INTRODUCTIO
MEASURE AND PROBABILITY

INTRODUCTION TO MEASURE AND PROBABILITY

BY

J. F. C. KINGMAN

Professor of Mathematics in the University of Oxford

AND

S. J. TAYLOR

Professor of Mathematics at Westfield College, University of London

CAMBRIDGE UNIVERSITY PRESS

CAMBRIDGE UNIVERSITY PRESS
Cambridge, New York, Melbourne, Madrid, Cape Town, Singapore, São Paulo, Delhi

Cambridge University Press
The Edinburgh Building, Cambridge CB2 8RU, UK

Published in the United States of America by Cambridge University Press, New York

www.cambridge.org
Information on this title: www.cambridge.org/9780521058889

First published 1966
Reprinted (with corrections) 1973
This digitally printed version 2008

A catalogue record for this publication is available from the British Library

Library of Congress Catalogue Card Number: 66–12308

ISBN 978-0-521-05888-9 hardback
ISBN 978-0-521-09032-2 paperback

CONTENTS

PREFACE

We decided to write this book largely as a result of experience in teaching at the Instructional Conference on Probability held at Durham in 1963 under the auspices of the London Mathematical Society. It seemed that a proper treatment of probability theory required an adequate background in the theory of finite measures in general spaces. The first part of the book attempts to set out this material in a form which not only provides an introduction for the intending specialist in measure theory, but also meets the needs of students of probability.

The theory of measure and integration is presented in the first instance for general spaces, though at each stage Lebesgue measure and the Lebesgue integral are considered as important examples, and their detailed properties are obtained. An introduction to functional analysis is given in Chapters 7 and 8; this contains not only the material (such as the various notions of convergence) which is relevant to probability theory, but also covers the basic theory of L_2-spaces important, for instance, in modern physics.

The second part of the book is an account of the fundamental theoretical ideas which underlie the applications of probability in statistics and elsewhere. The treatment leans heavily on the machinery developed in the first half of the book, and indeed some of the most important results are merely restatements of standard theorems of measure theory. No attempt has been made to give a detailed description of the applications of the theory, and in particular there is no account of the problem of statistical inference, but it is hoped that this part of the book may provide the framework for a rigorous mathematical study of these subjects.

The book is designed primarily for final year Honours mathematicians at British universities, but might also serve as a text for graduate courses in measure theory, or in probability theory. We have included a large number of examples; these are not an optional extra, but form an essential part of the development, and include many facts which might have been stated as theorems. We hope that the book will also be found useful by those statisticians, and other applied mathematicians, who find a need to acquire a basic working knowledge of the mathematical foundations for the techniques which they use.

The book is largely self-contained. The first two chapters contain the essential parts of set theory and point set topology; these could well be omitted by a reader already familiar with these subjects. Chapters 3 and 4 develop the theory of measure by the usual process of extension from 'simple sets' to those of a larger class, and the properties of Lebesgue measure are obtained. The integral is defined in Chapter 5, again by extending its definition stage by stage, using monotone sequences, Chapter 6 includes a discussion of product measures and a definition of measure in function space. Convergence in function space is considered in Chapter 7, and Chapter 8 includes a treatment of complete orthonormal sets in Hilbert space. Chapter 9 deals with special spaces; differentiation theory for real functions of real variable is developed and related to Lebesgue measure theory, and the Haar measure on a locally compact group is defined.

Chapter 10 introduces the idea of probability, and Chapter 11 the concepts of a random variable and its distribution. In Chapter 12 characteristic functions are defined and their properties established; these are then used in Chapter 13 where the classical results on sums of independent random variables are presented. Chapter 14 deals with joint distributions of several random variables. Finally, in Chapter 15, the idea of a stochastic process and a few examples of such processes are discussed.

Starred sections contain more advanced material and can be omitted at a first reading.

It will be clear to any reader familiar with the standard treatises that this book owes much to what has gone before. We make no attempt to establish precedence and make detailed acknowledgements; nor do we claim any particular originality for our treatment; but the form of presentation owes a great deal to our experience in undergraduate teaching—at Cornell University, the University of London, and the University of Cambridge—and we readily acknowledge the stimulus received from these sources. We are also grateful to Mr B. Fishel and Professor G. E. H. Reuter, who among others made helpful criticisms of an early draft, and to Professor D. G. Kendall for his encouragement in the early stages of this project.

J.F.C.K.

S.J.T.

PREFACE TO SECOND IMPRESSION

We have taken the opportunity of correcting the misprints and errors in the first edition but we have resisted the temptation to carry out a major revision of the text. We are grateful to the many friends and readers who pointed out to us errors and obscurities which we have tried to correct.

J.F.C.K.
S.J.T.

1

THEORY OF SETS

1.1 Sets

We do not want to become involved in the logical foundations of mathematics. In order to avoid these we will adopt a rather naïve attitude to set theory. This will not lead us into difficulties because in any given situation we will be considering sets which are all contained in (are subsets of) a fixed set or space or suitable collections of such sets. The logical difficulties which can arise in set theory only appear when one considers sets which are 'too big'—like the set of all sets, for instance. We assume the basic algebraic properties of the positive integers, the real numbers, and Euclidean spaces and make no attempt to obtain these from more primitive set theoretic notions. However, we will give an outline development (in Chapter 2) of the topological properties of these sets.

In a space X a set E is well defined if there is a rule which determines, for each *element* (or point) x in X, whether or not it is in E. We write $x \in E$ (read 'x belongs to E') whenever x is an element of E, and the negation of this statement is written $x \notin E$. Given two sets E, F we say that E is contained in F, or E is a subset of F, or F contains E and write $E \subset F$ if every element x in E also belongs to F. If $E \subset F$ and there is at least one element in F but not in E, we say that E is a proper subset of F.

Two sets E, F are *equal* if and only if they contain the same elements; i.e. if and only if $E \subset F$ *and* $F \subset E$. In this case we write $E = F$. This means that if we want to prove that $E = F$ we must prove both $x \in E \Rightarrow x \in F$ and $x \in F \Rightarrow x \in E$ (the symbol $\Rightarrow$ should be read 'implies').

Since a set is determined by its elements, one of the commonest methods of describing a set is by means of a defining sentence: thus E is the set of all elements (of X) which have the property P (usually delineated). The notation of 'braces' is often used in this situation

$$E = \{x \colon x \text{ has property } P\}$$

but when we use this notation we will always assume that only elements x in some fixed set X are being considered—as otherwise logical paradoxes can arise. When a set has only a finite number of elements we can write them down between braces $E = \{x, y, z, a, b\}$. In particular $\{x\}$ stands for the set containing the single element x. One must always distinguish between the element x and the set $\{x\}$, for example, the empty set $\varnothing$ defined below is not the same as the class $\{\varnothing\}$ containing the empty set.

Empty set (*or null set*)

The set which contains no elements is called the empty set and will be denoted by $\varnothing$. Clearly

$$\varnothing = \{x : x \neq x\}, \quad \text{and} \quad \varnothing \subset E \text{ for all sets } E.$$

In fact since $\varnothing$ contains no element, any statement made about the elements of $\varnothing$ is true (as well as its negative).

There are some sets which will be considered very frequently, and we consistently use the following notation:

Z, for the set of positive integers,
Q, for the set of rationals,
$\mathbf{R} = \mathbf{R}^1$, for the set of all real numbers,
C, for the set of complex numbers,
$\mathbf{R}^n$, for Euclidean n-dimensional space, i.e. the set of ordered n-tuples $(x_1, x_2, \ldots, x_n)$ where all the x_i are in **R**.

We assume that the reader is familiar with the algebraic and order properties of these sets. In particular we will use the fact that **Z** is *well ordered*, that is, that every non-empty set of positive integers has a least member: this is equivalent to the principle of mathematical induction.

We frequently have to consider sets of sets, and occasionally sets of sets of sets. It is convenient to talk of *classes* of sets and *collections* of classes to distinguish these types of set, and we will use italic capitals A, B, ... for sets, script capitals $\mathscr{A}$, $\mathscr{B}$, $\mathscr{C}$, ... for classes and Greek capitals Δ, Γ, ... for collections. Thus $C \in \mathscr{C}$ is read 'the set C belongs to the class $\mathscr{C}$'; and $\mathscr{A} \subset \mathscr{B}$ means that every set in the class $\mathscr{A}$ is also in the class $\mathscr{B}$.

Cartesian product

Given two sets E, F we define the Cartesian (or direct) product $E \times F$ to be the set of all *ordered* pairs (x, y) whose first element $x \in E$ and whose second element $y \in F$. This clearly extends immediately to the product $E_1 \times E_2 \times \ldots \times E_n$ of any finite number of sets. In particular it is immediate that $\mathbf{R}^n$, Euclidean n-space, is the Cartesian product of n copies of $\mathbf{R}$. For an infinite indexed class $\{E_i, i \in I\}$ of sets, the product $\prod_{i \in I} E_i$ is the set of elements of the form $\{a_i, i \in I\}$ with $a_i \in E_i$ for each $i \in I$.

Exercises 1.1

1. Describe in words the following sets:

(i) $\{t \in \mathbf{R}: 0 \leqslant t \leqslant 1\}$;
(ii) $\{(x, y) \in \mathbf{R}^2: x^2 + y^2 \leqslant 1\}$;
(iii) $\{k \in \mathbf{Z}: k = n^2 \text{ for some } n \in \mathbf{Z}\}$;
(iv) $\{k \in \mathbf{Z}: n | k \Rightarrow n = 1 \text{ or } k\}$;
(v) $\{\mathscr{A}: E \in \mathscr{A}\}$;
(vi) $\{B: B \subset E\}$.

2. Show that the relation $\subset$ is reflexive and transitive, but not in general symmetric.

3. The sets $X \times (Y \times Z)$ and $(X \times Y) \times Z$ are different but there is a natural correspondence between them.

4. Suppose x is an element of X and $A = \{x\}$. Which of the following statements are correct: $x \in A$, $x \in X$, $x \subset A$, $x \subset X$, $A \in X$, $A \subset X$, $A \subset x$?

5. Suppose $P(\alpha)$ and $Q(\alpha)$ are two propositions about the element such that $P(\alpha) \Rightarrow Q(\alpha)$. Show that $\{\alpha: P(\alpha)\} \subset \{\alpha: Q(\alpha)\}$.

1.2 Mappings

Suppose A and B are any two sets: a *function* from A to B is a rule which, for each element in A, determines a unique element in B. We talk of the function f and use the notation $f: A \to B$ to denote a function f defined on A and taking values in B. For any $x \in A$, $f(x)$ means the value of the function f at the point x and is therefore an element of the set B: we therefore avoid the terminology (common in older text books) 'the function $f(x)$'. The words *mapping* and *transformation* are often used as a synonym for function.

For a given function $f: A \to B$, we call A the *domain* of f and the subset of B consisting of the set of values $f(x)$ for x in A is called the *range* of f and may be denoted $f(A)$. When $f(A) = B$ we say that f is a function from A *onto* B. Given a function $f: A \to B$, by definition $f(x)$ is a uniquely determined element of B for each $x \in A$; if in addition for each y in $f(A)$ there is a *unique* $x \in A$ (we know there is at least one) with $y = f(x)$ we say that the function f is $(1, 1)$. Another shorter way of saying this is that $f: A \to B$ is $(1, 1)$ if and only if for $x_1, x_2 \in A$,

$$x_1 \neq x_2 \Rightarrow f(x_1) \neq f(x_2).$$

Given $f: A \to B$ there is an associated $f: \mathscr{A} \to \mathscr{B}$, where $\mathscr{A}$ is the class of all subsets of A and $\mathscr{B}$ is the class of all subsets of B, defined by

$$f(E) = \{y \in B: \exists x \in E \quad \text{with} \quad y = f(x)\}$$

for each $E \subset A$. (the symbol $\exists$ should be read, 'there exists': i.e. the set described by $\{x \in E: y = f(x)\}$ is not empty). There is also a function $f^{-1}: \mathscr{B} \to \mathscr{A}$ defined by

$$f^{-1}(F) = \{x \in A: f(x) \in F\},$$

for each $F \subset B$. The set $f^{-1}(F)$ is called the *inverse image* of F under f. Note that if $y \in B - f(A)$, then the inverse image $f^{-1}(\{y\})$ of the one point set $\{y\}$ is the empty set. If $f: A \to B$ is $(1, 1)$ and $y \in f(A)$, then it is clear that $f^{-1}(\{y\})$ is a one point subset of A, so that in this case (only) we can think of f^{-1} as a function from $f(A)$ to A. In particular, if $f: A \to B$ is $(1, 1)$ and onto there is a function $f^{-1}: B \to A$ called the *inverse function* of f such that $f^{-1}(y) = x$ if and only if $y = f(x)$.

Now suppose $f: A_1 \to B$, $g: A_2 \to B$ are functions such that $A_1 \supset A_2$ and $f(x) = g(x)$ for all x in A_2: under these conditions we say that f is an *extension* of g (from A_2 to A_1) and g is the *restriction* of f (to A_2). For example, if

$$g(x) = \cos x \quad (x \in \mathbf{R});$$

$$f(x+iy) = \cos x \cosh y + i \sin x \sinh y \quad (x+iy \in \mathbf{C});$$

then $f: \mathbf{C} \to \mathbf{C}$ is an extension of $g: \mathbf{R} \to \mathbf{C}$ from $\mathbf{R}$ to $\mathbf{C}$, and the usual convention of designating both f and g by 'cos' obscures the differences in their domains.

If we have two functions $f: A \to B$, $g: B \to C$ the result of applying the rule for g to the element $f(x)$ defines an element in C for all $x \in A$. Thus we have defined a function $h: A \to C$ which is called the *composition* of f and g and denoted $g \circ f$ or $g(f)$. Thus, for $x \in A$

$$h(x) = (g \circ f)x = g(f(x)) \in C.$$

Note that, if $f: A \to B$ is $(1, 1)$ and onto we could define the inverse function $f^{-1}: B \to A$ as the unique function from B to A such that

$$(f \circ f^{-1})(y) = y \quad \text{for all} \quad y \in B,$$

$$(f^{-1} \circ f)(x) = x \quad \text{for all} \quad x \in A.$$

Sequence

Given any set X a *finite sequence* of n points of X is a function from $\{1, 2, \ldots, n\}$ to X. This is usually denoted by $x_1, x_2, \ldots, x_n$ where $x_i \in X$ is the value of the function at the integer i. Similarly, an *infinite sequence* in X is a function from $\mathbf{Z}$ to X (where $\mathbf{Z}$ is the set of positive integers). This is denoted $x_1, x_2, \ldots$, or $\{x_i\}$ $(i = 1, 2, \ldots)$, or just $\{x_i\}$ where x_i is the value of the function at i, and is called the ith element of the sequence. Given a sequence $\{n_i\}$ of positive integers (that is, a function $f: \mathbf{Z} \to \mathbf{Z}$ where $f(i) = n_i$) such that $n_i > n_j$ for $i > j$, and a sequence $\{x_i\}$ of elements of X (a function $g: \mathbf{Z} \to X$) it is clear that the composite function $g \circ f: \mathbf{Z} \to X$ is again a sequence. Such a sequence is called a *subsequence* of $\{x_i\}$ and is denoted $\{x_{n_i}\}$ $(i = 1, 2, \ldots)$. Thus $\{x_{n_i}\}$ is a subsequence of $\{x_i\}$ if $n_i \in \mathbf{Z}$ for all $i \in \mathbf{Z}$, and $i > j \Rightarrow n_i > n_j$.

We can think of a sequence as a point in the product space $\prod_{i=1}^{\infty} X_i$ where $X_i = X$ for all i. More generally a point in the product space $\prod_{i \in I} X_i$ with $X_i = X$ for $i \in I$ can be identified as a function $f: I \to X$.

Exercises 1.2

1. Suppose $f: \mathbf{R} \to \mathbf{R}$ is defined by $f(x) = \sin x$. Describe each of the following sets:

$$f^{-1}\{0\},\quad f^{-1}\{1\},\quad f^{-1}\{2\},\quad f^{-1}\{y: 0 \leqslant y \leqslant \tfrac{1}{2}\}.$$

2. Suppose $f: A \to B$ is any function. Prove

(i) $E \subset f^{-1}(f(E))$, for each $E \subset A$;
(ii) $F \supset f(f^{-1}(F))$, for each $F \subset B$;

and give examples in which there is not equality in (i), (ii).

3. Suppose $f: A \to B$, $g: B \to C$ are functions and $h = g \circ f$: show that $h^{-1}(E) = f^{-1}[g^{-1}(E)]$ for each $E \subset C$.

4. If $A \subset B \subset C$, $f: A \to X$, $g: B \to X$, $h: C \to X$ are such that h is an extension of g and g is an extension of f, prove that f is the restriction of h to A.

5. Show that the restriction of a $(1, 1)$ mapping is $(1, 1)$.

6. Suppose $m, n \in \mathbf{Z}$, A is a set with m distinct elements and B is a set with n distinct elements. How many distinct functions are there from A to B?

1.3 Cardinal numbers

If there is a mapping $f: A \to B$ which is $(1, 1)$ and onto, then it is reasonable to say that there are the same number of elements in A as there are in B. In fact, for finite sets, the elementary process of counting sets up such a mapping from the set being counted to the integers $\{1, 2, \ldots, n\}$, and from experience we know that if the same finite set of objects is counted in different ways we always end up with the same integer n. (This fact can also be deduced from primitive axioms about the integers.) We say that the set A is equivalent to the set B, and write $A \sim B$ if there is a mapping $f: A \to B$ which is $(1, 1)$ and onto. It is clear that $\sim$ is an equivalence relation between sets in the sense that it is reflexive, symmetric and transitive, and we can therefore form equivalence classes of sets with respect to this relation. Such an equivalence class of sets is called a *cardinal number*, but by noting that the equivalence class is determined by any one of its members, we see that the easiest way to specify a cardinal number is to specify a representative set. Thus any set which can be mapped $(1, 1)$ onto the representative set will have the same cardinal. As is usual we shall use the following notation:

the cardinal of the empty set $\varnothing$ is 0;
the cardinal of the set of integers $\{1, 2, \dots n\}$ is n;
the cardinal of the set $\mathbf{Z}$ of positive integers is $\aleph_0$;
the cardinal of the set $\mathbf{R}$ of real numbers is c.

Since $\mathbf{Z}$ is ordered we can clearly order the cardinals of *finite* sets by saying that A has a smaller cardinal than B if A is equivalent to a proper subset of B. This definition does not work for infinite sets as the mappings

$$n \to 2^n \quad \text{or} \quad n \to n^2$$

map $\mathbf{Z}$ onto a proper subset of $\mathbf{Z}$ and are $(1, 1)$. Instead we say that the cardinal of a set A is less than the cardinal of the set B if there is a subset $B_1 \subset B$ such that $A \sim B_1$ but no subset $A_1 \subset A$ such that $A_1 \sim B$.

From this definition of ordering we consider the following statements, where m, n, p denote cardinals

(i) $m < n, n < p \Rightarrow m < p$;

(ii) at *most* one of the relations $m < n$, $m = n$, $n < m$ holds so that $m \leqslant n, n \leqslant m \Rightarrow m = n$.

(iii) at *least* one of the relations $m < n$, $m = n$, $n < m$ holds.

Now (i) follows easily from the definition, for let M, N, P be sets with cardinals m, n, p and suppose $N_1 \subset N$, $P_1 \subset P$ with $M \sim N_1$, $N \sim P_1$. The mapping $f: N \to P_1$ when restricted to N_1 gives an equivalence $N_1 \sim P_2 \subset P_1$ so that $M \sim P_2 \subset P$. Further if $P \sim M_1 \subset M$ the mapping $g: M \to N_1$ when restricted to M_1 shows $P \sim M_1 \sim N_2 \subset N$ which contradicts $n < p$. (ii) can also be deduced from the definition (see exercise 1.3 (5)), though this requires quite a complicated argument: (ii) is known as the Schröder–Bernstein theorem. However, the truth of (iii)—that all cardinals are comparable—cannot be proved without the use of an additional axiom (known as the axiom of choice) which we will discuss briefly in §1.6. If we assume the axiom of choice or something equivalent, then (iii) is also true.

A set of cardinal $\aleph_0$ is said to be *enumerable*. Thus such a set $A \sim \mathbf{Z}$ so that the elements of A can be 'enumerated' as a sequence $a_1, a_2, \ldots$ in which each element of A occurs once and only once. A set which has a cardinal $m \leqslant \aleph_0$ is said to be *countable*. Thus E is countable if there is a subset $A \subset \mathbf{Z}$ such that $E \sim A$, and a set is countable if it is either finite or enumerable.

Given any infinite set B we can choose, by induction, a sequence $\{b_i\}$ of distinct elements in B and if B_1 is the set of elements in $\{b_i\}$ the cardinal of B_1 is $\aleph_0$. Hence if m is an infinite cardinal we always have $m \geqslant \aleph_0$. By using the equivalence

$$b_i \leftrightarrow b_{2i}$$

between B_1 and the proper subset $B_2 \subset B_1$ where B_2 contains the even elements of $\{b_i\}$ and the identity mapping

$$b \leftrightarrow b \quad \text{for} \quad b \in B - B_1$$

we have an equivalence between $B = B_1 \cup (B - B_1)$ and $B_2 \cup (B - B_1)$, a proper subset of B. This shows that any infinite set B contains a proper subset of the same cardinal.

In order to see that some infinite sets have cardinal $> \aleph_0$ it is sufficient to recall that the set $\{x \in \mathbf{R}: 0 < x < 1\}$ cannot be arranged as a sequence.† Now $\pi^{-1} \tan^{-1} x + \frac{1}{2} = f(x)$, $x \in \mathbf{R}$ defines a mapping f: $\mathbf{R} \to (0, 1)$ which is $(1, 1)$ and onto so that $\mathbf{R}$ has the same cardinal as the interval $(0, 1)$ and we have $\mathbf{c} > \aleph_0$. It is worth remarking that a famous unsolved problem of mathematics concerns the existence or otherwise of cardinals m such that $\mathbf{c} > m > \aleph_0$. The axiom that no such exist, that is that $m > \aleph_0 \Rightarrow m \geqslant \mathbf{c}$ is known as the *continuum hypothesis*.

The fact that there are infinitely many different infinite cardinals follows from the next theorem, which compares the cardinal of a set E with the cardinal of the class of subsets of E.

***Theorem* 1.1.** *For any set E, the class $\mathscr{C} = \mathscr{C}(E)$ of all subsets of E has a cardinal greater than that of E.*

Proof. For sets E of finite cardinal n, one can prove directly that the cardinal of $\mathscr{C}(E)$ is 2^n, and an induction argument easily yields $n < 2^n$ for $n \in \mathbf{Z}$. However, the case of finite sets E is included in the general proof, so there is nothing gained by this special argument.

† See, for example, J. C. Burkill, *A First Course in Mathematical Analysis* (Cambridge, 1962).

Suppose $\mathscr{Q}$ is the class of one points sets $\{x\}$ with $x \in E$. Then $\mathscr{Q} \subset \mathscr{C}$ and $E \sim \mathscr{Q}$ because of the mapping $x \leftrightarrow \{x\}$. Therefore it is sufficient to prove by (ii) above, that $\mathscr{C}$ is equivalent to no subset $E_1 \subset E$. Suppose then that $\phi: \mathscr{C} \to E_1$ is $(1, 1)$ and onto and let $\chi: E_1 \to \mathscr{C}$ denote the inverse function. Let A be the subset of E_1 defined by

$$A = \{x \in E_1, \quad x \notin \chi(x)\}.$$

Then $A \in \mathscr{C}$ so that $\phi(A) = x_0 \in E_1$. Now if $x_0 \in A$, $\chi(x_0) = A$ does not contain x_0 which is impossible, while if $x_0 \notin A$, then x_0 is not in $\chi(x_0)$ so that $x_0 \in A$. In either case we have a contradiction. ∎

It is possible to build up systematically an arithmetic of cardinals. This will only be needed for finite cardinals and $\aleph_0$ in this book, so we restrict the results to these cases and discuss them in the next section.

Exercises 1.3

1. Show that $(0, 1] \sim (0, 1)$ by considering f defined by

$$\begin{aligned} f(x) &= \tfrac{3}{2} - x, \quad \text{for} \quad \tfrac{1}{2} < x \leqslant 1; \\ &= \tfrac{3}{4} - x, \quad \text{for} \quad \tfrac{1}{4} < x \leqslant \tfrac{1}{2}; \\ &= \tfrac{3}{8} - x, \quad \text{for} \quad \tfrac{1}{8} < x \leqslant \tfrac{1}{4}; \\ &= \frac{3}{2^n} - x, \quad \text{for} \quad \frac{1}{2^n} < x \leqslant \frac{1}{2^{n-1}} \quad (n = 1, 2, \ldots). \end{aligned}$$

Deduce that all intervals (a, b), $(a, b]$, $[a, b]$ or $[a, b)$ with $a < b$ have the same cardinal $\mathbf{c}$.

2. Every function f: $[a, b] \to \mathbf{R}$ which is monotonic, i.e.

$$a \leqslant x_1 < x_2 \leqslant b \Rightarrow f(x_1) \leqslant f(x_2),$$

is discontinuous at the points of a countable subset of $[a, b]$.

Hint. Consider the sets of points x where the size of the discontinuity $d(x) = f(x+0) - f(x-0)$ satisfies $1/(n+1) < d(x) \leqslant 1/n$ and prove this is finite for all n in $\mathbf{Z}$.

3. Show that $\mathbf{R}^2 \sim \mathbf{R}$.

Hint.

$$(\cdot a_1 a_2 a_3 \ldots, \cdot b_1 b_2 b_3 \ldots) \leftrightarrow \cdot a_1 b_1 a_2 b_2 \ldots$$

defines a $(1, 1)$ mapping between pairs of decimal expansions and single expansions of numbers in $(0, 1)$. Modify this mapping to eliminate the difficulty caused by the fact that decimal expansions are not quite unique.

4. Prove that a finite set E of cardinal m has 2^m distinct subsets.

5. Suppose $A_1 \subset A$, $B_1 \subset B$, $A_1 \sim B$ and $A \sim B_1$. Construct a mapping to show that $A \sim B$.

Hint. Suppose $f: A \to B_1$, $g: B \to A_1$ are (1, 1) and onto. Say x (in either A or B) is an ancestor of y if and only if y can be obtained from x by successive applications of f and g. Decompose A into 3 sets A_o, A_e, A_i according as to whether the element x has an odd, even or infinite number of ancestors and decompose B similarly. Consider the mapping which agrees with f on A_e and A_i, and with g^{-1} on A_o.

1.4 Operations on subsets

For two sets A, B we define the *union* of A and B (denoted $A \cup B$) to be the set of elements in either A or B or both. The *intersection* of A and B (denoted $A \cap B$) is the set of elements in both A and B.

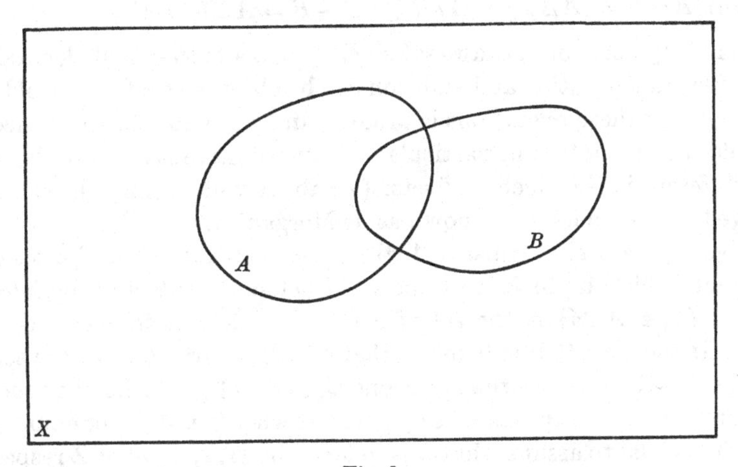

Fig. 1

If $A \subset X$, the *complement* of A with respect to X (denoted $X - A$) is the set of those elements in X which are not in A. We also use $(A - B)$ to denote the set of elements in A which are not in B for arbitrary sets A, B. For any two sets A, B the *symmetric difference* (denoted $A \triangle B$) is $(A - B) \cup (B - A)$, that is the set of elements which are in one of A, B but not in both. Note that $A \triangle B = B \triangle A$.

These finite operations on sets are best illustrated by means of a Venn diagram. In this some figure (like a rectangle) denotes the whole space X and suitable geometrical figures inside denote the subsets A, B, etc. It is well known that drawing does not prove a theorem, but the reader is advised to illustrate the results of the next paragraph by means of suitable Venn diagrams (see Figure 1).

The operations $\cup, \cap, \triangle$ satisfy algebraic laws, some of which are listed below. We assume the reader is familiar with these, so proofs are omitted.

(i) $A \cup B = B \cup A, \quad A \cap B = B \cap A$;

(ii) $(A \cup B) \cup C = A \cup (B \cup C), \quad (A \cap B) \cap C = A \cap (B \cap C)$;

(iii) $A \cap (B \cup C) = (A \cap B) \cup (A \cap C)$,
$A \cup (B \cap C) = (A \cup B) \cap (A \cup C)$;

(iv) $A \cup \varnothing = A, \quad A \cap \varnothing = \varnothing$;

(v) if $A \subset X$, then $A \cup X = X, \quad A \cap X = A$;

(vi) if $A \subset X, \quad B \subset X$, then $X - (A \cup B) = (X - A) \cap (X - B)$,
$X - (A \cap B) = (X - A) \cup (X - B)$;

(vii) $A \cup B = (A \triangle B) \triangle (A \cap B), \quad A - B = A \triangle (A \cap B)$.

A similarity between the laws satisfied by $\cap$, $\cup$ and the usual algebraic laws for multiplication and addition can be observed (in fact the older notation for these operations is product and sum) but the differences should also be noted: in particular the distributive laws, (iii) above, are different in the algebra of sets. (vi) above will be generalized and proved as a lemma—it is known as de Morgan's law.

Given a class $\mathscr{C}$ of subsets A, the *union* $\bigcup\{A; A \in \mathscr{C}\}$ is the set of elements which are in at least one set A belonging to $\mathscr{C}$ and the *intersection* $\bigcap\{A; A \in \mathscr{C}\}$ is the set of elements which are in every set A of $\mathscr{C}$. If the class $\mathscr{C}$ is indexed so that $\mathscr{C}$ consists precisely of the sets A_α, $(\alpha \in I)$, then we use the notations $\bigcup_{\alpha \in I} A_\alpha$, $\bigcap_{\alpha \in I} A_\alpha$ for the union and intersection of the class. In particular when $\mathscr{C}$ is finite or enumerable it is usual to assume that it is indexed by $\{1, 2, \ldots, n\}$ or $\mathbf{Z}$ respectively and the notation is

$$\bigcup_{i=1}^{n} A_i, \quad \bigcap_{i=1}^{n} A_i, \quad \bigcup_{i=1}^{\infty} A_i, \quad \bigcap_{i=1}^{\infty} A_i.$$

When the class $\mathscr{C}$ is empty, that is $I = \varnothing$, we adopt the conventions

$$\bigcup_{\alpha \in I} E_\alpha = \varnothing, \quad \bigcap_{\alpha \in I} E_\alpha = X, \quad \text{the whole space.}$$

This ensures that certain identities are valid without restriction on I.

Lemma. *Suppose $E_\alpha, \alpha \in I$ is a class of subsets of X, and E_1 is one set of the class, then*

(i) $\bigcap_{\alpha \in I} E_\alpha \subset E_1 \subset \bigcup_{\alpha \in I} E_\alpha$;

(ii) $X - \bigcup_{\alpha \in I} E_\alpha = \bigcap_{\alpha \in I} (X - E_\alpha)$;

(iii) $X - \bigcap_{\alpha \in I} E_\alpha = \bigcup_{\alpha \in I} (X - E_\alpha)$.

Proof. (i) This is immediate from the definition.

(ii) Suppose $x \in X - \bigcup_{\alpha \in I} E_\alpha$, then $x \in X$ and x is not in $\bigcup_{\alpha \in I} E_\alpha$, that is x is not in any E_α, $\alpha \in I$ so that $x \in X - E_\alpha$ for every α in I, and $x \in \bigcap_{\alpha \in I} (X - E_\alpha)$. Conversely if $x \in \bigcap_{\alpha \in I} (X - E_\alpha)$, then for every $\alpha \in I$, x is in X but not in E_α, so $x \in X$ but x is not in $\bigcup_{\alpha \in I} E_\alpha$; that is, $x \in X - \bigcup_{\alpha \in I} E_\alpha$.

(iii) Similar to (ii). ▮

Two sets A, B are said to be *disjoint* if they have no elements in common; that is, if $A \cap B = \varnothing$. A *disjoint class* is a class $\mathscr{C}$ of sets such that any two distinct sets of $\mathscr{C}$ are disjoint. The union of a disjoint class is sometimes called a *disjoint union.*

Lemma. *Given a finite or enumerable union of sets* $\bigcup_{i=1}^{p} E_i$ *(where* p *can be* $+\infty$*), there are subsets* $F_i \subset E_i$ *such that the sets* F_i *are disjoint and* $\bigcup_{i=1}^{p} E_i = \bigcup_{i=1}^{p} F_i$.

Proof. We write out the details for $p = \infty$. Only obvious changes are needed for $p \in \mathbf{Z}$. Put $C = \bigcup_{i=1}^{p} E_i$ and define $F_1 = E_1$,

$$F_n = E_n - \bigcup_{i=1}^{n-1} E_i \quad (n = 2, 3, \ldots).$$

Then $F_n \subset E_n$ for all n, and if $i > j$, F_i and E_j are disjoint, so that F_i, F_j must be disjoint. Further if $x \in C$, and n is the smallest integer (which exists because $\mathbf{Z}$ is well ordered) such that $x \in E_n$; then $x \in E_n$ but not to E_i for $i < n$. Thus $x \in F_n$ and so $x \in \bigcup_{i=1}^{\infty} F_j$. Thus

$$C \subset \bigcup_{i=1}^{\infty} F_i, \text{ and the reverse inclusion is immediate.} ▮$$

***Theorem* 1.2.** *The union of a countable class of countable sets is a countable set.*

Proof. By the process of the above lemma we can replace the countable union by a countable disjoint union of sets which are subsets of those in the original class—each of which is therefore countable. Each countable set can be enumerated as a finite or infinite sequence. So we have

$$C = \bigcup_{i=1}^{\infty} E_i \quad \text{a disjoint union,}$$

$$E_i = \{x_{ij}\} \quad (j = 1, 2, \ldots),$$

where the infinite union may be a finite one and some (or all) of the sequences $\{x_{ij}\}$ may be finite. Put $F_n = \{x_{ij} : i + j = n\}$, then F_n is a finite set containing at most $(n+1)$ elements. The sets F_n are disjoint, and C can be enumerated by first enumerating F_1, then F_2, and so on. If $F_n = \varnothing$ for $n \geqslant N$ then C is finite; otherwise it is enumerable. ∎

Corollary. *The set* Q *of rational numbers is enumerable.*

Proof. $\mathbf{Q} = \bigcup_{n=1}^{\infty} E_n$, where E_n is the set of real numbers of the form p/n where p is an integer. E_n is enumerable since

$$0, +1, -1, +2, -2, \ldots, +p, -p, \ldots$$

is an enumeration of the set of integers. ∎

For a sequence $E_1, E_2, \ldots$ of sets, we put

$$\limsup E_i = \bigcap_{n=1}^{\infty}\left(\bigcup_{i=n}^{\infty} E_i\right), \quad \liminf E_i = \bigcup_{n=1}^{\infty}\left(\bigcap_{i=n}^{\infty} E_i\right);$$

and if $\{E_i\}$ is such that $\limsup E_i = \liminf E_i$ we say that the sequence converges to the set $E = \limsup E_i = \liminf E_i$. For any sequence $\{E_i\}$, $\limsup E_i$ is the set of those elements which are in E_i for infinitely many i and $\liminf E_i$ is the set of those elements which are in all but a finite number of the sets E_i.

A sequence $\{E_i\}$ is said to be *increasing* if, for each positive integer n, $E_n \subset E_{n+1}$; it is said to be *decreasing* if, for each positive integer n, $E_n \supset E_{n+1}$. A monotone sequence of sets is one which is either increasing or decreasing. Note that any monotone sequence convergences to a limit for

(i) If $\{E_i\}$ is increasing,

$$\bigcup_{i=n}^{\infty} E_i = \bigcup_{i=1}^{\infty} E_i, \quad \bigcap_{i=n}^{\infty} E_i = E_n \quad \text{for all } n,$$

so that $\limsup E_i = \liminf E_i = \bigcup_{i=1}^{\infty} E_i$; while

(ii) If $\{E_i\}$ is decreasing,

$$\bigcup_{i=n}^{\infty} E_i = E_n, \quad \bigcap_{i=n}^{\infty} E_i = \bigcap_{i=1}^{\infty} E_i \quad \text{for all } n,$$

so that $\limsup E_i = \liminf E_i = \bigcap_{i=1}^{\infty} E_i$.

Indicator function

Given a subset E of a space X, the function $\chi_E: X \to \mathbf{R}$ defined by

$$\chi_E(x) = \begin{cases} 1 & \text{for } x \in E, \\ 0 & \text{for } x \in X - E \end{cases}$$

is called the indicator function of E (many books use the term 'characteristic function' for χ_E, but we will avoid this term because characteristic function has a different meaning in probability theory). The correspondence between subsets of X and indicator functions is clearly (1, 1) for

$$E = \{x\colon \chi_E(x) = 1\},$$

and we will use indicator functions as a convenient tool for carrying out operations on sets.

Exercises 1.4

1. Prove each of the following set identities:

$$A - B = A - (A \cap B) = A \cup B - B,$$
$$A \cap (B - C) = A \cap B - A \cap C,$$
$$(A - B) - C = A - (B \cup C),$$
$$A - (B - C) = (A - B) \cup (A \cap C),$$
$$(A - B) \cap (C - D) = A \cap C - B \cup D,$$
$$E \triangle (F \triangle G) = (E \triangle F) \triangle G,$$
$$E \cap (F \triangle G) = (E \cap F) \triangle (E \cap G),$$
$$E \triangle \varnothing = E, E \triangle X = X - E,$$
$$E \triangle E = \varnothing, E \triangle (X - E) = X,$$
$$E \triangle F = (E \cup F) - (E \cap F).$$

2. With respect to which of the operations $\triangle$, $\cup$, $\cap$ does the class of all subsets of X form a group?

3. Show that $E \subset F$ if and only if $X - F \subset X - E$.

4. Prove that $A \triangle B = C \triangle D$ if and only if $A \triangle C = B \triangle D$, by showing that either equality is equivalent to the statement that every point of X is in 0, 2 or 4 of the sets A, B, C, D.

5. Show that if $I_1 \subset I_2$, then

$$\bigcap_{\alpha \in I_1} E_\alpha \supset \bigcap_{\alpha \in I_2} E_\alpha, \quad \bigcup_{\alpha \in I_1} E_\alpha \subset \bigcup_{\alpha \in I_2} E_\alpha.$$

6. A real number is said to be algebraic if it is a zero of a polynomial $a_n x^n + a_{n-1} x^{n-1} + \ldots + a_0$ where the coefficients a_i are integers. Defining the 'height' of a polynomial to be the integer

$$h = n + |a_n| + |a_{n-1}| + \ldots + |a_0|,$$

show that there are only finitely many polynomials of height h, and deduce that the set of all algebraic real numbers is enumerable. Deduce that the set of transcendental numbers (real numbers which are not algebraic) has cardinal $> \aleph_0$.

7. Show that in $\mathbf{R}^n$, the set $\mathbf{Q}^n$ of points $(x_1, x_2, \ldots, x_n)$, where each coordinate x_i is rational, is an enumerable set.

Further, the class of all spheres with centres at points of $\mathbf{Q}^n$ and rational radii is an enumerable class.

8. Show that any sequence of disjoint sets converges to $\varnothing$. Show that $\{E_n\}$ is a convergent sequence if and only if there is no point x of X such that each of $x \in E_n$, $x \in X - E_n$ holds for infinitely many n.

Suppose

$$E_n = \begin{matrix} \{x : 0 < x \leqslant 1 - (1/n)\} & n \text{ odd}, \\ \{x : (1/n) \leqslant x < 1\} & n \text{ even}; \end{matrix}$$

show that $\{E_n\}$ converges but is not monotone.

9. For any sequence $\{E_n\}$ of sets prove

(i) $\limsup E_n$, $\liminf E_n$ are unaltered by the omission or alteration of any finite number of sets in the sequence.

(ii) for any set F,

$$F - \limsup E_n = \liminf (F - E_n),$$

$$F - \liminf E_n = \limsup (F - E_n).$$

10. If $E_n = A$ for n even, $E_n = B$ for n odd, show that $\limsup E_n = A \cup B$, $\liminf E_n = A \cap B$.

11. Can an uncountable union of distinct sets be countable?

12. If $\{E_n\}$ is a sequence of sets and

$$D_1 = E_1, \quad D_n = D_{n-1} \triangle E_n \quad \text{for} \quad n = 2, 3, \ldots,$$

show that the sequence $\{D_n\}$ converges to a limit if and only if $\lim E_n = \varnothing$.

13. Show that $\chi_E(x) \leqslant \chi_F(x)$ for all x in X if and only if $E \subset F$. Suppose $A = E \cup F$, $B = E \cap F$, $C = E \triangle F$: show that

$$\chi_B = \chi_E \cdot \chi_F, \quad \chi_A = \chi_E + \chi_F - \chi_B, \quad \chi_C = |\chi_E - \chi_F|.$$

Generalise the first two of these identities to finite unions and intersections.

14. If χ_n is the indicator function of E_n $(n = 1, 2, \ldots)$ and $A = \limsup E_n$, $B = \liminf E_n$, show that, for all x in X,

$$\chi_A(x) = \limsup_{n \to \infty} \chi_n(x), \quad \chi_B(x) = \liminf_{n \to \infty} \chi_n(x).$$

1.5 Classes of subsets

Up to the present our operations have been defined on the class $\mathscr{C}$ of all subsets of a given set X. This class is too large for many purposes and it is usual to restrict attention to subclasses of $\mathscr{C}$. However it is important that the subclasses considered have sufficient structure, and we now define various types of class starting with the simplest.

1. *Semi-ring*

A class $\mathscr{S}$ of subsets such that

(i) $\varnothing \in \mathscr{S}$;

(ii) $A, B \in \mathscr{S} \Rightarrow A \cap B \in \mathscr{S}$;

(iii) $A, B \in \mathscr{S} \Rightarrow A - B = \bigcup_{i=1}^{n} E_i$, where the E_i are disjoint sets in $\mathscr{S}$, is

called a semi-ring. (Note that many authors, following Von Neumann, who first defined the concept, have an additional condition in the definition of a semi-ring—instead of (iii) they assume that if A, $B \in \mathscr{S}$ and $B \subset A$ there is a finite class $C_0, C_1, \ldots, C_n$ of sets of $\mathscr{S}$ such that $B = C_0 \subset C_1 \subset \ldots \subset C_n = A$ and $D_i = C_i - C_{i-1} \in \mathscr{S}$ for $i = 1, 2, \ldots, n$. This stronger condition causes complications and we weaken it since it is unnecessary.) An important example of a semi-ring of subsets of $\mathbf{R}$ is the class $\mathscr{P} = \mathscr{P}^1$ of finite intervals $(a, b]$ which are open on the left and closed on the right. Similarly, $\mathscr{P}^n$ consisting of the rectangles in $\mathbf{R}^n$ of the form $\{(x_1, x_2, \ldots, x_n)\colon a_i < x_i \leqslant b_i\}$ is a semi-ring in $\mathbf{R}^n$.

2. *Ring*

This is any non-empty class $\mathscr{R}$ of subsets such that

$$A, B \in \mathscr{R} \Rightarrow A \cap B \in \mathscr{R} \quad \text{and} \quad A \triangle B \in \mathscr{R}.$$

Since $\varnothing = A \triangle A$, $A \cup B = (A \triangle B) \triangle (A \cap B)$, and $A - B = A \triangle (A \cap B)$ we see that a ring is a class of sets closed under the operations of union, intersection, and difference and $\varnothing \in \mathscr{R}$. Thus a ring is certainly also a semi-ring. As examples the system $\{\varnothing, X\}$ is a ring as is the class of all subsets of X. However, the class $\mathscr{P}$ of half-open intervals in $\mathbf{R}$ is not a ring, for it is not closed under the operation of difference.

3. *Field* (*or algebra*)

Any class $\mathscr{A}$ of subsets of X which is a ring and contains X is called a field. Thus a ring is a field if and only if it is closed under the operation of taking the complement. The class of all finite subsets of a space X is a ring, but is not a field unless X is finite. In $\mathbf{R}$ the class of all bounded subsets is a ring but not a field.

4. *Sigma ring* (σ*-ring*)

A ring $\mathscr{R}$ is called a σ-ring if it is closed under countable unions, i.e. if

$$A_i \in \mathscr{R} \quad (i = 1, 2, \ldots) \Rightarrow \bigcup_{i=1}^{\infty} A_i \in \mathscr{R}.$$

Now put $A = \bigcup_{i=1}^{\infty} A_i$ and use the identity $\bigcap_{i=1}^{\infty} A_i = A - \bigcup_{i=1}^{\infty} (A - A_i)$ to see that a σ-ring is also closed under countable intersections. Hence if $\mathscr{R}$ is a σ-ring and $\{A_n\}$ is a sequence of sets from $\mathscr{R}$ then $\limsup A_n$ and $\liminf A_n$ both belong to $\mathscr{R}$.

5. *Sigma-field* (σ*-field, Borel field,* σ*-algebra*)

Any class $\mathscr{F}$ of sets which contains the whole space X and is a σ-ring is called a σ-field. Alternatively, a σ-field can be defined as a field which is closed under countable unions. For any space X, the class of all countable subsets will be a σ-ring, but will only be a σ-field if X is countable.

6. *Monotone class*

Any class $\mathscr{M}$ of subsets such that, for any monotone sequence $\{E_n\}$ of sets in $\mathscr{M}$ we have $\lim E_n \in \mathscr{M}$ is called a monotone class. It is clear that a σ-ring is a monotone class, and any monotone class which is a ring is also a σ-ring since

$$E_i \in \mathscr{M} \Rightarrow \bigcup_{i=1}^{n} E_i \in \mathscr{M},$$

and $\bigcup_{i=1}^{n} E_i$ is monotone so that $\bigcup_{i=1}^{\infty} E_i = \lim \bigcup_{i=1}^{n} E_i$ is in $\mathscr{M}$.

We now use the term z-class to denote any one of the types 2, 3, 4, 5, 6 above (but not a semi-ring), and we consider a collection of z-classes.

Lemma. *If* $\mathscr{C}_\alpha$, *for* $\alpha \in I$ *is a z-class, then* $\mathscr{C} = \bigcap_{\alpha \in I} \mathscr{C}_\alpha$ *is a z-class.*

Proof. Each of these z-classes is defined in terms of closure with respect to specified operations. Since each $\mathscr{C}_\alpha$ is closed with respect to operations, the resulting subset will be in $\mathscr{C}_\alpha$ for all $\alpha \in I$ and therefore in $\mathscr{C}$, so that $\mathscr{C}$ is also a z-class. ∎

Note. The intersection of a collection of semi-rings need not be a semi-ring.

***Theorem* 1.3.** *Given any class* $\mathscr{C}$ *of subsets of* X *there is a unique z-class* $\mathscr{S}$ *containing* $\mathscr{C}$ *such that, if* $\mathscr{Q}$ *is any other z-class containing* $\mathscr{C}$ *we must have* $\mathscr{Q} \supset \mathscr{S}$.

Remark. The z-class $\mathscr{S}$ obtained in this theorem is called the z-class *generated by* $\mathscr{C}$. It is clearly the smallest z-class of subsets which contains $\mathscr{C}$.

Proof. The class of all subsets of X is a z-class containing $\mathscr{C}$. Put $\mathscr{S} = \bigcap \{\mathscr{Q}: \mathscr{Q} \supset \mathscr{C} \text{ and } \mathscr{Q} \text{ is a } z\text{-class}\}$. Then $\mathscr{S}$ is a z-class by the lemma and it clearly satisfies the conditions of the theorem. ∎

In certain special cases one can specify the nature of the z-class generated by a given class.

Theorem 1.4. *The ring $\mathscr{R}(\mathscr{S})$ generated by a semi-ring $\mathscr{S}$ consists precisely of the sets which can be expressed in the form*

$$E = \bigcup_{k=1}^{n} A_k$$

of a finite disjoint union of sets of $\mathscr{S}$.

Proof. (i) The ring $\mathscr{R}(\mathscr{S})$ certainly must contain all sets of this form, since it has to be closed under finite unions.

(ii) To see that the system $\mathscr{Q}$ of sets of this type form a ring suppose

$$A = \bigcup_{k=1}^{n} A_k, \quad B = \bigcup_{k=1}^{m} B_k$$

and put $C_{ij} = A_i \cap B_j \in \mathscr{S}$. Then since the sets C_{ij} are disjoint and

$$A \cap B = \bigcup_{i=1}^{n} \bigcup_{j=1}^{m} C_{ij}$$

the system $\mathscr{Q}$ is closed under intersections. Now from the definition of a semi-ring, an induction argument shows that

$$A_i = \bigcup_{j=1}^{m} C_{ij} \cup \bigcup_{k=1}^{r_i} D_{ik}, \quad (i = 1, \ldots, n);$$

$$B_j = \bigcup_{i=1}^{n} C_{ij} \cup \bigcup_{k=1}^{s_j} E_{kj}, \quad (j = 1, 2, \ldots, m);$$

where the finite sequences $\{D_{ik}\}\,(k = 1, \ldots, r_i)$ and $\{E_{kj}\}\,(k = 1, \ldots, s_j)$ consist of disjoint sets in $\mathscr{S}$. It follows now that

$$A \triangle B = \bigcup_{i=1}^{n} \left(\bigcup_{k=1}^{r_i} D_{ik} \right) \cup \bigcup_{j=1}^{m} \left(\bigcup_{k=1}^{s_j} E_{kj} \right)$$

so that the system $\mathscr{Q}$ is also closed under the operation of taking the symmetric difference. ∎

Example. We have already seen that $\mathscr{P}$, the class of intervals $(a, b]$ in $\mathbf{R}$, is a semi-ring. The generated ring is the class $\mathscr{E}$ of finite unions of disjoint half-open intervals. $\mathscr{E}$ is called the class of *elemen-*

tary figures in $\mathbf{R}$. Similarly, the elementary figures in $\mathbf{R}^n$ form the class $\mathscr{E}^n$ of finite disjoint unions of half-open rectangles from $\mathscr{P}^n$.

The next theorem is often important in proving that a given class is a σ-ring.

Theorem 1.5. *If $\mathscr{R}$ is any ring, the monotone class $\mathscr{M}(\mathscr{R})$ generated by $\mathscr{R}$ is the same as the σ-ring $\mathscr{S}(\mathscr{R})$ generated by $\mathscr{R}$.*

Corollary. *Any monotone class $\mathscr{M}$ which contains a ring $\mathscr{R}$ contains the σ-ring $\mathscr{S}(\mathscr{R})$ generated by $\mathscr{R}$.*

Proof. Since a σ-ring is always a monotone class and $\mathscr{S}(\mathscr{R}) \supset \mathscr{R}$ we must have

$$\mathscr{S}(\mathscr{R}) \supset \mathscr{M}(\mathscr{R}), \text{ denoted by } \mathscr{M}.$$

Hence it is sufficient to show that $\mathscr{M}$ is a σ-ring, and this will follow if we can prove that $\mathscr{M}$ is a ring. For any set F, let $\mathscr{Q}(F)$ be the class of sets E for which $E-F$, $F-E$, $E \cup F$ are all in $\mathscr{M}$. Then if $\mathscr{Q}(F)$ is not empty it is easy to check that it is a monotone class. It is clear that $\mathscr{Q}(F) \supset \mathscr{R}$ for any $F \in \mathscr{R}$ so that $\mathscr{Q}(F) \supset \mathscr{M}$. Hence, $E \in \mathscr{M}$, $F \in \mathscr{R} \Rightarrow E \in \mathscr{Q}(F) \Rightarrow F \in \mathscr{Q}(E)$ by the symmetry of the definition of the class $\mathscr{Q}$, and it follows that $\mathscr{M} \subset \mathscr{Q}(E)$ since $\mathscr{Q}$ is a monotone class. But the truth of this for every $E \in \mathscr{M}$ implies that $\mathscr{M}$ is a ring. ∎

In §1.2 we discussed mappings $f: X \to Y$ and saw that any such mapping induced a set mapping f^{-1} on the class of all subsets of Y. If f^{-1} is restricted to a special class $\mathscr{C}$ of subsets in Y, then the image of $\mathscr{C}$ under f^{-1} will be a class of subsets in X. The interesting thing is that the structure of the class $\mathscr{C}$ is often preserved by such a mapping f^{-1}.

Theorem 1.6. *Suppose $\mathscr{C}$ is a z-class of subsets of Y, $f: X \to Y$ is any mapping and $f^{-1}(\mathscr{C})$ denotes the class of subsets of X of the form $f^{-1}(E)$, $E \in \mathscr{C}$. Then $f^{-1}(\mathscr{C})$ is a z-class of subsets of X.*

Proof. It is easy to check that the mapping $f^{-1}: \mathscr{C} \to f^{-1}(\mathscr{C})$ commutes with each of the set operations union, symmetric difference, countable union and monotone limit. The closure of $\mathscr{C}$ with respect to any of these operations therefore implies the closure of $f^{-1}(\mathscr{C})$ with respect to the same operation. ∎

Exercises 1.5

1. Give an example of two semi-rings $\mathscr{S}_1$, $\mathscr{S}_2$ whose intersection is not a semi-ring.

2. Prove that any finite field is also a σ-field.

3. If $\mathscr{R}$ is a ring of sets and we define operations $\odot$ = multiplication and $\oplus$ = addition by

$$E \odot F = E \cap F, \quad E \oplus F = E \triangle F$$

show that $\mathscr{R}$ becomes a ring in the algebraic sense.

4. If $\mathscr{R}$ is a ring and $\mathscr{C}$ is the class of all subsets E of X such that either E or $(X-E)$ is in $\mathscr{R}$, show that $\mathscr{C}$ is a field.

5. What is the ring $\mathscr{R}(\mathscr{C})$ generated by each of the following classes:
(i) for a single fixed E, $\mathscr{C} = \{E\}$;
(ii) for a single E, $\mathscr{C}$ is class of all subsets of E;
(iii) $\mathscr{C}$ is class of all sets with precisely 2 points?

6. Prove that if A is any subset of a space X, $A \neq \varnothing$ or X, then the σ-field $\mathscr{F}(A)$ generated by the set A is the class $\{\varnothing, A, X-A, X\}$.

7. If $\mathscr{C}$ is a non-empty class of sets show that every set in the σ-ring generated by $\mathscr{C}$ is a subset of a countable union of sets of $\mathscr{C}$.

8. For each of the following classes $\mathscr{C}$ describe the σ-field, σ-ring and monotone glass generated by $\mathscr{C}$.

(i) P is any permutation of the points of X, i.e. any transformation from X to itself which is $(1,1)$ and onto, and $\mathscr{C}$ is the class of subsets of X left invariant by P.

(ii) X is $\mathbf{R}^3$, Euclidean 3-space, $\mathscr{C}$ is the class of all cylinders in X, i.e. sets E such that $(x, y, z_1) \in E \Rightarrow (x, y, z_2) \in E$ for all $z_2 \in \mathbf{R}$.

(iii) $X = \mathbf{R}^2$, the plane, $\mathscr{C}$ is class of all sets which are subsets of a countable union of horizontal lines.

9. Suppose X is the set of rational numbers in $0 \leqslant x \leqslant 1$, and let $\mathscr{Q}$ be the set of intervals of the form $\{x \in X;\ a < x \leqslant b\}$ where $0 \leqslant a \leqslant b \leqslant 1$; $a, b \in X$. Show that $\mathscr{Q}$ is a semi-ring and every set in $\mathscr{Q}$ is either empty or infinite.

Show that the σ-ring generated by $\mathscr{Q}$ contains all subsets of X.

10. Given a function $f: X \to Y$, and a class of subsets $\mathscr{A}$ of X, $f(\mathscr{A})$ will denote the class of subsets of Y of the form $f(A)$, $A \in \mathscr{A}$.

What is the relation between $f(A-B)$ and $f(A)-f(B)$? Give an example in which $f(A \cap B) \neq f(A) \cap f(B)$. Show that it is possible to have a ring $\mathscr{A}$ such that $f(\mathscr{A})$ is not a ring.

Give an example of a mapping $f: X \to Y$ and a semi-ring $\mathscr{S}$ in Y such that $f^{-1}(\mathscr{S})$ is not a semi-ring. For any class $\mathscr{N}$ of sets in Y show that

$$\mathscr{R}(f^{-1}(\mathscr{N})) = f^{-1}(\mathscr{R}(\mathscr{N})), \quad \mathscr{F}(f^{-1}(\mathscr{N})) = f^{-1}(\mathscr{F}(\mathscr{N})),$$

where $\mathscr{R}(\mathscr{C})$ is the ring generated by $\mathscr{C}$, and $\mathscr{F}(\mathscr{C})$ is the σ-field generated by $\mathscr{C}$.

1.6 Axiom of choice

Any non-empty set A contains at least one element x, and in the ordinary process of logic one can choose a particular element from a non-empty set. By using the principle of induction it follows that one can choose an element from each of a sequence of non-empty sets, but difficulty arises if one has to make the simultaneous choice of an element from each set of a non-countable class $\mathscr{C}$. The assumption that such a choice is possible can be formulated in the following equivalent forms, known as the axiom of choice:

(1) Given a non-empty class $\mathscr{C}$ of disjoint non-empty sets E_α, there is a set $G \subset \bigcup \{E_\alpha : E_\alpha \in \mathscr{C}\}$ such that $G \cap E_\alpha$ is a single point set for each $E_\alpha \in \mathscr{C}$.

(2) For a non-empty class $\mathscr{C}$ of non-empty sets E_α, there is a function (called a choice function) $f: \mathscr{C} \to \bigcup\{E_\alpha : E_\alpha \in \mathscr{C}\}$ such that, for each E_α in $\mathscr{C}$, $f(E_\alpha) \in E_\alpha$.

The difficulty in proofs using the axiom of choice is that only the existence of a choice function is postulated, and if $\mathscr{C}$ is uncountable, one has no information about its nature. However, we will find it convenient at times to use this axiom (or something equivalent). It has recently been shown that both the axiom of choice, and its negative, are consistent with the other axioms of set theory, so that one has to postulate this as an axiom. Although part of our theory will be valid without this axiom we will not trouble to discover how much and we will use the axiom of choice throughout when it is convenient.

There are a large number of other apparently different axioms which turn out to be logically equivalent to the axiom of choice. We will formulate just two of these, as they will be convenient later. Various new concepts will be needed before we can state them precisely.

Partial ordering

Suppose V is a set with elements $a, b, \ldots$ and $\prec$ is a relation defined between some but not necessarily all pairs $a, b \in V$ such that

(i) $\prec$ is transitive, i.e. $a \prec b,\ b \prec c \Rightarrow a \prec c$;

(ii) $\prec$ is reflexive, i.e. $a \prec a$ for all a in V;

(iii) $a \prec b,\ b \prec a \Rightarrow a = b$;

then V is said to be partially ordered by the relation $\prec$. V is said to be *simply* (or totally) *ordered* if,

(iv) for each pair $a, b \in V$ at least one of $a \prec b$, $b \prec a$ is valid.

Any partial ordering in a set V induces automatically a partial ordering in every subset of V. If $W \subset V$ and the induced ordering in W is a simple ordering, then W is said to be a chain in V.

For example, in $\mathbf{R}$ the usual $\leqslant$ relation defines a total ordering of $\mathbf{R}$. However, in $\mathbf{R}^2$, if we say $(x_1, y_1) \prec (x_2, y_2)$ if and only if $y_1 \leqslant y_2$ and $x_1 \leqslant x_2$ we have an example of a partial ordering which is not simple. A more useful example is the class $\mathscr{C}$ of all subsets of a fixed set X with $A \prec B$ meaning $A \subset B$.

A chain W in a partially ordered set V is called a *maximal* chain if it is not possible to obtain a larger chain by the addition of an element in $(V-W)$. We can now state

Kuratowski's lemma. *Every partially ordered set V contains a maximal chain.*

This means that there is a totally ordered subset $W \subset V$ such that for every $x \in V-W$, there is some element $y \in W$ such that neither of $x \prec y$, $y \prec x$ is true.

For a partially ordered set V, the element a is said to be an *upper bound* for the subset $C \subset V$ if $c \prec a$ for every $c \in C$. The element a is said to be the least upper bound or *supremum* of the subset C if

(i) a is an upper bound for C;

(ii) if b is an upper bound for C, then $a \prec b$.

It is easy to check that, in any partially ordered set V, it is impossible for two distinct elements a_1, a_2 to satisfy the above conditions (i), (ii) so that the supremum of a set C is unique when it exists. However, even when a set V is totally ordered, not all its subsets need have a supremum. With the usual ordering $\mathbf{R}$ has the property that any non-empty subset C which is bounded above has a supremum (this is known as the *least upper bound axiom*), but $\mathbf{Q}$ does not have this property.

Finally, we say that the element $m \in V$ is a maximal element of V if $m \prec a \Rightarrow m = a$. We can now state

Zorn's lemma. *If V is partially ordered and each chain W in V has a supremum, then V has a maximal element.*

Both Zorn's lemma and Kuratowski's lemma can be deduced from the axiom of choice,† but we will not give the details as these are complicated and outside the mainstream of our argument. However, the next theorem shows that, if we assume Zorn's lemma as an additional axiom, then both the axiom of choice and Kuratowski's lemma will be valid. This means that, in our subsequent work, we will assume whichever of these three results happens to be most convenient.

***Theorem* 1.6.** *The statements of (A) Kuratowski's lemma and (B) Zorn's lemma are equivalent. Either of them implies (C) the axiom of choice.*

Proof. $(A) \Rightarrow (B)$. Suppose W is a maximal chain in V, then by the hypothesis of (B) there is a supremum m for W so that $a \prec m$ for all $a \in W$. If m is not a maximal element of V, then there is a $b \in V$ such that $b \neq m$ and $m \prec b$. Then b is not in W as this would imply $b \prec m$,

† For a discussion of these and other axioms equivalent to axiom of choice see, for example, J. L. Kelley, *General Topology* (Van Nostrand, 1955).

and $b = m$. Hence we may add b to the chain W and the new set obtained is still a chain. This would contradict the fact that W is a maximal chain.

$(B) \Rightarrow (A)$. The chains in V form a class $\mathscr{C}$ which is partially ordered by inclusion. If now $\mathscr{W}$ is a chain in $\mathscr{C}$ with elements W (each of which is a chain in V), then the union $\bigcup \{W: W \in \mathscr{W}\}$ is a chain in V so that it is an element of $\mathscr{C}$ which can only be the supremum of $\mathscr{W}$. Hence by hypothesis $\mathscr{C}$ contains a maximal element, i.e. V contains a maximal chain.

$(B) \Rightarrow (C)$. We now suppose given a class $\mathscr{N}$ of sets E_α. There are clearly some subsets (in fact any finite subset) $\mathscr{N}_1 \subset \mathscr{N}$ on which it is possible to define a choice function $g: \mathscr{N}_1 \to \bigcup \{E_\alpha : E_\alpha \in \mathscr{N}_1\}$ such that $g(E_\alpha) \in E_\alpha$. The set V of all such functions g is therefore non-empty and it is partially ordered if we say $g_1 \prec g_2$ if g_1 is defined on $\mathscr{N}_1$, g_2 is defined on $\mathscr{N}_2$, $\mathscr{N}_1 \subset \mathscr{N}_2$ and $g_1(E_\alpha) = g_2(E_\alpha)$ for $E_\alpha \in \mathscr{N}_1$ (i.e. g_2 is an extension of g_1). If now W is a chain in V containing functions g_i defined on $\mathscr{N}_i$, the supremum of W is the function defined on $\bigcup \mathscr{N}_i$ which has the value $g_i(E_\alpha)$ on any set $E_\alpha \in \mathscr{N}_i$. If we now assume (B) it follows that the set V has a maximal element f. Then this function f must be defined on all the sets E_α, for otherwise if f is not defined on E_1 we could choose an element $x_1 \in E_1$, put $f(E_1) = x_1$ and this would be a proper extension of f and therefore contradict the fact that f is maximal.]

Exercises 1.6

1. Show that $\mathbf{Z}$ is partially ordered if $a < b$ means that a is a divisor of b.

2. Suppose α is a decomposition of the non-empty set X into disjoint subsets; $X = \bigcup A_i$ all the A_i disjoint. Show that the collection of such decompositions is partially ordered if $\alpha \prec \beta$ means that β is a refinement of α, i.e. if β is the decomposition $X = \bigcup B_j$ then each B_j is a subset of some A_i.

3. A partially ordered set V is said to be *well ordered* if each non-empty subset $W \subset V$ has a least element, i.e. there is a $w_0 \in W$ such that $w_0 \prec w$ for all $w \in W$. Show that, if V is well ordered, then it is simply ordered, and by considering the natural ordering of $\mathbf{R}$ show that there exist simply ordered sets which are not well ordered.

4. Assuming Zorn's lemma, show that any set X can be well ordered. *Hint.* Consider the class $\mathscr{C}$ of well ordered subsets $V \subset X$ with the partial ordering $V_1 \prec V_2$ if: (i) $V_1 \subset V_2$, (ii) the ordering in V_1 is the same as that induced by the ordering in V_2, (iii) V_1 is an initial segment of V_2 in the sense that $a \in V_1, b \in V_2, b \prec a \Rightarrow b \in V_1$. Show that each chain in $\mathscr{C}$ has a supremum and show that the maximal element V_0 in $\mathscr{C}$ must be X.

2

POINT SET TOPOLOGY

2.1 Metric space

In the first chapter we were concerned with abstract sets where no structure in the set was assumed or used. In practice, most useful spaces do have a structure which can be described in terms of a class of subsets called 'open'. By far the most convenient method of obtaining this class of open sets is to quantify the notion of nearness for each pair of points in the space. A non-empty set X together with a 'distance' function $\rho: X \times X \to \mathbf{R}$ is said to form a *metric space* provided that

(i) $\rho(y,x) = \rho(x,y) \geqslant 0$ for all $x, y \in X$;
(ii) $\rho(x,y) = 0$ if and only if $x = y$;
(iii) $\rho(x,y) \leqslant \rho(x,z) + \rho(y,z)$ for all $x, y, z \in X$.

The real number $\rho(x,y)$ should be thought of as the distance from x to y. Note that it is possible to deduce conditions (i), (ii) and (iii) from a smaller set of axioms: this has little point as all the conditions agree with the intuitive notion of distance. Condition (iii) for ρ is often called the triangle inequality because it says that the lengths of two sides of a triangle sum to at least that of the third. Condition (ii) ensures that ρ distinguishes distinct points of X, and (i) says that the distance from y to x is the same as the distance from x to y. When we speak of a metric space X we mean the set X together with a particular ρ satisfying conditions (i), (ii) and (iii) above. If there is any danger of ambiguity we will speak of the metric space (X, ρ).

In the set $\mathbf{R}$ of real numbers, it is not difficult to check (i), (ii) and (iii) for the usual distance function

$$\rho(x,y) = |x-y|,$$

and similarly in $\mathbf{R}^n$, $x = (x_1, \ldots, x_n)$, $y = (y_1, \ldots, y_n)$

$$\rho(x,y) = \left[\sum_{i=1}^{n} (x_i - y_i)^2\right]^{\frac{1}{2}}$$

(one always assumes the positive square root) the conditions for a metric are satisfied. Thus $\mathbf{R}$ and $\mathbf{R}^n$ are metric spaces with the usual Euclidean distance for ρ.

Open sphere

In a metric space (X, ρ), if $x \in X$, $r > 0$, then

$$S(x,r) = \{y : \rho(x,y) < r\};$$

the set consisting of those points of X whose distance from x is less than r is called an open sphere (spherical neighbourhood) centre x, radius r. Clearly, in $\mathbf{R}^n$, $S(x,r)$ is the inside of the usual Euclidean n-sphere centre x, radius r (for $n = 2$, the 'sphere' is the interior of a circle while for $n = 1$ it reduces to the interval $(x-r, x+r)$).

Open set

A subset E of a metric space X is said to be *open* if, for each point x in E there is an $r > 0$ such that the open sphere $S(x,r) \subset E$. Note that the open spheres defined above are examples of open sets since

$$y \in S(x,r) \Rightarrow \rho(x,y) = r_1 < r,$$

so that, for $0 < r_2 \leqslant r - r_1$, $S(y, r_2) \subset S(x,r)$.

***Theorem* 2.1.** *In a metric space X, the class $\mathscr{G}$ of open sets satisfies*

(i) $\varnothing, X \in \mathscr{G}$;

(ii) $A_1, A_2, \ldots, A_n \in \mathscr{G} \Rightarrow \bigcap_{i=1}^{n} A_i \in \mathscr{G}$;

(iii) $A_\alpha \in \mathscr{G}$ *for* α *in* $I \Rightarrow \bigcup_{\alpha \in I} A_\alpha \in \mathscr{G}$.

Proof. (i) Since any statement about the elements of $\varnothing$ is true, $\varnothing \in \mathscr{G}$, and it is clear that $S(x,r) \subset X$ for any $x \in X$, $r > 0$ so certainly $X \in \mathscr{G}$.

(ii) If $x \in \bigcap_{i=1}^{n} A_i$, then $x \in A_i$ for $i = 1, \ldots, n$ and each A_i is open so there are real numbers $r_i > 0$ for which $S(x, r_i) \subset A_i$. If we put $r = \min_{1 \leqslant i \leqslant n} r_i$, then $0 < r \leqslant r_i$ so that $S(x,r) \subset S(x,r_i) \subset A_i$ for $i = 1, \ldots, n$; and $S(x,r) \subset \bigcap_{i=1}^{n} A_i$.

(iii) For any $x \in \bigcup_{\alpha \in I} A_\alpha$, there must be a particular α in I such that $x \in A_\alpha$. Since this A_α is open, there is an $r > 0$ such that

$$S(x,r) \subset A_\alpha \subset \bigcup_{\alpha \in I} A_\alpha. \;\blacksquare$$

Remark. The condition (ii) says that $\mathscr{G}$ is closed for finite intersections, while (iii) says it is closed under arbitrary unions. One cannot extend (ii) to give closure for infinite intersections for, in $\mathbf{R}$ the intervals $(0, 1+(1/n))$ are open sets, but

$$\bigcap_{n=1}^{\infty}\left(0, 1+\frac{1}{n}\right) = \{x\colon 0 < x \leqslant 1\} = (0, 1]$$

is not open as it contains no open sphere centre 1.

It is more general to start with a set X and a class $\mathscr{G}$ of subsets of X satisfying (i), (ii), (iii) of theorem 2.1 and to call these 'the open sets' in X. Such a class $\mathscr{G}$ and set X are said to form a *topological space*, and $\mathscr{G}$ is said to determine the topology in X. A topological space $(X, \mathscr{G})$ is said to be *metrisable* if there is a distance function ρ defined on it which determines the class $\mathscr{G}$ for its open sets. Most topological spaces $(X, \mathscr{G})$ of interest satisfy the rather weak conditions which are sufficient to ensure metrisability, so that little is lost by assuming in the first place that we have a metric space (X, ρ). Of course two different metrics ρ_1, ρ_2 on a set X may define the same class $\mathscr{G}$ of open sets, so that even when a topological space is metrisable, the metric ρ is not uniquely determined—see exercise 2.4 (1).

In this chapter we will define most of the further concepts which depend on the toplogy of X in terms of the class $\mathscr{G}$ of open sets in X: this means that the definitions will make sense either in a metric space (X, ρ) or in a topological space $(X, \mathscr{G})$. However, when it simplifies the proof, we will assume that X has a metric ρ determining $\mathscr{G}$ and use this metric, so that some theorems will be stated and proved for metric spaces even though they are true more generally.

Closed set

A subset E of X is said to be closed if $(X-E)$ is open. If we apply this definition, with de Morgan's laws, to the conditions (i), (ii), (iii) of theorem 2.1 satisfied by the class of open sets, we see that the class $\mathscr{C}$ of closed sets satisfies

(i) $\varnothing, X \in \mathscr{C}$;

(ii) $A_1, A_2, \ldots, A_n \in \mathscr{C} \Rightarrow \bigcup_{i=1}^{n} A_i \in \mathscr{C}$;

(iii) $A_\alpha \in \mathscr{C}, \alpha$ in $I \Rightarrow \bigcap_{\alpha \in I} A_\alpha \in \mathscr{C}$;

so that the class $\mathscr{C}$ is closed for finite unions and arbitrary intersections.

In a metric space (X, ρ), for $x \in X$, $r > 0$ the set

$$\bar{S}(x, r) = \{y : \rho(x, y) \leqslant r\}$$

is called the *closed sphere* centre x, radius r. It is always a closed set according to our definition for

$$y \in G = X - \bar{S}(x,r) \Rightarrow \rho(x,y) = r_1 > r$$

so that $S(y,r_2) \subset G$ for $0 < r_2 \leqslant r_1 - r$.

Neighbourhood

In a topological space $(X, \mathscr{G})$, any open set containing $x \in X$ is said to be a neighbourhood of x.

Limit point of a set

Given a subset E of X, a point $x \in X$ is said to be a limit point (or point of accumulation) of E if every neighbourhood of x contains a point of E other than x. Note that the point x may or may not be in E. In a metric space it is easy to see that x is a limit point of E only if every neighbourhood N of x contains infinitely many points of E: for, if N contains only the points $x_1, x_2, \ldots, x_n$ of E (all different from x), then $S(x,r)$ where $r = \min_{1 \leqslant i \leqslant n} \rho(x, x_i)$ is a neighbourhood of x which contains no point of E other than x.

Lemma. *A set $E \subset X$ is closed if and only if E contains all its limit points.*

Proof. Suppose E is closed, then $X - E = G$ is open, so that if $x \in G$ there is a neighbourhood N of x with $N \subset G$. This means that N contains no point of E so that x is not a limit point of E. Conversely, if E is a set which contains its limit points and $x \in G = X - E$, then x is not a limit point of E so there is a neighbourhood N_x of x containing no point of E. Since N_x is open, so is $H = \bigcup_{x \in G} N_x$. But $N_x \subset G$ for all $x \in G$ so $H \subset G$, and every point x of G is in the corresponding N_x so $H \supset G$. Thus $H = G$ and G is open.]

Closure

For any set $E \subset X$, the closure of E, denoted by $\bar{E}$, is the intersection of all the closed subsets of X which contain E. It is immediate that $\bar{E}$ is a closed set, and $E = \bar{E}$ if and only if E is closed. Further since a closed set contains its limit points, $\bar{E}$ must contain all the limit points of E: in fact

$$\bar{E} = E \cup E',$$

where E' is the set of limit points of E, known as the derived set of E; for if $x \notin E \cup E'$, there is a neighbourhood N of x which contains no point of $E \cup E'$ so that $(X - N)$ is a closed set containing E and $x \notin \bar{E}$.

Limit of a sequence

Given a sequence $\{x_i\}$ of points in a metric space (X, ρ) we say that the sequence converges to the point $x \in X$ if each neighbourhood of x contains all but a finite number of points of the sequence. Thus $\{x_i\}$ converges to x if given $\epsilon > 0$, there is an integer N such that

$$i \geqslant N \Rightarrow \rho(x, x_i) < \epsilon.$$

We then write $x = \lim_{i \to \infty} x_i$ or $x = \lim x_i$ and say that x is the limit of the sequence $\{x_i\}$. Note that, in a metric space, the limit of a sequence is unique—see exercise 2.1 (7).

In a metric space X, given a point x and a set E, the *distance from x to E*, denoted by $d(x, E)$ is defined by

$$d(x, E) = \inf\{\rho(x, y) : y \in E\}.$$

This is always defined since $\{\rho(x, y) : y \in E\}$ is a set of non-negative real numbers. If $E \subset S(x, r)$ for some open sphere, we say that E is bounded and define the diameter of E, denoted diam (E), by

$$\operatorname{diam}(E) = \sup\{\rho(x, y) : x, y \in E\}.$$

If E is not bounded then the set $\{\rho(x, y) : x, y \in E\}$ is not bounded above and we put diam $(E) = +\infty$. Note that diam (E) is finite if and only if E is bounded. Finally, if E, F are two subsets of X, we define $d(E, F)$ by

$$\begin{aligned} d(E, F) &= \inf\{\rho(x, y) : x \in E, y \in F\} \\ &= \inf\{d(x, E), x \in F\} \end{aligned}$$

and call $d(E, F)$ the distance from E to F. Note that if $E \cap F \neq \varnothing$, then $d(E, F) = 0$ but there is no converse to this statement.

Remark

Many readers will be familiar with the concepts of this section for $\mathbf{R}$ and $\mathbf{R}^2$. Usually the proofs given in these special cases can be generalised to a general metric space, and often even to a topological space. The reader who has difficulty in working in an abstract situation should visualise the argument in the plane $\mathbf{R}^2$, but not use any of the special properties of $\mathbf{R}^2$.

Exercises 2.1

1. In any set X, the class $\mathscr{C}$ of all subsets of X satisfies the conditions for a class of open sets. Show that this topology can be generated by the metric

$$\rho(x,y) = 1 \quad \text{for} \quad x \neq y, \quad x, y \in X.$$

(This is called the *discrete* topology in X.) At the opposite end of the scale, the *indiscrete* topology in X is that for which $\mathscr{G} = \{\varnothing, X\}$: in this case the space is non-metrisable if X contains at least two points.

2. Show that in a metric space X containing at least 2 points

(i) single point sets $\{x\}$ are closed, but they are open only if

$$d(x, X-\{x\}) > 0;$$

(ii) finite sets are closed;

(iii) any open set G is the union of the class of open spheres contained in G;

(iv) any open set G is the union of the class of closed spheres contained in G.

3. In the topological space X, given a set $E \subset X$ a point x is said to be an *interior point* of E if there is a neighbourhood N of x with $N \subset E$. Prove

(i) the set E^0 consisting of the interior points of E is open;

(ii) E is open if and only if $E = E^0$.

4. In a topological space X show that

$$\overline{A_1 \cup A_2 \cup \ldots \cup A_n} = \bar{A}_1 \cup \bar{A}_2 \cup \ldots \cup \bar{A}_n,$$

but this does not extend to arbitrary unions. Give an example in which $\overline{E \cap F} \neq \bar{E} \cap \bar{F}$.

5. If X is a 2-point space $\{x_1, x_2\}$, $\rho(x_1, x_2) = 1$ show that (X, ρ) is a metric space in which the closure $\overline{S(x_1, 1)}$ of the open sphere $S(x_1, 1)$ is not the same as the closed sphere $\bar{S}(x_1, 1)$. However, if X is a normed linear space (see §2.6) then $\bar{S}(x, r) = \overline{S(x, r)}$.

6. Suppose $A \subset \mathbf{R}$ and A is closed and bounded below. Show that the infinum of A is an element of A.

7. In a metric space, suppose $\{x_n\}$ is a sequence converging to x and E is the set of points in this sequence. Show

(i) every subsequence $\{x_{n_i}\}$ converges to x;

(ii) either x is the only limit point of E or there is an integer N such that $x_n = x$ for $n \geqslant N$. Deduce that a sequence $\{x_n\}$ cannot converge to two different limit points.

8. Suppose E is the set of points in a sequence $\{x_n\}$ in a metric space and x is a limit point of E. Show there is a subsequence $\{x_{n_i}\}$ which converges to x.

9. For any set E in a metric space, show that

$$\bar{E} = \{x: d(x, E) = 0\}.$$

10. If E, F are subsets of a metric space X, $x, y \in X$, show

(i) $\rho(x,y) \geqslant |d(x,E) - d(y,E)|$;
(ii) $\rho(x,y) \leqslant d(x,E) + \text{diam}(E)$, if $y \in E$;
(iii) $|\rho(x,y_1) - \rho(x,y_2)| \leqslant \text{diam}(E)$, if $y_1, y_2 \in E$;
(iv) $d(E,F) = d(F,E)$.

Is d a metric on the space of subsets of X?

· **11.** In **R** show that a bounded open set is uniquely expressible as a countable union of disjoint open intervals.

Hint. For each $x \in E$, put $a = \inf\{y: (y,x) \subset E\}$, $b = \sup\{y: (x,y) \subset E\}$; and show that the open interval $I_x = (a,b)$ contains x, is contained in E and is such that any open interval I satisfying $x \in I \subset E$ satisfies $I \subset I_x$. Deduce that for $x_1, x_2 \in E$, either $I_{x_1} = I_{x_2}$ or $I_{x_1} \cap I_{x_2} = \varnothing$, so that $E = \bigcup_{x \in E} I_x$ is a disjoint union. Enumerate the intervals I_x by considering those of length greater than $1/n$ $(n = 1, 2, \ldots)$.

In $\mathbf{R}^n$ $(n \geqslant 2)$ show that a bounded open set can be expressed as a disjoint union of a countable number of half-open rectangles in $\mathscr{P}^n$ (but that this expression is never unique). Show that in general an open set in $\mathbf{R}^n$ $(n \geqslant 2)$ cannot be expressed as a disjoint union of open spheres, or of open rectangles.

2.2 Completeness and compactness

In a metric space (X, ρ) a sequence $\{x_n\}$ is said to be a *Cauchy sequence* if given $\epsilon > 0$, there is an integer N such that

$$n, m \geqslant N \Rightarrow \rho(x_n, x_m) < \epsilon.$$

It is immediate that any sequence $\{x_n\}$ in a metric space which converges to a point $x \in X$, is a Cauchy sequence.

Complete metric space

A metric space (X, ρ) is said to be complete if, for each Cauchy sequence $\{x_n\}$ in X, there is a point $x \in X$ such that $x = \lim x_n$. For example, the set **Q** of rationals is a metric space with the usual distance, but it is not complete for $\sqrt{2} \notin \mathbf{Q}$, but one can easily define a Cauchy sequence $\{x_n\}$ of rationals which converges to $\sqrt{2}$ (in **R**), and this sequence cannot converge to any rational. One of the important properties of the space **R** is that it is complete. This property is equivalent to the assumption that, in the usual ordering, every non-void subset of **R** which is bounded above has a supremum or least upper bound. We now give a proof of the completeness of **R** by a method which will turn out to be useful in more complicated situations.

Lemma. *The space* $\mathbf{R}$ *is complete.*

Proof. Let $\{x_n\}$ be a Cauchy sequence in $\mathbf{R}$. Define a sequence of integers $\{n_i\}$ by $n_0 = 1$; if n_{i-1} is defined, let n_i be such that $n_i > n_{i-1}$ and $n, m \geqslant n_i \Rightarrow |x_n - x_m| < 1/2^i$. Then the series

$$\sum_{i=1}^{\infty} (x_{n_i} - x_{n_{i-1}})$$

is absolutely convergent, and therefore convergent,† say to y. But

$$\sum_{i=1}^{p} (x_{n_i} - x_{n_{i-1}}) = x_{n_p} - x_1,$$

so the subsequence $\{x_{n_p}\}$ $(p = 1, 2, \ldots)$ must converge to $x = x_1 + y$. Given $\epsilon > 0$, choose integers $P, N \geqslant n_P$ such that

$$p \geqslant P \Rightarrow |x - x_{n_p}| < \tfrac{1}{2}\epsilon,$$

$$n, m \geqslant N \Rightarrow |x_n - x_m| < \tfrac{1}{2}\epsilon.$$

Now, if $m \geqslant N$, we can take $n_p \geqslant N$ with $p \geqslant P$ to obtain

$$|x - x_m| \leqslant |x - x_{n_p}| + |x_{n_p} - x_m| < \epsilon,$$

so the sequence $\{x_n\}$ must converge to x. ∎

Covering systems

A class $\mathscr{C}$ of subsets of X is said to *cover* the set $E \subset X$ or form a covering for E, if $E \subset \bigcup\{S : S \in \mathscr{C}\}$. If all the sets of $\mathscr{C}$ are open, and $\mathscr{C}$ covers E, then we say that $\mathscr{C}$ is an *open covering* of E.

Compact set

A subset E of X is said to be *compact* if, for each open covering $\mathscr{C}$ of E, there is a finite subclass $\mathscr{C}_1 \subset \mathscr{C}$ such that $\mathscr{C}_1$ covers E. For example, the celebrated Heine–Borel theorem states that any finite closed interval $[a, b]$ is compact. Though this is proved in most elementary text-books we include a proof which starts from the least upper bound property.

Lemma. *If* a, b *are real numbers, the closed interval*

$$[a, b] = \{x : a \leqslant x \leqslant b\}$$

is compact.

† The fact that absolute convergence implies convergence is a consequence of the least upper bound axiom for $\mathbf{R}$, that is, it follows from the assumption that every non-empty subset of $\mathbf{R}$ which is bounded above has a supremum. (See, for example, Burkill, *A First Course in Mathematical Analysis*, Cambridge, 1962.)

Proof. Let $\mathscr{C}$ be any open cover of $[a, b]$ and let c be the supremum of the set of x in $[a, b]$ for which some finite subfamily $\mathscr{C}_1 \subset \mathscr{C}$ covers $[a, x]$. (This set is non-void since it contains a.) Choose a set $G \in \mathscr{C}$ with $c \in G$ and choose a point $d \in (a, c)$ such that the closed interval $[d, c] \subset G$. Then there is a finite subfamily covering $[a, d]$ and the addition of G to this family gives a finite subfamily covering $[a, c]$. But unless $c = b$, since G is open, we have covered by a finite subfamily the interval $[a, e]$ for some $e > c$ which contradicts the choice of c. ∎

It is also possible to prove directly that any closed rectangle in $\mathbf{R}^n$ is compact, but we will be able to deduce this from theorem 2.6. We can use this to show that, in $\mathbf{R}^n$, every closed bounded set is compact. This will follow from the following:

Lemma. *If E is compact, and F is a closed subset of E, then F is compact.*

Proof. Suppose $\mathscr{C}$ is an open covering for F. Then $\mathscr{C}$, together with $(X - F)$, which is open, forms an open covering for E. This has a finite subcovering $\mathscr{C}_1$ of E and $\mathscr{C} \cap \mathscr{C}_1$ must be a finite subclass of $\mathscr{C}$ which covers F. ∎

It is not true in a general metric space that closed bounded subsets are compact—see exercise 2.2 (3). However, we can prove:

Lemma. *In a metric space X, every compact subset is closed and bounded.*

Proof. If E is not closed, there is a point $x_0 \in X$ which is a limit point of E but is not in E. For every $x \in E$, put $S_x = S(x, r)$ with $r = \frac{1}{2}\rho(x, x_0)$. Then the collection of all such open spheres covers E, but every finite subclass $S(x_1, r_1), \ldots, S(x_n, r_n)$ has a void intersection with $S(x_0, r)$, where $r = \min_{1 \leq i \leq n} r_i$ and so cannot cover E, for $S(x_0, r)$ contains points of E. On the other hand, if E is not bounded, the class of open spheres of radius 1 and centres in E covers E, but no finite subclass can cover E. ∎

Whenever the whole space X is compact, we talk of a compact space. The above lemma shows that $\mathbf{R}^n$ is not compact because it is not bounded. A space X is said to be *locally compact* if every point x in X has a neighbourhood N such that $\bar{N}$ is compact. It is clear that $\mathbf{R}^n$ is locally compact.

There are various other properties in a topological space which are equivalent to compactness under suitable conditions.

Weierstrass property

A set E is said to have property (W) if every infinite subset of E has at least one limit point.

Finite intersection property

A class $\mathscr{A}$ of subsets of E is said to have the finite intersection property if every finite intersection $\bigcap_{i=1}^{n} A_i$, where $A_i \in \mathscr{A}$, $(i = 1, 2, \ldots, n)$ is non-void.

***Theorem* 2.2.** (i) *A closed subset E of a topological space X is compact if and only if every class $\mathscr{A}$ of closed subsets of E with the finite intersection property has a non-void intersection.*

(ii) *In a metric space X, a subset E is compact if and only if it has property (W).*

Remark. In a general topological space, property (W) is equivalent to sequential compactness—the property that every countable open covering has a finite subcovering. A space in which arbitrary open coverings can be replaced by countable subcoverings is called *Lindelöf*. Thus in any Lindelöf space, property (W) is equivalent to compactness.

Proof. (i) Suppose E is compact and F_α, $\alpha \in I$ is a class of closed subsets of E with $\bigcap_{\alpha \in I} F_\alpha$ void. Then the class of sets $G_\alpha = X - F_\alpha$, $\alpha \in I$ is an open covering of E. Choose a finite subcovering

$$G_{\alpha_1}, G_{\alpha_2}, \ldots, G_{\alpha_n}; \quad \text{then} \quad \bigcap_{i=1}^{n} F_{\alpha_i} = \varnothing$$

so that the class of sets F_α, $\alpha \in I$ has not got the finite intersection property. This proves that compactness implies that any class $\mathscr{A}$ of closed subsets with the finite intersection property has a non-void intersection. Conversely suppose a closed set E is such that any class $\mathscr{A}$ of closed subsets with the finite intersection property has a non-void intersection, and suppose G_α, $\alpha \in I$ is an open covering of E, so that $\bigcap_{\alpha \in I} (X - G_\alpha) = \varnothing$. If E is closed $E \cap (X - G_\alpha)$, $\alpha \in I$ is a family of closed subsets of E, so there must be a finite set $\alpha_1, \alpha_2, \ldots, \alpha_n$ such that $\bigcap_{i=1}^{n} (X - G_{\alpha_i}) \cap E = \varnothing$. This means that $G_{\alpha_1}, \ldots, G_{\alpha_n}$ form a finite subcovering for E.

(ii) Suppose first that E has not got property (W). Let A be an infinite subset of E with no limit point. If A is not enumerable, choose an enumerable subset B of A. Then B is closed and $(X - B)$ is open. Enumerate B as a sequence of distinct points $\{x_i\}$, and for each x_i choose a neighbourhood N_i which contains x_i but no other point of B.

Then the sequence $\{N_i\}$ together with the set $(X-B)$ form an open covering of E, which has no finite subcovering as none of the open sets N_i can be omitted without 'uncovering' the corresponding x_i.

Conversely suppose E has property (W). Then there is a finite class $\mathscr{C}_n$ of spherical neighbourhoods of radius $1/n$ which covers E; for otherwise we could find an infinite subset of E all of whose points were distant more than $1/n$ apart and such a subset can have no limit point. Let $\mathscr{C} = \bigcup_{n=1}^{\infty} \mathscr{C}_n$; so that $\mathscr{C}$ is countable. Now if G is any open set, for each x in $G \cap E$ we can find a sphere $S \in \mathscr{C}$ containing x with $S \subset G$: for we can first choose $\delta > 0$ so that $S(x, \delta) \subset G$ and if $n > 2/\delta$, the sphere of $\mathscr{C}_n$ which contains x will be contained in $S(x, \delta)$. Given any open covering $\mathscr{D}$ of E, carry out the above process for each set D of $\mathscr{D}$ which intersects E, and each x in $D \cap E$, and let $\mathscr{C}' \subset \mathscr{C}$ be the countable collection of open spheres obtained. For each $S \in \mathscr{C}'$, choose *one* set $D \in \mathscr{D}$ with $D \supset S$, and let $\mathscr{D}'$ be the countable class of sets so obtained. Then, since $\mathscr{C}'$ covers E, the class $\mathscr{D}'$ is a countable subcovering. This means that, if we assume property (W), open coverings can be replaced by a countable subcovering.

Now suppose $E \subset \bigcup_{i=1}^{\infty} G_i$ where the sets G_i are open. Then, if there is no finite subcovering, for each integer n we can find a point

$$x_n \in E - \bigcup_{i=1}^{n} G_i$$

and the sequence $\{x_n\}$ must form an infinite set, so that there is a limit point $x_0 \in E$. But $x_0 \in G_k$ for some k, and G_k is open and therefore is a neighbourhood of x_0; this means we can find an $n > k$ such that

$$x_n \in G_k \subset \bigcup_{i=1}^{n} G_i,$$

which is a contradiction. ∎

Compactification

Many operations can be carried out more easily in compact spaces than in non-compact spaces. Given a non-compact space X a useful trick is to enlarge it to a topological space $X^* \supset X$ which is compact and such that the system G of open sets in X is obtained by taking the intersection $X \cap G$ with X of sets G which are open in X^*. This device is known as the compatification of X. For example, $\mathbf{R}$ is not compact, but if we adjoin two points $+\infty$, $-\infty$ to form the space $\mathbf{R}^*$ of *extended* *real numbers* we can show that $\mathbf{R}^*$ is compact if we call a set $E \subset \mathbf{R}^*$ open if $E \cap \mathbf{R}$ is open and

if $+\infty \in E$, there is a neighbourhood $\{x : a < x \leqslant +\infty\} \subset E$;

if $-\infty \in E$, there is a neighbourhood $\{x : -\infty \leqslant x < b\} \subset E$,

where a, $b \in \mathbf{R}$. Note that the extended real number system $\mathbf{R}^*$ is simply ordered if we put $-\infty < x < +\infty$ for all $x \in \mathbf{R}$.

In general, a non-compact topological space X may be compactified in many different ways. The simplest method is to adjoin a single point ∞ (which can be thought of as a point at infinity) to give the space $X^* = X \cup \{\infty\}$ and say that a subset G of X^* is open if either

(i) $G \subset X$ and G is open in X; or

(ii) $\infty \in G$ and $(X^* - G)$ is a closed compact subset of X. It is not difficult to verify that this collection of 'open' sets defines a topology in X^* in which X^* is compact. This process is called the *one-point* compactification of X. It is familiar in the theory of the complex plane, where it is usual to add a single point at infinity (with neighbourhoods of the form $|z| > R$) to make the resulting 'closed plane' compact. Note that, if G^* is the class of open sets in X^*, G is the class of sets of the form $X \cap E$, $E \in G^*$.

There are other, more sophisticated, methods of compactifying a topological space X, but we will not require these in the present book.

Exercises 2.2

1. If (X, ρ) is a compact metric space, show that

(i) X is complete;

(ii) for each $\epsilon > 0$, X can be covered by a finite class of open spheres of radius ϵ.

2. If (X, ρ) is a complete metric space which, for each $\epsilon > 0$, can be covered by a finite number of spheres of radius ϵ, show that X is compact.

3. The open interval $X = (0, 1) \subset \mathbf{R}$ with the usual metric

$$\rho(x_1, x_2) = |x_1 - x_2|$$

is a metric space. In (X, ρ) the set X is closed and bounded. Show that X is not compact (and therefore not complete by example 2).

4. Construct a covering of the closed interval $[0, 1]$ by a family of closed intervals such that there is no finite subcovering.

5. If A, B are compact subsets of a metric space X, show that there are points $x_0 \in A$, $y_0 \in B$ for which

$$\rho(x_0, y_0) = d(A, B).$$

Hint. Take sequences $\{x_i\}$ in A, $\{y_i\}$ in B with $\rho(x_i, y_i) < d(A,B)+1/i$, and apply property (W) to find convergent subsequences.

6. Give the details of the proof that the process of adjoining a point ∞ used to give the one-point compactification does yield a compact set X^*. If this process is applied to a space X which is already compact, show that the one point set $\{\infty\}$ is then both open and closed in X^*.

2.3 Functions

In Chapter 1 we defined the notion of a function $f: X \to Y$. When X and Y are topological spaces it is natural to enquire how the function f is related to these topologies. In particular do points which are 'close' in X map into points which are close in Y? We make this precise first for metric spaces.

Continuous function

If (X, ρ_X), (Y, ρ_Y) are metric spaces, a function $f: X \to Y$ is said to be continuous at $x = a$ if, given $\epsilon > 0$ there is a $\delta > 0$ such that

$$\rho_X(x,a) < \delta \Rightarrow \rho_Y(f(x), f(a)) < \epsilon.$$

If $E \subset X$, we say f is continuous on E if it is continuous at each point of E. In particular $f: X \to Y$ is said to be continuous (or continuous on X) if it is continuous at each point of X.

Lemma. *If (X, ρ_X), (Y, ρ_Y) are metric spaces a function $f: X \to Y$ is continuous if and only if $f^{-1}(G)$ is an open set in X for each open set G in Y.*

Proof. Suppose first that f is continuous and G is an open set in Y. If $f^{-1}(G)$ is void, then it is open. Otherwise, let $a \in f^{-1}(G)$, $f(a) \in G$ so that there is an $\epsilon > 0$ for which the sphere $S(f(a), \epsilon) \subset G$. But then we can find a $\delta > 0$ such that

$$\rho_X(x,a) < \delta \Rightarrow f(x) \in S(f(a), \epsilon) \subset G$$

so that the sphere $S(a, \delta) \subset f^{-1}(G)$.

Conversely consider f at a point a of X. For each $\epsilon > 0$, $S(f(a), \epsilon) = H$ is an open set in Y, so that if $f^{-1}(H)$ is open, we can find a $\delta > 0$ for which $S(a, \delta) \subset f^{-1}(H)$, that is such that

$$\rho_X(x,a) < \delta \Rightarrow \rho_Y(f(x), f(a)) < \epsilon. ▌$$

If $(X, \mathscr{G})$, $(Y, \mathscr{H})$ are topological spaces, the function $f: X \to Y$ is said to be *continuous* if $f^{-1}(H) \in \mathscr{G}$ for every H in $\mathscr{H}$. The lemma just proved shows that when the topologies in X, Y are determined by metrics this definition agrees with the one first given for mappings from one metric space to another.

Now if $f: X \to Y$ is continuous and E is a closed subset of Y, it follows that $f^{-1}(E)$ is closed in X. One has to be careful about the implications in the reverse direction. In general, it is not true for a continuous $f: X \to Y$, that A open in $X \Rightarrow f(A)$ open in Y. There is one important result of this kind which is valid:

***Theorem* 2.3.** *If $f: X \to Y$ is continuous, and A is a compact subset of X, then $f(A)$ is compact in Y.*

Proof. Suppose G_α, $\alpha \in I$ forms an open covering of $f(A)$. Then $f^{-1}(G_\alpha)$ is open for each α and the class $f^{-1}(G_\alpha)$, $\alpha \in I$ must cover A. Since A is compact, there is a finite subcovering $f^{-1}(G_1), \ldots, f^{-1}(G_n)$ which covers A, and this implies that $G_1, \ldots, G_n$ cover $f(A)$. ▌

Corollary. *If $f: X \to \mathbf{R}$ is continuous, and A is compact, the set $f(A)$ is bounded and the function f attains its bounds on A at points in A.*

Proof. $f(A)$ is compact, and so it must be closed and bounded. Hence $\sup\{x: x \in f(A)\}$ and $\inf\{x: x \in f(A)\}$ exist and belong to the set $f(A)$. Hence there are points $x_1, x_2 \in A$ for which $f(A) \subset [f(x_1), f(x_2)]$. ▌

Remark. The reader will recognise this corollary as a generalisation of the elementary theorem that a continuous function $f: [a, b] \to \mathbf{R}$ is bounded and attains its bounds.

It is important to notice that, in a metric space X, the distance function $\rho(x, y)$ is continuous for each fixed y considered as a function from X to $\mathbf{R}$. Further, for a fixed set A, $d(x, A)$ defines a continuous function from X to $\mathbf{R}$ since

$$\rho(x_1, x_2) \geqslant |d(x_1, A) - d(x_2, A)|.$$

This means that if E is compact, F is any set, the function $d(x, F)$ for $x \in E$ attains its lower bound so that there is an x_0 in E with

$$d(x_0, F) = d(E, F).$$

Now, if F is also compact

$$d(x_0, F) = \inf\{\rho(x_0, y): y \in F\}$$

is the lower bound on a compact set of another continuous function, so that there is a y_0 in F such that

$$d(x_0, F) = \rho(x_0, y_0) = d(E, F).$$

Thus we have proved a further corollary to theorem 2.3—which could have been proved by a different argument (see exercise 2.2 (5)).

Corollary. *If E, F are two compact subsets of a metric space (X,ρ), there are points $x_0 \in E$, $y_0 \in F$ such that*

$$\rho(x_0, y_0) = d(E,F).$$

Uniformly continuous function

A mapping $f: X \to Y$ from the metric space (X,ρ_X) to the metric space (Y,ρ_Y) is said to be uniformly continuous on the subset $A \subset X$ if given $\epsilon > 0$, there is a $\delta > 0$ for which

$$x, y \in A,\ \rho_X(x,y) < \delta \Rightarrow \rho_Y(f(x),f(y)) < \epsilon. \tag{2.3.1}$$

Clearly a function which is uniformly continuous on A is certainly continuous at each point of A, but the point of the condition (2.3.1) is that one can make $f(x)$ close to $f(y)$ in Y simply by making x close to y simultaneously for all $x, y \in A$. The choice of δ in (2.3.1) does not depend on x or y. In general, uniform continuity does not follow from continuity, but there is an important case in which it does:

Theorem 2.4. *If X, Y are metric spaces, and $f: X \to Y$ is continuous on A where A is a compact subset of X, then f is uniformly continuous on A.*

Proof. Given $\epsilon > 0$, for each $x \in A$, there is a $\delta_x > 0$ such that

$$\xi \in S(x, \delta_x) \cap A \Rightarrow f(\xi) \in S(f(x), \tfrac{1}{2}\epsilon).$$

For $x \in A$, the class of spheres $S_x = S(x, \frac{1}{2}\delta_x)$ form an open covering of A. Choose a finite subcovering $S_{x_1}, \ldots, S_{x_n}$ and put $\delta = \frac{1}{2}\min(\delta_{x_1}, \ldots, \delta_{x_n})$. Then if $\rho_X(\xi, \eta) < \delta$, ξ, $\eta \in A$, there must be a sphere S_{x_i} which contains ξ, and $S(x_i, \delta_{x_i})$ will then contain η. This implies

$$\rho_Y(f(\xi),f(\eta)) \leqslant \rho_Y(f(\xi),f(x_i)) + \rho_Y(f(\eta),f(x_i)) < \epsilon. \blacksquare$$

Remark. The reader will recognize the above theorem as a generalisation of the result that a continuous function f: $[a,b] \to \mathbf{R}$ is uniformly continuous.

Exercises 2.3

1. Consider the function f: $(0, \infty) \to \mathbf{R}$ given by $f(x) = \min(1, 1/x)$, for $x > 0$.

Show that it is continuous. Find the image $f(E)$ of

(i) the set $E = (\frac{1}{2}, 1)$;
(ii) the set Z of positive integers;

(i) shows that E can be open, $f(E)$ not open; (ii) shows that E can be closed, $f(E)$ not closed.

2. The function

$$f:(0,1) \to \mathbf{R} \text{ given by } f(x) = \frac{1}{x(1-x)}$$

is continuous on $(0, 1)$, but not bounded and not uniformly continuous, so theorems 2.3, 2.4 fail if the set is not closed. To see they also fail if the set is closed but not compact, examine

$$g: \mathbf{R} \to \mathbf{R} \quad \text{given by} \quad g(x) = \exp(x).$$

3. In the argument of the proof to theorem 2.4 why could we not have put

$$\delta = \tfrac{1}{2}\inf\{\delta_x : x \in A\}$$

before first restricting to a finite subset?

4. Suppose A is compact and $\{f_r\}$ is a monotone sequence of continuous functions $f_r: A \to \mathbf{R}$ converging to a continuous $f: A \to \mathbf{R}$. Show that the convergence must be uniform, and give an example to show that the condition that A be compact is essential.

5. Prove Lebesgue's covering lemma, which states that if $\mathscr{C}$ is an open cover of a compact set A in a metric space (X, ρ), then there is a $\delta > 0$, such that the sphere $S(x, \delta)$ is contained in a set of $\mathscr{C}$ for each $x \in X$.

2.4 Cartesian products

We have already defined the direct product of two arbitrary sets X, Y as the set of ordered pairs (x, y) with $x \in X$, $y \in Y$. If $(X, \mathscr{G})$ $(Y, \mathscr{H})$ are topological spaces, then there is a natural method of defining a topology in $X \times Y$. Let $\mathscr{A}$ be the class of *rectangle* sets $G \times H$ with $G \in \mathscr{G}$, $H \in \mathscr{H}$ and let $\mathscr{C}$ be the class of sets in $X \times Y$ which are unions of sets in $\mathscr{A}$ (finite or infinite unions): it is immediate that $\mathscr{C}$ satisfies the conditions (i), (ii), (iii) of theorem 2.1. This class $\mathscr{C}$ of 'open' sets in $X \times Y$ is said to define the *product topology*. This definition extends in an obvious manner to finite products $X_1 \times X_2 \times \ldots \times X_n$, and it is also possible to extend it to an arbitrary product of topological spaces—though we will not have occasion to consider a topology for infinite product spaces.

***Theorem* 2.5.** *If (X_i, ρ_i) $(i = 1, \ldots, n)$ are metric spaces then*

$$\rho((x_1, \ldots, x_n), (y_1, \ldots, y_n)) = \max_{1 \leqslant i \leqslant n} \rho_i(x_i, y_i)$$

defines a metric in the Cartesian product $X_1 \times \ldots \times X_n$ which generates the product topology.

Proof. It is clear that $\rho(x, y) = 0$ if and only if $\rho_i(x_i, y_i) = 0$ for each i

which implies $x_i = y_i$, $1 \leqslant i \leqslant n$ or $x = y$. Thus in order to show that ρ is a metric it is sufficient to prove the triangle inequality. But

$$\rho_i(x_i, y_i) \leqslant \rho_i(x_i, z_i) + \rho_i(y_i, z_i) \quad (i = 1, \ldots, n)$$

since the ρ_i are all metrics, so that

$$\max_{1 \leqslant i \leqslant n} \rho_i(x_i, y_i) \leqslant \max_{1 \leqslant i \leqslant n} \{\rho_i(x_i, z_i) + \rho_i(y_i, z_i)\}$$

$$\leqslant \max_{1 \leqslant i \leqslant n} \rho_i(x_i, z_i) + \max_{1 \leqslant i \leqslant n} \rho_i(y_i, z_i).$$

Now in the product space, the open sphere centre x, radius r has the form

$$\{(y_1, \ldots, y_n) : \rho_i(x_i, y_i) < r, \quad 1 \leqslant i \leqslant n\}$$

$$= S(x_1, r) \times S(x_2, r) \times \ldots \times S(x_n, r)$$

that is, it is the product of spheres in each of the component spaces. Thus the open spheres are open sets in the product topology and since every open set in the topology of a metric ρ is a union of the open spheres contained in it, each such set must be open in the product topology.

Conversely if G is an open set in the product topology and $x \in G$, there must be open sets $G_i \subset X_i$ such that $x \in G_1 \times \ldots \times G_n \subset G$. Choose $r_i > 0$ such that $S(x_i, r_i) \subset G_i$ and put $r = \min_{1 \leqslant i \leqslant n} r_i$. Then $r > 0$ and $S(x, r) \subset S(x_1, r_1) \times \ldots \times S(x_n, r_n) \subset G$. Thus any set G open in the product topology is also open in the topology of the metric ρ. ∎

Remark. The metric ρ defined in this theorem is by no means the only one which generates the product topology—see exercise 2.4 (1, 2).

***Theorem* 2.6.** *If X, Y are compact topological spaces, then $X \times Y$ is compact in the product topology.*

Remark. The proof which follows extends immediately to finite Cartesian products of compact sets. Actually the theorem is true for arbitrary products, and in this more general form is due to Tychonoff.

Proof. Suppose first that $R_\alpha = G_\alpha \times H_\alpha$, $\alpha \in I$ is a covering of $X \times Y$ by open rectangles. Then if x_0 is a fixed point of X and I_{x_0} is the set of indices α for which $(x_0, y) \in G_\alpha \times H_\alpha$ for some $y \in Y$, the class H_α, $\alpha \in I_{x_0}$ forms an open covering of the compact set Y. Hence, there is a finite set $J_{x_0} \subset I_{x_0}$ such that R_α, $\alpha \in J_{x_0}$ covers the set $\{x_0\} \times Y$. But if we put $A_{x_0} = \bigcap_{\alpha \in J_{x_0}} G_\alpha$, A_{x_0} is open, contains x_0, and the finite class R_α, $\alpha \in J_{x_0}$ must cover all of $A_{x_0} \times Y$. For each $x_0 \in X$, form such an open set A_{x_0}: the class of all sets of this form is an open covering of the compact set X, and so has a finite subcovering $A_{x_1}, A_{x_2}, \ldots, A_{x_n}$. It follows that R_α, $\alpha \in \bigcup_{i=1}^{n} J_{x_i}$ is a finite subcovering of $X \times Y$.

It remains to show that, in testing for compactness, it is sufficient to consider coverings by open rectangles $G \times H$ with G open in X, H open in Y. Suppose then that every covering of $X \times Y$ by open rectangles has a finite subcovering, and consider an arbitrary open covering $G_\alpha, \alpha \in I$. Each point $(x, y) \in X \times Y$ is an element of an open set G_α and therefore there is an open rectangle $R_{x,y}$ with $(x, y) \in R_{x,y} \subset G_\alpha$. The class of open rectangles $R_{x,y}$, $(x, y) \in X \times Y$ clearly covers $X \times Y$ and so we can find a finite subcovering $R_1, R_2, \ldots, R_n$. The corresponding sets $G_1, \ldots, G_n$ then form a finite subclass of the original covering class which covers $X \times Y$.]

Corollary. *Any bounded closed set in* $\mathbf{R}^n$ *is compact.*

Proof. The usual topology in $\mathbf{R}^n$ is given by the distance function

$$\rho(x, y) = \left\{\sum_{i=1}^{n} (x_i - y_i)^2\right\}^{\frac{1}{2}},$$

while theorem 2.5 shows that the product topology is given by the distance function

$$\tau(x, y) = \max_{1 \leqslant i \leqslant n} |x_i - y_i|.$$

Now $\tau(x, y) \leqslant \rho(x, y) \leqslant \sqrt{n}\, \tau(x, y)$ for all $x, y \in \mathbf{R}^n$ so that the topology of the usual metric ρ is the product topology.

If E is bounded, there is a real number K such that E is a subset of the Cartesian product of n intervals $[-K, K]$. Since each of these intervals is compact, the product is compact and therefore E is compact if it is closed.]

Exercises 2.4

1. If ρ_1, ρ_2 are two metrics in X such that $c\rho_1(x, y) \leqslant \rho_2(x, y) \leqslant k\rho_1(x, y)$ for all $x, y \in X$ where $c > 0$, $k > 0$; show that ρ_1 and ρ_2 generate the same topology in X.

2. Show that, if $X \times Y$ has the product topology, then the projection function p: $X \times Y \to X$ defined by $p(x, y) = x$ is continuous.

In a space X, if $\mathscr{G}_1, \mathscr{G}_2$ are two collections of 'open' sets defining topologies and $\mathscr{G}_1 \supset \mathscr{G}_2$ we say that the topology given by $\mathscr{G}_2$ is *coarser* than that given by $\mathscr{G}_1$. Show that the product topology in $X \times Y$ is the coarsest topology for which projections are continuous. (For an arbitrary Cartesian product $\prod_{\alpha \in I} X_\alpha$ of topological spaces $(X_\alpha, \mathscr{G}_\alpha)$ the projection p_β can be defined as $p_\beta(x_\alpha, \alpha \in I) = x_\beta \in X_\beta$ for any $\beta \in I$. One method of defining the product topology in the Cartesian product space is to say that it is the coarsest topology for which each of the projections is continuous.)

3. Suppose $X \times Y$ has the product topology and $A \subset X$, $B \subset Y$. Show that

$$\overline{A \times B} = \bar{A} \times \bar{B},$$

and prove that the product of closed sets is closed.

2.5 Further types of subset

In a topological space X, a subset $E \subset X$ is said to be *nowhere dense* if the closure $\bar{E}$ of E contains no non-void open set. If E is nowhere dense, and G is any non-void open set, the intersection $G \cap (X - \bar{E})$ is a non-void open subset disjoint from $\bar{E}$, and therefore from E. Conversely if $\bar{E}$ contains a non-void open set H then every non-void open subset of H is a neighbourhood of each of its points, and therefore contains points of E. Thus E is nowhere dense if and only if every non-void open set in X contains a non-void open set disjoint from E.

Category

A subset $E \subset X$ is said to be of the *first category* (in X) if there is a sequence $\{E_n\}$ of nowhere dense subsets of X such that $E = \bigcup_{i=1}^{\infty} E_i$. A set $E \subset X$ which cannot be expressed as a countable union of nowhere dense sets is said to be of the *second category*.

Before proving that complete metric spaces are necessarily of the second category, it is convenient to prove a lemma which again generalises a well-known result in **R** (about a decreasing sequence of closed intervals).

Lemma (*Cantor*). *In a complete metric space, given $\{A_n\}$ a decreasing sequence of non-empty closed sets such that* $\operatorname{diam}(A_n) \to 0$, $\bigcap_{n=1}^{\infty} A_n$ *is a one point set.*

Proof. For each integer n, choose a point $x_n \in A_n$. Then given $\epsilon > 0$ we can choose N so that

$$n \geqslant N \Rightarrow \operatorname{diam}(A_n) < \epsilon,$$

and, since $A_N \supset A_n$ for $n \geqslant N$,

$$n, m \geqslant N \Rightarrow x_n, x_m \in A_N \Rightarrow \rho(x_n, x_m) < \epsilon;$$

so that $\{x_n\}$ is a Cauchy sequence. Since the space is complete, there is a point x_0 such that $x_n \to x_0$ as $n \to \infty$. For each n, since A_n is closed and $x_i \in A_n$ for $i \geqslant n$ we must have $x_0 \in A_n$ so that $x_0 \in \bigcap_{n=1}^{\infty} A_n$. Now $\operatorname{diam}\left(\bigcap_{n=1}^{\infty} A_n\right) \leqslant \operatorname{diam}(A_n)$ for each n so that $\operatorname{diam}(\bigcap A_n) = 0$ and the set $\bigcap A_n$ cannot contain more than one point. ∎

***Theorem* 2.7** (*Baire*). *Every complete metric space X is of the second category.*

Proof. It is sufficient to show that if $\{A_n\}$ is any sequence of nowhere dense sets there are points $x \in X - \bigcup_{n=1}^{\infty} A_n$. Starting with such a sequence $\{A_n\}$, since we can find a non-void open set disjoint from A_1 there is a sphere $S(x_1, r_1)$ with $0 < r_1 < 1$ such that $\bar{S}(x_1, r_1) \cap A_1 = \varnothing$. Suppose we have found spheres $S(x_1, r_1) \supset S(x_2, r_2) \supset \ldots \supset S(x_{n-1}, r_{n-1})$ with $0 < r_{n-1} < 1/(n-1)$ such that $\bar{S}(x_i, r_i) \cap A_i = \varnothing$ for $1 \leqslant i \leqslant n-1$. Then $S(x_{n-1}, r_{n-1})$ is a non-void open set in X and A_n is nowhere dense. There must therefore be an open subset disjoint from A_n so that we can find a sphere $S(x_n, r_n)$ with

$$0 < r_n < 1/n, \quad S(x_n, r_n) \subset S(x_{n-1}, r_{n-1}) \quad \text{and} \quad \bar{S}(x_n, r_n) \cap A_n = \varnothing.$$

By the last lemma, there is a unique point $x_0 \in \bigcap_{n=1}^{\infty} \bar{S}(x_n, r_n)$ and this point x_0 cannot be in A_n for any integer n. This means that $X \neq \bigcup_{n=1}^{\infty} A_n$. ∎

Corollary. *$\mathbf{R}^n$ is of the second category. For $a < b$, the interval $[a, b] \subset \mathbf{R}$ is of the second category.*

Dense subset

If A, E are subsets of a topological space X, we say that A is *dense in E* if $\bar{A} \supset E$. This means that any open set G which contains a point of E also contains a point of A. In particular A is dense in X if $\bar{A} = X$, that is, if every non-void open set contains a point of A.

Separable space

A topological space X is said to be separable if there is a countable set $E \subset X$ such that E is dense in X. This implies the existence of a sequence $\{x_n\}$ in X such that every non-void open subset contains a point of the sequence. It is immediate that $\mathbf{R}$ and $\mathbf{R}^n$ are separable as the set of points with rational coordinates is dense and countable. Further, every compact metric space X is separable, since X can be covered by a finite class $\mathscr{C}_n$ of open spheres of radius $1/n$ for $n = 1, 2, \ldots$ and the (countable) set consisting of the centres of all these spheres is clearly dense in X.

Borel sets and Borelian sets

In any topological space X, we will call the σ-field $\mathscr{B}$ generated by the open sets the class of Borel sets, and the σ-field $\mathscr{K}$ generated by the compact sets the class of Borelian sets. (One must remark that some authors use the term Borel sets for $\mathscr{K}$.)

In a metric space the compact sets are closed, and therefore in $\mathscr{B}$ so that $\mathscr{K} \subset \mathscr{B}$. If $X = \bigcup_{i=1}^{\infty} K_i$ is a countable union of compact sets (in this case we can say that X is σ-compact), then $\mathscr{K} = \mathscr{B}$. In order to prove this it is sufficient to show that each open set $G \in \mathscr{K}$: but if G is open, $E = X - G$ is closed and so $E \cap K_i$ is compact for each i and this implies $E = \bigcup_{i=1}^{\infty} E \cap K_i \in \mathscr{K}$ and $G = X - E \in \mathscr{K}$. Now Euclidean n-space $\mathbf{R}^n$ is the union of the closed spheres $\bar{S}(0, k)$ $(k = 1, 2, \ldots)$ each of which is compact, so $\mathbf{R}^n$ is σ-compact. This means that, in $\mathbf{R}^n$ the Borel sets and the Borelian sets are the same.

Note that, by our definition, the class $\mathscr{B}^n$ of Borel sets in $\mathbf{R}^n$ is the σ-field generated by the open sets in $\mathbf{R}^n$. It is convenient to see that $\mathscr{B}^n$ can also be obtained as the σ-field generated by a simpler class of sets.

Lemma. *The class $\mathscr{P}^n$ of half-open intervals in $\mathbf{R}^n$ generates the σ-field $\mathscr{B}^n$ of Borel sets in $\mathbf{R}^n$.*

Proof. Let $\mathscr{F}^n$ be the σ-field generated by $\mathscr{P}^n$. Each set in $\mathscr{P}^n$,

$$\{(x_1, x_2, \ldots, x_n): a_i < x_i \leqslant b_i, \quad i = 1, 2, \ldots, n\}$$

can clearly be obtained as a countable intersection

$$\bigcap_{k=1}^{\infty} \left\{(x_1, \ldots, x_n): a_i < x_i < b_i + \frac{1}{k}, \quad i = 1, \ldots, n\right\}$$

of open rectangles, and is therefore in $\mathscr{B}^n$. Hence $\mathscr{B}^n \supset \mathscr{P}^n$ so that $\mathscr{B}^n \supset \mathscr{F}^n$.

On the other hand, each open set G in $\mathbf{R}^n$ is a union of those rectangles of $\mathscr{P}^n$ whose boundary points a_i, b_i are all rational. Since there are only countably many such sets, each G is a countable union of sets in $\mathscr{P}^n$ and so $\mathscr{P}^n \supset \mathscr{G}$. This implies that $\mathscr{F}^n \supset \mathscr{B}^n$. ∎

It is sometimes useful to be able to describe sets which can be obtained from a given class $\mathscr{C}$ by a countable operation. We say that

E is a $\mathscr{C}_\sigma$ set if it is possible to find sets $E_1, E_2 \ldots$ in $\mathscr{C}$ such that $E = \bigcup_{i=1}^{\infty} E_i$; and E is a $\mathscr{C}_\delta$ set if $E = \bigcap_{i=1}^{\infty} E_i$ for a sequence $\{E_n\}$ in $\mathscr{C}$. In particular, if $\mathscr{G}$ is the class of open sets in a space X, we see that $\mathscr{B} \supset \mathscr{G}_\delta \supset \mathscr{G}$, $\mathscr{G}_\sigma = \mathscr{G}$. Similarly, if $\mathscr{F}$ is the class of closed sets, $\mathscr{B} \supset \mathscr{F}_\sigma \supset \mathscr{F}$, and $\mathscr{F}_\delta = \mathscr{F}$.

Perfect set

A subset E of a topological space X is said to be perfect if E is closed, and each point of E is a limit point of E. For example, in $\mathbf{R}^n$, any closed sphere $\bar{S}(X, r)$, $r > 0$ is perfect and, in particular, the closed interval $[a, b]$ is perfect in $\mathbf{R}$ for any $a < b$. It is obvious that finite sets in a metric space cannot be perfect. In fact more is true—see exercise 2.5 (7).

Exercises 2.5

1. Show that, in $\mathbf{R}^n$, any countable set is of the first category. Give a category argument for the existence of irrational numbers.

2. Show that the class $\mathscr{N}$ of nowhere dense subsets of X is a ring, and the class $\mathscr{C}$ containing all sets of the first category is the generated σ-ring.

3. Show that in a complete metric space, a set of the first category contains no non-empty open set. Deduce that every non-empty open set is of the second category.

4. If G is an open set in a topological space, prove that $(\bar{G} - G)$ is nowhere dense.

5. In $\mathbf{R}$ show that the class $\mathscr{P}_0$ of half-open intervals with rational end-points generates the σ-field $\mathscr{B}$ of all Borel sets. Similarly in $\mathbf{R}^n$, show that $\mathscr{P}_0^n$ generates the Borel sets $\mathscr{B}^n$.

6. Show that a set E is perfect if and only if $E = E'$, where E' is the set of limit points of E.

7. Show that any non-empty perfect subset of a complete metric space is non-countable. *Hint.* Use theorem 2.7 and the fact that a closed subset of a complete metric space is complete.

2.6 Normed linear space

There are many abstract sets which have an algebraic structure as well as a topology. Thus if, in the set X there is a binary operation $+$ (called addition) and an operation in which elements of X can be multiplied by elements of the real number field **R** to give elements in X we say that X is a *real linear space* if for all $x, y, z \in X$, $a, b, \in \mathbf{R}$;

(i) $x+y = y+x$;

(ii) $x+(y+z) = (x+y)+z$;

(iii) $x+y = x+z \Rightarrow y = z$;

(iv) $a(x+y) = ax+ay$;

(v) $(a+b)x = ax+bx$;

(vi) $a(bx) = (ab)x$;

(vii) $1.x = x$.

It follows from these axioms that X has a unique zero element $0 = 0.y$ for all $y \in X$, and that subtraction can be defined in X by $x-y = x+(-1)y$.

In the present book we will only consider linear spaces over **R**. Most of our results can be extended, though sometimes with a little difficulty, to linear spaces over the number field **C**. We will not carry out this extension, nor do we consider any more general number fields.

It is immediate that $\mathbf{R}^n$ is a real linear space with vector addition and scalar multiplication for the two operations. The properties of linear spaces are studied at length in elementary courses on linear algebra.† We will not require many of these, but will develop the properties of linear independence when they are needed in Chapter 8.

Norm

If in a real linear space X there is a function $n: X \to \mathbf{R}$ satisfying

(i) $n(0) = 0$, $n(x) > 0$ if $x \neq 0$;

(ii) $n(x+y) \leqslant n(x)+n(y)$ for all $x, y \in X$;

(iii) $n(ax) = |a|\, n(x)$ for $a \in \mathbf{R}$, $x \in X$,

we say that X is a *normed linear space*. We will in this case use the usual notation $\|x\|$ for the value $n(x)$ of the norm function n at x.

In any normed linear space X,

$$\rho(x,y) = \|x-y\| = \rho(x-y, 0)$$

defines a metric, and in the topology determined by this metric, the algebraic operations are continuous in the sense that

(i) $x+y$ is continuous in the product topology of $X \times X$;

(ii) ax is continuous in the product topology of $X \times \mathbf{R}$.

† See, for example, G. Birkoff and S. MacLane. *A Survey of Modern Algebra*, (Macmillan, 1941).

It follows in particular that

(iii) $a \in \mathbf{R}$, $\lim x_n = 0 \Rightarrow \lim (ax_n) = 0$;

(iv) $x \in X$, $a_n \in R$, $\lim a_n = 0 \Rightarrow \lim (a_n x) = 0$.

(The reader is advised to check (i)–(iv) using the axioms.)

Special normed linear spaces will be studied in Chapters 7 and 8. At this stage we consider a few important examples of such spaces and examine the topological structure imposed by the norm.

M. Consider the set of those functions $x: [0,1] \to \mathbf{R}$ which are bounded. Define, for $t \in [0,1]$

$$(x+y)(t) = x(t)+y(t),$$
$$(ax)(t) = ax(t)$$

and check that this makes $\boldsymbol{M}$ a linear space. If we put

$$\|x\| = \sup_{0 \leqslant t \leqslant 1} |x(t)|$$

it is not hard to check that the conditions for a norm are also satisfied, so we have a normed linear space.

C. The set of those functions x: $[0,1] \to \mathbf{R}$ which are continuous is a subset $\boldsymbol{C}$ of $\boldsymbol{M}$. Since this subset $\boldsymbol{C}$ is closed under the operations of addition and scalar multiplication, it must be a normed linear space with the same norm

$$\|x\| = \sup_{0 \leqslant t \leqslant 1} |x(t)|.$$

s. The set of all sequences of real numbers $\{x_i\}$ is a linear space if we put

$$\{x_i\}+\{y_i\} = \{x_i+y_i\},$$
$$a\{x_i\} = \{ax_i\}.$$

Since for x, y real we have

$$\frac{|x+y|}{1+|x+y|} \leqslant \frac{|x|}{1+|x|} + \frac{|y|}{1+|y|}$$

it follows that $$\rho(\{x_i\}, \{y_i\}) = \sum_{i=1}^{\infty} \frac{1}{2^i} \frac{|x_i-y_i|}{1+|x_i-y_i|}$$

defines a metric in $\boldsymbol{s}$.

m. This is the set of all bounded sequences of real numbers with the same linear structure as $\boldsymbol{s}$. However this time it is more convenient to use the norm

$$\|\{x_i\}\| = \sup_i |x_i|,$$

to make $\boldsymbol{m}$ into a normed linear space.

c. This is the set of convergent sequences of real numbers with the same norm and linear structure as $\boldsymbol{m}$.

Each of the above spaces has a topology defined by the norm. We now obtain a few of the topological properties of these spaces, leaving the reader to determine the remainder.

Lemma. *The space $\boldsymbol{C}$ is complete.*

Proof. If $\{x_n\}$ is a Cauchy sequence in $\boldsymbol{C}$, then for each $t \in [0, 1]$ $\{x_n(t)\}$ is a Cauchy sequence in $\mathbf{R}$ which must converge to a real number $x_0(t)$. For each $\epsilon > 0$, there is an integer N such that, if

$$y_m(t) = |x_N(t) - x_m(t)| \quad (m \geqslant N),$$

then $\|y_m\| < \frac{1}{2}\epsilon$; that is,

$$0 \leqslant y_m(t) < \tfrac{1}{2}\epsilon \quad \text{for each } t \text{ in } [0, 1].$$

If we now let $m \to \infty$, it follows that

$$|x_N(t) - x_0(t)| \leqslant \tfrac{1}{2}\epsilon \quad \text{for all } t \text{ in } [0, 1]$$

so that, if $n \geqslant N$, $t \in [0, 1]$

$$|x_n(t) - x_0(t)| \leqslant |x_N(t) - x_n(t)| + |x_N(t) - x_0(t)| < \epsilon$$

and $\|x_n - x_0\| \to 0$ as $n \to \infty$. This means that x_0 is the uniform limit of a sequence of continuous functions and must therefore be continuous; that is, $x_0 \in \boldsymbol{C}$. ∎

Lemma. *The space $\boldsymbol{M}$ is not separable.*

Proof. For each $s \in [0, 1)$, let x_s be the function given by

$$x_s(t) = \begin{cases} 0 & \text{for } 0 \leqslant t \leqslant s, \\ 1 & \text{for } s < t \leqslant 1. \end{cases}$$

Then if $r, s \in [0, 1)$ and $r \neq s$ we must have $\|x_r - x_s\| = 1$. Now any dense set in $\boldsymbol{M}$ has to contain a point y_s such that $\|y_s - x_s\| < \frac{1}{2}$ for each $s \in [0, 1)$; and we cannot have $y_r = y_s$ with $r \neq s$ for then

$$1 = \|x_r - x_s\| \leqslant \|x_r - y_r\| + \|y_r - y_s\| + \|y_s - x_s\| < 1.$$

This means that any set dense in $\boldsymbol{M}$ must contain at least $\boldsymbol{c}$ points, and therefore $\boldsymbol{M}$ is not separable. ∎

Lemma. *The space $\boldsymbol{c}$ is not locally compact.*

Proof. A metric space X is locally compact if, for each $z \in X$, there is an $\epsilon > 0$ such that the closed sphere $\bar{S}(z, \epsilon)$ is compact. Now put

$$x_i^k = \begin{cases} 1 & \text{for } i = k, \\ 0 & \text{for } i \neq k, \end{cases}$$

and for each integer k,

$$x^k = \{x_i^k\}\ (i = 1, 2, \ldots) \in \boldsymbol{c} \quad \text{and} \quad k \neq j \Rightarrow \|x^k - x^j\| = 1.$$

Given $z = \{z_i\} \in \boldsymbol{c}$ and $\epsilon > 0$, put

$$z^k = z + \epsilon x^k$$

and all the points z^k are in $\bar{S}(z, \epsilon)$. But

$$\|z^k - z^j\| = \epsilon \quad \text{if} \quad k \neq j$$

so that $\{z^k\}\ (k = 1, 2, \ldots)$ forms an infinite set in $\bar{S}(z, \epsilon)$ with no limit point, and $\bar{S}(z, \epsilon)$ cannot be compact. ∎

Exercises 2.6

1. Show that $\boldsymbol{s}$ is bounded but not compact. If $x = \{x_i\} \in \boldsymbol{s}$, and

$$E = \{y \colon |y_i| \leqslant |x_i|\},$$

show that E is compact in $\boldsymbol{s}$, but show that $\boldsymbol{s}$ is not locally compact.

2. Show that each of the spaces $\boldsymbol{M}$, $\boldsymbol{c}$, $\boldsymbol{m}$, $\boldsymbol{s}$ is complete.

3. Show that each of the spaces $\boldsymbol{C}$, $\boldsymbol{c}$, $\boldsymbol{s}$ is separable, but that $\boldsymbol{m}$ is not separable.

Hint for $\boldsymbol{C}$. Consider the set of functions which take rational values at each of a finite set of rationals in $[0, 1]$ and are defined by linear interpolation between these points.

4. $\boldsymbol{C}^*(X)$ denotes the space of functions $f \colon X \to \mathbf{R}$ which are continuous and bounded. Show that $\boldsymbol{C}^*(\mathbf{R})$ is not separable by considering continuous functions which take the values $+1$, -1 on disjoints subsets Z_1, Z_2 of the set Z of positive integers and are defined elsewhere by linear interpolation. (The distance between any two such functions is 2, and there are $\boldsymbol{c}$ of them.)

However, let $\mathscr{D}$ be the subset of $\boldsymbol{C}^*(\mathbf{R})$ consisting of those functions f for which $\lim\limits_{x \to +\infty} f(x)$, $\lim\limits_{x \to -\infty} f(x)$ both exist. Prove that $\mathscr{D}$ is separable.

5. Let $\boldsymbol{l}_2$ be the subset of $\boldsymbol{s}$ such that $\sum\limits_{i=1}^{\infty} x_i^2$ converges. In the linear structure of $\boldsymbol{s}$ show that $\boldsymbol{l}_2$ is a linear space and that

$$\|x\| = \sum_{i=1}^{\infty} x_i^2$$

defines a norm. In the topology of this norm show that $\boldsymbol{l}_2$ is separable.

The subset $\{x \colon |x_i| < 1/i\}$ of $\boldsymbol{l}_2$ is known as the Hilbert cube: prove it is compact.

Hint. Starting with an infinite sequence in the Hilbert cube pick a subsequence in which the first coordinate converges, then successive subsequences in which the 2nd, 3rd, ..., nth coordinate converges. Show that the sequence to which these coordinates converge is in the Hilbert cube and is a limit point of the original set.

2.7 Cantor set

We now digress briefly from the study of general situations and consider the definition and properties of a special subset of **R** first considered by Cantor. This set and associated functions will be useful in the sequel to provide counter-examples to several conjectures which are plausible but false.

If we denote the open interval $((3r-2)/3^n, (3r-1)/3^n)$ by $E_{n,r}$, put

$$G_n = \bigcup_{r=1}^{3^{n-1}} E_{n,r} \quad G = \bigcup_{n=1}^{\infty} G_n;$$

it is clear that G is an open subset of $[0, 1]$. Its complement

$$C = [0,1]-G$$

is called the Cantor set. From its definition C is closed.

Lemma. *The Cantor set C is nowhere dense and perfect.*

Proof. If we express points $x \in [0, 1]$ in the form

$$x = \sum_{i=1}^{\infty} \frac{a_i}{3^i} \quad (a_i = 0, 1, 2), \tag{2.7.1}$$

then the set G_n above is the set of x for which $a_n = 1$. Hence, the set C consists of precisely those real numbers which have a representation in the form (2.7.1) with each $a_i = 0$ or 2. Given a point $x_1 \in C$, altering the nth term a_n (replacing 0, 2 by 2, 0 respectively) gives a new point x_2 in C such that

$$|x_1 - x_2| = 2.3^{-n}.$$

This shows that every point of C is a limit point of C, so that C is perfect.

If H is any open set with $H \cap [0, 1]$ not void then $H \cap [0, 1]$ contains an open interval I of length $\delta > 0$. If $\delta > 3^{1-n}$, then I must contain an interval $E_{n,r}$ so that H contains an open set disjoint from C. This proves that C is nowhere dense. ∎

From the above lemma and example 2.5 (7) we can deduce that C is not countable. However, one can prove that C must have the same cardinal as the continuum $[0, 1]$ by considering the following mapping:

$$\text{if} \quad x \in [0,1], \quad \text{put} \quad x = \sum_{i=1}^{\infty} \frac{b_i}{2^i} \quad (b_i = 0 \quad \text{or} \quad 1),$$

where the sequence $\{b_i\}$ does not satisfy $b_i = 1$ for $i \geq N$. Put

$$f(x) = \sum_{i=1}^{\infty} \frac{a_i}{3^i} \quad \begin{cases} a_i = 0 & \text{if} \quad b_i = 0, \\ a_i = 2 & \text{if} \quad b_i = 1. \end{cases}$$

Then $f: [0, 1] \to C$ is (1, 1) and maps $[0, 1]$ onto a (proper) subset of C. Since $C \subset [0, 1]$ the cardinal of C is $\mathbf{c}$.

We can think of f as a function on $[0, 1]$ to $[0, 1]$. It is easy to see that f is monotonic, that is,

$$x_1 < x_2 \Rightarrow f(x_1) < f(x_2)$$

so that, for each $y \in [0, 1]$,

$$f^{-1}[0, y] = [0, z]. \tag{2.7.2}$$

If z is defined by (2.7.2), then we say that

$$z = g(y).$$

This defines $g: [0, 1] \to [0, 1]$ as a monotonic function which is clearly constant on each of the sets $E_{n,r}$. In fact

$$\frac{3r-2}{3^n} \leqslant y \leqslant \frac{3r-1}{3^n} \Rightarrow g(y) = \frac{2r-1}{2^n}.$$

The function g is continuous and monotonic increasing, for

$$0 \leqslant y_1 - y_2 < 3^{-n-1} \Rightarrow 0 \leqslant g(y_1) - g(y_2) < 2^{-n}.$$

Since the function g is constant in each $E_{n,r}$ it follows that it is differentiable with zero derivative at each point of G. One can easily see that g increases at each point of C—and in fact the 'upper derivative' at points of C is $+\infty$.

Note that there is nothing magical about the integer 3 used in the construction of C. Similar constructions using expansions to a different base will give sets with similar properties.

Exercises 2.7

1. If $x = \sum_{i=1}^{\infty} \frac{c_i}{10^i}$ ($c_i = 0, 1, \ldots, 9$) is a decimal expansion of real numbers in $[0, 1]$ and T is the set of such x for which $c_i \neq 7$, show that T is perfect and nowhere dense.

2. Construct a set which is dense in $[0, 1]$ and yet the union of a countable class of nowhere dense perfect sets.

3. Show that the function $g: [0, 1] \to [0, 1]$ defined above satisfies a Lipschitz condition of order $\alpha = \log 2/\log 3$, but not of any order $\beta > \alpha$. (A function $h: I \to \mathbf{R}$ is said to satisfy a Lipschitz condition of order α at $x_0 \in I$ if $|h(x) - h(x_0)| \leqslant K|x - x_0|^\alpha$ for $x \in I$ and some suitable $K \in \mathbf{R}$.)

3

SET FUNCTIONS

3.1 Types of set function

We consider only functions $\mu:\mathscr{C} \to \mathbf{R}^*$, where $\mathscr{C}$ is a non-empty class of sets. Thus μ is a rule which determines, for each $E \in \mathscr{C}$, a unique element $\mu(E)$ which is either a real number or $\pm\infty$. We always assume that $\mathscr{C}$ contains the empty set $\varnothing$. $\mathbf{R}^*$ denotes the compactification of the real number field $\mathbf{R}$ by the addition of two points $+\infty$, $-\infty$, while $\mathbf{R}^+$ will denote the set of non-negative real numbers together with $+\infty$. It is not possible to arrange for $\mathbf{R}^*$ to be an algebraic field extending $\mathbf{R}$, though we will preserve as many of the algebraic properties of $\mathbf{R}$ as possible by adopting the convention that, for any $a \in \mathbf{R}$,

$$-\infty < a < +\infty, \quad -\infty \pm a = -\infty, \quad +\infty \pm a = +\infty;$$

$$\text{if } \; a > 0, \quad a(+\infty) = +\infty, \quad a(-\infty) = -\infty;$$

$$\text{if } \; a = 0, \quad a(\pm\infty) = 0;$$

$$\text{if } \; a < 0, \quad a(+\infty) = -\infty, \quad a(-\infty) = +\infty;$$

$$(+\infty)(+\infty) = +\infty, \quad (+\infty)(-\infty) = -\infty, \quad (-\infty)(-\infty) = +\infty.$$

Thus the operation of dividing by $(\pm\infty)$ is not allowed and

$$(+\infty)+(-\infty)$$

is not defined in $\mathbf{R}^*$. All these definitions are natural except for the convention $0(\pm\infty) = 0$.

Arbitrary set functions $\mu:\mathscr{C} \to \mathbf{R}^*$ are not of much interest. We adopt conditions on μ which correspond to our intuitive idea of mass for a physical object (we generalise the notion to allow negative masses). The first property we define corresponds to the notion that the mass of a pair of disjoint objects is the sum of the masses of the individual objects.

Additive set function

A set function $\mu:\mathscr{C} \to \mathbf{R}^*$ is said to be (finitely) additive if

(i) $\mu(\varnothing) = 0$,

(ii) for every finite collection $E_1, E_2, \ldots, E_n$ of disjoint sets of $\mathscr{C}$ such that $\bigcup_{i=1}^{n} E_i \in \mathscr{C}$ we have

$$\mu\left(\bigcup_{i=1}^{n} E_i\right) = \sum_{i=1}^{n} \mu(E_i). \tag{3.1.1}$$

In this definition condition (i) is almost redundant for it will be implied by (ii) provided there is at least one $E \in \mathscr{C}$ with $\mu(E)$ finite. Note also that we do not in the definition assume that $\mathscr{C}$ is closed under finite disjoint unions so that, in testing a given set function $\mu: \mathscr{C} \to \mathbf{R}^*$ to see whether or not it is additive, we can only use sets $E_i \in \mathscr{C}$ which are disjoint and have their union in $\mathscr{C}$. However, the definition is taken to imply that the right-hand side of (3.1.1) has a unique meaning in $\mathbf{R}^*$ so that in particular there are no sets E, $F \in \mathscr{C}$ such that $E \cap F = \varnothing$, $E \cup F \in \mathscr{C}$, $\mu(E) = +\infty$, $\mu(F) = -\infty$.

The natural domain of definition for an additive set function μ is a ring since, if $\mathscr{C}$ is a ring,

$$E_i \in \mathscr{C} \quad (i = 1, 2, \ldots, n) \Rightarrow \bigcup_{i=1}^{n} E_i \in \mathscr{C}.$$

For a ring $\mathscr{C}$ it is worth noticing that $\mu: \mathscr{C} \to \mathbf{R}^*$ is additive if and only if $\mu(\varnothing) = 0$ and

$$E, F \in \mathscr{C}, \quad E \cap F = \varnothing \Rightarrow \mu(E \cup F) = \mu(E) + \mu(F),$$

since in this case the general result (3.1.1) can be obtained from the result for two sets by a simple induction argument.

When $\mathscr{C}$ is a finite class of sets it is easy to give examples of set functions defined on $\mathscr{C}$ which are additive. We now give a number of less trivial examples which will be useful for illustrating our later definitions. In each case the reader should check that $\mu: \mathscr{C} \to \mathbf{R}^*$ is additive.

Example 1. Ω any space with infinitely many points, $\mathscr{C}$ the class of all subsets of Ω. Define μ by

$$\mu(E) = \text{number of points in } E, \text{ if } E \text{ is finite;}$$
$$\mu(E) = +\infty, \text{ if } E \text{ is infinite.}$$

Example 2. Ω any topological space, $\mathscr{C}$ the class of all subsets of Ω. Put

$$\mu(E) = 0, \text{ if } E \text{ is of the first category in } \Omega;$$
$$\mu(E) = +\infty, \text{ if } E \text{ is of the second category in } \Omega.$$

Example 3. $\Omega = \mathbf{R}$, $\mathscr{C}$ the class of all finite intervals of $\mathbf{R}$. For $E = [a, b]$ or $[a, b)$ or $(a, b]$ or (a, b), put

$$\mu(E) = b - a.$$

Example 4. Ω is any space with at least two distinct points n, s; $\mathscr{C}$ is the class of all subsets of Ω.

$\mu(E) = 0$, if E contains neither or both of n, s;

$\mu(E) = 1$, if E contains n but not s;

$\mu(E) = -1$, if E contains s but not n.

Example 5. $\Omega = (0, 1]$, the set of real numbers x with $0 < x \leqslant 1$, $\mathscr{C}$ the class of half-open intervals $(a, b]$ where $0 \leqslant a \leqslant b \leqslant 1$.

$$\mu(a, b] = b - a \quad \text{if} \quad a \neq 0;$$

$$\mu(0, b] = +\infty.$$

Example 6. Ω is any infinite space, $\mathscr{C}$ the class of all its subsets. Let $x_1, x_2, \ldots, x_n, \ldots$ be an enumerable sequence of distinct points of Ω, and suppose $p_1, p_2, \ldots$ is a sequence of real numbers such that $\sum_{i=1}^{\infty} p_i$ either converges absolutely or is properly divergent to $+\infty$ or $-\infty$ (the case Σp_i convergent, $\Sigma |p_i|$ divergent is not allowed: why?). Put

$$\mu(E) = \sum_{x_i \in E} p_i,$$

where the sum extends over all integers $i = 1, 2, \ldots$ for which $x_i \in E$. Any set function which can be defined as in example 6 is called *discrete*. Note that example (4) can be thought of as a special case of example (6).

Although it is not sufficient to restrict our attention to set functions $\mu: \mathscr{C} \to \mathbf{R}$ which are finite valued, the condition of additivity which is usually assumed prevents μ from taking both the values $+\infty$, $-\infty$ at least when $\mathscr{C}$ is a ring. This is one of the results in the next theorem.

***Theorem* 3.1.** *Suppose $\tau: \mathscr{C} \to \mathbf{R}^*$ is an additive set function defined on a ring $\mathscr{C}$ and $E, F \in \mathscr{C}$. Then*

(i) *if $E \supset F$ and $\tau(F)$ is finite*

$$\tau(E - F) = \tau(E) - \tau(F);$$

(ii) *if $E \supset F$ and $\tau(F)$ is infinite*

$$\tau(E) = \tau(F);$$

(iii) *if $\tau(E) = +\infty$, then $\tau(F) \neq -\infty$.*

Proof. (i) Since $\mathscr{C}$ is a ring, $E - F \in \mathscr{C}$ and additivity implies, since $F \cap (E - F) = \varnothing$,

$$\tau(E) = \tau(E - F) + \tau(F). \tag{3.1.2}$$

Subtracting the finite real number $\tau(F)$ gives the result.

(ii) If $\tau(F) = +\infty$, then (3.1.2) can only have a meaning if $\tau(E-F) \neq -\infty$, and this implies $\tau(E) = +\infty$. The case $\tau(F) = -\infty$ is similar.

(iii) Since $E \cap F$, $E-F$, $F-E$ are disjoint sets of $\mathscr{C}$

$$\tau(E) = \tau(E \cap F) + \tau(E-F) = +\infty,$$

$$\tau(F) = \tau(E \cap F) + \tau(F-E) = -\infty$$

could only have meaning if $\tau(E \cap F)$ is finite. But this would imply

$$\tau(E-F) = +\infty,\ \tau(F-E) = -\infty, \text{ and then, since } E \triangle F \in \mathscr{C},$$

$$\tau(E \triangle F) = \tau(E-F) + \tau(F-E) = +\infty + (-\infty)$$

which is impossible.]

Our definition of additivity means that for $\mu: \mathscr{C} \to \mathbf{R}^*$ to be additive any set $E_0 \in \mathscr{C}$ which can be split into a finite number of disjoint subsets in $\mathscr{C}$ must be such that $\mu(E_0)$ is the same as the sum of the values of μ on the 'pieces'. We often want this to be true for a dissection of E_0 into a countably infinite collection of subsets in $\mathscr{C}$.

σ-additive set function

A set function $\mu: \mathscr{C} \to \mathbf{R}^*$ is said to be σ-additive (sometimes called completely additive, or countably additive) if

(i) $\mu(\varnothing) = 0$,

(ii) for any disjoint sequence $E_1, E_2, \ldots$ of sets of $\mathscr{C}$ such that

$$E = \bigcup_{i=1}^{\infty} E_i \in \mathscr{C},$$

$$\mu(E) = \sum_{i=1}^{\infty} \mu(E_i). \tag{3.1.3}$$

As before the condition (i) is redundant if μ takes any finite values. Since we may assume that all but a finite number of the sequence $\{E_i\}$ are void it is clear that any set function which is σ-additive is also additive. To see that the converse is not true it is sufficient to consider example (5) on p. 53. Put

$$E = (0, 1], \quad E_n = \left(\frac{1}{n+1}, \frac{1}{n}\right] \quad (n = 1, 2, \ldots);$$

then $\{E_n\}$ is a disjoint sequence in $\mathscr{C}$ whose union E is in $\mathscr{C}$ but

$$+\infty = \mu(E) \neq 1 = \sum_{n=1}^{\infty} \left(\frac{1}{n} - \frac{1}{n+1}\right) = \sum_{n=1}^{\infty} \mu(E_n).$$

Notice further that even when $\mathscr{C}$ is a ring it does not follow that

$$E_i \in \mathscr{C} \quad (i = 1, 2, \ldots) \Rightarrow E = \bigcup E_i \in \mathscr{C};$$

so that in testing (3.1.3) we can only use those sets $E \in \mathscr{C}$ which can be split into a countable sequence of disjoint subsets in $\mathscr{C}$. In particular if $\mathscr{C}$ is a finite class of sets then additivity for $\mu: \mathscr{C} \to \mathbf{R}^*$ implies σ-additivity. We also interpret (3.1.3) to mean that the right-hand side is uniquely defined and independent of the order of the sets E_i; thus if μ is σ-additive, and we have such a decomposition $E = \bigcup_{i=1}^{\infty} E_i$, we cannot have $\mu(E_i) = +\infty$, $\mu(E_j) = -\infty$, nor can the series in (3.1.3) converge conditionally.

It is easy to check that each of the set functions in examples (1), (2), (4), (6) on p. 52 is σ-additive, and the set function of example (3) is also σ-additive though the proof of this fact is non-trivial. This proof will be given in detail in § 3.4, as it is an essential step in the definition of Lebesgue measure in $\mathbf{R}$.

Measure

Any non-negative set function $\mu: \mathscr{C} \to \mathbf{R}^+$ which is σ-additive is called a measure on $\mathscr{C}$, ($\mathbf{R}^+ = \{x \in \mathbf{R}^*: x \geqslant 0\}$).

We should remark that there is not general agreement in the literature as to which set functions ought to be called measures. According to our definition the set functions in examples (1), (2) and (3) are measures, those in (4) and (6) are not because μ can take negative values while the set function in (5) is not because it is not σ-additive.

The natural domain of definition of a measure, or indeed of any σ-additive set function, is a σ-ring since then

$$E_i \in \mathscr{C} \quad (i = 1, 2, \ldots) \Rightarrow \bigcup_{i=1}^{\infty} E_i \in \mathscr{C}.$$

However, we will not restrict our consideration to σ-additive set functions already defined on a σ-ring.

Given a set function $\mu: \mathscr{C} \to \mathbf{R}^*$ where $\mathscr{C}$ is a ring it is usually quite easy to check whether or not μ is additive for one only has to check (3.1.1) for $n = 2$. In order to check that it is also σ-additive it is useful to have a characterisation of σ-additive μ in terms of a continuity condition for monotone sequences of sets. Since we have seen already (theorem 3.1) that such set functions cannot attain both values $+\infty$, $-\infty$ there will be no loss of generality in assuming that

$$-\infty < \mu(E) \leqslant +\infty \quad \text{for all} \quad E \in \mathscr{C}.$$

Continuity

Suppose $\mathscr{R}$ is a ring and $\mu: \mathscr{R} \to \mathbf{R}^*$ is additive with $\mu(E) > -\infty$ for all $E \in \mathscr{R}$. Then for any $E \in \mathscr{R}$ we say that:

(i) μ is continuous from below at E if

$$\lim_{n\to\infty} \mu(E_n) = \mu(E) \tag{3.1.4}$$

for every monotone increasing sequence $\{E_n\}$ of sets in $\mathscr{R}$ which converges to E;

(ii) μ is continuous from above at E if (3.1.4) is satisfied for any monotone decreasing sequence $\{E_n\}$ in $\mathscr{R}$ with limit E which is such that $\mu(E_n) < \infty$ for some n;

(iii) μ is continuous at E if it is continuous at E from below and from above (when $E = \varnothing$ the first requirement is trivially satisfied).

***Theorem* 3.2.** *Suppose $\mathscr{R}$ is a ring and $\mu: \mathscr{R} \to \mathbf{R}^*$ is additive with $\mu(E) > -\infty$ for all $E \in \mathscr{R}$.*

(i) *If μ is σ-additive, then μ is continuous at E for all $E \in \mathscr{R}$;*

(ii) *if μ is continuous from below at every set $E \in \mathscr{R}$, then μ is σ-additive;*

(iii) *if μ is finite and continuous from above at $\varnothing$, then μ is σ-additive.*

Proof. (i) If $\mu(E_n) = +\infty$ for $n = N$ and $\{E_n\}$ is monotone increasing then $\mu(E) = +\infty$ and $\mu(E_n) = +\infty$ for $n \geqslant N$ by theorem 3.1 (ii), where $E = \lim E_n$. Thus in this case $\mu(E_n) \to \mu(E)$ as $n \to \infty$. On the other hand, if $\mu(E_n) < \infty$ for all n and $\{E_n\}$ increases to E, then

$$E = E_1 \cup \bigcup_{n=1}^{\infty} (E_{n+1} - E_n)$$

is a disjoint decomposition of E and

$$\mu(E) = \mu(E_1) + \sum_{n=1}^{\infty} \mu(E_{n+1} - E_n)$$

$$= \mu(E_1) + \lim_{N\to\infty} \sum_{n=1}^{N} \mu(E_{n+1} - E_n) = \lim_{N\to\infty} \mu(E_N),$$

since μ is additive on the ring $\mathscr{R}$. Thus μ is continuous from below at E. Now suppose $\{E_n\}$ decreases to E and $\mu(E_N) < +\infty$. Put

$$F_n = E_N - E_n \quad \text{for} \quad n \geqslant N.$$

Then, by theorem 3.1 (ii), $\mu(F_n) < \infty$ and the sequence $\{F_n\}$ is monotone increasing to $E_N - E$. Hence, as $n \to \infty$,

$$\mu(F_n) \to \mu(E_N - E) = \mu(E_N) - \mu(E).$$

But $\mu(F_n) = \mu(E_N) - \mu(E_n)$ so that $\mu(E_n) \to \mu(E)$ as $n \to \infty$, since $\mu(E_N)$ is finite, and μ is also continuous from above at E.

(ii) Suppose $E \in \mathscr{R}$, $E_i \in \mathscr{R}$ $(i = 1, 2, \ldots)$ are such that $E = \bigcup_{i=1}^{\infty} E_i$ and the sets E_i are disjoint. Put

$$F_n = \bigcup_{i=1}^{n} E_i \in \mathscr{R} \quad (n = 1, 2, \ldots),$$

and $\{F_n\}$ is a monotone increasing sequence of sets in $\mathscr{R}$ which converges to $E \in \mathscr{R}$. If μ is continuous from below at E

$$\sum_{i=1}^{n} \mu(E_i) = \mu(F_n) \to \mu(E) \quad \text{as} \quad n \to \infty$$

so that
$$\mu(E) = \sum_{i=1}^{\infty} \mu(E_i),$$
and μ is σ-additive.

(iii) In the notation of (ii) put

$$G_n = E - F_n \in \mathscr{R} \quad (n = 1, 2, \ldots).$$

Then $\{G_n\}$ is a monotone decreasing sequence converging to $\varnothing$ and, for $n = 1, 2, \ldots$

$$\mu(E) = \sum_{i=1}^{n} \mu(E_i) + \mu(G_n).$$

If μ is finite and continuous from above at $\varnothing$ we must have

$$\mu(G_n) \to 0 \quad \text{as} \quad n \to \infty$$

so that again
$$\mu(E) = \sum_{i=1}^{\infty} \mu(E_i). \blacksquare$$

Remark 1. In our definition of continuity from above we only require to have $\mu(E_n) \to \mu(E)$ for those sequences $\{E_n\}$ which decrease to E for which $\mu(E_n)$ is finite for some n. To see that we could not relax this finiteness condition, consider example (2) on p. 52 which we have already seen to be σ-additive with $\Omega = (0, 1)$. Then if

$$E_n = \left(0, \frac{1}{n}\right) \quad (n = 1, 2, \ldots)$$

we have a sequence decreasing to $\varnothing$ such that $\mu(E_n) = +\infty$ for all n since E_n is of the second category.

Remark 2. The condition that μ be finite cannot be omitted in theorem 3.2 (iii). Consider example (5) on p. 53 which we saw was additive but not σ-additive. Actually the class $\mathscr{C}$ of sets on which μ is defined is a semi-ring rather than a ring, but its definition can easily be extended to the ring of finite disjoint unions of sets in $\mathscr{C}$ by using theorem 3.4. It is easy to check that it will remain continuous from above at $\varnothing$, but not σ-additive.

Part (iii) of theorem 3.2 will prove very useful in practice, especially for finite valued set functions $\mu: \mathscr{R} \to \mathbf{R}^+$ which are non-negative and additive on a ring $\mathscr{R}$. In order to prove that such a μ is a measure it is sufficient to show that, if $\{E_n\}$ is any sequence of sets in $\mathscr{R}$ decreasing to $\varnothing$,

$$\mu(E_n) \to 0 \quad \text{as} \quad n \to \infty. \tag{3.1.5}$$

If (3.1.5) is false for some such sequence $\{E_n\}$ then, since $\mu(E_n)$ is monotone decreasing we must have

$$\mu(E_n) \to \delta > 0 \quad \text{as} \quad n \to \infty. \tag{3.1.6}$$

If we can establish a contradiction by assuming (3.1.6), then (3.1.5) will be proved and we will have deduced that μ is a measure.

When we come to consider particular set functions one of our objectives will be to define μ on as large a class $\mathscr{C}$ as possible. We will also want μ to be σ-additive. It would be desirable to define μ on the class of all subsets of Ω, but unfortunately this is not possible if Ω is not countable and μ is to have an interesting structure. In particular it has been shown, using the continuum hypothesis, that it is impossible to define a measure μ on all subsets of the real line such that (*a*) sets consisting of a single point have zero measure (this eliminates discrete set functions like examples (1), (4), (6) on pp. 52–3); (*b*) every set of infinite measure has a subset of finite positive measure (this eliminates example (2)); (*c*) the measure of the whole space is not zero. In practice the method used is to define μ with desired properties on a restricted class of sets $\mathscr{C}$ (as in examples (3) or (5)) and then extend the definition to a larger class $\mathscr{D} \supset \mathscr{C}$.

Extension

Given two classes $\mathscr{C} \subset \mathscr{D}$ of subsets of Ω and set functions $\mu: \mathscr{C} \to \mathbf{R}^*$, $\nu: \mathscr{D} \to \mathbf{R}^*$ we say that ν is an extension of μ if, for all $E \in \mathscr{C}$

$$\nu(E) = \mu(E);$$

under the same conditions we say that μ is the *restriction* of ν to $\mathscr{C}$.

It is sometimes appropriate (as in probability theory) to work with set functions μ which are finite. However, most of the theorems which can be proved for finite σ-additive set functions can also be obtained with a condition slightly weaker than finiteness.

σ-finite set function

A set function $\mu:\mathscr{C} \to \mathbf{R}^*$ is said to be σ-finite if, for each $E \in \mathscr{C}$, there is a sequence of sets $C_i\ (i = 1, 2, \ldots) \in \mathscr{C}$ such that $E \subset \bigcup_{i=1}^{\infty} C_i$ and $\mu(C_i)$ is finite for all i.

In our examples, on p. 52, the set functions in (3), (4), and (5) are all finite, (1) gives a σ-finite measure if and only if Ω is countable, (2) is not σ-finite if Ω is of the second category, (6) is finite if $\Sigma|p_i|$ converges and otherwise it is σ-finite.

Sometimes it is useful to relax the condition of additivity in order to be able to define μ on the class of all subsets. The most common example of this is in the concept of outer measure.

Outer measure

If $\mathscr{C}$ is the class of all subsets of Ω, then $\mu:\mathscr{C} \to \mathbf{R}^+$ is called an outer measure on Ω if

(i) $\mu(\varnothing) = 0$;

(ii) μ is monotone in the sense that $E \subset F \Rightarrow \mu(E) \leqslant \mu(F)$;

(iii) μ is countably subadditive in the sense that for any sequence $\{E_i\}$ of sets,

$$E \subset \bigcup_{i=1}^{\infty} E_i \Rightarrow \mu(E) \leqslant \sum_{i=1}^{\infty} \mu(E_i). \tag{3.1.7}$$

Note that every measure on the class of all subsets of a space Ω is an outer measure on Ω. However, it is not difficult to give examples of outer measures which are not measures.

Example 7. Ω any space with more than one point. Put

$$\mu(\varnothing) = 0, \quad \mu(E) = 1 \quad \text{for all} \quad E \neq \varnothing.$$

In this book we do not study the properties of outer measures for their own sake, but we will use them as a tool to extend the definition of measures.

Exercises 3.1

1. If $\Omega = [0, 1)$ and $\mathscr{C}$ consists of the 6 sets

$$\varnothing, \quad \Omega, \quad [0, \tfrac{1}{2}), \quad [0, \tfrac{1}{4}), \quad [0, \tfrac{3}{4}), \quad [\tfrac{1}{4}, \tfrac{3}{4}),$$

$$\mu(\varnothing) = 0, \quad \mu[0, \tfrac{1}{4}) = 2, \quad \mu[0, \tfrac{1}{2}) = 2,$$

$$\mu[0, \tfrac{3}{4}) = 4, \quad \mu[\tfrac{1}{4}, \tfrac{3}{4}) = 2, \quad \mu(\Omega) = 4,$$

show that μ is additive on $\mathscr{C}$. Can μ be extended to an additive set function on the ring generated by $\mathscr{C}$?

2. Show that if $\mathscr{R}$ is any finite ring of subsets of Ω and μ is additive on $\mathscr{R}$ then μ is σ-additive on $\mathscr{R}$.

3. A set function $\mu: \mathscr{C} \to \mathbf{R}^*$ is said to be monotone if $\mu(\varnothing) = 0$ and $E \subset F, E, F \in \mathscr{C} \Rightarrow \mu(E) \leqslant \mu(F)$. Show that monotone set functions are non-negative, and if $\mathscr{C}$ is a ring, show that an additive non-negative set function is monotone. Of the set functions in examples (1)–(7), which are monotone?

4. $\mathbf{Z}$ is the space of positive integers and $\sum_{n=1}^{\infty} a_n$ is a convergent series of positive terms.

If E is a finite subset of $\mathbf{Z}$, put $\tau(E) = \sum_{n \in E} a_n$; if E is an infinite subset of $\mathbf{Z}$, put $\tau(E) = +\infty$.

Show that τ is additive, but not σ-additive on the class of all subsets of $\mathbf{Z}$.

5. $\mathbf{Z}$ is the space of positive integers; for $E \subset \mathbf{Z}$ let $r_n(E)$ be the number of integers in E which are not greater than n. Let $\mathscr{C}$ be the class of subsets E for which

$$\lim_{n\to\infty} \frac{r_n(E)}{n} = \tau(E)$$

exists. Show that τ is finitely additive, but not σ-additive on $\mathscr{C}$, but that $\mathscr{C}$ is not even a semi-ring.

6. If μ is finitely additive on a ring $\mathscr{R}$; $E, F, G \in \mathscr{R}$ show

$$\mu(E) + \mu(F) = \mu(E \cup F) + \mu(E \cap F),$$

$$\mu(E) + \mu(F) + \mu(G) + \mu(E \cap F \cap G)$$
$$= \mu(E \cup F \cup G) + \mu(E \cap F) + \mu(F \cap G) + \mu(G \cap E).$$

State and prove a relationship of this kind for n subsets of $\mathscr{R}$.

7. Suppose $\mathscr{S}$ is a σ-ring of subsets of Ω, μ is a measure on $\mathscr{S}$. Show that the class of sets $E \in \mathscr{S}$ with $\mu(E)$ finite forms a ring, and the class with $\mu(E)$ σ-finite forms a σ-ring.

8. If E is a set in $\mathscr{S}$ of σ-finite μ-measure (where μ is a measure on $\mathscr{S}$) and $\mathscr{D} \subset \mathscr{S}$ where $\mathscr{D}$ is a class of disjoint subsets of E show that the subclass of those $D \in \mathscr{D}$ for which $\mu(D) > 0$ is countable.

9. State and prove a version of theorem 3.2 (i) for set functions μ defined on a semi-ring $\mathscr{C}$.

10. To show that the finiteness condition in the definition of 'continuous from above' in theorem 3.2 (i) cannot be relaxed, consider any infinite space Ω and put

$\tau(E) =$ number of points in E, if E is finite;

$\tau(E) = +\infty$, if E is infinite.

Then τ is a measure on the class of all subsets of Ω, but for any sequence $\{E_n\}$ of infinite sets which decreases to $\varnothing$, we do not have $\lim \tau(E_n) = 0$.

11. Suppose $\mathscr{P}$ is the semi-ring of half-open intervals $(a, b]$, Q is the set of rationals in $(0, 1]$ and $\mathscr{P}_Q$ is the semi-ring of sets of the form $(a, b] \cap Q$. Put

$$\mu\{(a, b] \cap Q\} = b - a \quad \text{if} \quad 0 \leqslant a \leqslant b \leqslant 1.$$

Show that μ is additive on $\mathscr{P}_Q$ and is continuous above and below at every set in $\mathscr{P}_Q$, but is not σ-additive. This shows that theorem 3.2 (ii), (iii) is not true for semi-rings.

12. Show that if μ is an outer measure on Ω, E_0 any fixed subset then $\mu_0(E) = \mu(E \cap E_0)$ defines another outer measure on Ω.

13. Show that if μ, ν are outer measures on Ω, so is τ defined by

$$\tau(E) = \max[\mu(E), \nu(E)].$$

14. Suppose $\{\phi_n\}$ is a sequence of σ-additive set functions defined on a σ-ring $\mathscr{S}$ and that $\lim \phi_n(E) = \phi(E)$ exists for all E in $\mathscr{S}$. Show that ϕ is finitely additive on $\mathscr{S}$. If either

(i) $\phi_n(E) \to \phi(E)$ uniformly on $\mathscr{S}$ with $\phi(E) > -\infty$ for all $E \in \mathscr{S}$; or

(ii) $\phi_1(E) > -\infty$, $\phi_n(E)$ monotone increasing for all $E \in \mathscr{S}$;

show that ϕ is σ-additive on $\mathscr{S}$.

3.2 Hahn–Jordan decompositions

When discussing σ-additive set functions we will usually restrict our attention to the non-negative ones (which we call measures). The present section justifies this procedure by showing that, under reasonable conditions a 'signed' set function $\mu: \mathscr{C} \to \mathbf{R}^*$ which is completely additive can be expressed as the difference of two measures. This means that properties of completely additive set functions can be deduced from the corresponding properties of measures. There are also versions of the decomposition theorem for finitely additive set functions, but we will not consider these.

We have already seen (theorem 3.1 (iii)) that an additive set function defined on a ring cannot take both the values $+\infty$, $-\infty$. If $\mathscr{S}$ is a σ-ring and $\mu: \mathscr{S} \to \mathbf{R}^*$ is completely additive then for any sequence $\{E_i\}$ of disjoint sets in $\mathscr{S}$,

$$\mu\left(\bigcup_{i=1}^{\infty} E_i\right) = \sum_{i=1}^{\infty} \mu(E_i).$$

Since $\bigcup_{i=1}^{\infty} E_i$ is independent of the order of the sets in the sequence, it follows that the series on the right-hand side must be either absolutely convergent or properly divergent. In the case of example (6) the set function

$$\mu(E) = \sum_{x_i \in E} p_i$$

can be decomposed $\quad \mu(E) = \mu_+(E) - \mu_-(E),$

where $\quad \mu_+(E) = \sum_{x_i \in E} \max(0, p_i), \quad \mu_-(E) = - \sum_{x_i \in E} \min(0, p_i)$

so that μ_+, μ_- are measures of which at least one is finite. Further if we put $P = \{x; \mu\{x\} > 0\}$, $N = \Omega - P$ we have $\mu_+ E = \mu_+(P \cap E)$, $\mu_- E = -\mu(N \cap E)$ for all $E \subset \Omega$, so that the decomposition into the difference of two measures can also be obtained by splitting Ω into two subsets P, N such that μ is non-negative on every subset of P and non-positive on every subset of N. These two aspects of the decomposition are true in general, provided $\mathscr{S}$ is a σ-field.

***Theorem* 3.3.** *Given a completely additive $\tau: \mathscr{F} \to \mathbf{R}^*$ defined on a σ-field $\mathscr{F}$, there are measures τ_+* and *τ_- defined on $\mathscr{F}$ and subsets P, N in $\mathscr{F}$ such that $P \cup N = \Omega$, $P \cap N = \varnothing$ and for each $E \in \mathscr{F}$,*

$$\tau_+(E) = \tau(E \cap P) \geqslant 0, \quad \tau_-(E) = -\tau(E \cap N) \geqslant 0,$$
$$\tau(E) = \tau_+(E) - \tau_-(E);$$

so that τ is the difference of two measures τ_+, τ_- on $\mathscr{F}$. At least one of τ_+, τ_- is finite and, if τ is finite or σ-finite so are both τ_+, τ_-.

Proof. Since τ can take at most one of the values $+\infty$, $-\infty$ we may assume without loss of generality that, for all $E \in \mathscr{F}$,

$$-\infty < \tau(E) \leqslant +\infty.$$

We first prove that, if $E \in \mathscr{F}$ and

$$\lambda(E) = \inf_{B \subset E,\, B \in \mathscr{F}} \tau(B), \tag{3.2.1}$$

then $\lambda(\Omega) \neq -\infty$. If this is false then there is a set $B_1 \in \mathscr{F}$ for which $\tau(B_1) < -1$. At least one of $\lambda(B_1)$, $\lambda(\Omega - B_1)$ must be $-\infty$; since $\lambda(A \cup B) \geqslant \lambda(A) + \lambda(B)$ if A, B are disjoint sets of F. Put A_1 equal to B_1 if $\lambda(B_1) = -\infty$ and $(\Omega - B_1)$ otherwise. Proceed by induction. For each positive integer n, choose $B_{n+1} \subset A_n$ such that

$$\tau(B_{n+1}) < -(n+1).$$

If $\lambda(B_{n+1}) = -\infty$, put $A_{n+1} = B_{n+1}$; otherwise put $A_{n+1} = A_n - B_{n+1}$. Then $\lambda(A_{n+1}) = -\infty$.

There are two possible cases:

(i) for infinitely many integers n, $A_n = A_{n-1} - B_n$;

(ii) for $n \geqslant n_0$, $A_n = B_n$.

In case (i) there is a subsequence $\{B_{n_i}\}$ of disjoint sets and

$$\tau\left(\bigcup_{i=1}^{\infty} B_{n_i}\right) = \sum_{i=1}^{\infty} \tau(B_{n_i}) \leqslant \sum_{i=1}^{\infty} -(n_i+1) = -\infty,$$

so that τ takes the value

$$-\infty \quad \text{on} \quad E = \bigcup_{i=1}^{\infty} B_{n_i} \in \mathscr{F},$$

contrary to assumption. In case (ii), the set

$$E = \bigcap_{n=n_0}^{\infty} B_n \in \mathscr{F}$$

and since $\{B_n\}$ is then a decreasing sequence of sets we have

$$\tau(E) = \lim_{n\to\infty} \tau(B_n) = -\infty$$

again giving a contradiction.

Since $\tau(\varnothing) = 0$, $\lambda(\Omega) \leqslant 0$ so that $\lambda = \lambda(\Omega)$ is finite and we can find a sequence $\{C_n\}$ of sets in $\mathscr{F}$ for which

$$\tau(C_n) < \lambda + 2^{-n}.$$

Now consider the set $C_n \cap C_{n+1}$. By noting that

$$C_n \cup C_{n+1} = (C_n - C_n \cap C_{n+1}) \cup (C_{n+1} - C_n \cap C_{n+1}) \cup (C_n \cap C_{n+1})$$

is a decomposition into disjoint sets, it follows that

$$\begin{aligned}\tau(C_n \cap C_{n+1}) &= \tau(C_n) + \tau(C_{n+1}) - \tau(C_n \cup C_{n+1}),\\ &< \lambda + 2^{-n} + \lambda + 2^{-n-1} - \lambda = \lambda + 2^{-n} + 2^{-n-1}.\end{aligned}$$

This argument can be repeated to the pair of sets $(C_n \cap C_{n+1})$ and C_{n+2}: by induction it can be proved that

$$\tau\left(\bigcap_{r=n}^{p} C_r\right) < \lambda + \sum_{r=n}^{p} 2^{-r} < \lambda + 2^{1-n}.$$

If we put $D_n = \bigcap_{r=n}^{\infty} C_r$ we have $D_n \in \mathscr{F}$ and, by theorem 3.2 (i),

$$\tau(D_n) \leqslant \lambda + 2^{1-n}.$$

But now $\{D_n\}$ is a monotone increasing sequence of sets in $\mathscr{F}$ so that

$$N = \lim_{n\to\infty} D_n = \liminf_{n\to\infty} C_n \in \mathscr{F},$$

and $$\tau(N) = \lambda,$$

by theorem 3.2 (i).

Finally, put $P = \Omega - N$. If $E \in \mathscr{F}$, $E \subset P$ we must have $\tau(E) \geqslant 0$ for, if $\tau(E) < 0$, then $\tau(E \cup N) = \tau(E) + \tau(N) < \lambda$. Also if $E \in \mathscr{F}$, $E \subset N$ we must have $\tau(E) \leqslant 0$ for, if $\tau(E) > 0$, then

$$\tau(N - E) = \tau(N) - \tau(E) < \lambda.$$

If we now put

$$\tau_+(E) = \tau(E \cap P), \quad \tau_-(E) = \tau(E \cap N),$$

it is clear that all the conditions of the theorem are satisfied. ∎

Remark. It is usual to call the decomposition $\tau = \tau_+ - \tau_-$, of τ into the difference of two measures, the Jordan decomposition while the decomposition of Ω into positive and negative sets P and N is called the Hahn decomposition. It is not difficult to show that the Jordan decomposition is unique while the sets P, N are not uniquely determined by τ unless $\tau(E) \neq 0$ for all $E \in \mathscr{F}$ such that $\mu(E) \neq 0$ and $\mu(F) = 0$ or $\mu(E)$ for every $F \subset E$ with $F \in \mathscr{F}$. It is further clear that

$$\left.\begin{aligned} \tau_-(E) &= -\lambda(E), \\ \tau_+(E) &= \sup_{B \subset E,\, B \in \mathscr{F}} \tau(B) \end{aligned}\right\} \tag{3.2.2}$$

under the conditions of theorem 3.3, where $\lambda(E)$ is given by (3.2.1).

If one is given a σ-additive $\mu: \mathscr{S} \to \mathbf{R}^*$ defined on a σ-ring $\mathscr{S}$ which is not a σ-field then it is not, in general, possible to obtain the Hahn decomposition, but the Jordan decomposition is still possible, using (3.2.1), (3.2.2) as the definition of τ_-, τ_+.

Exercises 3.2

1. If $\phi: \mathscr{S} \to \mathbf{R}^*$ is σ-additive on a σ-ring $\mathscr{S}$, show that, for any $E \in \mathscr{S}$, there are sets $A \subset E$, $B \subset E$ with $A, B \in \mathscr{S}$ such that

$$\phi(A) = \inf_{C \subset E,\, C \in \mathscr{S}} \phi(C), \quad \phi(B) = \sup_{C \subset E,\, C \in \mathscr{S}} \phi(C).$$

2. Show that, given a (finitely) additive $\mu: \mathscr{R} \to \mathbf{R}^*$ defined on a σ-ring $\mathscr{R}$ and taking finite values, there is a decomposition

$$\mu(E) = \mu_+(E) - \mu_-(E)$$

of μ into the difference of two non-negative additive set functions on $\mathscr{R}$.

3. The set $E_0 \in \mathscr{C}$ is said to be an *atom* of a set function $\phi: \mathscr{C} \to \mathbf{R}^*$ if $\phi(E_0) \neq 0$ and for every $E \subset E_0$, $E \in \mathscr{C}$; $\phi(E) = 0$ or $\phi(E_0)$. Write down the atoms of the set functions of examples (4) and (6) on page 53.

4. A set function $\phi: \mathscr{C} \to \mathbf{R}^*$ is said to be *non-atomic* if it has no atoms. Suppose $\phi: \mathscr{F} \to \mathbf{R}^*$ is σ-additive on the σ-field $\mathscr{F}$, non-atomic, and finite valued. Show that for any $A \in \mathscr{F}$, ϕ takes every real value between $-\phi_-(A)$ and $\phi_+(A)$ for some subset $E \subset A$.

3.3 Additive set functions on a ring

In order to simplify the arguments we now consider only non-negative set functions $\mu:\mathscr{C} \to \mathbf{R}^+$. It is often possible, for a given ring $\mathscr{R}$ to find a semi-ring $\mathscr{C} \subset \mathscr{R}$ such that $\mathscr{R}$ is the ring generated by $\mathscr{C}$. We saw (see § 1.5) that the sets of $\mathscr{R}$ can then be expressed in terms of the sets of $\mathscr{C}$, so it is natural to ask whether in these circumstances a set function $\mu:\mathscr{C} \to \mathbf{R}^+$ can be extended to $\mu:\mathscr{R} \to \mathbf{R}^+$. We now prove that, if μ is additive on $\mathscr{C}$, this is always possible and that the result is unique.

***Theorem* 3.4.** *If $\mu:\mathscr{C} \to \mathbf{R}^+$ is a non-negative additive set function defined on a semi-ring $\mathscr{C}$, then there is a unique additive set function ν defined on the generated ring $\mathscr{R} = \mathscr{R}(\mathscr{C})$ such that ν is an extension of μ. ν is non-negative on $\mathscr{R}$, and is called the extension of μ from $\mathscr{C}$ to $\mathscr{R}(\mathscr{C})$.*

Proof. Suppose A is any set of $\mathscr{R} = \mathscr{R}(\mathscr{C})$, then by theorem 1.4, $A = \bigcup_{k=1}^{n} E_k$ where the sets E_k are disjoint and $E_k \in \mathscr{C}$. Define

$$\nu(A) = \sum_{k=1}^{n} \mu(E_k). \tag{3.3.1}$$

Since for any $a, b \in \mathbf{R}^+$, $a+b$ is always defined, the right-hand side of (3.3.1) defines a number in $\mathbf{R}^+$. ν is thus defined on $\mathscr{R}$ provided we can show that (3.3.1) gives the same result for any two decompositions of A into disjoint subsets in $\mathscr{C}$.

Suppose

$$A = \bigcup_{k=1}^{n} E_k = \bigcup_{j=1}^{m} F_j,$$

where $F_j \in \mathscr{C}$ and are disjoint. Put $H_{kj} = E_k \cap F_j$. Then since $\mathscr{C}$ is a semi-ring the sets $H_{kj} \in \mathscr{C}$, are disjoint and

$$E_k = \bigcup_{j=1}^{m} H_{kj} \quad (k = 1, 2, \ldots, n);$$

$$F_j = \bigcup_{k=1}^{n} H_{kj} \quad (j = 1, 2, \ldots, m);$$

so that, since μ is additive on $\mathscr{C}$,

$$\sum_{k=1}^{n} \mu(E_k) = \sum_{k=1}^{n} \left(\sum_{j=1}^{m} \mu(H_{kj}) \right) = \sum_{j=1}^{m} \left(\sum_{k=1}^{n} \mu(H_{kj}) \right) = \sum_{j=1}^{m} \mu(F_j)$$

and it makes no difference which decomposition is used with (3.3.1) to define $\nu(A)$.

If A_1, A_2 are disjoint sets of $\mathscr{R}$, and

$$A_1 = \bigcup_{k=1}^{n} E_k, \quad A_2 = \bigcup_{i=1}^{m} F_j,$$

then $A_1 \cup A_2$ is a set of $\mathscr{R}$ with a possible decomposition into disjoint subsets of $\mathscr{C}$ given by

$$A_1 \cup A_2 = \bigcup_{k=1}^{n} E_k \cup \bigcup_{i=1}^{m} F_i.$$

Hence
$$\nu(A_1 \cup A_2) = \sum_{k=1}^{n} \mu(E_k) + \sum_{i=1}^{m} \mu(F_j)$$
$$= \nu(A_1) + \nu(A_2),$$

since ν is uniquely determined by (3.3.1). Thus ν is finitely additive on $\mathscr{R}$ (since $\mathscr{R}$ is a ring). It is obvious that ν is non-negative.

Now let τ be any extension of μ from $\mathscr{C}$ to $\mathscr{R}$ which is additive. If $A \in \mathscr{R}$ and $A = \bigcup_{k=1}^{n} E_k$ is a decomposition into disjoint sets of $\mathscr{C}$,

$$\tau(A) = \sum_{k=1}^{n} \tau(E_k), \quad \text{since } \tau \text{ is additive;}$$

$$= \sum_{k=1}^{n} \mu(E_k), \quad \text{since } \tau \text{ is an extension;}$$

$$= \nu(A) \quad \text{by} \quad (3.3.1).$$

Thus ν is the unique additive extension of μ from $\mathscr{S}$ to $\mathscr{R}$. ∎

If we start with a measure $\mu: \mathscr{C} \to \mathbf{R}^+$ on a semi-ring $\mathscr{C}$, then μ is clearly a non-negative finitely additive set function, and so possesses a unique additive extension to the generated ring $\mathscr{R}$. What can we say about this extension?

***Theorem* 3.5.** *If $\mu: \mathscr{C} \to \mathbf{R}^+$ is a measure defined on a semi-ring $\mathscr{C}$, then the (unique) additive extension of μ to the generated ring $\mathscr{R}(\mathscr{C})$ is also a measure.*

Proof. In the last theorem we discovered the form of the unique additive extension ν of μ from $\mathscr{C}$ to $\mathscr{R}$. It is sufficient to show that ν is σ-additive on $\mathscr{R}$. Suppose $E \in \mathscr{R}$, $E_k\ (k = 1, 2, \ldots) \in \mathscr{R}$ and are disjoint, and $E = \bigcup_{k=1}^{\infty} E_k$.

Put
$$E = \bigcup_{r=1}^{n} A_r, \quad A_r \text{ disjoint sets of } \mathscr{C};$$

$$E_k = \bigcup_{i=1}^{n_k} B_{ki}, \quad B_{ki} \text{ disjoint sets of } \mathscr{C}.$$

Put

$$C_{rki} = A_r \cap B_{ki} \quad (r = 1, 2, \ldots, n;\ k = 1, 2, \ldots;\ i = 1, 2, \ldots, n_k),$$

then $\{C_{rki}\}$ forms a disjoint collection of sets in $\mathscr{C}$, and

$$A_r = \bigcup_{k=1}^{\infty} \bigcup_{i=1}^{n_k} C_{rki}, \quad B_{ki} = \bigcup_{r=1}^{n} C_{rki}$$

are disjoint decompositions into sets of $\mathscr{C}$. Since μ is additive on $\mathscr{C}$,

$$\mu(B_{ki}) = \sum_{r=1}^{n} \mu(C_{rki});$$

and since μ is σ-additive on $\mathscr{C}$

$$\mu(A_r) = \sum_{k=1}^{\infty} \sum_{i=1}^{n_k} \mu(C_{rki}).$$

Since the order of summation of double series of non-negative terms makes no difference, we have

$$\nu(E) = \sum_{r=1}^{n} \mu(A_r) = \sum_{r=1}^{n} \left(\sum_{k=1}^{\infty} \sum_{i=1}^{n_k} \mu(C_{rki}) \right)$$

$$= \sum_{k=1}^{\infty} \left(\sum_{i=1}^{n_k} \sum_{r=1}^{n} \mu(C_{rki}) \right)$$

$$= \sum_{k=1}^{\infty} \sum_{i=1}^{n_k} \mu(B_{ki}) = \sum_{k=1}^{\infty} \nu(E_k). \blacksquare$$

The above theorem gives one method of obtaining a measure on a ring—it is sufficient to define a measure on any semi-ring which generates the given ring. The extension to the generated ring is easily carried out, is unique, and gives a measure. There are circumstances in which one can define a set function μ directly on a ring so that it is easy to see that μ is non-negative and additive. Under these circumstances one can often use theorem 3.2 as a criterion for determining whether or not μ is a measure. Another useful criterion is given by the following theorem.

***Theorem* 3.6.** *Suppose $\mu: \mathscr{R} \to \mathbf{R}^+$ is non-negative and additive on a ring $\mathscr{R}$. Then*

(i) *if $E \in \mathscr{R}$, and $\{E_i\}$ is a sequence of disjoint sets of $\mathscr{R}$ such that $E \supset \bigcup_{i=1}^{\infty} E_i$*

$$\mu(E) \geqslant \sum_{i=1}^{\infty} \mu(E_i);$$

(ii) *μ is a measure if and only if for any sequence $\{E_i\}$ of sets in $\mathscr{R}$ such that $\bigcup_{i=1}^{\infty} E_i \supset E \in \mathscr{R}$,*

$$\mu(E) \leqslant \sum_{i=1}^{\infty} \mu(E_i).$$

Proof (i). For each positive integer n,

$$E \supset \bigcup_{i=1}^{n} E_i \in \mathscr{R} \quad \text{so that} \quad \mu(E) \geq \mu\left(\bigcup_{i=1}^{n} E_i\right) = \sum_{i=1}^{n} \mu(E_i),$$

since μ is additive. Hence

$$\mu(E) \geq \sum_{i=1}^{\infty} \mu(E_i).$$

(ii) First, suppose that μ is a measure. Put

$$F_i = E \cap E_i \; (i = 1, 2, \ldots); \; G_1 = F_1,$$

and
$$G_n = F_n - \bigcup_{i=1}^{n-1} F_i \quad (n = 2, 3, \ldots.).$$

Then $\{G_n\}$ is a sequence of disjoint sets of $\mathscr{R}$ such that

$$G_n \subset F_n \subset E_n, \quad \bigcup_{n=1}^{\infty} G_n = \bigcup_{n=1}^{\infty} F_n = E.$$

Thus
$$\mu(E) = \mu\left(\bigcup_{i=1}^{\infty} G_i\right) = \sum_{i=1}^{\infty} \mu(G_i) \leq \sum_{i=1}^{\infty} \mu(E_i),$$

since μ is σ-additive and non-negative.

Conversely if it is known that μ is additive and $E = \bigcup_{i=1}^{\infty} E_i$ is a disjoint decomposition of $E \in \mathscr{R}$ into sets in $\mathscr{R}$, by (i)

$$\mu(E) \geq \sum_{i=1}^{\infty} \mu(E_i);$$

and if the condition in (ii) is satisfied,

$$\mu(E) \leq \sum_{i=1}^{\infty} \mu(E_i)$$

so that we must have
$$\mu(E) = \sum_{i=1}^{\infty} \mu(E_i)$$

and μ is a measure on $\mathscr{R}$.▌

Exercises 3.3

1. If $\Omega = \{1, 2, 3, 4, 5,\}$, show that $\mathscr{C}$ consisting of $\varnothing$, Ω, $\{1\}$, $\{2, 3\}$, $\{1, 2, 3,\}$, $\{4, 5\}$ is a semi-ring and that 0, 3, 1, 1, 2, 1 defines a set of values for an additive set function μ on $\mathscr{C}$. What is the ring $\mathscr{R}$ generated by $\mathscr{C}$? Find the additive extension of μ to $\mathscr{R}$, and show that it is a measure.

2. Suppose $\mathscr{R}$ is any ring of subsets, $\phi: \mathscr{R} \to \mathbf{R}^+$ is non-negative, finitely additive on $\mathscr{R}$, and $\mu: \mathscr{R} \to \mathbf{R}^+$ is a measure on $\mathscr{R}$ such that, for any sequence of sets $\{A_n\}$ in $\mathscr{R}$

$$\mu(A_n) \to 0 \Rightarrow \phi(A_n) \to 0 \quad \text{as} \quad n \to \infty;$$

show that ϕ is completely additive.

3. If $\mu: \mathscr{R} \to \mathbf{R}^+$ is finitely additive on a ring $\mathscr{R}$ and $E, F \in \mathscr{R}$ are such that $\mu(E \triangle F) = 0$, we say that $E \sim F$. Show that $\sim$ is an equivalence relation in $\mathscr{R}$ and that

$$E \sim F \Rightarrow \mu(E) = \mu(F) = \mu(E \cup F) = \mu(E \cap F).$$

Is the class of all sets $E \in \mathscr{R}$ for which $E \sim \varnothing$ a ring?

4. In the notation of question 3, put $\rho(E,F) = \mu(E \triangle F)$ and show that $\rho(E,F) \geqslant 0$, $\rho(E,F) = \rho(F,E)$, $\rho(E,F) \leqslant \rho(E,G) + \rho(G,F)$. If $E_1 \sim E_2$, $F_1 \sim F_2$ are all sets in $\mathscr{R}$, show that $\rho(E_1,F_1) = \rho(E_2,F_2)$. Does ρ define a metric in $\mathscr{R}$?

3.4 Length, area and volume of elementary figures

In §1.5 we saw that:

(i) In $\mathbf{R} = \mathbf{R}^1$ (Euclidean 1-space) the class $\mathscr{P} = \mathscr{P}^1$ of half-open intervals $(a,b]$ forms a semi-ring which generates the ring $\mathscr{E}$ of elementary figures (sets E of the form $E = \bigcup_{i=1}^{n} (a_i, b_i]$ with $b_i < a_{i+1}$ $(i = 1, 2, \ldots, n-1)$.

(ii) In $\mathbf{R}^k$ the half-open intervals have the form $\{(x_1, x_2, \ldots, x_k)$: $a_i < x_i \leqslant b_i, i = 1, 2, \ldots, k\}$ and they again form a semi-ring $\mathscr{P}^k$ which generates the ring $\mathscr{E}^k$ of elementary figures (sets which can be expressed as a finite union of disjoint sets of $\mathscr{P}^k$).

Instead of using the terms length (for $k = 1$), area (for $k = 2$) and volume (for $k \geqslant 3$) of an interval we will use the same word 'length' in each case. Thus the 'length' of an interval of $\mathscr{P}^k$ will be the product of the lengths of k perpendicular edges.

$$\mu(a,b] = b - a,$$

$$\mu\{(x_1, \ldots, x_k): a_i < x \leqslant b_i, i = 1, 2, \ldots, k\} = \prod_{i=1}^{k} (b_i - a_i).$$

Thus for each k we have defined a set function

$$\mu: \mathscr{P}^k \to \mathbf{R}^+$$

which has the usual physical meaning of length, area or volume. Historically this set function and its extension to a larger class of subsets of $\mathbf{R}^k$ was the first to be studied; it leads quickly to the definition of Lebesgue measure in $\mathbf{R}^k$. Our object in the present section is to show that the set function obtained by extending μ from $\mathscr{P}^k$ to $\mathscr{E}^k$ is a measure on $\mathscr{E}^k$. There are essentially two distinct methods of doing this, and both will work for each k. In both it is necessary to show that μ is additive on $\mathscr{P}^k$ so that it has a unique extension to an additive set function in $\mathscr{E}^k$. Then one can either make use of the continuity theorem 3.2 to show that $\mu: \mathscr{E}^k \to \mathbf{R}^+$ is a measure on $\mathscr{E}^k$, or one can prove directly that μ is a measure on $\mathscr{P}^k$ and appeal to theorem 3.5 to deduce that its extension is also a measure. We illustrate by applying the first method to the case $k = 1$, and the second method to the case $k = 2$.

$k = 1$

For each $(a, b] \in \mathscr{P}$ we put $\mu(a, b] = b - a$. It follows that μ is additive on $\mathscr{P}$ for if $(a, b] = \bigcup_{i=1}^{n} (a_i, b_i]$ and the $(a_i, b_i]$ are disjoint we may assume that these intervals are ordered so that $b_i \leqslant a_{i+1}$ $(i = 1, 2, \ldots, n-1)$. It follows that we must have $a_1 = a, b_n = b$ and $b_i = a_{i+1}$ $(i = 1, 2, \ldots, n-1)$ so that, if $a_{n+1} = b_n$,

$$\sum_{i=1}^{n} \mu(a_i, b_i] = \sum_{i=1}^{n} (b_i - a_i) = \sum_{i=1}^{n} (a_{i+1} - a_i)$$
$$= (b - a) = \mu(a, b].$$

By theorem 3.4 there is a unique additive extension $\mu: \mathscr{E} \to \mathbf{R}^+$ since $\mathscr{E}$ is the smallest ring containing the semi-ring $\mathscr{P}$. Since μ is finite on $\mathscr{E}$ it will follow from theorem 3.2 (iii) that μ is a measure, if we can prove that μ is continuous from above at $\varnothing$.

Suppose this is false; then there is a monotone sequence $\{E_n\}$ of sets in $\mathscr{E}$ for which $\lim E_n = \varnothing$ but $\mu(E_n) \to \delta > 0$ as $n \to \infty$. Now E_1 consists of a finite number of intervals of $\mathscr{P}$. Let F_1 be a set of $\mathscr{E}$ obtained by taking away short half-open intervals of $\mathscr{P}$ from the left-hand end of each of the intervals of E_1 in such a way that

$$F_1 \subset \bar{F}_1 \subset E_1; \quad \mu(F_1) > \mu(E_1) - \delta/2^2.$$

We now proceed by induction. Suppose we have obtained $F_n \in \mathscr{E}$ such that

$$F_n \subset \bar{F}_n \subset E_n \cap F_{n-1}$$

and

$$\mu(F_n) > \mu(E_n) - \sum_{r=1}^{n} \frac{\delta}{2^{r+1}}. \tag{3.4.1}$$

Then $F_n \cap E_{n+1} \in \mathscr{E}$ and

$$\mu(F_n \cap E_{n+1}) \geqslant \mu(E_{n+1}) - \mu(E_n - F_n) \geqslant \mu(E_{n+1}) - \sum_{r=1}^{n} \frac{\delta}{2^{r+1}}. \tag{3.4.2.}$$

We can again remove small half-open intervals from the left-hand end of each interval of $F_n \cap E_{n+1}$ to give a set $F_{n+1} \in \mathscr{E}$ such that

$$\mu(F_{n+1}) > \mu(E_{n+1} \cap F_n) - \delta/2^{n+2} \tag{3.4.3}$$

and

$$F_{n+1} \subset \bar{F}_{n+1} \subset E_{n+1} \cap F_n.$$

By (3.4.2) and (3.4.3) we deduce that

$$\mu(F_{n+1}) > \mu(E_{n+1}) - \sum_{r=1}^{n+1} \frac{\delta}{2^{r+1}}.$$

Thus by induction we can establish (3.4.1) for all n. Since $\mu(E_n) \geqslant \delta$ for all n, we have

$$\mu(F_n) > \tfrac{1}{2}\delta, \quad \text{for all } n$$

so that all the sets F_n are non-void. Hence $\{\bar{F}_n\}$ is a decreasing sequence of non-empty bounded closed sets. Hence $\bigcap_{n=1}^{\infty} \bar{F}_n$ is not void. But

$$\bigcap_{n=1}^{\infty} \bar{F}_n \subset \bigcap_{n=1}^{\infty} E_n = \varnothing,$$

so we obtain a contradiction.

$k = 2$

Suppose $C = \{(x,y): a < x \leqslant b,\ c < y \leqslant d\}$ is a set of $\mathscr{P}^2$, and $\mu(C) = (b-a)(d-c)$. In order to prove that μ is additive on $\mathscr{P}^2$, suppose that $C = \bigcup_{i=1}^{n} C_i$ is a decomposition of C into disjoint rectangles in each of which one of the sides (say $(c, d]$) remains the same. Then the other sides $(a_i, b_i]$ must be disjoint and satisfy

$$(a, b] = \bigcup_{i=1}^{n} (a_i, b_i]$$

so that by the corresponding result in $\mathscr{P}^1$, μ is additive in this case. More generally if

$$C = \bigcup_{i=1}^{n} C_i, \quad C_i = \{(x,y): a_i < x \leqslant b_i,\quad c_i < y \leqslant d_i\}$$

is a decomposition of C into a finite number of disjoint rectangles, use the infinite lines $x = a_i$, $x = b_i$, $(i = 1, 2, \ldots, n)$ to decompose each C_i into a finite number of pieces C_{ik} each with the same bounds for the y-coordinate. Hence

$$\sum_{i=1}^{n} \mu(C_i) = \sum_i \sum_k \mu(C_{ik}),$$

and we can sum the right-hand side by first summing over the rectangles whose x-coordinate is bounded by a pair of contiguous a_i, b_j and then summing over these intervals in x. Thus by repeated application of additivity in $\mathscr{P}^1$ we get

$$\mu(C) = \sum_{i=1}^{n} \mu(C_i),$$

as required. (The reader should draw a picture.)

Now suppose $C = \bigcup_{i=1}^{\infty} C_i$ is an infinite decomposition of C into disjoint sets of $\mathscr{P}^2$. We must show that μ is completely additive on $\mathscr{P}^2$. Since $\mathscr{P}^2$ is a semi-ring it follows by induction that, for each n.

$$C - \bigcup_{i=1}^{n} C_i$$

can be expressed as a finite union of sets of $\mathscr{P}^2$. Since μ is non-negative, this implies that

$$\mu(C) \geqslant \sum_{i=1}^{n} \mu(C_i), \quad \text{for all } n,$$

so that

$$\mu(C) \geqslant \sum_{i=1}^{\infty} \mu(C_i).$$

Suppose if possible that μ is not σ-additive, then there will be such a set C for which

$$\mu(C) = \sum_{i=1}^{\infty} \mu(C_i) + 2\delta \quad (\delta > 0). \tag{3.4.4}$$

We now use another form of compactness argument to obtain a contradiction. Suppose $e > 0$ is small enough to ensure that, if

$$F_0 = \{(x, y): a + e < x \leqslant b, c + e < y \leqslant d\},$$

then

$$\mu(F_0) > \mu(C) - \delta;$$

and $e_i > 0$ are small enough to ensure that, if

$$F_i = \{(x, y): a_i < x \leqslant b_i + e_i, c_i < y \leqslant d_i + e_i\},$$

then

$$\mu(F_i) < \mu(C_i) + \delta 2^{-n} \quad (i = 1, 2, \ldots). \tag{3.4.5}$$

Then $\overline{F}_0 \subset C$ and $C_i \subset F_i^0$, the interior of F_i $(i = 1, 2, \ldots)$; so that

$$\overline{F}_0 \subset \bigcup_{i=1}^{\infty} F_i^0.$$

Since $\overline{F}_0$ is compact and the sets F_i^0 are open it follows that, for some integer n, we have

$$\overline{F}_0 \subset \bigcup_{i=1}^{n} F_i^0 \quad \text{so that} \quad F_0 \subset \bigcup_{i=1}^{n} F_i.$$

By the finite additivity of μ on $\mathscr{P}^2$ this implies

$$\mu(F_0) \leqslant \sum_{i=1}^{n} \mu(F_i)$$

so that, by (3.4.5)

$$\mu(C) - \delta < \sum_{i=1}^{\infty} \mu(C_i) + \sum_{i=1}^{\infty} \delta 2^{-i}.$$

Which contradicts (3.4.4).

Thus μ is a measure on $\mathscr{P}^2$ and by theorem 3.5 the unique additive extension $\mu: \mathscr{E}^2 \to \mathbf{R}^+$ is also a measure. Either form of argument clearly extends to the class of elementary figures in $\mathbf{R}^k$, so we have proved:

***Theorem* 3.7.** *Suppose $\mathscr{E}^k$ is the class of elementary figures in $\mathbf{R}^k$, that is, the class of those sets*

$$E = \bigcup_{i=1}^{n} C_i$$

where the C_i are disjoint half-open intervals in $\mathbf{R}^k$. If we put $\mu(C_i)$ = length of C_i = product of lengths of the sides of the interval C_i and

$$\mu(E) = \sum_{i=1}^{n} \mu(C_i),$$

then μ is uniquely defined on $\mathscr{E}^k$ and is a measure.

4

CONSTRUCTION AND PROPERTIES OF MEASURES

4.1 Extension theorem; Lebesgue measure

Measure was defined as a non-negative σ-additive set function defined on a class of sets $\mathscr{C}$. In testing τ for σ-additivity we only needed

$$\tau(E) = \sum_{i=1}^{\infty} \tau(E_i)$$

for sequences $\{E_i\}$ of disjoint sets of $\mathscr{C}$ for which $E = \bigcup_{i=1}^{\infty} E_i \in \mathscr{C}$. This is an artificial restriction as the condition of additivity does not apply to a sequence $\{E_i\}$ unless the union set $\bigcup_{i=1}^{\infty} E_i$ happens to belong to $\mathscr{C}$. For this reason the natural domain of definition for a measure $\tau: \mathscr{C} \to \mathbf{R}^+$ is a σ-ring, and in practice most useful measures are defined on σ-rings.

In the last chapter we considered properties of measures defined on a ring $\mathscr{R}$, so our first objective in the present chapter will be to prove that these can always be extended to a measure on the σ-ring $\mathscr{S}$ generated by $\mathscr{R}$. This extension is unique provided the measure on $\mathscr{R}$ is σ-finite. We introduce an (unnecessary) simplifying assumption—that the generated σ-ring $\mathscr{S}$ is also a σ-field, i.e. that it contains the whole space Ω. Even with this simplification the main extension theorem is somewhat involved. The main idea is that of introducing an outer approximating set function, defined in terms of the measure on $\mathscr{R}$, and then restricting this to a class of sets on which it is σ-additive. The relevant set function turns out to be an outer measure, so it is convenient first to obtain a theorem about all outer measures.

Measurable set

Suppose μ^* is an outer measure defined for all subsets of Ω: that is, μ^* is non-negative, countably subadditive and monotone (see p. 59). A subset E is said to be measurable with respect to μ^* if, for every set $A \subset \Omega$,

$$\mu^*(A) = \mu^*(A \cap E) + \mu^*(A - E). \tag{4.1.1}$$

It is important to stress that the concept of measurability for a set depends on the outer measure μ^*. The same set E may well be measurable with respect to μ_1^* and non-measurable with respect to μ_2^*. It helps ones intuition to realise that (4.1.1) states that if one divides the set A using E and its complement, then the outer measure of the 'pieces' adds up correctly. Thus a set E is μ^*-measurable if and only if it breaks up no set A into two subsets on which μ^* is not additive. The measurability of E depends on what the set E does to the outer measure of all the other subsets.

The reader may find the above explanation of condition (4.1.1) still inadequate to provide the definition of measurability with much intuitive content. This is a case where the definition is justified by the result—it turns out that, for suitable outer measures, a wide class of sets is measurable and the class of measurable sets has got the right kind of structural properties. The definition is therefore justified ultimately by the elegance and usefulness of the theory which results from it.

Note that, because of the subadditivity condition on outer measures, we always have

$$\mu^*(A) \leqslant \mu^*(A \cap E) + \mu^*(A - E)$$

for all sets A, E. Hence E is μ^*-measurable if and only if

$$\mu^*(A) \geqslant \mu^*(A \cap E) + \mu^*(A - E) \tag{4.1.2}$$

for every set $A \subset \Omega$. Since (4.1.2) is automatically satisfied for sets A with $\mu^*(A) = +\infty$, E is μ^*-measurable if and only if (4.1.2) is satisfied for every $A \subset \Omega$ with $\mu^*(A) < \infty$.

It is worth remarking that many of the early discussions of measurability use the concept of *inner measure*. If $\mu^*(\Omega) < \infty$, this can be defined for all subsets E by

$$\mu_*(E) = \mu^*(\Omega) - \mu^*(\Omega - E).$$

In this method of procedure a set E is said to be measurable if $\mu_*(E) = \mu^*(E)$. This apparently weaker definition of measurability can be shown to be equivalent to the one we have adopted provided the outer measure μ^* is *regular*. (An outer measure is said to be regular if, for every $A \subset \Omega$, there is a measurable cover $E \supset A$ such that $\mu^*(E) = \mu^*(A)$.) This means in particular that, under these circumstances, it is sufficient to use the single test set Ω for A in (4.1.1.). We do not use the concept of inner measure in our development.

***Theorem* 4.1.** *Let μ^* be an outer measure on Ω, and let $\mathcal{M}$ be the class of sets of Ω which are measurable with respect to μ^*, Then $\mathcal{M}$ is a σ-field and the restriction of μ^* to $\mathcal{M}$ defines a measure on $\mathcal{M}$.*

Proof. We first show that any finite union of sets of $\mathscr{M}$ is in $\mathscr{M}$. It is clearly sufficient to prove that $E_1 \cup E_2 \in \mathscr{M}$ for any $E_1, E_2 \in \mathscr{M}$. For any set A, since E_1 is measurable,

$$\mu^*(A) = \mu^*(A \cap E_1) + \mu^*(A - E_1). \tag{4.1.3}$$

Now use $(A - E_1)$ as a test set for the measurable E_2

$$\mu^*(A - E_1) = \mu^*((A - E_1) \cap E_2) + \mu^*(A - E_1 - E_2),$$
$$\mu^*(A - E_1) = \mu^*((A - E_1) \cap E_2) + \mu^*(A - (E_1 \cup E_2)). \tag{4.1.4}$$

But $$[(A - E_1) \cap E_2] \cup (A \cap E_1) = A \cap (E_1 \cup E_2),$$

so that if we substitute (4.1.4) into (4.1.3) and use the subadditivity of μ^*, we obtain

$$\begin{aligned}\mu^*(A) &= \mu^*(A \cap E_1) + \mu^*((A - E_1) \cap E_2) + \mu^*(A - (E_1 \cup E_2)) \\ &\geqslant \mu^*(A \cap (E_1 \cup E_2)) + \mu^*(A - (E_1 \cup E_2))\end{aligned}$$

so that, by (4.1.2), $E_1 \cup E_2 \in \mathscr{M}$.

Now, since $A \cap E = A - (\Omega - E)$, the equation for the measurability of $(\Omega - E)$ is the same as that for the measurability of E. Hence, $(\Omega - E)$ is measurable if and only if E is measurable.

Since

$$\bigcap_{i=1}^{n} E_i = \Omega - \bigcup_{i=1}^{n} (\Omega - E_i), \quad \text{(see §1.4)},$$

it follows that the class $\mathscr{M}$ is also closed under finite intersections so that $\mathscr{M}$ is a field. In order to show that $\mathscr{M}$ is a σ-field it is sufficient to show that $E = \bigcup_{k=1}^{\infty} E_k \in \mathscr{M}$ for any sequence $\{E_k\}$ of sets of $\mathscr{M}$. There is no loss in generality in assuming that the sets E_k are disjoint for, since $\mathscr{M}$ is a ring, any countable union can be replaced by a countable disjoint union of subsets in $\mathscr{M}$. Put

$$F_n = \bigcup_{k=1}^{n} E_k \quad (n = 1, 2, \ldots),$$

and let $\mathscr{H}_n$ be the hypothesis that, for any A,

$$\mu^*(A \cap F_n) = \sum_{k=1}^{n} \mu^*(A \cap E_k).$$

Clearly $\mathscr{H}_1$ is true. Use $A \cap F_{n+1}$ as a test set for the measurability of F_n: then

$$\mu^*(A \cap F_{n+1}) = \mu^*(A \cap F_n) + \mu^*(A \cap E_{n+1}).$$

Hence $\mathscr{H}_n \Rightarrow \mathscr{H}_{n+1}$ so that, by induction $\mathscr{H}_n$ is true for all positive integers n.

Since μ^* is monotonic, for each n

$$\mu^*(A \cap E) \geqslant \mu^*(A \cap F_n) = \sum_{k=1}^{n} \mu^*(A \cap E_k),$$

so that

$$\mu^*(A \cap E) \geqslant \sum_{k=1}^{\infty} \mu^*(A \cap E_k),$$

and the subadditivity of μ^* now implies that

$$\mu^*(A \cap E) = \sum_{k=1}^{\infty} \mu^*(A \cap E_k). \tag{4.1.5}$$

Thus, for any A, and any n,

$$\mu^*(A) = \mu^*(A \cap F_n) + \mu^*(A - F_n) \geqslant \sum_{k=1}^{n} \mu^*(A \cap E_k) + \mu^*(A - E)$$

using $\mathscr{H}_n$ and the monotonicity of μ^*. Thus, by (4.1.5),

$$\mu^*(A) \geqslant \mu^*(A \cap E) + \mu^*(A - E),$$

and this implies $E \in \mathscr{M}$ by (4.1.2).

Now the restriction of μ^* to $\mathscr{M}$ is a non-negative set function. Further (4.1.5) with A replaced by Ω shows that μ^* is σ-additive on $\mathscr{M}$ and is therefore a measure on $\mathscr{M}$. ∎

We can now prove the basic extension theorem. In order to simplify the formulation we will assume that the ring $\mathscr{R}$ of subsets of Ω is such that there is a sequence of sets $\{E_n\}$ in $\mathscr{R}$ such that $\Omega = \bigcup_{n=1}^{\infty} E_n$. We then say that Ω is σ-$\mathscr{R}$. This condition implies that the σ-ring generated by $\mathscr{R}$ is a σ-field. Theorem 4.2 is true without this restriction, but the proof would then require the consideration of outer measures defined on a suitable class of subsets of Ω, rather than on all subsets. Since this generalisation also causes complications in the definition of the integral, and the extra generality is rarely needed, we will keep the condition that Ω be σ-$\mathscr{R}$.

***Theorem* 4.2.** *Suppose $\mathscr{R}$ is a ring of subsets of Ω such that Ω is σ-$\mathscr{R}$ and $\mu: \mathscr{R} \to \mathbf{R}^+$ is a measure defined on $\mathscr{R}$. Then there is an extension of μ to a measure ν defined on $\mathscr{S}(\mathscr{R})$, the σ-ring generated by $\mathscr{R}$. If μ is σ-finite on $\mathscr{R}$, then the extension is unique, and is σ-finite on $\mathscr{S}$.*

Proof. Let $\mathscr{C}$ be the class of all subsets of Ω. Since Ω is σ-$\mathscr{R}$, any $E \in \mathscr{C}$ can be covered by a countable sequence of sets of $\mathscr{R}$. Put

$$\mu^*(E) = \inf \sum_{i=1}^{\infty} \mu(F_i),$$

the infimum being taken over all sequences of sets $\{F_i\}$ in $\mathscr{R}$ such that $E \subset \bigcup_{i=1}^{\infty} F_i$. It is clear that $\mu^*: \mathscr{C} \to \mathbf{R}^+$ is non-negative, monotone and that $\mu^*(\varnothing) = 0$. Suppose now that $E \subset \bigcup_{i=1}^{\infty} E_i$. Then, if $\mu^*(E_i)$ is infinite for some i,

$$\mu^*(E) \leqslant \sum_{i=1}^{\infty} \mu^*(E_i) \tag{4.1.6}$$

is immediate. If $\mu^*(E_i) < \infty$ for all i; for any $\epsilon > 0$, choose sets F_{ik} $(k = 1, 2, \ldots)$ in $\mathscr{R}$ such that

$$E_i \subset \bigcup_{k=1}^{\infty} F_{ik} \quad \text{and} \quad \sum_{k=1}^{\infty} \mu(F_{ik}) < \mu^*(E_i) + \frac{\epsilon}{2^i} \quad (i = 1, 2, \ldots).$$

The countable collection $\{F_{ik}\}$ will now cover E, and

$$\mu^*(E) \leqslant \sum_{i=1}^{\infty} \sum_{k=1}^{\infty} \mu(F_{ik}) \leqslant \sum_{i=1}^{\infty} \left(\mu^*(E_i) + \frac{\epsilon}{2^i} \right).$$

Since ϵ is arbitrary, (4.1.6) now follows, and we have proved that $\mu^*: \mathscr{C} \to \mathbf{R}^+$ is an outer measure. Let $\mathscr{M}$ be the class of subsets of Ω which are measurable with respect to μ^*.

We first want to show that $\mathscr{M} \supset \mathscr{R}$. If $E \in \mathscr{R}$ and $\mu^*(A) < \infty$ (the case $\mu^*(A) = +\infty$ is unimportant as (4.1.2) is then trivially satisfied), choose a sequence $\{E_i\}$ of sets of $\mathscr{R}$ such that $A \subset \bigcup_{i=1}^{\infty} E_i$ and

$$\begin{aligned} \mu^*(A) + \epsilon &\geqslant \sum_{i=1}^{\infty} \mu(E_i) = \sum_{i=1}^{\infty} [\mu(E_i \cap E) + \mu(E_i - E)] \\ &\geqslant \mu^*(A \cap E) + \mu^*(A - E), \end{aligned}$$

by the subadditivity of μ^*. Since ϵ is arbitrary, we have again proved (4.1.2), so that $E \in \mathscr{M}$. By theorem 4.1, $\mathscr{M}$ is a σ-ring, so that $\mathscr{M} \supset \mathscr{S}$, the σ-ring generated by $\mathscr{R}$. But the restriction of μ^* to $\mathscr{M}$ is a measure, so that its further restriction ν to $\mathscr{S}$ is also a measure.

If $E \in \mathscr{R}$ it is clear that $\mu^*(E) \geqslant \mu(E)$ because of theorem 3.6 (i), and since E is a covering of itself, $\mu^*(E) \leqslant \mu(E)$. Hence, for all sets $E \in \mathscr{R}$, we have $\nu(E) = \mu^*(E) = \mu(E)$, so that ν is an extension of μ from $\mathscr{R}$ to $\mathscr{S}$.

If we now assume that μ^* is σ-finite on $\mathscr{R}$, it follows that $\Omega = \bigcup_{i=1}^{\infty} E_i$ with $\{E_i\}$ an increasing sequence of sets in $\mathscr{R}$ and $\mu(E_i)$ finite, $i = 1, 2, \ldots$. For a fixed integer n, consider the ring $\mathscr{R}_n$ consisting of sets of the form $E_n \cap E$ with $E \in \mathscr{R}$. Suppose μ_1 and μ_2 are any two extensions of μ from $\mathscr{R}_n$ to $\mathscr{S}_n = \mathscr{S}(\mathscr{R}_n)$. Then all the subsets in $\mathscr{S}_n$ are contained in the set E_n so that μ_1 and μ_2 are finite on $\mathscr{S}_n$. Now let $\mathscr{T}_n$ be the subclass of those sets E of $\mathscr{S}_n$ for which $\mu_1(E) = \mu_2(E)$. Since finite measures are continuous from above and below, it follows that $\mathscr{T}_n$ is a monotone class. By theorem 1.5, since $\mathscr{T}_n \supset \mathscr{R}_n$, it follows that $\mathscr{T}_n \supset \mathscr{S}_n$ and we must have $\mathscr{T}_n = \mathscr{S}_n$. Thus the extension of μ to $\mathscr{S}_n$ is unique for every n. But, for any $E \in \mathscr{S}$ we have

$$E = \lim_{n \to \infty} E \cap E_n$$

so that a further application of the continuity theorem shows that the extension of μ to $\mathscr{S}$ must be unique. ∎

Theorem 4.2 can be applied to any measure defined on a ring $\mathscr{R}$. In 3.4 we saw that the concept of length in $\mathbf{R}^1$, area in $\mathbf{R}^2$ and volume in $\mathbf{R}^k$ ($k \geqslant 3$) could be precisely formulated on the ring $\mathscr{E}^k$ of elementary figures to define a measure on $\mathscr{E}^k$. It is clear that $\mathbf{R}^k$ is σ-$\mathscr{E}^k$, and the measure is actually finite on $\mathscr{E}^k$. The σ-ring generated by $\mathscr{E}^k$ is the class $\mathscr{B}^k$ of Borel sets in $\mathbf{R}^k$ (proved in §2.5). Thus if we apply the statement of theorem 4.2 to this measure μ: $\mathscr{E}^k \to \mathbf{R}^+$ we obtain a unique extension to a measure ν: $\mathscr{B}^k \to \mathbf{R}^+$ which is σ-finite on $\mathscr{B}^k$. It is worth noticing that in the proof of theorem 4.2 the extension was actually carried out to a class of measurable sets containing $\mathscr{B}^k$. This class is denoted by $\mathscr{L}^k$ and can be shown to be larger than $\mathscr{B}^k$. A set $E \subset \mathbf{R}^k$ is said to be *Lebesgue measurable* if and only if it is in the class $\mathscr{L}^k$. In particular all Borel sets in $\mathbf{R}^k$ are Lebesgue measurable. The set function ν: $\mathscr{L}^k \to \mathbf{R}^+$ is called Lebesgue measure in k-space and should be thought of as a generalisation of the notion of k-dimensional volume to a very wide class of sets. We will examine the properties of this set function in some detail in §4.4, and it will then become clear that many of our intuitive ideas of length, area, and volume can be precisely formulated and remain valid for Lebesgue measure.

It is worth noticing that the outer measure obtained by covering as in theorem 4.2 is always a regular outer measure. For, if $\mu^*(E) < \infty$, choose sets $T_{n,r} \in \mathscr{R}$ ($r = 1, 2, \ldots$) such that

$$E \subset \bigcup_{r=1}^{\infty} T_{n,r}, \quad \mu^*(E) + \frac{1}{n} > \sum_{r=1}^{\infty} \mu(T_{n,r}).$$

Then
$$A = \bigcap_{n=1}^{\infty} \bigcup_{r=1}^{\infty} T_{n,r} \supset E, \quad A \in \mathscr{S},$$

and $\mu^*(A) = \mu^*(E)$. This means that the approach through inner measure will lead to the same class of measurable sets and the same extension to this class. In particular the Lebesgue measure can be obtained by this method provided one considers subsets of a fixed bounded interval (of finite measure) in the first instance and then allows the interval to expand to the whole Euclidean space.

Exercises 4.1

1. Suppose μ^* is an outer measure on $\Omega = \lim E_k$ where $\{E_k\}$ is a monotone increasing sequence of sets. Show that if a set E is such that $E \cap E_k$ is measurable (μ^*) for all sufficiently large k, then E is measurable (μ^*).

2. Show that if μ^* is a regular outer measure on Ω and $\mu^*(\Omega) < \infty$, then a necessary and sufficient condition for E to be measurable (μ^*) is that

$$\mu^*(\Omega) = \mu^*(E) + \mu^*(\Omega - E).$$

3. In each of the following cases, show that μ^* is an outer measure, and determine the class of measurable sets

(i) $\mu^*(\varnothing) = 0$, $\mu^*(E) = 1$ for all $E \neq \varnothing$.

(ii) $\mu^*(\varnothing) = 0$, $\mu^*(E) = 1$ for $E \neq \varnothing$ or Ω, $\mu^*(\Omega) = 2$.

(iii) Ω is not countable; $\mu^*(E) = 0$ if E is countable, $\mu^*(E) = 1$ if E is not countable.

4. Show that any outer measure which is (finitely) additive is σ-additive.

5. Suppose μ^* is an outer measure on Ω and E, F are two subsets, at least one of which is measurable (μ^*). Show that

$$\mu^*(E) + \mu^*(F) = \mu^*(E \cup F) + \mu^*(E \cap F).$$

6. Suppose $\{E_n\}$ is a sequence of sets in a σ-ring $\mathscr{S}$, and μ is a measure on $\mathscr{S}$. Show that

(i) $\mu(\liminf E_n) \leqslant \liminf \mu(E_n)$;

(ii) provided $\bigcup_{k=n}^{\infty} E_k$ has finite measure for some n,

$$\mu(\limsup E_n) \geqslant \limsup \mu(E_n).$$

If $\sum_{n=1}^{\infty} \mu(E_n) < \infty$, show that $\mu(\limsup E_n) = 0$.

7. Show that, if μ is a discrete measure on Ω (as in example (6) of §3.1 with $p_i \geqslant 0$), then the operation of extending it to an outer measure and restricting this extension to the class of measurable sets as in theorem 4.2 yields nothing new.

8. Suppose $\mathscr{M}$ is the σ-ring of μ^*-measurable sets in Ω. Then if $\{E_n\}$ is a monotone increasing sequence of sets in $\mathscr{M}$ and A is any set

$$\mu^*(\lim_{n\to\infty} A \cap E_n) = \lim_{n\to\infty} \mu^*(A \cap E_n).$$

Prove a corresponding result for a decreasing sequence (which needs an additional condition).

9. If μ^* is a regular outer measure, show that $\mu^*(\lim A_n) = \lim \mu^*(A_n)$ for any increasing sequence $\{A_n\}$.

10. Suppose in theorem 4·2 that μ is known only to be finitely additive on $\mathscr{R}$; then the same procedure yields an outer measure μ^* and a restriction

$\bar{\mu}$ of μ^* to the μ^*-measurable sets. Show that $\bar{\mu}$ is a measure but is not necessarily an extension of μ.

11. Suppose $\mathscr{R}$ is a ring of subsets of a countable set Ω such that every set in $\mathscr{R}$ is either empty or infinite, but the generated sigma-ring $\mathscr{S}(\mathscr{R})$ contains all subsets of Ω (see exercise 1.5(8)). Put $\mu_1(E) =$ number of points in E, $\mu_2(E) = 2\mu_1(E)$ for all subsets $E \subset \Omega$. Then μ_1, μ_2 agree on $\mathscr{R}$ but not on $\mathscr{S}(\mathscr{R})$ so that the uniqueness assertion of theorem 4.2 requires μ to be σ-finite.

12. Suppose $h(t)$ is any continuous monotonic increasing function defined on $(0, y)$, $y > 0$ with $\lim_{t \to 0+} h(t) = 0$. If Ω is any metric space, let

$$h - m^*(E) = \lim_{\delta \to 0} \left[\inf \sum_{i=1}^{\infty} h\{\text{diam}\,(C_i)\} \right],$$

where the infimum is taken over all sequences $\{C_i\}$ of sets of diameter $< \delta$ which cover E (if there are no such coverings then the inf is $+\infty$). Show that $h - m^*(E)$ defines an outer measure in Ω. (It is called the Hausdorff measure with respect to $h(t)$.)

4.2 Complete measures

If we again think of measure as a mass distribution in the space Ω, it is clear that any subset of a set of zero mass should have the mass zero assigned to it. The present section seeks to make this notion precise.

Given a measure $\tau : \mathscr{C} \to \mathbf{R}^+$ we say that the class $\mathscr{C}$ is complete with respect to τ if

$$E \subset F, \quad F \in \mathscr{C}, \quad \tau(F) = 0 \Rightarrow E \in \mathscr{C}.$$

(Note that since τ is monotone it follows that $\tau(E) = 0$.) If $\tau : \mathscr{C} \to \mathbf{R}^+$ is such that $\mathscr{C}$ is complete with respect to τ we also say that τ is a *complete measure*.

All measures μ which are obtained (as in theorem 4.1) by restricting an outer measure μ^* to the class $\mathscr{M}$ of sets which are measurable (μ^*) are complete measures. For, since outer measures are monotone, non-negative,

$$E \subset F, \quad \mu^*(F) = 0 \Rightarrow \mu^*(E) = 0,$$

and all sets E of zero μ^*-measure are measurable μ^* by (4.1.2) since

$$\mu^*(A) \geqslant \mu^*(A - E) = \mu^*(A - E) + \mu^*(A \cap E).$$

In particular Lebesgue measure defined on the class $\mathscr{L}^k$ is a complete measure.

Given any measure μ on a σ-ring $\mathscr{S}$, there is a simple method of extending it to a complete measure on a larger σ-ring—called the *completion* of $\mathscr{S}$ with respect to μ.

Theorem 4.3. *Given a measure μ on a σ-ring $\mathscr{S}$, let $\overline{\mathscr{S}}$ be the class of all sets of the form $E \triangle N$ where $E \in \mathscr{S}$ and $N \subset F \in \mathscr{S}$ with $\mu(F) = 0$. Then $\overline{\mathscr{S}}$ is a σ-ring and if we put*

$$\overline{\mu}(E \triangle N) = \mu(E),$$

then $\overline{\mu}: \overline{\mathscr{S}} \to \mathbf{R}^+$ is a (uniquely) defined extension of μ from $\mathscr{S}$ to $\overline{\mathscr{S}}$, and $\overline{\mu}$ is a complete measure on $\overline{\mathscr{S}}$.

Proof. Let $E_0 = E \triangle N$, where $E \in \mathscr{S}$, $N \subset F \in \mathscr{S}$, $\mu(F) = 0$. Put $E_1 = E - F$, then $E_1 \subset E_0$, $E_1 \in \mathscr{S}$ and $\mu(E_1) = \mu(E)$. If

$$N_1 = E_0 - E_1,$$

then E_1, N_1 are disjoint and $E_0 = E_1 \cup N_1$. Further, since

$$E_0 \subset E \cup F = (E - F) \cup F,$$

we have $N_1 \subset F$ and $\mu(F) = 0$. Thus the class $\overline{\mathscr{S}}$ is the same as the class of sets $E \cup N$ with $E \in \mathscr{S}$, $N \subset F \in \mathscr{S}$, $\mu(F) = 0$ and $E \cap N = \varnothing$. A similar argument shows that $\overline{\mathscr{S}}$ is also the same as the class of sets of the form $E - N$ with $E \in \mathscr{S}$, $N \subset F \in \mathscr{S}$, $\mu(F) = 0$ and $N \subset E$.

It is now easy to check that $\overline{\mathscr{S}}$ is a ring. Suppose $E_1, E_2 \in \overline{\mathscr{S}}$; first express them as $E_1 = X_1 - N_1$, $E_2 = X_2 - N_2$, $N_1 \subset X_1$, $N_2 \subset X_2$ where $N_1 \subset F_1$, $N_2 \subset F_2$ and $\mu(F_1) = \mu(F_2) = 0$. Then

$$E_1 \cap E_2 = X_1 \cap X_2 - (N_1 \cup N_2),$$

and $X_1 \cap X_2 \in \mathscr{S}$, $N_1 \cup N_2 \subset F_1 \cup F_2 \in \mathscr{S}$, $\mu(F_1 \cup F_2) = 0$; so that $E_1 \cap E_2 \in \overline{\mathscr{S}}$. Now put

$$E_1 = X_3 \cup N_3, \; E_2 = X_2 - N_2, \; N_3 \cap X_3 = \varnothing, \quad N_3 \subset F_3 \text{ with } \mu(F_3) = 0.$$

Then

$$E_1 - E_2 = (X_3 - X_2) \cup (N_3 - X_2) \cup (N_2 \cap E_1) = (X_3 - X_2) \cup N_5,$$

where $N_5 \subset F_3 \cup F_2$ and $\mu(F_3 \cup F_2) = 0$. Finally

$$E_1 = X_3 \cup N_3, \quad E_2 = X_4 \cup N_4, \quad \text{where } X_4 \cap N_4 = \varnothing,$$

and $N_4 \subset F_4$ with $\mu(F_4) = 0$. Then

$$E_1 \cup E_2 = (X_3 \cup X_4) \cup (N_3 \cup N_4 - X_3 \cup X_4) = (X_3 \cup X_4) \cup N_6,$$

where $N_6 \subset F_3 \cup F_4$ and $\mu(F_3 \cup F_4) = 0$. Thus $\bar{\mathscr{S}}$ is closed under the finite operations of intersection, difference, union so it is a ring. To prove it is a σ-ring, put

$$E_i = X_i \cup N_i, \quad N_i \subset F_i, \quad \mu(F_i) = 0 \quad (i = 1, 2, \ldots);$$

then
$$\bigcup_{i=1}^{\infty} E_i = \bigcup_{i=1}^{\infty} X_i \cup \bigcup_{i=1}^{\infty} N_i = X \cup N,$$

where
$$N \subset \bigcup_{i=1}^{\infty} F_i = F \quad \text{and} \quad \mu(F) = 0.$$

Hence
$$\bigcup_{i=1}^{\infty} E_i \in \bar{\mathscr{S}}.$$

To see that $\bar{\mu}$ is uniquely defined on $\bar{\mathscr{S}}$, let

$$E_1 \triangle N_1 = E_2 \triangle N_2$$

be two representations of the same set. Then (see exercise 1.4 (5))

$$E_1 \triangle E_2 = N_1 \triangle N_2$$

and $N_1 \triangle N_2 \subset F \in \mathscr{S}$ with $\mu(F) = 0$. Hence

$$\mu(E_1 - E_2) = \mu(E_2 - E_1) = 0,$$

and
$$\mu(E_1) = \mu(E_1 \cap E_2) = \mu(E_2).$$

Thus if we define $\bar{\mu}$ on $\bar{\mathscr{S}}$ by

$$\bar{\mu}(E_0) = \mu(E_1) \quad \text{if} \quad E_0 = E_1 \triangle N_1,$$

$\bar{\mu}$ is uniquely defined.

It only remains to show that $\bar{\mathscr{S}}$ is complete with respect to $\bar{\mu}$. Suppose E is any set of $\bar{\mathscr{S}}$ with $\bar{\mu}(E) = 0$. Then $E = X \cup N$ where $X \in \mathscr{S}$, $\mu(X) = 0$, $N \subset F \in \mathscr{S}$, $\mu(F) = 0$. Thus, if $G \subset E$, we have $G \subset X \cup F$ with $\mu(X \cup F) = 0$ and $X \cup F \in \mathscr{S}$; so that

$$G = \varnothing \cup G \in \bar{\mathscr{S}},$$

and $\bar{\mu}(G) = 0$. ∎

We already saw that if μ was a σ-finite measure defined on a ring $\mathscr{R}$, then it had only one extension to a measure on the generated σ-ring $\mathscr{S}$. If we now complete $\mathscr{S}$ to obtain the measure $\bar{\mu}$ defined on $\bar{\mathscr{S}}$ so that $\bar{\mathscr{S}}$ is now complete with respect to the extension $\bar{\mu}$ of μ, then we have extended μ from $\mathscr{R}$ to $\bar{\mathscr{S}}$. Since the extension from $\mathscr{S}$ to $\bar{\mathscr{S}}$ is also unique, it follows that there is only one extension of μ from $\mathscr{R}$ to $\bar{\mathscr{S}}$. There is a sense in which, in general, this is as far as one can get with extensions while still preserving uniqueness, though it may be possible to extend μ further to a larger σ-field; see theorem 6.11.

It should also be noticed that in the extension theorem 4.2, the class $\mathcal{M}$ of μ^*-measurable sets is none other than $\bar{\mathcal{S}}$ the completion of the σ-ring $\mathcal{S}$ with respect to μ. For, in the first place, $\mathcal{M} \supset \mathcal{S}$ and $\mathcal{M}$ is complete, hence $\mathcal{M} \supset \bar{\mathcal{S}}$. Secondly, if E is any set of $\mathcal{M}$ such that $\mu(E) < \infty$, we can cover it by $F \in \mathcal{S}$ such that $\mu^*(F) = \mu^*(E)$. Then $F - E \in \mathcal{M}$ and has zero measure, so that it can be covered by a $G \in \mathcal{S}$ with $\mu(G) = 0$, and

$$E = (F - G) \cup (E \cap G) \in \bar{\mathcal{S}}.$$

Since μ is σ-finite on $\mathcal{M}$, and $\bar{\mathcal{S}}$ is a σ-ring, it now follows that $\mathcal{M} \subset \bar{\mathcal{S}}$.

In particular, Lebesgue measure on $\mathcal{L}^k$ is the unique extension of the concept of length from the semi-ring $\mathcal{P}^k$ to the σ-ring $\mathcal{L}^k$ which is the completion of $\mathcal{B}^k$.

Exercises 4.2

1. Suppose μ is a measure on a σ-ring $\mathcal{Q}$ and $\bar{\mu}$ on $\bar{\mathcal{Q}}$ is its completion. Show that if $A, B \in \mathcal{Q}$ with $A \subset E \subset B$, $\mu(B - A) = 0$ then $E \in \bar{\mathcal{Q}}$, and $\bar{\mu}(E) = \mu(A) = \mu(B)$.

2. Given a σ-finite measure μ on a ring $\mathcal{R}$ the extension given by theorem 4.2 yields a complete measure on the class $\mathcal{M}$ of μ^*-measurable sets which is the completion of $\mathcal{S}$ the generated σ-ring. The following example shows that this is not true if the hypothesis of σ-finiteness is omitted: Let Ω be non-countable, $\mathcal{S}$ the ring (also a σ-ring) of all sets which are countable or have countable complements, $\mu(E)$ = number of points in E for $E \in \mathcal{S}$. Then $\mathcal{S}$ is complete with respect to μ, but applying theorem 4.2 yields a complete measure on the class of all subsets (as every subset is measurable).

4.3 Approximation theorems

We have seen how the definition of a measure can be extended from a ring $\mathcal{R}$ to the generated σ-ring $\mathcal{S}$, and its completion $\bar{\mathcal{S}}$. It is convenient to think of the sets of $\mathcal{R}$ as having a simple structure, so that it becomes interesting to see that the sets of $\bar{\mathcal{S}}$ can always be approximated in measure with arbitrary accuracy by sets in the original ring $\mathcal{R}$.

***Theorem* 4.4.** *Suppose $\mathcal{R}$ is a ring for which Ω is σ-$\mathcal{R}$, and the σ-finite measure $\mu: \mathcal{R} \to \mathbf{R}^+$ has been extended (uniquely) to the completion $\bar{\mathcal{S}}$ of the σ-ring $\mathcal{S}$ generated by $\mathcal{R}$. Then for any $\epsilon > 0$, any set $E \in \bar{\mathcal{S}}$ with $\mu(E) < \infty$, there is a set $F \in \mathcal{R}$ such that*

$$\mu(E \triangle F) < \epsilon$$

Proof. First, find a set $E_1 \in \mathscr{S}$ such that

$$\mu(E \triangle E_1) = 0.$$

Then $\mu(E_1) = \mu(E) < \infty$, so that by the construction of theorem 4.2, we have

$$\mu(E_1) = \mu^*(E_1) = \inf_{\substack{E_1 \subset \cup T_i \\ T_i \in \mathscr{R}}} \Sigma \mu(T_i),$$

so that we can choose a sequence of disjoint sets $\{T_i\}$ of $\mathscr{R}$ such that

$$E_1 \subset \bigcup_{i=1}^{\infty} T_i \quad \text{and} \quad \mu^*(E_1) + \tfrac{1}{2}\epsilon > \sum_{i=1}^{\infty} \mu(T_i).$$

Now choose a finite integer n such that

$$\sum_{n+1}^{\infty} \mu(T_i) < \tfrac{1}{2}\epsilon,$$

and put

$$F = \bigcup_{i=1}^{n} T_i \in \mathscr{R}.$$

Then

$$E_1 - F \subset \bigcup_{i=n+1}^{\infty} T_i, \quad \text{so that} \quad \mu(E_1 - F) < \tfrac{1}{2}\epsilon;$$

and

$$F - E_1 \subset \bigcup_{i=1}^{\infty} T_i - E_1 \quad \text{so that} \quad \mu(F - E_1) < \tfrac{1}{2}\epsilon.$$

Hence

$$\mu(E \triangle F) = \mu(E_1 \triangle F) < \epsilon. ▌$$

Remark. The condition $\mu(E) < \infty$ cannot be omitted from the above theorem, since it is possible for a finite measure μ on $\mathscr{R}$ to have an extension to $\mathscr{S}$ which is σ-finite but not finite (for example, Lebesgue measure).

It is also worth noticing that the sets E of $\overline{\mathscr{S}}$ can be approximated exactly in measure by sets in $\mathscr{S}$, by theorem 4.3. We noticed earlier that the outer measure μ^* generated by the process of theorem 4.2 is always regular. This means that an arbitrary set $E \subset \Omega$ is always contained in a set $F \in \mathscr{S}$ for which $\mu^*(E) = \mu(F)$, so that every set can be approximated from the outside by a set of $\mathscr{S}$ of the same measure. If E is not μ^*-measurable (i.e. not in $\overline{\mathscr{S}}$) then two-sided approximation is not possible.

Up to the present we have only considered general approximation theorems valid in any abstract space. If the measure is defined in a topological space, then it is of interest to obtain approximation theorems which connect the measure properties to the topology of the space. We do not, however, discuss this problem in general: instead we consider Euclidean space with the usual topology, and Lebesgue measure.

Regular measure

Suppose $\mathscr{S}$ is a σ-ring of subsets of a topological space Ω which includes the open and the closed subsets of Ω, and $\mu:\mathscr{S}\to\mathbf{R}^+$ is a measure. Then the measure μ is said to be regular if, for each $\epsilon > 0$,

(i) given $E\in\mathscr{S}$, there is an open $G\supset E$ with $\mu(G-E) < \epsilon$;

(ii) given $E\in\mathscr{S}$, there is a closed $F\subset E$ with $\mu(E-F) < \epsilon$.

Since the class $\mathscr{B}$ of Borel sets in Ω is the σ-ring generated by the open sets, the condition that $\mathscr{S}$ includes the open sets implies $\mathscr{S}\supset\mathscr{B}$. If μ is regular on $\mathscr{S}$, then $\overline{\mathscr{B}}\supset\mathscr{S}$, where $\overline{\mathscr{B}}$ denotes the completion of $\mathscr{B}$ with respect to μ; for if δ_n is a sequence of positive numbers decreasing to zero one can find for any E in $\mathscr{S}$ an open set G_n and a closed set F_n such that

$$\mu(G_n-F_n) < \delta_n \quad\text{and}\quad G_n\supset E\supset F_n,$$

and
$$G=\bigcap_{n=1}^{\infty} G_n,\quad F=\bigcup_{n=1}^{\infty} F_n$$

will then be Borel sets with $G\supset E\supset F$ and $\mu(G-F)=0$.

Metric outer measure

An outer measure μ^* defined on a metric space Ω and such that μ^* is additive on separated sets, i.e.

$$d(E,F) > 0 \Rightarrow \mu^*(E\cup F)=\mu^*(E)+\mu^*(F),$$

is said to be a metric outer measure. It can be proved that, for any metric outer measure, the class $\mathscr{M}$ of measurable sets contains the open sets (and therefore contains $\mathscr{B}$), and that, if μ^* is also σ-finite, the restriction of μ^* to $\mathscr{M}$ is regular. Since Lebesgue measure is generated by a metric outer measure, this general theory would allow us to deduce that Lebesgue measure is regular. However, we prefer instead to prove the result only for the special case of Lebesgue measure.

Theorem 4.5. *Lebesgue k-dimensional measure, defined on the class $\mathscr{L}^k$ of Lebesgue measurable sets in $\mathbf{R}^k$, is a regular measure.*

Proof. We give the details of the proof for $k=1$; only obvious alterations are needed for general k. Suppose $E\in\mathscr{L}=\mathscr{L}^1$; then $E\cap[n,n+1]=E_n\in\mathscr{L}$ for every integer n, and $\mu(E_n)\leqslant 1<\infty$. By the construction of theorem 4.2, there is a countable covering $\{C_{ni}\}$ of E_n by $\frac{1}{2}$-open intervals of $\mathscr{P}$ such that

$$\mu(E_n)+\frac{1}{4}\frac{\epsilon}{2^{|n|}} > \sum_{i=1}^{\infty}\mu(C_{ni}).$$

Enlarge each of these intervals C_{ni} to an open interval G_{ni} such that

$$\mu(G_{ni}-C_{ni}) < \frac{1}{4}\frac{\epsilon}{2^{|n|+i}}.$$

Then $Q_n = \bigcup_{i=1}^{\infty} G_{ni}$ is an open set which contains E_n and satisfies

$$\mu(Q_n - E_n) < \frac{1}{2}\frac{\epsilon}{2^{|n|}}.$$

If we now put $Q = \bigcup_{n=-\infty}^{\infty} Q_n$, then Q is open, $Q \subset E$, and $\mu(Q-E) < \epsilon$. This proves condition (i) for regularity.

For any $E \in \mathscr{L}$, $\Omega - E \in \mathscr{L}$, and we can apply the above argument to obtain an open $R \supset \Omega - E$ such that $\mu(R-(\Omega-E)) < \epsilon$. Then $F = \Omega - R$ is closed, $F \subset E$ and $\mu(E-F) = \mu(R \cap E) < \epsilon$, so that the second condition for regularity is also satisfied.]

Corollary. *Given any set $E \in \mathscr{L}^k$, there is a $\mathscr{G}_\delta$-set Q and an $\mathscr{F}_\sigma$-set R such that*

$$Q \supset E \supset R \quad \text{and} \quad \mu(Q-R) = 0.$$

Proof. Note that $\mathscr{G}_\delta$ and $\mathscr{F}_\sigma$ sets were defined in §2.5. For each integer n, take an open set $G_n \supset E$ and a closed set $F_n \subset E$ such that

$$\mu(G_n - E) < \frac{1}{n}, \quad \mu(E-F_n) < \frac{1}{n}.$$

The sets

$$Q = \bigcap_{n=1}^{\infty} G_n \quad \text{and} \quad R = \bigcup_{n=1}^{\infty} F_n$$

then satisfy the conditions of the corollary.]

This corollary strengthens the result that any set in $\mathscr{L}^k$ can be approximated exactly in measure by a set in $\mathscr{B}^k$—which follows from the fact that $\mathscr{L}^k$ is the completion of $\mathscr{B}^k$ with respect to Lebesgue measure.

Exercises 4.3

1. Suppose $\mathscr{Q}$ is the σ-ring generated by a ring $\mathscr{R}$ and μ, ν are two σ-finite measures on $\mathscr{R}$. Show that if $E \in \mathscr{Q}$ is such that both $\mu(E)$, $\nu(E)$ are finite then, for any $\epsilon > 0$, there is a set $E_0 \in \mathscr{R}$ for which

$$\mu(E \,\triangle\, E_0) < \epsilon, \quad \nu(E \,\triangle\, E_0) < \epsilon.$$

2. Suppose Ω is a metric space and μ^* is an outer measure on Ω such that every Borel set is μ^*-measurable. Show that μ^* is a metric outer measure, i.e. that for $E_1, E_2 \subset \Omega$,

$$d(E_1, E_2) > 0 \Rightarrow \mu^*(E_1 \cup E_2) = \mu^*(E_1) + \mu^*(E_2).$$

Hint. Take an open set $G \supset E_1$, $G \cap E_2 = \varnothing$ and use $E_1 \cup E_2$ as a test set for the measurability of G.

3. Suppose μ^* is a metric outer measure on a metric space Ω. Show that if E is a subset of an open subset G and $E_n = E \cap \{x: d(x, \Omega - G) \geqslant 1/n\}$ then $\lim_{n \to \infty} \mu^*(E_n) = \mu^*(E)$.

Hint. $\{E_n\}$ is a monotonic increasing sequence of sets whose limit is E. Put $E_0 = \varnothing$, $D_n = E_{n+1} - E_n$ and notice that if neither D_{n+1} nor E_n is empty then $d(D_{n+2}, D_n) > 0$ so that

$$\mu^*(E_{2n+1}) \geqslant \sum_{i=1}^{n} \mu^*(D_{2i}), \quad \mu^*(E_{2n}) \geqslant \sum_{i=1}^{n} \mu^*(D_{2i-1}).$$

If either of these series diverges, then $\mu^*(E_n) \to \infty = \mu^*(E)$. If both converge, use

$$\mu^*(E) \leqslant \mu^*(E_{2n}) + \sum_{i=n}^{\infty} \mu^*(D_{2i}) + \sum_{i=n}^{\infty} \mu^*(D_{2i+1}).$$

4. If μ^* is a metric outer measure, show that all open sets (and therefore all Borel sets) are μ^*-measurable.

Hint. If G is open, A any subset, use notation of (3) applied to $E = A \cap G$. Then $d(E_n, A \cap -G) > 0$ so

$$\mu^*(A) \geqslant \mu^*\{E_n \cup (A \cap -G)\} = \mu^*(E_n) + \mu^*(A \cap -G).$$

4.4* Geometrical properties of Lebesgue measure

We have now defined Lebesgue measure in Euclidean space and considered some of its measure-theoretic properties. However, the justification for studying Lebesgue measure is that it makes precise our intuitive notion of length, area, volume in Euclidean space and generalises these notions to sets where our intuition breaks down. In the present section we want to show that Lebesgue measure has got the properties which geometrical intuition would lead us to expect.

It is convenient to adopt the notation $|E|$ for the Lebesgue measure of any set $E \in \mathscr{L}^k$, so that for sets $E \in \mathscr{L}^1$, $|E|$ is a generalisation of length; for sets $E \in \mathscr{L}^2$, $|E|$ is a generalisation of area; for sets $E \in \mathscr{L}^k$ $(k \geqslant 3)$, $|E|$ is a generalisation of volume.

Since the set consisting of a single point x can be enclosed in an interval of $\mathscr{P}$ of arbitrarily small length, it follows that

$$|\{x\}| = 0 \quad \text{for} \quad x \in \mathbf{R}^k.$$

In particular, in $\mathbf{R}^1$,

$$|[a, b]| = |(a, b)| = |(a, b]| = |[a, b)| = b - a$$

so that the Lebesgue measure of any interval on the line is its length. Any countable set in $\mathbf{R}^k$ is the union of its single points, and is therefore of zero measure. In particular the set of points in $\mathbf{R}^k$ with rational coordinates forms a set of zero measure (even though this set is dense in the whole space).

In $\mathbf{R}^k$ $(k \geq 2)$, any segment of length l of a straight line can be covered by $[nl]+1$ cubes of $\mathscr{P}^k$ of side $1/n$ so that the Lebesgue measure of such a segment must be less than ($[x]$ denotes the largest integer not greater than x)

$$\left(\frac{1}{n}\right)^k \{[nl]+1\} = O\left(\frac{1}{n^{k-1}}\right) \quad \text{as} \quad n \to \infty,$$

and so $|L| = 0$ for any segment L of finite length. Any infinite straight line in $\mathbf{R}^k$, $k \geq 2$, is the countable union of segments of finite length so that $|L| = 0$ for any straight line L in $\mathbf{R}^k$ $(k \geq 2)$. It follows that, if we are calculating the measure of any geometrical figure in the plane which is bounded by a countable collection of straight lines, then the area will be the same whether all, some or none of the boundary lines are included in the set.

The above argument shows that there are sets E in $\mathbf{R}^k$ $(k \geq 2)$ which are not countable, but such that $|E| = 0$. The question arises whether or not such sets exist in $\mathbf{R}^1$. This is easily answered by the Cantor set

$$C = \bigcap_{n=0}^{\infty} F_n,$$

defined in § 2.7 where $F_0 = [0, 1]$ and F_n is obtained from F_{n-1} by replacing each closed interval of F_{n-1} by two closed intervals obtained by removing an open interval of one third its length from the centre. We proved that C was perfect and therefore non-countable. But

$$|F_n| = \tfrac{2}{3}|F_{n-1}| = (\tfrac{2}{3})^n |F_0| = (\tfrac{2}{3})^n,$$

so that

$$|C| = \lim_{n\to\infty} |F_n| = 0.$$

It is worth remarking that it is also possible for perfect nowhere dense sets in $\mathbf{R}$ to have positive measure—see exercises 4.4 (2, 3).

We now consider what happens to the Lebesgue measure of sets under elementary transformations of the space.

(i) *Translation*

Suppose $\mathbf{x} \in \mathbf{R}^k$ and $E \subset \mathbf{R}^k$. Put

$$E(\mathbf{x}) = \{\mathbf{z}: \mathbf{z} = \mathbf{x}+\mathbf{y}, \quad \mathbf{y} \in E\}.$$

For the intervals $I \in \mathscr{P}^k$, it is immediate that

$$|I(\mathbf{x})| = |I|$$

so that the outer measure μ^* is invariant under translations, and Lebesgue measure must therefore also be invariant provided measurability is preserved. Suppose $E \in \mathscr{L}^k$, and A is a test set for $E(\mathbf{x})$. Then since E is measurable, using $A(-\mathbf{x})$ as a test set,

$$\mu^*(A(-\mathbf{x})) = \mu^*(A(-\mathbf{x}) \cap E) + \mu^*(A(-\mathbf{x}) - E)$$

so that
$$\mu^*(A) = \mu^*(A \cap E(\mathbf{x})) + \mu^*(A - E(\mathbf{x}))$$

and $E(\mathbf{x})$ must also be measurable.

(ii) *Reflexion in a plane perpendicular to an axis*

(For $k = 1$ this means reflexion in a point, for $k = 2$ this means reflexion in a line parallel to an axis.) It is clear that μ^* is invariant under such a reflexion because the reflexion of the covering sets of $\mathscr{P}^k$ again gives $\frac{1}{2}$-open intervals of the same measure. A similar argument to that used in (i) shows that measurability is preserved, so that Lebesgue measure is invariant under such reflexions.

(iii) *Uniform magnification*

For $p > 0$, the transformation of $\mathbf{R}^k$ obtained by putting $\mathbf{y} = p\mathbf{x}$ for all $\mathbf{x} \in \mathbf{R}^k$ will be called a magnification by the factor p, and pE denotes the result of applying this magnification to the set E. If $I \in \mathscr{P}^k$, then it is clear that

$$pI \in \mathscr{P}^k \quad \text{and} \quad |pI| = p^k|I|.$$

Hence, if μ^* denotes the outer measure generated by Lebesgue measure on $\mathscr{P}^k$,

$$\mu^*(pE) = p^k \mu^*(E)$$

for all sets E. A similar argument to that used in (i) shows that measurability is preserved by magnification, so that if E is Lebesgue measurable, so is pE and

$$|pE| = p^k|E|.$$

(iv) *Rotation about the origin*

Lebesgue measure is invariant in this case also, but rather more work is needed to prove it. The key idea needed for the proof is that an open sphere centre O is invariant under rotation about O. Suppose I is a fixed interval of $\mathscr{P}^k$

$$I = \{\mathbf{x}: a_i < x_i \leqslant b_i, \quad i = 1, 2, \dots k\}.$$

Then for any $\mathbf{x} \in \mathbf{R}^k$ $(p > 0)$, $(pI)(\mathbf{x})$ is an interval of $\mathbf{R}^k$ similar, and similarly situated to I. If χ denotes the transformation of $\mathbf{R}^k$ consisting of a fixed rotation about O, then

$$\chi(pI)(\mathbf{x}) = (p\chi I)(\chi\mathbf{x}).$$

By (i) and (ii)

$$|\chi(pI)(x)| = p^k|\chi I|, \quad |(pI)(x)| = p^k|I|,$$

so that

$$|\chi(pI)(x)| = \frac{|\chi I|}{|I|}|(pI)(x)|$$

for all $p > 0$, $x \in \mathbf{R}^k$. This means that, for a given χ and I, the effect on the measure is the same for all intervals of the form $(pI)(x)$.

Now any open set G can be expressed as a countable union of disjoint sets of the form $(pI)(x)$. In particular the unit open sphere S centre the origin, can be expressed this way

$$S = \bigcup_{i=1}^{\infty} (p_i I)(x_i),$$

and

$$|S| = \sum_{i=1}^{\infty} |(p_i I)(x_i)|.$$

But $\chi S = S$, so that

$$\sum_{i=1}^{\infty} |(p_i I)(x_i)| = |S| = |\chi S| = \sum_{i=1}^{\infty} = |\chi(p_i I)\, x_i| = \frac{|\chi I|}{|I|} \sum_{i=1}^{\infty} |(p_i I)\, x_i|,$$

which implies that $|\chi I| = |I|$. This argument is valid for any interval $I \in \mathscr{P}^k$.

We can now use arguments similar to those in (i) to show that, for any set $E \subset \mathbf{R}^k$

$$\mu^*(\chi E) = \mu^*(E)$$

and measurability is preserved under χ. Thus if $E \in \mathscr{L}^k$, χE is also in $\mathscr{L}^k$ and

$$|\chi E| = |E|.$$

Note finally that reflexion in an arbitrary plane can be obtained by successively applying the operations (iv), (ii), (i), (iv). We have thus proved

***Theorem* 4.6.** *The class $\mathscr{L}^k$ of Lebesgue measurable subsets of $\mathbf{R}^k$, and Lebesgue measure on $\mathscr{L}^k$ are invariant under translations, reflexions and rotations. If E and F are two subsets of $\mathbf{R}^k$ which are congruent in the sense of Euclid and E is measurable, then so is F and*

$$|E| = |F|.$$

For $p > 0$, if pE denotes the set of vectors x of the form py, $y \in E$, then $E \in \mathscr{L}^k \Rightarrow pE \in \mathscr{L}^k$, and $|pE| = p^k|E|$.

If k, l, r are positive integers and $k + l = r$, then the Euclidean space $\mathbf{R}^r$ can be thought of as a Cartesian product $\mathbf{R}^k \times \mathbf{R}^l$. We have defined Lebesgue measure independently in each dimension, but the measure of the primary sets $\mathscr{P}^r$ could have been obtained as a product of the measures of corresponding sets in $\mathscr{P}^k$, $\mathscr{P}^l$. It is therefore not surprising that this is true of a wider class of sets.

***Theorem* 4.7.** *If $E \in \mathscr{L}^k$, $F \in \mathscr{L}^l$ then the Cartesian product $E \times F \in \mathscr{L}^{k+l}$ and*

$$|E \times F| = |E|.|F|.$$

Proof. We use μ^* to denote the outer measure generated by Lebesgue measure in the space where the set lies. Suppose first that E, F are bounded so that there are finite open intervals J, K such that $E \subset J, F \subset K$. We can then cover E and F by countable collections of open intervals such that

$$E \subset \bigcup_{i=1}^{\infty} Q_i \subset J, \quad F \subset \bigcup_{j=1}^{\infty} R_j \subset K,$$

$$\sum_{i=1}^{\infty} |Q_i| < |E| + \epsilon, \quad \sum_{j=1}^{\infty} |R_j| < |F| + \epsilon.$$

Then $E \times F \subset \bigcup_{i,j} Q_i \times R_j$, so that

$$\mu^*(E \times F) \leqslant \sum_{i,j} |Q_i \times R_j| = \sum_{i,j} |Q_i|\,|R_j|$$

$$= \sum_{i=1}^{\infty} |Q_i| \sum_{j=1}^{\infty} |R_j| < (|E| + \epsilon)\,(|F| + \epsilon).$$

Since ϵ is arbitrary, it follows that

$$\mu^*(E \times F) \leqslant |E|.|F|. \tag{4.4.1}$$

But

$$J \times K = E \times F \cup (J - E) \times F \cup E \times (K - F) \cup (J - E) \times (K - F),$$

and the subadditivity of μ^* gives, with (4.4.1),

$$\mu^*(J \times K) \leqslant |E|.|F| + |J - E|.|F| + |E|.|K - F| + |J - E|.|K - F|.$$

But $J \times K$ is an open rectangle and therefore in $\mathscr{L}^{k+l}$, and

$$\mu^*(J \times K) = |J|.|K| = (|E| + |J - E|)\,(|F| + |K - F|).$$

It follows that all the inequalities of type (4.4.1) must be equalities. In particular

$$\mu^*(E \times F) = |E|.|F|. \tag{4.4.2.}$$

By the corollary to theorem 4.5, we can find sequences $\{A_n\}$, $\{B_n\}$ of disjoint closed sets such that

$$A = \bigcup_{n=1}^{\infty} A_n \subset E, \quad B = \bigcup_{m=1}^{\infty} B_m \subset F,$$

$$|E - A| = 0, \quad |F - B| = 0.$$

Since $A \times B$ is an $\mathscr{F}_\sigma$-set in $\mathbf{R}^{k+l}$ it is measurable and

$$\mu^*(A \times B) = |A \times B| = |A|.|B| = |E|.|F|.$$

But $A \times B \subset E \times F$, and Lebesgue measure is complete so that we must have $E \times F$ measurable and

$$|E \times F| = \mu^*(E \times F) = |E|.|F|.$$

In order to remove the restriction of boundedness, apply the above to $E \cap S_n$, $F \cap S_n'$, where S_n S_n' are spheres of radius n centre the origin in k-space, l-space respectively. This shows that, for each n,

$$(E \cap S_n) \times (F \cap S_n') \in \mathscr{L}^{k+l},$$

$$|(E \cap S_n) \times (F \cap S_n')| = |E \cap S_n|\,|F \cap S_n'|$$

and the result follows from the continuity of measures on letting $n \to \infty$. ∎

Non-measurable sets

We have now seen that Lebesgue measure can be defined on $\mathscr{L}^k$, a large class of subsets of $\mathbf{R}^k$, in such a way as to preserve the intuitive geometrical ideas of volume. We also remarked earlier that it is impossible to define such a measure on all subsets of $\mathbf{R}^k$, so we now demonstrate the existence of at least one subset which is not in $\mathscr{L}^k$. Again we carry out the construction for $k = 1$. Consider subsets $E \subset (0, 1]$ and for $x \in (0, 1]$ let $E(x)$ be the set of real numbers z such that

$$z = x+y, \quad y \in E \quad \text{and} \quad x+y \leqslant 1,$$

or

$$z = x+y-1, \quad y \in E \quad \text{and} \quad x+y > 1;$$

that is, $E(x)$ is the result of translating E a distance x and then taking the non-integer part. From property (i), it follows immediately that

$$E \in \mathscr{L} \Rightarrow E(x) \in \mathscr{L}, \quad |E| = |E(x)|.$$

Now let Z be the set of rationals in $(0, 1]$. Two sets $Z(x_1)$, $Z(x_2)$ will be disjoint if $(x_1 - x_2)$ is irrational and identical if $(x_1 - x_2)$ is rational. Let $\mathscr{C}$ be the class of disjoint sets of the form $Z(x)$. By the axiom of choice (see §1.6) there is a set T containing precisely one point from each of the sets in $\mathscr{C}$. If Z is the set $(r_1, r_2, \ldots)$, we put

$$Q_n = T(r_n) \quad (n = 1, 2, \ldots).$$

Then

$$\bigcup_{n=1}^{\infty} Q_n = (0, 1],$$

since every point $x \in (0,1]$ is in $Z(x_1)$ for some x_1 and if $q \in Z(x_1) \cap T$, we have $q - x_1 = r_n$ so that $x \in Q_n$. Also the sets Q_n are disjoint as T contains only one point from each set in $\mathscr{C}$ and therefore cannot contain two points differing by a rational. If $T \in \mathscr{L}$, then $Q_n \in \mathscr{L}$ $(n = 1, 2, \ldots)$ and

$$|T| = |Q_n| \quad (n = 1, 2, \ldots).$$

But then

$$1 = |(0,1]| = \sum_{n=1}^{\infty} |Q_n|$$

and this equation cannot possibly be satisfied either by $|Q_n| = 0$ or $|Q_n| = c > 0$ for all n. The only possibility is that the set T is not measurable.

It is worth remarking that there are many more Lebesgue sets than there are Borel sets. The number of sets in $\mathscr{L}^k$ is not more than the number of subsets of $\mathbf{R}^k$, i.e. not more than $2^{\mathbf{c}}$. However it is at least $2^{\mathbf{c}}$ for it contains all subsets of the Cantor set (perfect with $\mathbf{c}$ points in it), so that the cardinality of $\mathscr{L}^k$ must be $2^{\mathbf{c}}$. However the cardinality of the class $\mathscr{B}^k$ of Borel subsets of $\mathbf{R}^k$ is $\mathbf{c}$ and $\mathbf{c} < 2^{\mathbf{c}}$ (see §1.3) so that there must be some sets which are in $\mathscr{L}^k$ but not in $\mathscr{B}^k$; this means that the class $\mathscr{B}^k$ is not complete with respect to Lebesgue measure. In order actually to exhibit a set in $\mathscr{L}^k$ but not in $\mathscr{B}^k$ one has to work a bit harder so we do not include such an example.

Exercises 4.4

1. Show that the set of points in $[0,1]$ whose binary expansion has zero in all the even places is a Lebesgue measurable set of zero measure. Is it a Borel set?

2. By changing the lengths of the extracted intervals in the construction of the Cantor set, show how to obtain a nowhere dense perfect set of measure $\frac{1}{2}$.

3. Generalise (2) to show that for any $\epsilon > 0$ there is a nowhere dense, perfect subset of $[0,1]$ with measure greater than $1 - \epsilon$.

4. Consider a union of sets of (3) to obtain a subset of $[0,1]$ of full measure which is of the first category, and another subset of $[0,1]$ of zero measure which is of the second category.

5 Show that any bounded set in Euclidean space $\mathbf{R}^k$ has finite Lebesgue outer measure. Is the converse of this statement true?

6. Suppose X is the circumference of a unit circle in $\mathbf{R}^2$. Show that there is a unique measure μ defined on Borel subsets of X such that $\mu(X) = 1$ and μ is invariant under all rotations of X into itself.

7. By considering suitable approximating polygons (finite unions of rectangles will do), show that the area of the plane region bounded by $x = 1$, $y = 0$, $y = x^3$ is $\frac{1}{4}$. Generalise to the case $y = x^k$, where $k > 0$ but need not be an integer.

8. Show that a subset E of a bounded interval $I \subset \mathbf{R}^k$ is measurable if, for any $\epsilon > 0$ there are elementary figures Q_1, $Q_2 \in \mathscr{E}^k$ such that $Q_1 \supset E$, $Q_2 \supset I - E$ and

$$|Q_1| + |Q_2| < |I| + \epsilon.$$

9. Suppose X is the unit square $\{(x, y): 0 \leqslant x \leqslant 1,\ 0 \leqslant y \leqslant 1\}$. If $E \subset [0, 1]$ put $\hat{E} = \{(x, y): x \in E,\ 0 \leqslant y \leqslant 1]$ and let $\mathscr{S}$ be the class of sets $\hat{E}$ such that E is $\mathscr{L}^1$-measurable. Put $\mu(\hat{E}) = |E|$, and show that the subset $M = \{(x, y): 0 \leqslant x \leqslant 1,\ y = \frac{1}{2}\}$ is not measurable with respect to the outer measure μ^* generated by μ on the class of all subsets of X. Show that

$$\mu^*(M) = 1, \quad \mu^*(X - M) = 1.$$

4.5 Lebesgue–Stieltjes measure

There are other measures in $\mathbf{R}^k$ which are of importance in probability theory. Suppose $F: \mathbf{R} \to \mathbf{R}$ is a monotone increasing real valued function of a real variable which is everywhere continuous on the right. Such a function is called a *Stieltjes measure function*. Put

$$\mu_F(a, b] = F(b) - F(a)$$

for each $(a, b] \in \mathscr{P}$. Then μ_F is non-negative and (finitely) additive on $\mathscr{P}$—the proof used for the length function in §3.4 can be easily adapted to show this (the length function corresponds to $F(x) = x$). By applying theorem 3.4 we can extend μ_F uniquely to an additive set function on $\mathscr{E}$, the ring of elementary figures. As in §3.4 we again have at least two methods of showing that μ_F is a measure on $\mathscr{E}$. By theorem 3.2 (iii) if μ_F is not a measure, then there is a monotone decreasing sequence $\{E_n\}$ of sets of $\mathscr{E}$ such that $\lim E_n = \varnothing$, but $\lim \mu_F(E_n) = \delta > 0$. The argument used in the Lebesgue case for $k = 1$ can be modified by using the fact that, for any $\epsilon > 0$, if

$$\mu_F(a, b] > 0,$$

we can always find a $y > 0$ such that

$$(a + y, b] \subset [a + y, b] \subset (a, b]$$

and
$$\mu_F(a + y, b] > \mu_F(a, b] - \epsilon,$$

since F is continuous on the right at a. This leads us to a contradiction which establishes that μ_F is a measure on $\mathscr{E}$.

For $k \geqslant 2$, we must start with a function $F: \mathbf{R}^k \to \mathbf{R}$ which is continuous on the right in each variable separately and such that, for $I \in \mathscr{P}^k$,

$$\mu_F(I) = \sum_{i=1}^{2k} y_i F(V_i) \geqslant 0, \tag{4.5.1}$$

where V_i are the 2^k vertices of the set $I \in \mathscr{P}^k$ and $y_i = +1$ for the vertex in which each co-ordinate is largest and $y_i = (-1)^r$ if the vertex V_i is such that r of its coordinates are at the lower bound (and $(k-r)$ at the upper bound). Any such function F is called a *k-dimensional Stieltjes measure function*. With a little care it is not difficult to show that, under these conditions, μ_F is a non-negative additive set function on $\mathscr{P}^k$ and that it therefore has a unique extension to $\mathscr{E}^k$. Either of the arguments given in §3.4 can now be modified to show that μ_F is a measure on $\mathscr{E}^k$.

We can now apply theorem 4.2 to this measure μ_F to extend it to the σ-ring $\mathscr{B}^k$ of Borel sets in $\mathbf{R}^k$. As in the case of Lebesgue measure, this extension automatically defines μ_F on the completion $\mathscr{L}_F^k$ of $\mathscr{B}^k$ with respect to μ_F. The class $\mathscr{L}_F^k$ is called the class of sets which are *Lebesgue–Stieltjes measurable* for the function F. The class clearly depends on the function F—for in the particular case $F \equiv c$, $\mathscr{L}_F^k$ contains all subsets of $\mathbf{R}^k$ as $\mu_F(\mathbf{R}^k) = 0$ and μ_F is complete; while if $F(x_1, x_2, \ldots x_k) = x_1 x_2 \ldots x_k$, then μ_F is the length function and $\mathscr{L}_F^k$ is the Lebesgue class $\mathscr{L}^k$.

Each of these measures $\mu_F: \mathscr{L}_F^k \to \mathbf{R}^+$ is regular. The proof given in theorem 4.5 can easily be modified to show this (we again do the case $k = 1$) by using the fact that, for any $\epsilon > 0$, if $(a, b] \in \mathscr{P}$, there is a $y > 0$ such that

$$(a, b+y] \supset (a, b+y) \supset (a, b]$$

and
$$\mu_F(a, b+y) \leqslant \mu_F(a, b+y] < \mu_F(a, b] + \epsilon,$$

to obtain economical coverings by open intervals.

Probability measure

Given a σ-field $\mathscr{F}$ of subsets of Ω, any measure $\mathsf{P}: \mathscr{F} \to \mathbf{R}^+$ such that $\mathsf{P}(\Omega) = 1$ is called a probability measure on $\mathscr{F}$. If in addition $\mathscr{F}$ is complete with respect to P we will say that the triple $(\Omega, \mathscr{F}, \mathsf{P})$ form a probability space.

Distribution function

A function $F: \mathbf{R} \to \mathbf{R}$ is called a distribution function if

(i) F is monotonic increasing, continuous on the right;

(ii) $F(x) \to 0$ as $x \to -\infty$, $F(x) \to 1$ as $x \to +\infty$.

A function $F: \mathbf{R}^k \to \mathbf{R}$ is called a (k-dimensional) distribution function if

(i) F is continuous on the right in each variable;

(ii) $\mu_F(I) \geqslant 0$ for all $I \in \mathscr{P}^k$, where μ_F is defined by (4.5.1),

(iii) $F(x_1, x_2, \ldots, x_k) \to 0$ as any one of $x_1, x_2, \ldots, x_k \to -\infty$,

$$F(x_1, x_2, \ldots, x_k) \to 1 \quad \text{as} \quad x_1, x_2, \ldots, x_k \quad \text{all} \quad \to +\infty.$$

It is immediate from our definitions that any distribution function F can be used to define a Lebesgue–Stieltjes measure μ_F on the σ-field $\mathscr{L}_F^k$. Further $\mu_F(\mathbf{R}^k) = 1$ and μ_F is complete, so that every distribution function determines a probability measure and $(\mathbf{R}^k, \mathscr{L}_F^k, \mu_F)$, is a probability space. There is a sense in which these are the only interesting probability measures on $\mathbf{R}^k$.

***Theorem* 4.8.** *Suppose $\mathscr{S}$ is a σ-field of sets in $\mathbf{R}$, $\mathscr{S}$ contains the open sets and $\mu: \mathscr{S} \to \mathbf{R}^+$ is a complete measure which is finite on bounded sets in $\mathscr{S}$. Then there is a Stieltjes measure function $F: \mathbf{R} \to \mathbf{R}$ such that $\mathscr{S} \supset \mathscr{L}_F$ and μ coincides with μ_F on $\mathscr{L}_F$. If $(\mathbf{R}, \mathscr{S}, \mu)$ is a probability space, then F can be chosen to be a distribution function.*

Proof. Since $\mathscr{S}$ contains the open sets and is a σ-field, it must contain $\mathscr{B}$, the Borel sets and in particular $\mathscr{S} \supset \mathscr{P}$, the class of half-open intervals. Define F by

$$F(x) = \begin{cases} \mu(0, x] & \text{for} \quad x \geqslant 0, \\ -\mu(x, 0] & \text{for} \quad x < 0. \end{cases}$$

Then $F: \mathbf{R} \to \mathbf{R}$ is clearly defined and is monotonic increasing for all real x (note that $F(0) = 0$). By theorem 3.2 (i), if $\{x_n\}$ is any monotonic sequence decreasing to x, $\lim_{n\to\infty} F(x_n) = F(x)$; since

$$\text{if} \quad x \geqslant 0, \quad \lim\,(0, x_n] = (0, x],$$

$$\text{if} \quad x < 0, \quad \lim\,(x_n, 0] = (x, 0].$$

Thus F is continuous on the right, and must therefore be a Stieltjes measure function.

Now

$$\text{if} \quad a \geqslant 0, \quad \mu(a, b] = \mu(0, b] - \mu(0, a] = F(b) - F(a);$$

$$\text{if} \quad a < 0 \leqslant b, \quad \mu(a, b] = \mu(a, 0] + \mu(0, b] = F(b) - F(a);$$

$$\text{if} \quad b < 0, \quad \mu(a, b] = \mu(a, 0] - \mu(b, 0] = F(b) - F(a);$$

so that μ coincides with μ_F on $\mathscr{P}$. By uniqueness of the extension of a measure to the generated σ-field and its completion, we have $\mu = \mu_F$ on $\mathscr{L}_F$ and $\mathscr{S} \supset \mathscr{L}_F$.

If μ is a probability measure on $\mathscr{S}$, we must have

$$\lim_{x \to +\infty} F(x) - \lim_{x \to -\infty} F(x) = \lim \mu(-n, n] = 1,$$

so that

$$F_1(x) = F(x) - \lim_{x \to -\infty} F(x)$$

will be a distribution function generating the same Stieltjes measure as F.]

Remark. The case where μ is a probability measure could have been done directly by defining

$$F_1(x) = \mu(-\infty, x].$$

It is clear that this case extends immediately to $\mathbf{R}^k$ since if we put

$$F(x_1, x_2, \ldots, x_k) = \mu\{(\xi_1, \ldots, \xi_k): \xi_i \leqslant x_i, i = 1, \ldots, k\}$$

it is easy to check that F is a k-dimensional distribution.

Discrete probability

There is a special case of a probability measure in which all the probability is concentrated on a countable set $E_0 \subset \Omega$. This can be defined by specialising example (6) of §3.1. If $\{x_n\}$ is any sequence in Ω, and $\{p_n\}$ is a sequence of positive real numbers with $\sum_{n=1}^{\infty} p_n = 1$, then it is clear that

$$\mathsf{P}(E) = \sum_{x_n \in E} p_n$$

defines a probability measure on the class of all subsets of Ω. When $\Omega = \mathbf{R}$, this measure can be obtained from the distribution function

$$F(x) = \sum_{x_n \leqslant x} p_n$$

so that, in $\mathbf{R}$ (or in $\mathbf{R}^k$ for that matter) a discrete probability measure can be expressed as the Lebesgue–Stieltjes measure of a suitable distribution function.

Exercises 4.5

1. To see that condition (4.5.1) for k-dimensional Stieltjes measure functions is not implied by the condition that F be monotonic increasing in each variable separately consider

$$F(x_1, x_2) = \begin{cases} \max(0, x_1 + x_2 + 1) & \text{for } x_1 + x_2 < 0, \\ 1 & \text{for } x_1 + x_2 \geqslant 0. \end{cases}$$

Does this condition (4.5.1) imply that F is monotonic in each variable?

2. If $F: \mathbf{R} \to \mathbf{R}$ is a Stieltjes measure function, show that

$$\mu_F(a,b) = F(b-0) - F(a), \quad \mu_F[a,b] = F(b) - F(a-0)$$

and determine μ_F for intervals of the form

$$[a,b), \quad (-\infty, a), \quad (a, \infty).$$

3. If F is a Stieltjes measure function in $\mathbf{R}$ which generates the Stieltjes measure μ_F, show that $F(x)$ is continuous if and only if $\mu_F\{x\} = 0$ for all single point sets $\{x\}$. What is the corresponding continuity condition in $\mathbf{R}^k$?

4. Consider Lebesgue measure on $\mathscr{L}^1$-subsets of $[0,1]$ and let E_0 be a subset of $[0,1]$ which is non-measurable, such that the Lebesgue outer measure of E_0 and $([0,1]-E_0)$ are both 1. Let $\mathscr{Q}$ be the smallest σ-field of subsets of $[0,1]$ containing E_0 and $\mathscr{L}^1$. Show that $\mathscr{Q}$ consists of sets of the form

$$E = A \cap E_0 + B \cap ([0,1]-E_0)$$

for $A, B \in \mathscr{L}^1$ and that $\mu(E) = |A \cap [0,1]|$ defines a probability measure on the σ-field $\mathscr{Q}$. By applying theorem 4.8 to this probability measure show that, in general it is not possible to deduce in theorem 4.8 that $\mathscr{S} = \mathscr{L}_F$.

5. Suppose

$$F(x) = \begin{cases} 0 & \text{for } x < 0, \\ 1 & \text{for } x \geqslant 0. \end{cases}$$

Show that

$$\mu_F(-1, 0) < F(0) - F(-1).$$

6. Give an example of a right-continuous monotone F such that

$$\mu_F(a,b) < F(b) - F(a) < \mu_F[a,b].$$

7. Show that, if F, G are distribution functions in $\mathbf{R}^k$, then $aF + bG$ is a distribution function for any $a \geqslant 0$, $b \geqslant 0$, $a+b = 1$.

8. In $\mathbf{R}^2$,

$$F(x_1, x_2) = \begin{cases} 1 & \text{for } x_1 \geqslant 0, \quad x_2 \geqslant 0, \\ 0 & \text{for all other points.} \end{cases}$$

Show that this F is a distribution function describing a unit mass at 0.

9. State and prove an n-dimensional form of theorem 4.8.

10. We can obtain completely additive set functions in $\mathbf{R}^1$ which are not necessarily non-negative by the following method. Suppose $F: \mathbf{R} \to \mathbf{R}$ is continuous on the right everywhere and of bounded variation in each finite interval and $F(b) - F(a)$ is bounded below for all $a < b$ and define

$$\tau_F(a,b] = F(b) - F(a).$$

Show that τ_F is additive on $\mathscr{P}$ and can be extended to $\mathscr{E}$. By an extension of theorem 4.2, τ_F can then be extended to a σ-additive set function on $\mathscr{B}$. Now apply theorem 3.3 to express τ_F as the difference of two measures. Finally, the argument of theorem 4.8 shows that τ_F is the difference of two Stieltjes measures.

5

DEFINITIONS AND PROPERTIES OF THE INTEGRAL

5.1 What is an integral?

Historically the concept of integration was first considered for real functions of a real variable where either the notion of 'the process inverse to differentiation' or the notion of 'area under a curve' was the starting point. In the first case a real number was obtained as the difference of two values of the 'indefinite' integral, while the second case corresponds immediately to the 'definite' integral. The so-called 'fundamental theorem of the integral calculus' provided the link between the two ideas. Our discussion of the operation of integration will start from the notion of a definite integral, though in the first instance the 'interval' over which the function is integrated will be the whole space. Thus, for 'suitable' functions $f: \Omega \to \mathbf{R}^*$ we want to define the integral $\mathscr{I}(f)$ as a real number. The 'suitable' functions will be called integrable and $\mathscr{I}(f)$ will be called the integral of f.

Before defining such an operator $\mathscr{I}$, we examine the sort of properties $\mathscr{I}$ should have before we would be justified in calling it an 'integral'. Suppose then that $\mathscr{A}$ is a class of functions $f: \Omega \to \mathbf{R}^*$, and $\mathscr{I}: \mathscr{A} \to \mathbf{R}$ defines a real number for every $f \in \mathscr{A}$. Then we want $\mathscr{I}$ to satisfy:

(i) $f \in \mathscr{A}, f(x) \geqslant 0$ all $x \in \Omega \Rightarrow \mathscr{I}(f) \geqslant 0$, that is $\mathscr{I}$ preserves positivity;

(ii) $f, g \in \mathscr{A}$, $\alpha, \beta \in \mathbf{R} \Rightarrow \alpha f + \beta g \in \mathscr{A}$ and

$$\mathscr{I}(\alpha f + \beta g) = \alpha \mathscr{I}(f) + \beta \mathscr{I}(g),$$

that is $\mathscr{I}$ is linear on $\mathscr{A}$;

(iii) $\mathscr{I}$ is continuous on $\mathscr{A}$ in some sense, at least we would want to have $\mathscr{I}(f_n) \to 0$ as $n \to \infty$ for any sequence $\{f_n\}$ of functions in $\mathscr{A}$ which is monotone decreasing with $f_n(x) \to 0$ for all x in Ω.

These conditions are satisfied by the elementary integration process, but the Riemann integral does not satisfy the following strengthened form of (iii):

(iii)* If $\{f_n\}$ is an increasing sequence of functions in $\mathscr{A}$, and

$$f_n(x) \to f(x) \quad \text{for all} \quad x \in \Omega,$$

then $f \in \mathscr{A}$ and $\mathscr{I}(f_n) \to \mathscr{I}(f)$ as $n \to \infty$.

This is the most serious limitation of the Riemann integral for, with this definition of integration, it is necessary in (iii)* to postulate $f_n(x) \to f(x)$ uniformly in x before one can conclude that $f \in \mathscr{A}$ and $\mathscr{I}(f_n) \to \mathscr{I}(f)$. Now conditions about the continuity of $\mathscr{I}$ are really essential if the operation is to be a useful tool in analysis—there would not be much of analysis left if one could not carry out at least sequential limiting operations. One of our main objectives, therefore, is to define an operator $\mathscr{I}$ which satisfies (iii)*.

One method of studying integration theory (essentially due to P. J. Daniell) is to start with a restricted class $\mathscr{A}_0$ of functions with a simple structure, define $\mathscr{I}:\mathscr{A}_0 \to \mathbf{R}$ to satisfy (i), (ii) and (iii) and then extend $\mathscr{A}_0$ and the functional $\mathscr{I}$ step by step until $\mathscr{I}:\mathscr{A} \to \mathbf{R}$ is defined on a sufficiently large class while (i), (ii) and (iii)* are satisfied. Using this approach one can deduce a measure on a suitable σ-ring of subsets of Ω by putting

$$\mu(E) = \mathscr{I}(\chi_E)$$

for those sets E for which $\chi_E \in \mathscr{A}$. Condition (i) then implies that μ is non-negative, condition (ii) that it is additive and condition (iii) that it is σ-additive provided the domain of definition is a ring. We will give details of this approach in § 9.4, but for the present we will regard the measure as the primary concept and define the integral in terms of a given measure. We will, however, obtain an operator $\mathscr{I}:\mathscr{A} \to \mathbf{R}$ which has the above properties and moreover in defining $\mathscr{I}$ we will continually have these desired properties in mind. Thus out of many possible ways of obtaining the integral starting from a measure, we choose the method of definition by limits of monotone sequences of 'simple' functions.

5.2 Simple functions; measurable functions

We now assume given $(\Omega, \mathscr{F}, \mu)$ where Ω is a space, $\mathscr{F}$ a σ-field of subsets of Ω and μ a measure on $\mathscr{F}$. All the concepts we now define are relative to $(\Omega, \mathscr{F}, \mu)$. It is worth remarking that our definitions can be modified to apply to the case where $\mathscr{F}$ is a σ-ring rather than a σ-field, but this results in additional complications in proofs. The additional labour involved does not seem justified for the small gain in generality.

Our object is to define an operation, called integration, having the properties discussed in § 5.1 on a suitable class of functions $f: \Omega \to \mathbf{R}^*$. Ultimately we want this domain of definition for the integral to be as large as possible. In the present section we obtain the properties of certain classes of functions which will be important later.

Dissection

If $\Omega = \bigcup_{i=1}^{n} E_i$ and the sets E_i are disjoint, then $E_1, E_2, \ldots, E_n$ are said to form a (finite) dissection of Ω. They are said to form an $\mathscr{F}$-dissection if, in addition, $E_i \in \mathscr{F}$ $(i = 1, 2, \ldots, n)$.

Simple function

A function $f: \Omega \to \mathbf{R}$ is called $\mathscr{F}$-simple if it can be expressed as

$$f(x) = \sum_{i=1}^{n} c_i \chi_{E_i}(x),$$

where $E_1, E_2, \ldots, E_n$ form an $\mathscr{F}$-dissection of Ω and

$$c_i \in \mathbf{R} \; (i = 1, 2, \ldots, n).$$

Thus an $\mathscr{F}$-simple function is one which takes a constant value c_i on the set E_i where the sets E_i are disjoint sets of $\mathscr{F}$. The additional condition implied by our definition that $\Omega = \bigcup_{i=1}^{n} E_i$ is not important (and is omitted by many authors), since if

$$E_{n+1} = \Omega - \bigcup_{i=1}^{n} E_i \neq \varnothing$$

we can always put $c_{n+1} = 0$ and write

$$f = \sum_{i=1}^{n+1} c_i \chi_{E_i}$$

to see that the function is $\mathscr{F}$-simple. If there is only one σ-field $\mathscr{F}$ under consideration we will talk of *simple* functions rather than $\mathscr{F}$-simple functions.

Lemma. *The sum, difference and product of two simple functions is a simple function.*

Proof. Suppose we have the representations

$$f = \sum_{i=1}^{n} c_i \chi_{E_i}, \quad g = \sum_{j=1}^{m} d_j \chi_{A_j};$$

then the sets $H_{ij} = E_i \cap A_j$ $(i = 1, 2, \ldots, n;\, j = 1, 2, \ldots, m)$ are in $\mathscr{F}$ and form a dissection of Ω. Further

$$f(x) = c_i \quad \text{and} \quad g(x) = d_j \quad \text{for} \quad x \in H_{ij}, \quad \chi_{H_{ij}} = \chi_{E_i} \cdot \chi_{A_j}$$

so that $\quad (f \pm g)(x) = c_i \pm d_j, \quad (fg)(x) = c_i d_j \quad \text{for} \quad x \in H_{ij}$

and $$f \pm g = \sum_{i=1}^{n} \sum_{j=1}^{m} (c_i \pm d_j) \chi_{H_{ij}}, \quad fg = \sum_{i=1}^{n} \sum_{j=1}^{m} c_i d_j \chi_{H_{ij}}. \; \blacksquare$$

Note that the constant functions

$$f(x) = c \quad \text{all} \quad x \in \Omega$$

are simple, so that by this lemma it also follows that cf is simple if f is and the class of simple functions forms a linear space over the reals. One should regard simple functions as a generalisation of 'step' functions, but it is clear that they form a very restricted class since the image of Ω under a simple function is a finite subset of $\mathbf{R}$.

In defining measurability we will want to consider functions $f: \Omega \to \mathbf{R}^*$ with extended real values. It is possible to define a topology in $\mathbf{R}^*$ and to define the class of Borel sets in $\mathbf{R}^*$ in terms of this topology. However, we adopt the simpler procedure of defining the class $\mathscr{B}^*$ of Borel sets in $\mathbf{R}^*$ directly. We say that a set $B \subset \mathbf{R}^*$ is a Borel set in $\mathbf{R}^*$ if it is the union of a set in $\mathscr{B}^1$ (the class of Borel sets in $\mathbf{R}$) with any subset of $\mathbf{R}^* - \mathbf{R} = \{-\infty, +\infty\}$.

Measurable function

A function $f: \Omega \to \mathbf{R}^*$ is said to be $\mathscr{F}$-measurable if and only if

$$f^{-1}(B) \in \mathscr{F}$$

for every $B \in \mathscr{B}^*$. If there is only one σ-field $\mathscr{F}$ under discussion we may say that f is a measurable function.

From the definition it appears at first sight that one has to work hard to check that a given function is $\mathscr{F}$-measurable. However, in practice it is sufficient to check that $f^{-1}(E) \in \mathscr{F}$ for a suitable class of subsets which generates the σ-field $\mathscr{B}^*$. The most important such class is given by the next theorem.

Theorem 5.1. *In order that $f: \Omega \to \mathbf{R}^*$ be $\mathscr{F}$-measurable each of the following conditions is necessary and sufficient:*

(i) $\{x: f(x) \leqslant c\} \in \mathscr{F}$ *for all* $c \in \mathbf{R}$;

(ii) $\{x: f(x) > c\} \in \mathscr{F}$ *for all* $c \in \mathbf{R}$;

(iii) $\{x: f(x) \geqslant c\} \in \mathscr{F}$ *for all* $c \in \mathbf{R}$;

(iv) $\{x: f(x) < c\} \in \mathscr{F}$ *for all* $c \in \mathbf{R}$.

Proof (i). Since $[-\infty, c] \in \mathscr{B}^*$, it is clear that the given condition is necessary. If we suppose that the condition is satisfied, and put

$$E_c = \{x: f(x) \leqslant c\} = f^{-1}[-\infty, c],$$

then $E_c \in \mathscr{F}$, for all $c \in \mathbf{R}$. But the sets $I_c = [-\infty, c]$, $c \in \mathbf{R}$ generate the σ-field $\mathscr{B}^*$, so that, for each $B \in \mathscr{B}^*$ (see exercise 1.5.(10)), the set $f^{-1}(B)$

is in the σ-field of subsets of Ω generated by the sets E_c, $c \in R$. Since $\mathscr{F}$ is a σ-field we have

$$f^{-1}(B) \in \mathscr{F}$$

for all $B \in \mathscr{B}^*$.

(ii), (ii) *and* (iv). A similar proof can be constructed for each case. Alternatively, it is easy to prove directly that each of (i), (ii), (iii), (iv) is equivalent to each of the others.]

Corollary. *Any $\mathscr{F}$-simple function is $\mathscr{F}$-measurable.*

Proof. If

$$f = \sum_{i=1}^{n} c_i \chi_{E_i}, \quad \text{then} \quad E_c = \{x : f(x) \leqslant c\}$$

is the finite union of those sets $E_i (\in \mathscr{F})$ for which $c_i \leqslant c$, and is therefore in $\mathscr{F}$. By condition (i) of the theorem, this implies that f is measurable.]

The next theorem examines further the relationship between simple functions and measurable functions. It is both important and somewhat surprising.

***Theorem* 5.2.** *Any non-negative measurable function $f : \Omega \to \mathbf{R}^+$ is the limit of a monotone increasing sequence of non-negative simple functions.*

Proof. For each positive integer s, let

$$Q_{p,s} = \left\{x : \frac{p-1}{2^s} \leqslant f(x) < \frac{p}{2^s}\right\} \quad (p = 1, 2, \ldots, 2^{2s});$$

$$Q_{0,s} = \Omega - \bigcup_{p=1}^{2^{2s}} Q_{p,s} = \{x : f(x) \geqslant 2^s\}.$$

Then, since f is $\mathscr{F}$-measurable, $Q_{p,s} \in \mathscr{F}$ and the sets

$$Q_{p,s} \,(p = 0, 1, \ldots, 2^{2s})$$

form an $\mathscr{F}$-dissection of Ω. The function

$$f_s(x) = \frac{p-1}{2^s} \quad \text{for} \quad x \in Q_{p,s} \quad (p = 1, 2, \ldots, 2^{2s});$$

$$= 2^s \quad \text{for} \quad x \in Q_{0,s}$$

is a simple function and it is immediate that

$$0 \leqslant f_s \leqslant f.$$

If $x \in Q_{p,s}$, then either $x \in Q_{2p-1,s+1}$ or $x \in Q_{2p,s+1}$ so that either

$$f_s(x) = f_{s+1}(x) \quad \text{or} \quad f_s(x) + \frac{1}{2^{s+1}} = f_{s+1}(x).$$

Further, if $x \in Q_{0,s}$, then $f_s(x) = 2^s \leqslant f(x)$ so that either $x \in Q_{0,s+1}$, or $x \in Q_{p,s+1}$ for some $p \geqslant 2^{2s+1}+1$; and in either case $f_{s+1}(x) \geqslant f_s(x)$. Thus for each integer s

$$f_{s+1}(x) \geqslant f_s(x) \quad \text{for all} \quad x \in \Omega;$$

and the sequence $\{f_s\}$ of simple functions is monotone increasing.

If x is such that $f(x)$ is finite, then, if $2^s > f(x)$ we have

$$0 \leqslant f(x) - f_s(x) < 2^{-s}$$

so that in this case $f_s(x) \to f(x)$ as $s \to \infty$. On the other hand, if

$$f(x) = +\infty, \quad \text{then} \quad f_s(x) = 2^s,$$

so that again $\qquad f_s(x) \to f(x) \quad \text{as} \quad s \to \infty.$]

For any function $f: \Omega \to \mathbf{R}^*$ we define the positive and negative parts f_+, f_- of f by

$$f_+(x) = \max[0, f(x)],$$
$$f_-(x) = -\min[0, f(x)].$$

Then clearly for all x,

$$f(x) = f_+(x) - f_-(x),$$
$$|f(x)| = f_+(x) + f_-(x),$$

and each of the functions f_+, f_- is non-negative. It follows immediately from theorem 5.1 that, for any measurable f, f_+ and f_- are both measurable. An application of theorem 5.2 now shows that any measurable function can be expressed as a limit of simple functions. Our next step is to show that the class of measurable functions $f: \Omega \to \mathbf{R}^*$ is closed both for finite algebraic operations and for countable limiting operations. A minor difficulty arises in that $\mathbf{R}^*$ is not an algebraic field so that, for example, $(f+g)(x)$ is not defined at points where $f(x) = +\infty$, $g(x) = -\infty$. In the following theorem therefore, we assume that the functions are such that the algebraic operations are possible.

***Theorem* 5.3.** *If f and g are measurable functions: $\Omega \to \mathbf{R}^*$ and $k \in \mathbf{R}$, then each of the functions:*

$$f+k, \quad kf, \quad f+g, \quad f^2, \quad fg, \quad 1/f \quad (\textit{where } (1/f)(x) = +\infty \textit{ if } f(x) = 0),$$
$$\max(f, g), \quad \min(f, g), \quad f_+, \quad f_-, \quad |f|;$$

which is defined, is measurable.

Proof. The measurability of the first two functions $f+k$, kf follows immediately from any part of theorem 5.1. Consider now the function $(f+g)$. Let $\{r_i\}$ be a sequence containing all the rationals in $\mathbf{R}$. Then, since $\{r_i\}$ is dense in $\mathbf{R}$, for any $c \in \mathbf{R}$,

$$\{x: f(x) + g(x) > c\} = \bigcup_{i=1}^{\infty} \{x: f(x) > r_i\} \cap \{x: g(x) > c - r_i\}.$$

By theorem 5.1 each of the sets on the right-hand side is in $\mathscr{F}$ so that, because $\mathscr{F}$ is a σ-field, the set on the left is also in $\mathscr{F}$, and $(f+g)$ must be measurable.

Now

$$\{x: (f(x))^2 \leqslant c\} = \begin{cases} \varnothing & \text{if } c < 0, \\ \{x: f(x) = 0\} & \text{if } c = 0, \\ \{x: -c \leqslant f(x) \leqslant c\} & \text{if } c > 0, \end{cases}$$

and each of these sets is in $\mathscr{F}$, so that f^2 is measurable. Further

$$\{x: (1/f)(x) < c\} = \begin{cases} \{x: 1/c < f(x) < 0\} & \text{if } c < 0, \\ \{x: -\infty < f(x) < 0\} & \text{if } c = 0, \\ \{x: -\infty \leqslant f(x) < 0\} \cup \{x: f(x) > 1/c\} & \text{if } c > 0, \end{cases}$$

and each of these sets is in $\mathscr{F}$, so that $1/f$ is measurable.

We have already seen that f_+ and f_- are measurable, so that $|f| = f_+ + f_-$ is also measurable. The measurability of the remaining functions now follow from the identities:

$$fg = \tfrac{1}{2}[(f+g)^2 - f^2 - g^2],$$

$$\max(f, g) = \tfrac{1}{2}[f + g - |f-g|],$$

$$\min(f, g) = f + g - \max(f, g). \blacksquare$$

It is clear that the above theorem extends immediately to functions obtained by carrying out a finite number of algebraic operations on any finite collection of measurable functions.

***Theorem* 5.4.** *Suppose $\{f_n\}$, $n = 1, 2, \ldots$ is a sequence of measurable functions: $\Omega \to \mathbf{R}^*$; then*

(i) *the functions $\sup_n f_n$ and $\inf_n f_n$ are measurable;*

(ii) *the functions $\limsup_{n\to\infty} f_n$, $\liminf_{n\to\infty} f_n$ are measurable;*

(iii) *if the sequence $\{f_n\}$ converges and in particular if it is monotone, $\lim_{n\to\infty} f_n$ is measurable.*

Proof. (i)

$$\{x: \sup_n f_n > c\} = \bigcup_{n=1}^{\infty} \{x: f_n(x) > c\}.$$

Since $\mathscr{F}$ is a σ-field, an application of theorem 5.1 now shows that $\sup_n f_n$ is measurable. The case of $\inf_n f_n$ can be proved similarly or it may be deduced from

$$\inf_n f_n = -\sup_n (-f_n).$$

Suppose now that $\{f_n\}$ is monotone increasing; then

$$\lim_{n\to\infty} f_n = \sup_n f_n$$

and is therefore measurable. Similarly, if $\{f_n\}$ is decreasing,

$$\lim_{n\to\infty} f_n = \inf_{n\to\infty} f_n.$$

(ii) If

$$g_n = \sup_{r\geqslant n} f_r, \quad h_n = \inf_{r\geqslant n} f_r,$$

then $\{g_n\}$, $\{h_n\}$ are monotone sequences, and

$$\limsup_{n\to\infty} f_n = \lim_{n\to\infty} g_n, \quad \liminf_{n\to\infty} f_n = \lim_{n\to\infty} h_n$$

so that both are measurable.

(iii) If $\{f_n\}$ converges its limit will be measurable because it is the common value of the measurable functions $\limsup f_n$, $\liminf f_n$. ▌

It should be remarked that the class of measurable functions is not closed for non-countable operations of the above type. Thus, if A is non-countable and $f_\alpha\colon \Omega \to \mathbf{R}^*$ is measurable for each $\alpha \in A$, there is no reason for

$$f(x) = \sup_{\alpha\in A} f_\alpha(x)$$

to be measurable. For example, let A be a subset of $[0, 1]$ which is not $\mathscr{L}$-measurable (see § 4.4), and put

$$\begin{aligned} f_\alpha(x) &= 1 \quad \text{if} \quad x = \alpha; \\ &= 0 \quad \text{if} \quad x \neq \alpha. \end{aligned}$$

Then for each $\alpha \in A$, f_α is $\mathscr{L}$-measurable (it is actually $\mathscr{L}$-simple) but

$$\chi_A(x) = \sup_{\alpha\in A} f_\alpha$$

is certainly not $\mathscr{L}$-measurable. In practice when one needs to consider non-countable suprema (as in the theory of stochastic processes with continuous time parameter) one tries to replace the index set A by a countable subset giving the same supremum for the family (at least except for a special subset of Ω of zero measure). If this procedure is impossible for any reason, then there are very serious difficulties in using non-countable suprema.

In the special case where Ω is a topological space and $\mathscr{B}$ is the σ-field of Borel sets in Ω, there is a special name for $\mathscr{B}$-measurable functions.

Borel measurability

If $\mathscr{B}$ is the class of Borel sets in Ω and $f\colon \Omega \to \mathbf{R}^*$ is $\mathscr{B}$-measurable, then we say that f is a *Borel measurable* function on Ω.

Lemma. *Any continuous function $f\colon \Omega \to \mathbf{R}$ on a topological space Ω is Borel measurable.*

Proof. Since, for continuous f, the inverse image of an open set in $\mathbf{R}^*$ is open in Ω it follows that $\{x: f(x) < c\}$ is open for all $c \in \mathbf{R}$ and is therefore in $\mathscr{B}$. ▌

If $\mathscr{F}$, $\mathscr{Q}$ are any two σ-fields of subsets of Ω such that $\mathscr{F} \supset \mathscr{Q}$, it is immediate that any function $f: \Omega \to \mathbf{R}^*$ which is $\mathscr{Q}$-measurable is also $\mathscr{F}$-measurable. In particular if $\mathscr{F} \supset \mathscr{B}$, then any continuous function on a topological space Ω is $\mathscr{F}$-measurable. If $\Omega = \mathbf{R}^k$ (Euclidean k-space) then we know that the class $\mathscr{L}^k$ of Lebesgue measurable sets, and the class $\mathscr{L}_F^k$ of sets which are measurable with respect to the Lebesgue–Stieltjes measure defined by F each contain $\mathscr{B}^k$, the Borel sets in $\mathbf{R}^k$. Hence, all continuous functions on $\mathbf{R}^k$ are Borel measurable and therefore $\mathscr{L}_F^k$-measurable for any F (in particular they are $\mathscr{L}^k$-measurable which we call *Lebesgue measurable*). Functions which normally occur in real analysis are usually obtainable from continuous functions and simple functions by the operations of the following types:

(i) finite algebraic operations;
(ii) countable limiting operations;
(iii) composition.

We have already seen that operations of types (i) and (ii) preserve measurability so that we should consider whether composition operations can be carried out within the class of measurable functions.

Lemma. *Suppose* $f: \mathbf{R}^* \to \mathbf{R}^*$ *is Borel measurable and* $g: \Omega \to \mathbf{R}^*$ *is* $\mathscr{F}$*-measurable, then the composite function* $f \circ g: \Omega \to \mathbf{R}^*$ *is* $\mathscr{F}$*-measurable.*

Proof. If A is any Borel set in $\mathbf{R}^*$, then since f is Borel measurable, the set $f^{-1}(A)$ is also a Borel set in $\mathbf{R}^*$. Now

$$\{x: f(g(x)) \in A\} = \{x: g(x) \in B\} \in \mathscr{F}$$

since $B = f^{-1}(A) \in \mathscr{B}^*$. ▌

Remark. In the above lemma, it is not sufficient to assume that $f: \mathbf{R}^* \to \mathbf{R}^*$ is Lebesgue measurable —see exercise 5.2 (9).

This means that, for most of the functions which normally occur in analysis, it is immediately obvious that they are $\mathscr{L}_F^k$-measurable for every F, and in particular that they are Lebesgue measurable.

Almost everywhere (*a.e.*)

It is convenient to have a way of describing the behaviour of a function $f: \Omega \to \mathbf{R}^*$ outside an (unspecified) set of zero measure. If P is some property describing the behaviour of $f(x)$ at a particular point x, then we say that $f(x)$ has a property P almost everywhere with respect to μ, if there is some set $E \in \mathscr{F}$ with $\mu(E) = 0$ such that $f(x)$ has property P for all $x \in \Omega - E$. We then write

$$f(x) \text{ has property } P \text{ a.e. } (\mu).;$$

and, if there is no ambiguity about the measure being considered, (μ) can be omitted.

Lemma. *If $\mathscr{F}$ is complete with respect to μ, and $f = g$ a.e., then f is measurable if and only if g is measurable.*

Proof. For any $c \in \mathbf{R}$ the set

$$\{x: f(x) < c\} \,\triangle\, \{x: g(x) < c\} \subset \{x: f(x) \neq g(x)\}$$

so that $\{x: f(x) < c\}$ differs from $\{x: g(x) < c\}$ by a subset of a set of zero measure. If $\mathscr{F}$ is complete with respect to μ, all such sets are in $\mathscr{F}$ so that

$$\{x: f(x) < c\} \in \mathscr{F} \quad \text{if and only if} \quad \{x: g(x) < c\} \in \mathscr{F}. \blacksquare$$

Exercises 5.2

1. In theorem 5.1, show that the condition $\{x: f(x) \leqslant c\} \in \mathscr{F}$ for all rational c is sufficient to imply that $f: \Omega \to \mathbf{R}^*$ is $\mathscr{F}$-measurable.

2. Suppose $\{f_n\}$ is a sequence of functions: $\Omega \to \mathbf{R}^*$ each of which is finite a.e. Show that, for almost all x in Ω, $f_n(x)$ is finite for all n.

3. Suppose G is an open set in $\mathbf{R}$ and $\{f_n\}$ is a convergent sequence of functions: $\Omega \to \mathbf{R}$. Show that

$$\{x: \lim_{n \to \infty} f_n(x) \in G\} = \bigcup_{m=1}^{\infty} \bigcup_{k=1}^{\infty} \bigcap_{n=k}^{\infty} \left\{x: d(f_n(x), \mathbf{R} - G) > \frac{1}{m}\right\},$$

where $d(y, E)$ denotes the distance from y to E (defined in §2.1).

4. Show that, in theorem 5.2, the condition $f \geqslant 0$ can be deleted provided we do not require monotonicity for the sequence $\{f_n\}$ of simple functions converging to f. Show that if f is unbounded above and below it is impossible to arrange for the sequence $\{f_n\}$ to be monotone.

5. An elementary function is one which assumes a countable set of values each on a measurable subset of Ω. Show that, if $f: \Omega \to \mathbf{R}$ is measurable then it is the *uniform* limit of a monotone sequence of elementary functions, but that if f is not bounded it is not the uniform limit of simple functions.

6. If $\mathscr{F}$ is a finite field of subsets of Ω, show that $f: \Omega \to \mathbf{R}$ is $\mathscr{F}$-measurable if and only if it is $\mathscr{F}$-simple.

7. If Ω is a topological space, give examples to show that, for $f: \Omega \to \mathbf{R}$, the condition

'f is continuous a.e. in Ω'

neither implies nor is implied by the condition

'there is a continuous $g: \Omega \to \mathbf{R}$ for which $f = g$ a.e.'

8. Suppose Ω is a topological space, $\mathscr{F} \supset \mathscr{B}$ and $(\Omega, \mathscr{F}, \mu)$ is such that $\mathscr{F}$ is complete with respect to μ. Show that any function f which is continuous a.e. is $\mathscr{F}$-measurable. Give an example of a measurable function which cannot be made continuous by altering its values on any set of zero measure.

9. If $\mathscr{L}$ is the class of Lebesgue measurable sets in $\mathbf{R}$, give an example of an $\mathscr{L}$-measurable function $f: \mathbf{R} \to \mathbf{R}$ and an $\mathscr{L}$-measurable set $E \subset \mathbf{R}$ for which $f^{-1}(E)$ is not $\mathscr{L}$-measurable.

Hint. Use a suitable subset of the Cantor set (see §2.7).

5.3 Definition of the integral

Our method is to define the operation of integration first for non-negative simple functions, and then extend the definition step-by-step showing at each stage that the desirable properties discussed in §5.1 are obtained. If we think of measure as a mass distribution in Ω, and integration as a means of averaging a given function f with respect to this mass distribution it is clear that there is only one reasonable definition for the integral of

(1) *A non-negative simple function*

If

$$f(x) = \sum_{i=1}^{n} c_i \chi_{E_i}(x), \tag{5.3.1}$$

where $c_i \geqslant 0$ $(i = 1, 2, \ldots, n)$ we define

$$\int f d\mu = \sum_{i=1}^{n} c_i \mu(E_i).$$

This sum is always defined since each of the terms is non-negative. It is called the integral of f with respect to μ. (Note that if $c_i = 0$, $\mu(E_i) = +\infty$ our convention is that $c_i \mu(E_i) = 0$.) Since the representation of a simple function in the form (5.3.1.) is not unique we must first see that our definition of the integral does not depend on the particular representation used. Suppose

$$f = \sum_{i=1}^{n} c_i \chi_{E_i} = \sum_{j=1}^{m} d_j \chi_{F_j},$$

then since both systems of sets are dissections of Ω

$$\mu(E_i) = \sum_{j=1}^{m} \mu(E_i \cap F_j) \quad \text{and} \quad \mu(F_j) = \sum_{i=1}^{n} \mu(E_i \cap F_j). \tag{5.3.2}$$

Also if $E_i \cap F_j$ is not empty, it will contain a point x and $f(x) = c_i = d_j$. Thus

$$\sum_{i=1}^{n} c_i \mu(E_i) = \sum_{i=1}^{n} \sum_{j=1}^{m} c_i \mu(E_i \cap F_j) = \sum_{i=1}^{n} \sum_{j=1}^{m} d_j \mu(E_i \cap F_j)$$

$$= \sum_{j=1}^{m} d_j \mu(F_j).$$

Now consider two non-negative simple functions

$$f = \sum_{i=1}^{n} c_i \chi_{E_i}, \quad g = \sum_{j=1}^{m} d_i \chi_{F_j}$$

and use the representations

$$f = \sum_{i=1}^{n} \sum_{j=1}^{m} c_i \chi_{E_i \cap F_j}, \quad g = \sum_{j=1}^{m} \sum_{i=1}^{n} d_j \chi_{E_i \cap F_j}$$

in terms of the dissection $E_i \cap F_j$. Then the simple function $(f+g)$ has the representation

$$f+g = \sum_{i=1}^{n} \sum_{j=1}^{m} (c_i + d_j) \chi_{E_i \cap F_j},$$

and $$\int (f+g)\, d\mu = \sum_{i=1}^{n} \sum_{j=1}^{m} (c_i + d_j) \mu(E_i \cap F_j)$$

$$= \sum_{i=1}^{n} \sum_{j=1}^{m} c_i \mu(E_i \cap F_i) + \sum_{i=1}^{n} \sum_{j=1}^{m} d_j \mu(E_i \cap F_j)$$

$$= \sum_{i=1}^{n} c_i \mu(E_i) + \sum_{j=1}^{m} d_j \mu(F_j), \quad \text{using} \quad (5.3.2)$$

$$= \int f d\mu + \int g d\mu.$$

It is now immediate from the definition that if $\alpha \geqslant 0$, $\beta \geqslant 0$ and f, g are non-negative simple functions then

$$\int (\alpha f + \beta g)\, d\mu = \alpha \int f d\mu + \beta \int g d\mu$$

so that our operator is linear on the class of non-negative simple functions. It is also clear that it is order preserving; that is, if f, g are simple functions and $f \geqslant g$ then $\int f d\mu \geqslant \int g d\mu$.

These properties allow us to extend our definition to:

(2) *Non-negative measurable functions*

Given a measurable $f: \Omega \to \mathbf{R}^+$, by theorem 5.2 there is a monotone increasing sequence f_n of simple functions such that $f_n \to f$. Since

$\int f_n d\mu$ is defined for all n, and is monotone increasing it has a limit in $\mathbf{R}^+$ (which may be $+\infty$). We define

$$\int f\,d\mu = \lim_{n\to\infty} \int f_n\,d\mu. \tag{5.3.3}$$

Since there are many possible monotone sequences of simple functions which converge to a given non-negative measurable f, we must show that the integral $\int f\,d\mu$ defined in this way is independent of the particular sequence used.

Suppose $\{f_n\}$ is an increasing sequence of non-negative simple functions and $f = \lim_{n\to\infty} f_n \geqslant g$, where g is non-negative simple. The first (and main) step in showing that our definition (5.3.3) is proper is to show that, in these circumstances

$$\lim_{n\to\infty} \int f_n\,d\mu \geqslant \int g\,d\mu. \tag{5.3.4}$$

Put
$$g = \sum_{i=1}^{k} c_i \chi_{E_i},$$

then if $\int g\,d\mu = +\infty$, there must be an integer i, $1 \leqslant i \leqslant k$ such that $c_i > 0$, $\mu(E_i) = +\infty$. Then for any fixed ϵ such that $0 < \epsilon < c_i$, the sequence of sets $\{A_n \cap E_i\}$ $(n = 1, 2, \ldots)$ is monotone increasing to E_i where

$$A_n = \{x : f_n + \epsilon > g\}. \tag{5.3.5}$$

Hence $\mu(A_n \cap E_i) \to +\infty$ as $n \to \infty$, by theorem 3.2. But

$$\int f_n\,d\mu \geqslant \int f_n \chi_{A_n \cap E_i}\,d\mu \geqslant (c_i - \epsilon)\,\mu(A_n \cap E_i) \to +\infty \quad \text{as} \quad n \to \infty.$$

Thus (5.3.4) is established, if $\int g\,d\mu = +\infty$. Now assume that $\int g\,d\mu$ is finite and put

$$A = \{x : g(x) > 0\} = \bigcup_{c_i > 0} E_i.$$

Since g is simple, $c = \min_{c_i > 0} c_i$ is positive. and $\mu(A) < \infty$. We now suppose $\epsilon > 0$ and again define A_n by (5.3.5). Then

$$\int f_n\,d\mu \geqslant \int f_n \chi_{A_n \cap A}\,d\mu \geqslant \int (g - \epsilon)\,\chi_{A_n \cap A}\,d\mu$$
$$= \int g \chi_{A_n \cap A}\,d\mu - \epsilon\mu(A_n \cap A) \geqslant \int g \chi_{A_n \quad A}\,d\mu - \epsilon\mu(A).$$

Since $\mu(A_n \cap E_i) \to \mu(E_i)$ for each i, we can evaluate the integrals as finite sums and find an integer $n_0 = n_0(\epsilon)$ such that

$$\int f_n\,d\mu \geqslant \int g\chi_A\,d\mu - \epsilon - \epsilon\mu(A) \quad \text{for} \quad n \geqslant n_0,$$

so that we have established (5.3.4) also in the case $\int g\,d\mu < \infty$.

We can now suppose given two sequences of simple functions

$$0 \leqslant f_1 \leqslant f_2 \leqslant \dots \leqslant f_n \leqslant \dots \to f,$$
$$0 \leqslant g_1 \leqslant g_2 \leqslant \dots \leqslant g_m \leqslant \dots \to f$$

each monotone increasing and convergent to f. Then for each fixed m we have, by (5.3.4), since

$$f = \lim_{n\to\infty} f_n \geqslant g_m,$$

$$\lim_{n\to\infty} \int f_n\,d\mu \geqslant \int g_m\,d\mu.$$

If we now let $m \to \infty$
$$\lim_{n\to\infty} \int f_n\,d\mu \geqslant \lim_{m\to\infty} \int g_m\,d\mu.$$

Since the situation is symmetrical, the opposite inequality is similarly proved and we must have

$$\lim_{n\to\infty} \int f_n\,d\mu = \lim_{m\to\infty} \int g_m\,d\mu.$$

Thus the operation of integration is properly defined for non-negative measurable functions. Because of the corresponding result for non-negative simple functions, it now follows that, if f, g are two non-negative measurable functions and $\alpha \geqslant 0$, $\beta \geqslant 0$ then

$$\int (\alpha f + \beta g)\,d\mu = \alpha \int f\,d\mu + \beta \int g\,d\mu.$$

By our definition, for $f \geqslant 0$ and measurable, $\int f\,d\mu$ may be finite or $+\infty$. A non-negative measurable function f is said to be *integrable* with respect to the measure μ if $\int f\,d\mu$ is finite.

There are clearly two possible reasons for such an f to fail to be integrable. Either there is a simple function $g \leqslant f$ for which $\int g\,d\mu = \infty$, which would imply the existence of a $c > 0$ for which $\mu\{x : f(x) > c\} = +\infty$, or alternatively it is possible that $\int g\,d\mu$ is finite for all simple functions $g \leqslant f$ (which implies $\mu\{x : f(x) > c\} < \infty$, all $c > 0$) but, for any sequence g_n of simple functions converging to f, $\int g_n\,d\mu \to +\infty$ as $n \to \infty$.

We can now define the integral for:

(3) *Integrable measurable functions*

We know that if $f : \Omega \to \mathbf{R}^*$ is measurable, then so are

$$f_+, f_- \quad \text{and} \quad f = f_+ - f_-.$$

If both f_+ and f_- are integrable, then we say that f is integrable and define

$$\int f d\mu = \int f_+ d\mu - \int f_- d\mu.$$

Thus our operation of integration is now well-defined on the class $\mathscr{A}$ of integrable functions. We will show in the next section that all the desirable properties discussed in §5·1 are satisfied by this operation.

Finally, we define:

(4) *Integral of a function f over a set A*

This can be considered only for sets A in $\mathscr{F}$. Put

$$\int_A f d\mu = \int f\chi_A d\mu$$

provided $\int f\chi_A d\mu$ is defined. Thus $\int_A f d\mu$ will be defined if either (i) $f\chi_A$ is non-negative and measurable, or (ii) $f\chi_A$ is measurable and integrable. We say that f is integrable over A (with respect to μ) if the function $f\chi_A$ is integrable. It is clear that

$$\int_\Omega f d\mu = \int f d\mu$$

and we will usually continue to omit the set Ω when we are integrating over the whole space.

Note that, if $E \in \mathscr{F}$ and $\mu(E) = 0$, then any function $f: \Omega \to \mathbf{R}^*$ is integrable over E with

$$\int_E f d\mu = 0.$$

Exercises 5.3

1. Show that a simple function

$$f(x) = \sum_{i=1}^{n} c_i \chi_{E_i}(x)$$

is integrable if and only if $c_i = 0$ for each integer i such that $\mu(E_i) = +\infty$.

2. Let Ω be a finite set, $\mu(E)$ the number of points in E. Show that all functions on Ω are simple functions and that the theory of integration reduces to the theory of finite sums.

3. If $f: \Omega \to \mathbf{R}^*$ is integrable (μ) show that, for any $\epsilon > 0$

$$\mu\{x: |f(x)| \geqslant \epsilon\} < \infty.$$

4. Suppose μ_1 and μ_2 are two measures defined on $\mathscr{F}$ and $\nu = \mu_1+\mu_2$. Show that if f is integrable with respect to μ_1 and μ_2 over a set E, then it is integrable with respect to ν and

$$\int_E f\,d\nu = \int_E f\,d\mu_1 + \int_E f\,d\mu_2.$$

5. Suppose $f: \Omega \to \mathbf{R}^+$ is a non-negative measurable function. Show that

$$\int_E f d\mu = \sup\left[\sum_{k=1}^{n} \mu(E_k) \inf\{f(x): x\in E_k\}\right],$$

where the supremum is taken over the collection of all finite classes of disjoint measurable sets with

$$E = \bigcup_{k=1}^{n} E_k.$$

(This is a possible way of defining $\int_E f\,d\mu$ which leads to the same class of integrable functions).

6. Suppose $\mu(E) < \infty$ and $f: E \to \mathbf{R}$ is a measurable finite-valued function defined on $E \in \mathscr{F}$. Put

$$S_n(E) = \sum_{k=-\infty}^{\infty} \frac{k}{2^n}\mu\left\{x: x\in E, \frac{k}{2^n} \leqslant f(x) < \frac{k+1}{2^n}\right\}.$$

Show that this series is absolutely convergent for all n if it is absolutely convergent for any one $n \in \mathbf{Z}$. Show that f is integrable on E if and only if the series converges absolutely for all n and in this case

$$\int_E f d\mu = \lim_{n\to\infty} S_n(E).$$

Show that this is not valid if $\mu(E) = +\infty$.

(This is another possible way of defining $\int_E f d\mu$.)

5.4 Properties of the integral

We have now defined the operation of integration with respect to a measure μ on a class of integrable functions. The first objective must be to show that our operation has the properties outlined in §5.1. These are of two types: those involving only a finite number of functions, and operations involving a countable class of functions. We will obtain various closure properties of the class $\mathscr{A}$ while we are examining the integration operation.

***Theorem* 5.5.** *Suppose $(\Omega, \mathscr{F}, \mu)$ is a measure space, A, B are disjoint sets in $\mathscr{F}$ and $f: \Omega \to \mathbf{R}^*$, $g: \Omega \to \mathbf{R}^*$ are two functions integrable* *(over Ω) with respect to μ. Then f is integrable over A, $f+g$ and $|f|$ are integrable (over Ω) and*

(i) $\int_{A\cup B} f d\mu = \int_A f d\mu + \int_B f d\mu;$

(ii) *f is finite a.e.;*

(iii) $\int (f+g)\, d\mu = \int f d\mu + \int g d\mu;$

(iv) $|\int f d\mu| \leqslant \int |f|\, d\mu;$

(v) *for any $c \in \mathbf{R}$, cf is integrable and* $\int cf\, d\mu = c \int f d\mu;$

(vi) $f \geqslant 0 \Rightarrow \int f d\mu \geqslant 0;\ f \geqslant g \Rightarrow \int f d\mu \geqslant \int g d\mu;$

(vii) *if $f \geqslant 0$ and $\int f d\mu = 0$, then $f = 0$ a.e.;*

(viii) $f = g$ *a.e.* $\Rightarrow \int f d\mu = \int g d\mu;$

(ix) *if $h: \Omega \to \mathbf{R}^*$ is $\mathscr{F}$-measurable and $|h| \leqslant f$, then h is integrable.*

Corollary. *Any function $f: \Omega \to \mathbf{R}^*$ which is bounded, $\mathscr{F}$-measurable, and zero outside a set E in $\mathscr{F}$ of finite μ-measure is integrable (over Ω) with respect to μ.*

Proof. If $f: \Omega \to \mathbf{R}^+$ is non-negative measurable and integrable (over Ω) and $0 \leqslant g \leqslant f$ with $g: \Omega \to \mathbf{R}^+$ measurable, it follows immediately from the definition of the integral of non-negative measurable functions that g is integrable. Since for any $A \in \mathscr{F}$, χ_A is measurable,

$$0 \leqslant f_+ \chi_A \leqslant f_+ \quad \text{and} \quad 0 \leqslant f_- \chi_A \leqslant f_-,$$

and a function f which is integrable over Ω is also integrable over any measurable set A.

(i) If A, B are disjoint,

$$\chi_{A\cup B} = \chi_A + \chi_B,$$

so that

$$f_+\chi_{A\cup B} = f_+\chi_A + f_+\chi_B,$$

$$f_-\chi_{A\cup B} = f_-\chi_A + f_-\chi_B;$$

and since property (i) is already known for non-negative measurable functions we must have

$$\int_{A\cup B} f d\mu = \int f_+\chi_{A\cup B} d\mu - \int f_-\chi_{A\cup B} d\mu$$

$$= \int f_+\chi_A d\mu - \int f_-\chi_A d\mu + \int f_+\chi_B d\mu - \int f_-\chi_B d\mu;$$

and

$$\int_{A\cup B} f d\mu = \int_A f d\mu + \int_B f d\mu, \tag{5.4.1}$$

since all the terms are finite.

(ii) If f is not finite a.e., then at least one of the sets

$$A_1 = \{x : f(x) = +\infty\}, \quad A_2 = \{x : f(x) = -\infty\}$$

has positive measure. Suppose $\mu(A_1) > 0$. Then it follows immediately from the definition that $\int f_+ d\mu = +\infty$ which means that f is not integrable (over Ω).

(iii) This has already been proved for non-negative functions f, g. If f_1, f_2 are non-negative and $f = f_1 - f_2$, then $f_1 + f_- = f_2 + f_+$ and applying (iii) for non-negative functions gives

$$\int f_1 d\mu + \int f d\mu = \int f_2 d\mu + \int f_+ d\mu$$

so that

$$\int f d\mu = \int f_1 d\mu - \int f_2 d\mu.$$

Now the general result follows since, for finite f, g

$$f + g = (f_+ + g_+) - (f_- + g_-),$$

so that

$$\int (f+g)\, d\mu = \int (f_+ + g_+)\, d\mu - \int (f_- + g_-)\, d\mu$$

$$= \int f_+ d\mu - \int f_- d\mu + \int g_+ d\mu - \int g_- d\mu$$

$$= \int f d\mu + \int g d\mu.$$

Finally apply (iii) to the function $|f| = f_+ + f_-$ to deduce that $|f|$ is integrable and

$$\int |f|\, d\mu = \int f_+ d\mu + \int f_- d\mu.$$

(iv) This now follows immediately as

$$\left| \int f d\mu \right| = \left| \int f_+ d\mu - \int f_- d\mu \right| \leqslant \int f_+ d\mu + \int f_- d\mu.$$

(v) If $c = 0$, $cf = 0$ and $\int cf d\mu = 0 = c \int f d\mu$. If $c > 0$ then

$$(cf)_+ = cf_+, \quad (cf)_- = cf_-$$

and the result follows since it has already been proved for non-negative functions (p. 113). Similarly, if $c < 0$

$$(cf)_+ = -cf_-, \quad (cf)_- = -cf_+,$$

$$\int (cf)_+ d\mu - \int (cf)_- d\mu = (-c)\int f_- d\mu + c\int f_+ d\mu = c\int f d\mu.$$

(vi) The first statement follows from the definition. If $f \geqslant g$, then $f = g + (f-g)$, and $(f-g) \geqslant 0$. By (iii), we now have

$$\int f d\mu = \int g d\mu + \int (f-g)\, d\mu \geqslant \int g\, d\mu.$$

(vii) If $\{x: f(x) > 0\}$ has positive measure, then by theorem 3.2 there is an integer n such that, if

$$A = \{x: f(x) > 1/n\},$$

$\mu(A) > 0$. But $n^{-1}\chi_A \leqslant f\chi_A \leqslant f$, so that

$$\int f d\mu \geqslant \frac{1}{n}\int \chi_A\, d\mu = \frac{1}{n}\mu(A) > 0.$$

Hence, if $f \geqslant 0$ and $\int f d\mu = 0$, we must have $\mu\{x: f(x) > 0\} = 0$.

(viii) If $f = g$ a.e., then $f_+ = g_+$, $f_- = g_-$ a.e. In the construction of theorem 5.2, the sets $Q_{p,s}$ for the two functions f_+ and g_+ will all have the same measure. Hence, there are simple functions $f_n \to f_+$, $g_n \to g_+$ such that

$$\int f_n d\mu = \int g_n d\mu \quad (n = 1, 2, \ldots),$$

and it follows that $\int f_+ d\mu = \int g_+ d\mu$. Similarly $\int f_- d\mu = \int g_- d\mu$.

(ix) If $|h| \leqslant f$ then $0 \leqslant h_+ \leqslant f$, $0 \leqslant h_- \leqslant f$. From (vi) it now follows that each of $\int h_+ d\mu$, $\int h_- d\mu$ is finite, and h is therefore integrable.

Proof of corollary. If $|f| \leqslant K$, then the simple function $K\chi_E$ is integrable and the integrability of f now follows from (ix). ∎

Remark. If $\mathscr{F}$ is complete with respect to μ, then (viii) can clearly be strengthened as follows:

If $f: \Omega \to \mathbf{R}^*$ is integrable, and $g: \Omega \to \mathbf{R}^*$ is such that $f = g$ a.e., then g is integrable and $\int f d\mu = \int g d\mu$. There is also a converse to this remark: if f and g are integrable functions such that

$$\int_E f d\mu = \int_E g d\mu \quad \text{for all} \quad E \in \mathscr{F},$$

then $f = g$ a.e. For, suppose not, so that $\mu\{x: f(x) \neq g(x)\} > 0$. Then at least one of $\{x:f(x) > g(x)\}$, $\{x:f(x) < g(x)\}$ has positive measure. By theorem 3.2 there must be an integer n such that, for

$$E_n = \left\{x:f(x) \geqslant g(x) + \frac{1}{n}\right\}, \quad \mu(E_n) > 0.$$

But then
$$\int_{E_n} f d\mu - \int_{E_n} g d\mu > \frac{1}{n}\mu(E_n) > 0$$

which is a contradiction establishing the required result.

We can now consider theorems about the continuity of the integration operator.

***Theorem* 5.6.** *Suppose $\{f_n\}$ is a monotone increasing sequence of non-negative measurable functions: $\Omega \to \mathbf{R}^+$ and $f_n(x) \to f(x)$ for all $x \in \Omega$: then*

$$\lim_{n\to\infty} \int f_n \, d\mu = \int f d\mu,$$

in the sense that, if f is integrable, the integrals $\int f_n d\mu$ converge to $\int f d\mu$; while if f is not integrable either f_n is integrable for all n and $\int f_n d\mu \to +\infty$ as $n \to \infty$, or there is an integer N such that f_N is not integrable so that $\int f_n d\mu = +\infty$ for $n \geqslant N$.

Proof. For each $n = 1, 2, \ldots$ choose an increasing sequence

$$\{f_{n,k}\} \, (k = 1, 2, \ldots)$$

of non-negative simple functions converging to f_n, and put

$$g_k = \max_{n \leqslant k} [f_{n,k}].$$

Then $\{g_k\}$ is a non-decreasing sequence of non-negative simple functions and

$$g = \lim_{k\to\infty} g_k$$

is non-negative measurable. But

$$f_{n,k} \leqslant g_k \leqslant f_k \leqslant f \quad \text{for} \quad n \leqslant k, \tag{5.4.2}$$

so that
$$f_n \leqslant g \leqslant f;$$

and, if we now let $n \to \infty$, we see that $f = g$. Using the order property (vi) of theorem 5.5 and (5.4.2) gives

$$\int f_{n,k} d\mu \leqslant \int g_k d\mu \leqslant \int f_k d\mu \quad \text{for} \quad n \leqslant k.$$

For fixed n, let $k \to \infty$; from the definition of the integral,

$$\int f_n d\mu \leqslant \int g d\mu \leqslant \lim_{k\to\infty} \int f_k d\mu.$$

If we now let $n \to \infty$, we obtain

$$\lim_{n\to\infty} \int f_n d\mu \leqslant \int g d\mu \leqslant \lim_{n\to\infty} \int f_k d\mu.$$

Since the two extremes of the inequality are the same, we must have

$$\lim_{n\to\infty} \int f_n d\mu = \int g d\mu = \int f d\mu.]$$

Corollary (absolute continuity). *If f is integrable (over Ω) then, for $A \in \mathscr{F}$,*

$$\int_A f d\mu \to 0 \quad \text{as} \quad \mu(A) \to 0.$$

Proof. Put

$$\begin{aligned} f_n &= f \quad \text{if} \quad |f| \leqslant n \\ &= n \quad \text{if} \quad |f| \geqslant n. \end{aligned}$$

Then $|f_n|$ is monotonic increasing to $|f|$ as $n \to \infty$ Since, by theorem 5.5, $|f|$ is integrable, we have

$$\int |f_n| \, d\mu \to \int |f| \, d\mu \quad \text{as} \quad n \to \infty.$$

Given $\epsilon > 0$, choose N such that

$$\int |f| \, d\mu < \int |f_n| \, d\mu + \tfrac{1}{2}\epsilon \quad \text{for} \quad n \geqslant N.$$

Then if $A \in \mathscr{F}$ is such that $\mu(A) < \epsilon/2N$, we have, by theorem 5.5,

$$\begin{aligned} \left| \int_A f d\mu \right| \leqslant \int_A |f| \, d\mu = \int_A |f_N| \, d\mu \\ + \int_A |f| - |f_N| d\mu < \tfrac{1}{2}\epsilon + \int_\Omega (|f| - |f_N|) \, d\mu < \epsilon.] \end{aligned}$$

Remark. The notion of absolute continuity for a set function ν: $\mathscr{F} \to \mathbf{R}$ will be considered more fully in §6.4.

Theorem 5.7 (*Fatou*). *If $\{f_n\}$ is a sequence of measurable functions which is bounded below by an integrable function, then*

$$\int \liminf_{n\to\infty} f_n d\mu \leqslant \liminf_{n\to\infty} \int f_n d\mu$$

Remark. The operation $\liminf_{n\to\infty}$ picks out the small values of a sequence. This theorem says that if this is done point-wise and the result is then integrated the answer will be not greater than the result of first integrating and then applying the operation.

Proof. Since $\{f_n\}$ is bounded below by an integrable function g we may assume without loss of generality that $f_n \geqslant 0$ for all n. For $h_n = f_n - g \geqslant 0$ a.e.† and

$$\int h_n \, d\mu = \int f_n \, d\mu - \int g \, d\mu, \quad \liminf h_n = \liminf f_n - g \text{ a.e.}$$

Put $g_n = \inf_{k \geqslant n} f_k$, then g_n is an increasing sequence of measurable functions and
$$\lim_{n\to\infty} g_n = \liminf_{n\to\infty} f_n.$$

Since $f_n \geqslant g_n$, for all n

$$\liminf_{n\to\infty} \int f_n \, d\mu \geqslant \lim_{n\to\infty} \int g_n \, d\mu = \int \lim_{n\to\infty} g_n \, d\mu = \int \liminf_{n\to\infty} f_n \, d\mu,$$

by theorem 5.6. ∎

Corollary. *If $\{f_n\}$ is a sequence of measurable functions which is bounded above by an integrable function, then*

$$\int \limsup_{n\to\infty} f_n \, d\mu \geqslant \limsup_{n\to\infty} \int f_n \, d\mu.$$

Proof. This can be proved directly by a method similar to that of theorem 5.7, or it can be deduced from that theorem by putting $g_n = -f_n \; (n = 1, 2, \ldots)$. ∎

***Theorem* 5.8** (*Lebesgue*). (i) *If $g: \Omega \to \mathbf{R}^+$ is integrable, $\{f_n\}$ is a sequence of measurable functions $\Omega \to \mathbf{R}^*$ such that $|f_n| \leqslant g \; (n = 1, 2, \ldots)$ and $f_n \to f$ as $n \to \infty$, then f is integrable and*

$$\int f_n \, d\mu \to \int f \, d\mu \quad \textit{as} \quad n \to \infty.$$

(ii) *Suppose $g: \Omega \to \mathbf{R}^+$ is integrable, $-\infty \leqslant a < b \leqslant +\infty$, and for each $t \in (a, b)$, f_t is a measurable function Ω to $\mathbf{R}^*$. Then if $|f_t| \leqslant g$ for all $t \in (a, b)$ and $f_t \to f$ as $t \to a+$ or $t \to b-$, then f is integrable and*

$$\int f_t \, d\mu \to \int f \, d\mu.$$

Proof. (i) We first prove the special case of the theorem where

† Since g is integrable the set $\{x: |g(x)| = +\infty\}$ has zero measure, so that the operation $f_n(x) - g(x)$ can be carried out at least outside the set $\{x: |g(x)| = +\infty\}$. We put in a.e. to cover the possible exceptional set of zero measure where $(f_n - g)$ is not defined. By theorem 5.5 (viii) such exceptional sets do not effect the value of the integrals, and we could arbitrarily define $(f_n - g)$ to be zero at the points where $f_n = g = \pm\infty$.

$f_n \geq 0$ and $f_n \to 0$ as $n \to \infty$. In this case we can apply theorem 5.7 and corollary to give

$$\limsup \int f_n \, d\mu \leq \int \limsup f_n \, d\mu = \int 0 \, d\mu = 0$$

$$= \int \liminf f_n \, d\mu \leq \liminf \int f_n \, d\mu \leq \limsup \int f_n \, d\mu.$$

Hence all the inequalities must be equalities, $\lim \int f_n \, d\mu$ exists, and has the value zero.

In the general case, put $h_n = |f_n - f|$; then $0 \leq h_n \leq 2g$, $2g$ is integrable and h_n is measurable with $h_n \to 0$ as $n \to \infty$. But then

$$\left| \int f_n \, d\mu - \int f \, d\mu \right| \leq \int |f_n - f| \, d\mu \quad \to 0 \quad \text{as} \quad n \to \infty,$$

and f is integrable by theorem 5.5(ix).

(ii) Suppose, for example, that $f_t \to f$ as $t \to a+$, then we can apply the sequence form of the theorem to $f_n = f_{t_n}$, where $\{t_n\}$ is any sequence in (a, b) converging to a. Since $f = \lim f_n$ we must have

$$\int f_n \, d\mu \to \int f \, d\mu.$$

But the right-hand side is now independent of the particular sequence $\{t_n\}$ chosen so that $\int f_t \, d\mu$ must approach the limit $\int f \, d\mu$ as $t \to a$ through values in (a, b). ❚

Exercises 5.4

1. Suppose $f: \Omega \to \mathbf{R}$ is measurable, $A \in \mathscr{F}$, $\mu(A) < \infty$ and

$$f(x) = 0 \quad \text{for} \quad x \in \Omega - A,$$

$$m \leq f(x) \leq M \quad \text{for} \quad x \in A, \quad \text{where} \quad m, M \in \mathbf{R}.$$

Show that f is integrable and

$$m\mu(A) \leq \int f \, d\mu \leq M\mu(A).$$

2. Prove that, if f and g are integrable functions,

$$\min \left[\int f \, d\mu, \quad \int g \, d\mu \right] \geq \int \min(f, g) \, d\mu.$$

If the two sides of this inequality are equal, what deduction can be made about the relation between f and g?

3. Prove that, for any $\epsilon > 0$, if f is integrable over E there is a subset $E_0 \subset E$ such that $\mu(E_0) < \infty$, and

$$\left| \int_E f \, d\mu - \int_{E_0} f \, d\mu \right| < \epsilon.$$

4. Show that $f: \Omega \to \mathbf{R}^*$ is integrable if and only if for any $\epsilon < 0$, there exist integrable functions g and h with $g \geqslant f \geqslant h$ and $\int (g-h)\, d\mu < \epsilon$.

5. If $E = \bigcup_{r=1}^{\infty} E_r$ is a countable union of disjoint sets of $\mathscr{F}$, and f is integrable over E, then

$$\int_E f d\mu = \sum_{r=1}^{\infty} \int_{E_r} f d\mu$$

and the series converges absolutely.

6. Suppose $\mathbf{Z}$ is the set of positive integers, $\mathscr{F}$ is the class of all subsets of $\mathbf{Z}$ and $\mu(E)$ denotes the number of points in E. Show that any $f: \mathbf{Z} \to \mathbf{R}^*$ is $\mathscr{F}$-measurable and that f is integrable if and only if $\sum_{n=1}^{\infty} f(n)$ converges absolutely. Deduce that the sum of an absolutely convergent series is unaffected by any rearrangement of the terms.

7. Suppose $\{f_n\}$ is a sequence of integrable functions and

$$\sum_{n=1}^{\infty} \int |f_n|\, d\mu < \infty.$$

Show that the series $\sum_{n=1}^{\infty} f_n(x)$ converges absolutely a.e. to an integrable function f and that

$$\int f d\mu = \sum_{n=1}^{\infty} \int f_n d\mu.$$

8. Suppose $\{E_n\}$ is a sequence of sets in $\mathscr{F}$, m is a fixed positive integer, and G is the set of points which are in E_n for at least m integers n. Then G is measurable and

$$m\mu(G) \leqslant \sum_{n=1}^{\infty} \mu(E_n).$$

9. Show that a measurable function f is integrable over a measurable set E if and only if

$$\Sigma \mu[E \cap \{x: |f(x)| \geqslant n\}]$$

converges.

10. Suppose f is measurable, g is integrable and $\alpha, \beta \in \mathbf{R}$ with $\alpha \leqslant f(x) \leqslant \beta$ a.e. Then there is a real γ such that $\alpha \leqslant \gamma \leqslant \beta$ and

$$\int f|g|\, d\mu = \gamma \int |g|\, d\mu.$$

Show by an example that we cannot replace $|g|$ by g in this equation.

11. Suppose μ is Lebesgue measure in $\mathbf{R}$ and put

$$\begin{aligned} f_n(x) &= -n^2 \quad \text{for} \quad x \in (0, 1/n), \\ &= 0 \qquad \text{otherwise.} \end{aligned}$$

Then $\liminf f_n = \lim f_n = 0$ for all x, but

$$\int f_n d\mu = -n.$$

This shows that theorem 5.7 is not valid without the restriction that $\{f_n\}$ be bounded below by an integrable g.

12. State and prove a version of Fatou's lemma (theorem 5.7) for a family f_t, $t \in (a, b)$, of non-negative measurable functions.

13. Is it true that, for measurable $f, g: \Omega \to \mathbf{R}^*$,

$$f^2 \text{ and } g^2 \text{ integrable} \Rightarrow fg \text{ integrable?}$$

Show that, if
$$\left[\int fg\, d\mu\right]^2 = \int f^2 d\mu \int g^2 d\mu,$$

then f and g are essentially proportional : that is, there is a real α such that $f = \alpha g$ a.e., or $g = 0$ a.e.

5.5 Lebesgue integral; Lebesgue–Stieltjes integral

We have defined the operation of integration on an abstract measure space $(\Omega, \mathcal{F}, \mu)$. Historically this method of integration was first defined on $(\mathbf{R}, \mathcal{L}, \mu)$, where μ denotes Lebesgue measure on the σ-field $\mathcal{L}$ of Lebesgue measurable sets. We have made the definition in the general case since no more work is involved, but we must now specialise it to obtain the Lebesgue integral.

If E is a Lebesgue measurable set in $\mathbf{R}$, μ denotes Lebesgue measure in $\mathbf{R}$, f is $\mathcal{L}$-measurable, then it is usual to use the notation

$$\int_E f(x)\, dx \quad \text{for} \quad \int_E f d\mu.$$

In particular, if E is an interval with end-points a, b we use the notation

$$\int_a^b f(x)\, dx \quad \text{for} \quad \int_E f dx,$$

where $E = [a, b]$ or (a, b) or $[a, b)$ or $(a, b]$. Note that, since the Lebesgue measure of a single point is zero, it makes no difference whether the interval is open or closed.

In the above notation a may be $-\infty$ and b may be $+\infty$ so that

$$\int_{-\infty}^{\infty} f(x)\, dx \quad \text{means} \quad \int_{\mathbf{R}} f d\mu = \int f d\mu.$$

It is worth remarking that the integral over an infinite interval is defined directly (an infinite interval is a measurable set) and not as the limit of integrals over finite intervals.

In $\mathbf{R}^k$ similar notations are used $\int_E \ldots \int f(x)\,dx$ means a 'multiple integral' of f over the set $E \in \mathscr{L}^k$ in Euclidean k-space with respect to Lebesgue measure.

If instead of using Lebesgue measure we use a Lebesgue–Stieltjes measure (defined in §4.5) given a point function F in $\mathbf{R}^k$, this is equivalent to working in the measure space $(\mathbf{R}^k, \mathscr{L}^k_F, \mu_F)$. We use the notation

$$\int_E f(x)\,dF(x) \quad \text{for} \quad \int_E f\,d\mu_F.$$

In this case we do not, in general, obtain the same result when we integrate over $E_1 = [a,b]$ and $E_2 = (a,b)$ so we will not use the notation

$$\int_a^b f(x)\,dF(x)$$

unless we know that F is continuous for all x. (This condition is sufficient to imply that the μ_F measure of single point sets is zero, so that the integrals over E_1 and E_2 are the same).

Because $\mathscr{L}^k$ is complete with respect to Lebesgue measure ($\mathscr{L}^k_F$ is complete with respect to μ_F) we see that if f: $\mathbf{R}^k \to \mathbf{R}^*$ is integrable and $f = g$ a.e., then g: $\mathbf{R}^k \to \mathbf{R}^*$ is also integrable.

The theorems of §5.4 were proved for any measure space $(\Omega, \mathscr{F}, \mu)$ so they are true in particular for Lebesgue measure in $\mathbf{R}^k$. Thus the Lebesgue integral is an order preserving linear operation on the class $\mathscr{A}$ of Lebesgue integrable functions. It is also a continuous operator in the following senses.

***Theorem* 5.6 A.** *If $\{f_n\}$ $(n = 1, 2, \ldots)$ is a monotone increasing sequence of non-negative Lebesgue measurable functions on $\mathbf{R}^k \to \mathbf{R}^+$ and*

$$f = \lim f_n,$$

then

$$\int_{-\infty}^{\infty} f(x)\,dx = \lim_{n\to\infty} \int_{-\infty}^{\infty} f_n(x)\,dx.$$

Corollary. *If $\{f_n\}$ is any sequence of non-negative Lebesgue measurable functions on $\mathbf{R}^k \to \mathbf{R}^+$ and $f = \sum_{n=1}^{\infty} f_n$, then*

$$\int_{-\infty}^{\infty} f(x)\,dx = \sum_{n=1}^{\infty} \int_{-\infty}^{\infty} f_n(x)\,dx.$$

***Theorem* 5.8 A.** *If g is Lebesgue integrable and $\{f_n\}$ is a sequence of Lebesgue measurable functions $\mathbf{R}^k \to \mathbf{R}$ such that $f_n \to f$ a.e. as $n \to \infty$*

and $|f_n| \leqslant g$ *a.e. for each* n; *then the functions* f_n, f *are Lebesgue integrable and*

$$\lim_{n\to\infty} \int_{-\infty}^{\infty} f_n(x)\,dx = \int_{-\infty}^{\infty} f(x)\,dx.$$

Corollary. *If* $E \in \mathscr{L}^k$ *and* $|E|$ *is finite, then for any sequence* $\{f_n\}$ *of* $\mathscr{L}^k$*-measurable functions* $\mathbf{R}^k \to \mathbf{R}$ *such that* $|f_n(x)| \leqslant a < \infty$ *for all* n, *all* $x \in E$, $f_n \to f$ *a.e. in* E *we have*

$$\int_E f(x)\,dx = \lim_{n\to\infty} \int_E f_n(x)\,dx.$$

It is clear that theorem 5.8A can also be translated to give a corresponding result for series. It is also worth remarking that the theorems corresponding to theorems 5.6A, 5.8A for the Riemann integral can only be proved by using some additional assumption that ensures that f is integrable: for example, it is sufficient to assume that $f_n \to f$ uniformly.

Exercises 5.5

1. From first principles calculate the Lebesgue integrals

(i) $\displaystyle\int_0^1 x^p\,dx \quad (p > -1)$; (ii) $\displaystyle\int_1^\infty x^q\,dx \quad (q < -1)$;

(iii) $\displaystyle\int_S f\,d\mu$, where μ is Lebesgue measure in $\mathbf{R}^2$, $f(x,y) = xy$ and S is the unit square $0 \leqslant x \leqslant 1$, $0 \leqslant y \leqslant 1$.

2. Suppose f: $\mathbf{R} \to \mathbf{R}^*$ is Lebesgue integrable and

$$F(x) = \int_{-\infty}^{x} f(t)\,dt.$$

Show that F is a uniformly continuous function.

3. Show that if $\{f_n\}$ is a sequence of integrable functions $E \to \mathbf{R}^*$ such that

$$\Sigma \int_E |f_n(x)|\,dx < \infty,$$

then $f_n(x) \to 0$ for almost all $x \in E$.

4. Show that if $|f_n(x)| \leqslant 1/n^2$ for all integers $n, x \in E$, and each f_n is measurable and g is integrable over E, then

$$\int_E \sum_{n=1}^{\infty} f_n(x)\,g(x)\,dx = \sum_{n=1}^{\infty} \int_E f_n(x)\,g(x)\,dx.$$

5. Carathéodory defines the Lebesgue integral of a non-negative measurable function in $\mathbf{R}$ as the Lebesgue measure of the ordinate set in $\mathbf{R}^2$

$$\int_E f(x)\,dx = |\{(x,y): x \in E, \quad 0 \leqslant y \leqslant f(x)\}|.$$

Show that this definition is equivalent to the one we have given.

6. Suppose $\{x_i\}$ is a sequence of points in $\mathbf{R}$ and $p_i \geqslant 0$, $\sum_{i=1}^{\infty} p_i < \infty$. $F(x)$ is defined by

$$F(x) = \sum_{x_i \leqslant x} p_i$$

and μ_F denotes the Lebesgue–Stieltjes measure with respect to F. Show that all functions $f: \mathbf{R} \to \mathbf{R}^*$ are measurable, and that f is integrable if and only if $\sum_{i=1}^{\infty} p_i f(x_i)$ converges absolutely.

7. Show that the function $f(x) = 1/x^2$ is integrable with respect to Lebesgue measure over $[1, \infty)$, but not with respect to the Lebesgue–Stieltjes measure generated by $F(x) = x^3$.

8. Show that the function $f(x) = x^2$ is integrable with respect to the Lebesgue–Stieltjes measure generated by

$$F(x) = \begin{cases} 0, & x \leqslant 0, \\ 1 - \dfrac{1}{(x+1)^4}, & x > 0. \end{cases}$$

9. Show that, for non-negative measurable functions $f: \mathbf{R} \to \mathbf{R}^+$, the Cauchy definition of the integral over an infinite interval

$$\int_0^\infty f\,d\mu = \lim_{t \to \infty} \int_0^t f(x)\,dx$$

is equivalent to the Lebesgue definition.

By considering the function $f(x) = \sin x/x$, show that this equivalence does not extend to all measurable f.

10. Show that if $f: [a,b] \to \mathbf{R}$ is continuous and $t \in (a,b)$

$$\lim_{y \to t} \frac{1}{y-t}\left[\int_a^y f(x)\,dx - \int_a^t f(x)\,dx\right] = f(t);$$

thus the Lebesgue indefinite integral can be differentiated at points where the integrand is continuous.

5.6* Conditions for integrability

The strength of the integration operator we have defined is that it works on a very wide class of functions. Provided the σ-field $\mathscr{F}$ is large, the restriction that f has to be $\mathscr{F}$-measurable is not a serious one, for we have seen that in a topological space Ω, if $\mathscr{F}$ contains the open sets, then any function which can be obtained from continuous functions or simple functions by countable operations will be $\mathscr{F}$-measurable. The only additional restriction for integrability of f is on the size (that is, the measure) of the sets where $|f|$ is large. It should be emphasised that our operation could be called 'absolute integration' for f is integrable if and only if $|f|$ is and we do not allow the large negative values of f to 'cancel out' the large positive values to give a finite integral unless each of f_+ and f_- is separately integrable (see exercise, 5.5 (9)).

If we restrict our consideration now to the Lebesgue integral on $\mathbf{R}$, these general comments still apply, but here it is worth comparing the Lebesgue integral with the Riemann integral over finite intervals. Since we want to compare integration operators, for the present section (only) we will use

$$\mathscr{L}\int_a^b f(x)\,dx \quad \text{to denote the Lebesgue integral,}$$

$$\mathscr{R}\int_a^b f(x)\,dx \quad \text{to denote the Riemann integral.}$$

It is easy to give examples of functions which are $\mathscr{L}$-integrable but not $\mathscr{R}$-integrable. There are two kinds of bad behaviour which can prevent a function from being $\mathscr{R}$-integrable. These are illustrated by:

(1) bounded functions which are badly discontinuous but still $\mathscr{L}$-measurable. For example

$$f(x) = \begin{cases} 1 & \text{when} \quad x \text{ is rational,} \\ 0 & \text{when} \quad x \text{ is irrational,} \end{cases}$$

is discontinuous everywhere. For any $a < b$, it is clear that

$$\mathscr{R}\int_a^b f(x)\,dx$$

cannot exist. However, the set of rational points is countable, and therefore $\mathscr{L}$-measurable with zero measure, so that $f(x)$ is an $\mathscr{L}$-simple function and

$$\mathscr{L}\int_a^b f(x)\,dx = 0.$$

(2) Functions which are unbounded in (a, b) cannot be $\mathscr{R}$-integrable even if they are continuous everywhere. For example, $f(x) = x^{-\alpha}$ $(0 < \alpha < 1)$ is not $\mathscr{R}$-integrable over $(0, 1)$, although an elementary

calculation shows that it is $\mathscr{L}$-integrable. If the points of unboundedness of f (as in the above case) are finite in number, it is sometimes possible to use the 'Cauchy–Riemann' process to define the integral. Thus

$$\lim_{\epsilon \to 0+} \mathscr{R} \int_\epsilon^1 f(x)\, dx$$

is defined in the above case and could be used as a definition of $\int_0^1 f(x)\, dx$. Provided the Cauchy–Riemann integral of $|f(x)|$ exists, it is not difficult to show that, if the Cauchy–Riemann process for $f(x)$ works, then f is $\mathscr{L}$-integrable to the same value. This is not true without the condition that the process works for $|f(x)|$, since the $\mathscr{L}$-integral is an absolute integral.

We know (corollary to theorem 5.5) that any function $f: [a, b] \to \mathbf{R}$ which is $\mathscr{L}$-measurable and bounded is $\mathscr{L}$-integrable. For the existence of the $\mathscr{R}$-integral it is necessary for f to be bounded, but the condition of measurability does not give sufficient smoothness. In fact the natural way of characterising functions which are $\mathscr{R}$-integrable over a finite interval is in terms of the measure of the set of points where the function is discontinuous.

***Theorem* 5.9.** *A bounded function f: $[a, b] \to \mathbf{R}$ is Riemann integrable if and only if the set E of points in $[a, b]$ at which f is discontinuous satisfies $|E| = 0$. Any $f: [a, b] \to \mathbf{R}$ which is Riemann integrable is Lebesgue integrable to the same value.*

Proof. We use the following definition for the Riemann integral of f over $[a, b]$ (this is not the usual one but can easily be seen to be equivalent by using the basic theory of the $\mathscr{R}$-integral). For any positive integer n, divide $I_0 = (a, b]$ into 2^n equal half-open intervals

$$I_{n,i} = (a_{n,i-1}, a_{n,i}] \quad (i = 1, 2, \ldots, 2^n);$$

put

$$m_{n,i} = \inf\{f(x): a_{n,i-1} < x \leqslant a_{n,i}\},$$

$$M_{n,i} = \sup\{f(x): a_{n,i-1} < x \leqslant a_{n,i}\},$$

$$g_n(x) = \begin{cases} m_{n,i} & \text{for } x \in I_{n,i}, \\ 0 & \text{for } x \notin I_0; \end{cases}$$

$$h_n(x) = \begin{cases} M_{n,i} & \text{for } x \in I_{n,i}, \\ 0 & \text{for } x \notin I_0. \end{cases}$$

Then for each integer n, $x \in I_0$

$$g_n(x) \leqslant f(x) \leqslant h_n(x);$$

$\{g_n\}$ is a monotone increasing sequence of simple functions, and $\{h_n\}$ is a monotone decreasing sequence of simple functions. If we put

$$g = \lim_{n\to\infty} g_n, \quad h = \lim_{n\to\infty} h_n,$$

then $g \leqslant f \leqslant h$. Further, by definition,

$$\mathscr{L}\int_a^b g(x)\,dx = \lim_{n\to\infty} \mathscr{L}\int_a^b g_n(x)\,dx$$

$$= \lim_{n\to\infty} \frac{b-a}{2^n} \sum_{i=1}^{2^n} m_{n,i} = \lim_{n\to\infty} s_n, \quad \text{say;}$$

$$\mathscr{L}\int_a^b h(x)\,dx = \lim \mathscr{L}\int_a^b h_n(x)\,dx$$

$$= \lim_{n\to\infty} \frac{b-a}{2^n} \sum_{i=1}^{2^n} M_{n,i} = \lim_{n\to\infty} S_n, \quad \text{say.}$$

We say that f is $\mathscr{R}$-integrable over $[a,b]$ if, only if

$$\lim_{n\to\infty} s_n = \lim_{n\to\infty} S_n \quad \text{and} \quad \mathscr{R}\int_a^b f(x)\,dx$$

is then the common value of the limit.

Now notice that if f is continuous at $x \in (a,b)$ then $g(x) = h(x)$. Conversely if $g(x) = h(x)$ and x is not a dyadic point (that is, $x \notin D$, where D is the countable set of end-points of intervals $I_{n,i}$), then f is continuous at x.

If $\mathscr{R}\int_a^b f(x)\,dx$ exists, since $g \leqslant f \leqslant h$,

$$\mathscr{L}\int_a^b g(x)\,dx = \mathscr{R}\int_a^b f(x)\,dx = \mathscr{L}\int_a^b h(x)\,dx$$

so that, by theorem 5.5 (vii) $g = h$ a.e. Since the set E of points where f is discontinuous is contained in $D \cup \{x: g(x) \neq h(x)\}$ it follows that $|E| = 0$. Further, since Lebesgue measure is complete, f is $\mathscr{L}$-measurable and, by theorem 5.5 (viii),

$$\mathscr{L}\int_a^b f(x)\,dx = \mathscr{L}\int_a^b g(x)\,dx = \mathscr{R}\int_a^b f(x)\,dx.$$

Conversely if the set E satisfies $|E| = 0$, this implies $g(x) = h(x)$ a.e., which gives, by theorem 5.5 (viii)

$$\mathscr{L}\int_a^b g(x)\,dx = \mathscr{L}\int_a^b h(x)\,dx$$

so that f is $\mathscr{R}$-integrable. ▌

Theorem 5.9 shows that $\mathscr{R}$-integrable functions have to be continuous at most points. We have many examples of $\mathscr{L}$-integrable functions which are continuous nowhere. However, there is a sense in which even $\mathscr{L}$-integrable functions have to be approximable by continuous functions—in fact by functions which are arbitrarily smooth, that is, functions that can be differentiated arbitrarily often.

***Theorem* 5.10.** *Given any $\mathscr{L}$-integrable function $f: \mathbf{R} \to \mathbf{R}^*$ and any $\epsilon > 0$ there is a finite interval (a, b), and a bounded function $g: \mathbf{R} \to \mathbf{R}$ such that $g(x)$ vanishes outside (a, b), is infinitely differentiable for all real x and*

$$\mathscr{L}\int_{-\infty}^{\infty} |f(x)-g(x)|\,dx < \epsilon.$$

Proof. We carry out the approximation in 4 stages.

(i) First, find a finite interval $[a, b]$ and a bounded measurable function f_1 which vanishes outside $[a, b]$ and is such that

$$\mathscr{L}\int_{-\infty}^{\infty} |f(x)-f_1(x)|\,dx < \tfrac{1}{4}\epsilon.$$

This can be done by considering the sequence of functions

$$g_n(x) = \begin{cases} f(x) & \text{if } x \in [-n, n] \text{ and } |f(x)| \leqslant n, \\ n & \text{if } x \in [-n, n] \text{ and } f(x) > n, \\ -n & \text{if } x \in [-n, n] \text{ and } f(x) < -n, \\ 0 & \text{if } x \notin [-n, n]. \end{cases}$$

Then $g_n(x) \to f(x)$ for all x and $|g_n| \leqslant |f|$. By theorem 5.8 it follows that

$$\int_{-\infty}^{\infty} |f(x)-g_n(x)|\,dx \to 0 \quad \text{as} \quad n \to \infty$$

so that we can fix a sufficiently large N and put $f_1(x) = g_N(x)$.

(ii) The next step is to approximate f_1 by an $\mathscr{L}$-simple function f_2 which vanishes outside $[a, b]$ and satisfies

$$\int_{-\infty}^{\infty} |f_1(x)-f_2(x)|\,dx < \tfrac{1}{4}\epsilon.$$

This is clearly possible since we defined the integral as a limit of the integrals of simple functions.

(iii) Now a simple function is a finite sum of multiples of indicator functions. If each indicator function can be approximated by the indicator function of a finite number of disjoint intervals, then it will follow that f_2 can be approximated by f_3, a step function of the form

$$f_3(x) = \sum_{i=1}^{n} q_i \chi_{J_i}(x),$$

where each J_i is a finite interval and

$$\int_{-\infty}^{\infty} |f_2(x)-f_3(x)|\,dx < \tfrac{1}{4}\epsilon.$$

To see that this is possible start with a bounded $\mathscr{L}$-measurable set E and $\eta > 0$. Find an open set $G \supset E$ such that $|G-E| < \frac{1}{2}\eta$ and from the countable union of disjoint open intervals making up G pick a finite number to form G_0 such that $|G-G_0| < \frac{1}{2}\eta$. It will then follow that $|E \triangle G_0| < \eta$ so that

$$\int_{-\infty}^{\infty} |\chi_E(x)-\chi_{G_0}(x)|\,dx < \eta.$$

(iv) In order to obtain the required infinitely differentiable function g for which

$$\int_{-\infty}^{\infty} |f_3(x)-g(x)|\,dx < \tfrac{1}{4}\epsilon$$

it is now sufficient to find a function for one of the components $\chi_{J_i}(x)$ of f_3.

Suppose $J = (a, b)$ and $0 < 2\eta < b-a$. Put

$$\phi_{a,\eta}(x) = \begin{cases} \exp\,[(x-a)^2-\eta^2]^{-1} & \text{for} \quad |x-a| < \eta, \\ 0 & \text{for} \quad |x-a| \geqslant \eta. \end{cases}$$

$$\text{If} \quad c_\eta^{-1} = \int_{-\infty}^{\infty} \phi_{a,\eta}(x)\,dx, \qquad \text{let} \quad h(x) = c_\eta \int_{-\infty}^{x} \{\phi_{a,\eta}(t)-\phi_{b,\eta}(t)\}\,dt.$$

It is easy to check that h is infinitely differentiable and

$$\int |\chi_J(x)-h(x)|\,dx < 4\eta,$$

since $0 \leqslant h(x) \leqslant 1$ for all x and $\{x: \chi_J(x) \neq h(x)\}$ is contained in the two intervals $(a-\eta, a+\eta)$ and $(b-\eta, b+\eta)$. ∎

Remark 1. We stated our approximation theorem in $\mathbf{R}^1$. It is also true in $\mathbf{R}^k$ for every k, and in this case we can require the approximating function to have partial derivatives of all orders everywhere. Our proof requires only minor modifications to give the corresponding theorem in $\mathbf{R}^k$.

Remark 2. If Ω is a topological space, and $\mathscr{F}$ includes the Baire sets in Ω then theorem 5.10 can be generalized to $(\Omega, \mathscr{F}, \mu)$ to show that any integrable function can be approximated by a continuous function.

(The Baire sets are the sets in the σ-ring generated by sets

$$\{x: f(x) > 0\} \quad \text{where} \quad f: \Omega \to \mathbf{R}$$

is continuous and vanishes outside a compact set).

Exercises 5.6

1. In theorem 5.10 it was shown that any integrable function f could be appproximated by a step function g in the sense that

$$\int |f(x)-g(x)|\,dx < \epsilon.$$

Show that in general it is not possible to arrange at the same time that $g \leqslant f$.

2. Show if $f: \mathbf{R} \to \mathbf{R}^*$ is integrable, then

$$\int |f(x+h)-f(x)|\,dx \to 0 \quad \text{as} \quad h \to 0.$$

Hint. Approximate to f by a function which is uniformly continuous and zero outside a bounded interval.

3. If $f_n(x) = e^{-nx} - 2e^{-2nx}$ show that f_n is integrable over $[0, +\infty)$ but that

$$\int_0^\infty \left(\sum_{n=1}^\infty f_n(x)\right) dx \neq \sum_{n=1}^\infty \int_0^\infty f_n(x)\,dx.$$

4. Put
$$g(x) = \begin{cases} x^2 \sin 1/x^3 & (x \neq 0), \\ 0 & (x = 0), \end{cases}$$

and
$$g'(x) = f(x) \qquad \text{for all} \quad x \in \mathbf{R}.$$

Show that $f(x)$ is finite for all x, but unbounded near $x = 0$. Show that $f(x)$ is not $\mathscr{R}$-integrable over $(0, 1)$, but that it is Cauchy–Riemann integrable (evaluate its integral). Is $f(x)$ $\mathscr{L}$-integrable over $(0, 1)$?

6

RELATED SPACES AND MEASURES

6.1 Classes of subsets in a product space

In the last few chapters we have defined all our concepts in a single abstract space Ω and usually we have at any time considered only one measure defined on a fixed class of subsets of Ω. In applications one often requires to consider more than one measure, and the relationship between the spaces and measures involved become important. We first consider measures defined on the Cartesian product of two measure spaces. Before considering the definition of such measures we must examine, in the present section, the structure of the relevant classes of subsets.

In §1.1 we defined the Cartesian product $X \times Y$ of two spaces X, Y to be the set of all ordered pairs (x, y) with $x \in X, y \in Y$.

Rectangle

Any set in $X \times Y$ of the form $E \times F$ with $E \subset X$, $F \subset Y$ is called a rectangle (set).

Product of classes

If $\mathscr{C}$, $\mathscr{D}$ denote classes of subsets in X, Y respectively, then $\mathscr{C} \times \mathscr{D}$ denotes the class of all rectangles $E \times F$ with $E \in \mathscr{C}$, $F \in \mathscr{D}$.

Product ring, field, σ-field

If z-class again denotes any one of ring, field, σ-ring, σ-field and $\mathscr{C}, \mathscr{D}$ are z-classes in X, Y, respectively, then the product z-class is the z-class in $X \times Y$ generated by $\mathscr{C} \times \mathscr{D}$.

Lemma. *If $\mathscr{C},\mathscr{D}$ are semi-rings in X, Y respectively, then $\mathscr{C} \times \mathscr{D}$ is a semi-ring in $X \times Y$.*

Proof. It is immediate that $\mathscr{C} \times \mathscr{D}$ is closed for finite intersections, so that we have only to prove that

$$E_1 \times F_1 - E_2 \times F_2$$

can be expressed as a union of disjoint sets of $\mathscr{C} \times \mathscr{D}$ for any

$$E_1, E_2 \in \mathscr{C}; \quad F_1, F_2 \in \mathscr{D}.$$

Since $\mathscr{C}, \mathscr{D}$ are semi-rings we have

$$E_1 - E_2 = \bigcup_{i=3}^{n} E_i, \quad F_1 - F_2 = \bigcup_{j=3}^{m} F_j,$$

where the sets $E_i\ (i = 3, 4, \ldots, n)$ are disjoint sets of $\mathscr{C}$, $F_j\ (j = 3, 4, \ldots, m)$ are disjoint sets of $\mathscr{D}$. Now

$$\begin{aligned} E_1 \times F_1 - E_2 \times F_2 &= (E_1 \cap E_2) \times (F_1 - F_2) \cup (E_1 - E_2) \times (F_1 \cap F_2) \\ &\qquad \cup (E_1 - E_2) \times (F_1 - F_2) \\ &= \bigcup_{j=3}^{m} (E_1 \cap E_2) \times F_j \cup \bigcup_{i=3}^{n} E_i \times (F_1 \cap F_2) \cup \bigcup_{i=3}^{n} \bigcup_{j=3}^{m} E_i \times F_j, \end{aligned}$$

and these are all disjoint sets in $\mathscr{C} \times \mathscr{D}$.]

It is important to notice that this lemma does not extend to any of the z-classes as $\mathscr{C} \times \mathscr{D}$ is not closed under the operation of union. In particular, if $\mathscr{C}, \mathscr{D}$ are σ-fields then $\mathscr{C} \times \mathscr{D}$ will not be a σ-field. We will use the notation $\mathscr{C} * \mathscr{D}$ for the σ-field in $X \times Y$ generated by $\mathscr{C} \times \mathscr{D}$. We also need some operations which are effective in the opposite direction, from the product space to the components.

Section

Given any set $E \subset X \times Y$ and any point $x \in X$, the subset

$$E_x = \{y\colon (x, y) \in E\}$$

of Y is called the section of E at x. Similarly, for $y \in Y$, the subset $E^y = \{x\colon (x, y) \in E\}$ of X is the section of E at y.

Projection

Given any set $E \subset X \times Y$ the sets $\{x\colon \text{there exists } y \text{ with } (x, y) \in E\}$, $\{y\colon \text{there exists } x \text{ with } (x, y) \in E\}$ are called the projections of E into the spaces X, Y, respectively.

Although the product σ-field $\mathscr{C} * \mathscr{D}$ of two σ-fields $\mathscr{C}, \mathscr{D}$ contains more than the rectangle sets $E \times F$, $E \in \mathscr{C}$, $F \in \mathscr{D}$, one can deduce an important restriction on the sections of its sets.

***Theorem* 6.1.** *If $\mathscr{F}$, $\mathscr{G}$ are σ-fields in X, Y, respectively, and $\mathscr{H}$ is the product σ-field in $X \times Y$, then all sets E in $\mathscr{H}$ have the property that $E_x \in \mathscr{G}$ for all $x \in X$ and $E^y \in \mathscr{F}$ for all $y \in Y$.*

Proof. Let $\mathscr{C}$ be the class of subsets E of $X \times Y$ with the property that every section of E is in the appropriate σ-field. Since rectangle sets certainly have this property, it is immediate that $\mathscr{C} \supset \mathscr{F} \times \mathscr{G}$. Moreover, it is not difficult to verify that $\mathscr{C}$ satisfies all the axioms of a σ-field. Hence, $\mathscr{C} \supset \mathscr{H}$ the σ-field generated by $\mathscr{F} \times \mathscr{G}$.]

Sections of functions

Given any function $f: X \times Y \to \mathbf{R}^*$; for each fixed $x \in X$,

$$f_x(y) = f(x, y) \quad \text{for} \quad y \in Y$$

defines a function $Y \to \mathbf{R}^*$ and for each fixed $y \in Y$, $f_y(x) = f(x, y)$ for $x \in X$ defines a function on X to $\mathbf{R}^*$. These functions f_x, f_y are called the *sections* of f at x, y, respectively.

Corollary. *Under the conditions of theorem* 6.1, *given any* $\mathscr{H}$*-measurable function* $f: X \times Y \to \mathbf{R}^*$, *each of the sections* $f_x(y)$ *is* $\mathscr{G}$*-measurable and each of the sections* $f_y(x)$ *is* $\mathscr{F}$*-measurable.*

Proof. Suppose x_0 is a fixed point in X and M is a Borel set in $\mathbf{R}^*$; then

$$\{y: f_{x_0}(y) \in M\} = \{y: f(x_0, y) \in M\} = \{(x, y): f(x, y) \in M\}_{x_0}$$

so that the test set is the section at x_0 of a set in $\mathscr{H}$. ∎

The results of theorem 6.1 and corollary can be extended in an obvious way to finite Cartesian products $X_1 \times X_2 \times \ldots \times X_n$; there is no difficulty in making the required modifications to the definitions and proofs. It is not quite so immediate that they can also be extended to arbitrary Cartesian products $\prod_{i \in I} X_i$.

Let us recall that a point in $\prod_{i \in I} X_i$ can be thought of as a function $f: I \to \bigcup_{i \in I} X_i$ such that $f(i) \in X_i$ for each $i \in I$. Suppose then that we have a collection $\{X_i, i \in I\}$ of spaces and σ-fields $\mathscr{F}_i$ of subsets of X_i.

Cylinder set

If $i_1, i_2, \ldots, i_n$ is any finite subset of I and $E_{i_k} \in \mathscr{F}_{i_k}$, $k = 1, \ldots, n$; the set of points $f \in \Pi X_i$ such that

$$f(i_k) \in E_{i_k} \quad (k = 1, 2, \ldots, n),$$

is said to be a (finite dimensional) cylinder set in ΠX_i. When we say that f is in a cylinder set C, the values of f are restricted only on a finite set of indices. The class of all such cylinder sets in ΠX_i will be denoted $\mathscr{C}(\mathscr{F}_i, i \in I)$.

Lemma. *The class* $\mathscr{C}(\mathscr{F}_i, i \in I)$ *of cylinder sets is a semi-ring of subsets of* $\prod_{i \in I} X_i$.

Proof. We can think of $\mathscr{F}_i$ as a semi-ring in X_i which contains the whole space X_i. Then if two sets

$$A = \{f: f(i) \in E_i, i \in J\},$$

$$B = \{f: f(i) \in F_i, i \in K\},$$

are in $\mathscr{C}(\mathscr{F}_i, i \in I)$, J and K must be finite subsets of I, and each of the sets E_i, F_i must be in the relevant σ-field $\mathscr{F}_i$. The set $J \cup K = L$ is also a finite subset of I and, if we put

$$E_i = X_i \quad \text{for} \quad i \in K - J, \quad F_i = X_i \quad \text{for} \quad i \in J - K,$$

then $$A = \{f : f(i) \in E_i, \, i \in L\}, \quad B = \{f : f(i) \in F_i, i \in L\},$$

are now cylinder sets in which the same finite subset L of indices are restricted. Since we know that any finite Cartesian product of semi-rings is a semi-ring, we can deduce that $A - B$ is a finite disjoint union of sets of this type and $A \cap B$ is a set of this type. Hence $\mathscr{C}(\mathscr{F}_i, i \in I)$ is a semi-ring. ▌

Note. The case $I = \mathbf{Z}$ is important. The cylinder sets in $\prod_{i=1}^{\infty} X_i$ then reduce to sets of the form $E_1 \times E_2 \times \ldots \times E_n \times \prod_{i=n+1}^{\infty} X_i$ with

$$E_i \in \mathscr{F}_i \, (i = 1, 2, \ldots, n).$$

The results corresponding to theorem 6.1 for $\prod_{i=1}^{\infty} X_i$ are formulated as examples for the reader to prove.

Exercises 6.1

1. If $\mathscr{R}$ is a ring of subsets of X, $\mathscr{S}$ is a ring of subsets of Y, show that the product ring consists of those sets in $X \times Y$ which are finite unions of disjoint rectangles in $\mathscr{R} \times \mathscr{S}$.

2. If $A_1, A_2 \subset X$, B_1, $B_2 \subset Y$ and $A_1 \times B_1 = A_2 \times B_2$ is not null, prove that $A_1 = A_2$, $B_1 = B_2$.

3. Suppose $E = A \times B$, $E_1 = A_1 \times B_1$ and $E_2 = A_2 \times B_2$ are all non-empty rectangles in $X \times Y$. Show that E is a disjoint union of E_1 and E_2 if and only if *either* A is a disjoint union of A_1, A_2 and $B = B_1 = B_2$, *or* B is a disjoint union of B_1, B_2 and $A = A_1 = A_2$.

4. If $\mathscr{S}$, $\mathscr{T}$ are σ-rings in X, Y, respectively, then the product σ-ring in $X \times Y$ is a σ-field if and only if both $\mathscr{S}$ and $\mathscr{T}$ are σ-fields.

5. Show that the intersection of a class of rectangles is a rectangle.

6. Suppose $X = Y$ is any uncountable set and $\mathscr{S} = \mathscr{T}$ is the class of subsets which are either countable or have countable complement. Determine the product σ-field of $\mathscr{S}$ and $\mathscr{T}$. If $D = \{(x, y) : x = y\}$ is the diagonal in $X \times Y$ show that every section of D is in $\mathscr{S}$ or $\mathscr{T}$ but D is not in the product σ-field. This shows that theorem 6.1 has no converse.

7. Suppose $\mathscr{F}$, $\mathscr{G}$ are σ-fields in X, Y; then a rectangle set $E \times F$ is in the product σ-field if and only if $E \in \mathscr{F}$, $F \in \mathscr{G}$.

8. Suppose $\mathscr{H}$ is the product σ-field of two σ-fields $\mathscr{F}$, $\mathscr{G}$. Show that any function on $X \times Y \to \mathbf{R}$ which is $\mathscr{H}$-simple has all its sections $\mathscr{F}$-simple or $\mathscr{G}$-simple.

9. Suppose $\mathscr{F}$ is the product σ-field of two σ-fields $\mathscr{F}_1, \mathscr{F}_2$. Show that the projection of a set in $\mathscr{F}$ on an axis need not be in $\mathscr{F}_1$, $\mathscr{F}_2$, respectively.

10. Suppose $\mathscr{F}_i$ is a σ-field in X_i $(i = 1, 2, \ldots)$ and the σ-field generated by cylinder sets $\mathscr{C}(\mathscr{F}_n, \mathscr{F}_{n+1}, \ldots)$ in $\prod_{i=n}^{\infty} X_i$ is denoted by $\mathscr{S}_n$. Then given any set E in $\prod_{i=1}^{\infty} X_i$ the (finite dimensional) section of E at $x_1, x_2, \ldots, x_k$ is the set $\left(\text{in} \prod_{i=k+1}^{\infty} X_i\right)$ of points $(x_{k+1}, x_{k+2}, \ldots)$ such that $(x_1, x_2, \ldots) \in E$. Then if $E \in \mathscr{S}_1$ the product σ-field in $\prod_{i=1}^{\infty} X_i$, all its k-dimensional sections belong to $\mathscr{S}_{k+1}$.

6.2 Product measures

We now assume that $(X_1, \mathscr{F}_1, \mu_1)$ and $(X_2, \mathscr{F}_2, \mu_2)$ are measure spaces and μ_1, μ_2 are σ-finite measures. The product σ-field $\mathscr{H}$ in $X_1 \times X_2$ was defined as the smallest σ-field containing the class $\mathscr{F}_1 \times \mathscr{F}_2$ which is known to be a semi-ring since each of $\mathscr{F}_1, \mathscr{F}_2$ are semi-rings. In Chapters 3 and 4 we developed a general method of extending a measure from a semi-ring to the generated σ-ring. Since the semi-ring $\mathscr{F}_1 \times \mathscr{F}_2$ contains the whole space $X_1 \times X_2$ this generated σ-ring must be a σ-field and is therefore $\mathscr{F}_1 * \mathscr{F}_2$, the product σ-field. Thus if we use theorems 3.5 and 4.2 we can extend any σ-finite measure on $\mathscr{F}_1 \times \mathscr{F}_2$ to a σ-finite measure on $\mathscr{F}_1 * \mathscr{F}_2$ in a unique way.

Suppose $E_1 \times E_2$ is any rectangle set in $\mathscr{F}_1 \times \mathscr{F}_2$ and put

$$\mu(E_1 \times E_2) = \mu_1(E_1)\,\mu_2(E_2),$$

with the usual convention that $0.\infty = \infty.0 = 0$. Then μ is a non-negative set function on $\mathscr{F}_1 \times \mathscr{F}_2$ which is easily seen to be σ-finite. Our first objective is to show that μ is a measure on the semi-ring $\mathscr{F}_1 \times \mathscr{F}_2$. First, suppose that

$$E \times F = \bigcup_{i=1}^{n} (E_i \times F_i)$$

with the sets $E_i \times F_i$ disjoint. Define the functions $f_i\colon X_1 \to \mathbf{R}^+$ by $f_i(x) = \mu_2(F_i)\,\chi_{E_i}(x)$ $(i = 1, 2, \ldots, n)$. Then f_i is a non-negative function or possibly a function which takes the value $+\infty$ on a measurable set E_i and zero outside it: in any case

$$\int f_i\, d\mu_1 = \mu_1(E_i)\,\mu_2(F_i) \quad (i = 1, 2, \ldots, n).$$

Similarly, if $f(x) = \mu_2(F)\,\chi_E(x)$ we have

$$\int f\,d\mu_1 = \mu_1(E)\,\mu_2(F).$$

Now for each fixed x in X_1 we have

$$(E \times F)_x = \bigcup_{i=1}^{n} (E_i \times F_i)_x$$

with the sets $(E_i \times F_i)_x$ disjoint. Since μ_2 is (finitely) additive it follows that

$$f(x) = \sum_{i=1}^{n} f_i(x).$$

If we now use (finite) additivity for integrals of non-negative simple functions we have

$$\mu_1(E)\,\mu_2(F) = \int f\,d\mu_1 = \int \sum_{i=1}^{n} f_i\,d\mu_1 = \sum_{i=1}^{n} \int f_i\,d\mu_1 = \sum_{i=1}^{n} \mu_1(E_i)\,\mu_2(F_i).$$

This shows that the set function μ we have defined is finitely additive on $\mathscr{F}_1 \times \mathscr{F}_2$. The same argument extends without difficulty to countable unions of disjoint rectangles

$$\bigcup_{i=1}^{\infty} (E_i \times F_i) = E \times F$$

because all the functions $f_i(x)$ are non-negative measurable, so that the monotone convergence theorem 5.6 justifies the inversion of integration and summation. Thus μ is a measure on the semi-ring $\mathscr{F}_1 \times \mathscr{F}_2$. It can be extended uniquely by theorem 3.5 to the generated ring, and then, by theorem 4.2, to the generated σ-ring which is the product σ-field $\mathscr{F}_1 * \mathscr{F}_2$. The result is called the *product measure* on $\mathscr{F}_1 * \mathscr{F}_2$. We have thus proved

***Theorem* 6.2.** *Given two measure spaces $(X_1, \mathscr{F}_1, \mu_1)$, $(X_2, \mathscr{F}_2, \mu_2)$ such that μ_1, μ_2 are σ-finite, there is a unique measure μ defined on the product σ-field $\mathscr{F}_1 * \mathscr{F}_2$ in $X_1 \times X_2$ such that*

$$\mu(E_1 \times E_2) = \mu_1(E_1)\,\mu_2(E_2) \quad \text{for} \quad E_1 \in \mathscr{F}_1, \quad E_2 \in \mathscr{F}_2.$$

The above theorem clearly extends immediately to any finite Cartesian product of σ-finite measure spaces. Difficulties arise with the Cartesian product of an enumerable collection of measure spaces unless we arrange that the infinite products of real numbers occurring converge. The easiest way to ensure this is to restrict the discussion to countable products of measure spaces $(X_i, \mathscr{F}_i, \mu_i)$ with $\mu_i(X_i) = 1$.

It is possible to define product measures on arbitrary product spaces $\prod_{i \in I} X_i$ such that $\mu_i(X_i) = 1$ by exactly the method used below. We carry out the construction only for enumerable products as, in applications, it is not usually appropriate to consider the product measure for non-countable products. In §6.6 we will give a general construction for a measure in $\prod_{i \in I} X_i$, an arbitrary product space—this construction could clearly be specialized to give the results of the remainder of this section, but it is simpler to deal with the case of product measures in countable product spaces first. We will set up our measure on the product σ-field by a slightly different procedure.

Let $\mathscr{C} = \mathscr{C}(\mathscr{F}_1, \mathscr{F}_2, \ldots)$ be the semi-ring of cylinder sets in $\prod_{i=1}^{\infty} X_i$. We define μ on $\mathscr{C}$ by $\mu(E) = \mu_1(E_1)\,\mu_2(E_2)\ldots\mu_n(E_n)$, if

$$E = E_1 \times E_2 \times \ldots \times E_n \times \prod_{i=n+1}^{\infty} X_i; \quad E_i \in \mathscr{F}_i, \quad (i = 1, 2, \ldots, n).$$

It is clear that $0 \leqslant \mu(E) \leqslant 1$ for all E in $\mathscr{C}$. To see that μ is finitely additive on $\mathscr{C}$ it is sufficient to see that, in any finite collection of cylinder sets, only a finite number of coordinates are involved so that, if $C = \bigcup_{j=1}^{m} C_j$ is a dissection of $C \in \mathscr{C}$ into disjoint sets of $\mathscr{C}$, there is an integer N such that C and $C_j\,(j = 1, \ldots, m)$ can all be expressed in the form

$$E_1 \times E_2 \times \ldots \times E_N \times \prod_{i=N+1}^{\infty} X_i.$$

We can then apply theorem 6.2 to the finite products to see that

$$\mu(C) = \sum_{j=1}^{m} \mu(C_j).$$

By theorem 3.4, μ has a unique additive extension to the ring $\mathscr{R}$ of finite unions of cylinder sets. In order to apply theorem 4.2 we must show that μ is a measure on $\mathscr{R}$. This can be done by using the continuity theorem 3.2. It is sufficient to show that any monotone decreasing sequence $\{A_n\}$ of sets in $\mathscr{R}$ such that

$$\mu(A_n) \to \epsilon > 0$$

has a non-void intersection.

Let $Y_n = \prod_{i=n+1}^{\infty} X_i$. Then by the above procedure we can define product set functions $\nu^{(n)}$ on the class $\mathscr{R}^{(n)}$ of finite unions of cylinder sets $\mathscr{C}(\mathscr{F}_{n+1}, \mathscr{F}_{n+2}, \ldots)$. It is clear that, for each integer n, we can obtain μ on $\mathscr{C}$ by taking the product of the measures μ_i $(i = 1, 2, \ldots, n)$ and $\nu^{(n)}$. Let

$$A_n(x_1) = \{y: y \in Y_1, \quad (x_1, y) \in A_n\}$$

be the section of A_n at $x_1 \in X_1$. It is clear that, for each $x_1 \in X_1$, $A_n(x_1) \in \mathscr{R}^{(1)}$ and if

$$B_{n,1} = \{x_1: \nu^{(1)}(A_n(x_1)) > \tfrac{1}{2}\epsilon\}$$

then $B_{n,1}$ is a finite union of sets in $\mathscr{F}_1$ and is therefore in $\mathscr{F}_1$: further we must have

$$\mu_1(B_{n,1}) + \tfrac{1}{2}\epsilon(1 - \mu_1(B_{n,1})) \geqslant \mu(A_n) \geqslant \epsilon,$$

by considering $A_n \cap (B_{n,1} \times Y_1)$ and $A_n \cap (X_1 - B_{n,1}) \times Y_1$. It follows that

$$\mu_1(B_{n,1}) > \tfrac{1}{2}\epsilon \quad (n = 1, 2, \ldots).$$

But $\{A_n\}$ is monotone decreasing so $\{B_{n,1}\}$ must also decrease with n and

$$\mu_1\left\{\bigcap_{n=1}^{\infty} B_{n,1}\right\} \geqslant \tfrac{1}{2}\epsilon.$$

Since μ_1 is a measure on $\mathscr{F}_1$, it follows that there must be at least one point $x_1 \in X_1$ for which

$$\nu^{(1)}(A_n(x_1)) > \tfrac{1}{2}\epsilon \quad \text{for all } n.$$

We now suppose x_1 is fixed as such a point in $\bigcap B_{n,1}$ and repeat the argument to the sequence of sets $\{A_n(x_1)\}$ in the space Y_1. This gives a point $x_2 \in X_2$ such that

$$\nu^{(2)}(A_n(x_1, x_2)) > \epsilon/2^2 \quad \text{for all } n.$$

By an induction argument we obtain a point $(x_1, x_2, \ldots,)$ in $\prod_{i=1}^{\infty} X_i$ such that, for any k, n

$$A_n(x_1, x_2, \ldots, x_k) \neq \varnothing.$$

But each set A_n has only a finite number of coordinates restricted so the point $(x_1, x_2, \ldots)$ must be in A_n for all n. This completes the proof that μ is continuous from above at $\varnothing$.

Since μ is now seen to be a finite measure on the ring $\mathscr{R}$ it has a unique extension to the generated σ-ring which is also the product σ-field in $\prod_{i=1}^{\infty} X_i$. This extension is called the *product measure*. Thus we have proved

***Theorem* 6.3.** *If* $(X_i, \mathscr{F}_i, \mu_i)$ *are measure spaces with*

$$\mu_i(X_i) = 1 \quad (i = 1, 2, \ldots);$$

then there is a unique measure μ defined on the product σ-field $\mathscr{F}$ of subsets of $X = \prod_{i=1}^{\infty} X_i$ which is generated by the cylinder sets of the form

$$E_1 \times E_2 \times \ldots \times E_n \times \prod_{i=n+1}^{\infty} X_i \quad (E_i \in \mathscr{F}_i, i = 1, 2, \ldots),$$

such that

$$\mu\left(E_1 \times \ldots \times E_n \times \prod_{i=n+1}^{\infty} X_i\right) = \mu_1(E_1)\,\mu_2(E_2) \ldots \mu_n(E_n).$$

Exercises 6.2

1. Given 3 σ-finite measure spaces $(X_1, \mathscr{F}_1, \mu_1)$, $(X_2, \mathscr{F}_2, \mu_2)$, $(X_3, \mathscr{F}_3, \mu_3)$ let τ be the product measure of μ_1, μ_2 in $X_1 \times X_2$ and ν the product measure of μ_2, μ_3 in $X_2 \times X_3$. Show that, in the space $X_1 \times X_2 \times X_3$ the product measure of τ and μ_3 is the same as the product measure of μ_1 and ν.

2. Suppose $(X_i, \mathscr{F}_i, \mu_i)$ $(i = 1, 2, \ldots)$ is a sequence of measure spaces with $\mu_i(X_i) = 1$. Let μ be the product measure of theorem 6.3 on $\prod_{i=1}^{\infty} X_i$ and suppose τ_n is the corresponding product measure of $\prod_{i=n+1}^{\infty} X_i$. Show that μ is the same as the product measure of $\mu_1, \mu_2, \ldots, \mu_n, \tau_n$ on the finite Cartesian product

$$X_1 \times X_2 \times \ldots \times X_n \times \left(\prod_{i=n+1}^{\infty} X_i\right).$$

3. The product measure of two σ-finite complete measures need not be complete. As an example take $X_1 = X_2 =$ unit interval with Lebesgue measure. Suppose M is a non-measurable set in X_1, and consider the set $M \times \{y\}$; use exercise 6.1 (7).

4. Suppose $\prod_{i=1}^{\infty} X_i$ is a product space with $\mu_i(X_i) = 1$. Let $E_i \in \mathscr{F}_i$, $(i = 1, 2, \ldots)$. Then the set $\prod_{i=1}^{\infty} E_i$ is in the product σ-field and $\mu(E) = \prod_{i=1}^{\infty} \mu(E_i)$.

5. If a cylinder set $E_1 \times E_2 \times \ldots \times E_n \times \prod_{n+1}^{\infty} X_i$ is in the product σ-field $\mathscr{F}$ generated by $\mathscr{C}(\mathscr{F}_1, \mathscr{F}_2, \ldots)$, then it is in $\mathscr{C}(\mathscr{F}_1, \mathscr{F}_2, \ldots)$; in fact $E_i \in \mathscr{F}_i$ $(i = 1, 2, \ldots, n)$.

6.3 Fubini's theorem

Given two measure spaces $(X, \mathscr{F}, \mu)$, $(Y, \mathscr{G}, \nu)$ we have now seen how to define a product measure on the product σ-field in $X \times Y$. Given a function $f: X \times Y \to \mathbf{R}^*$ there are sections $f_x: Y \to \mathbf{R}^*$ defined for every $x \in X$. Our objective in the present section is to compare the integral of $f(x, y)$ with respect to the product measure with the iterated integral obtained by first integrating $f_x(y)$ with respect to ν for each fixed x, and then integrating the resulting function of x with respect to the measure μ. Because of our method of defining the integral the general result will follow easily from the special case of simple functions. The essential step towards this case is given by the next theorem.

***Theorem* 6.4.** *Given $(X, \mathscr{F}, \mu)$, $(Y, \mathscr{G}, \nu)$ two σ-finite measure spaces, let λ be the product measure defined on the product σ-field $\mathscr{F} * \mathscr{G}$. Then for all $A \in \mathscr{F} * \mathscr{G}$, $\nu(A_x)$ is $\mathscr{F}$-measurable and $\mu(A^y)$ is $\mathscr{G}$-measurable; and*

$$\lambda(A) = \int \mu(A^y)\, d\nu = \int \nu(A_x)\, d\mu.$$

Proof. Suppose first that $\mu(X)$, $\nu(Y)$ are both finite. Let $\mathscr{M}$ be the class of subsets of $X \times Y$ for which the conclusions of the theorem are valid. Then $\mathscr{M} \supset \mathscr{F} \times \mathscr{G}$ since if $A = E_1 \times E_2$, $E_1 \in \mathscr{F}$, $E_2 \in \mathscr{G}$

$\nu(A_x)$ is $\mathscr{F}$-simple as a function of x,

$\mu(A^y)$ is $\mathscr{G}$-simple as a function of y,

and both these functions integrate to $\lambda(A)$ by the definition of λ on $\mathscr{F} \times \mathscr{G}$. It follows that $\mathscr{M}$ contains the ring $\mathscr{R}$ of finite unions of rectangle sets of $\mathscr{F} \times \mathscr{G}$. Since the limit of a monotone sequence of measurable functions is measurable, and theorem 5.6 applies to the integrals, it follows immediately that $\mathscr{M}$ is a monotone class. Hence, by theorem 1.5, $\mathscr{M}$ is a σ-ring. But clearly $\mathscr{M}$ contains $X \times Y$ so that $\mathscr{M}$ is a σ-field and $\mathscr{M} \supset \mathscr{F} * \mathscr{G}$. The restriction $\mu(X) < \infty$, $\nu(Y) < \infty$ can now be removed by the usual device of taking measurable sequences $\{A_n\}$ increasing to X and $\{B_n\}$ increasing to Y for which $\mu(A_n) < \infty$, $\nu(B_n) < \infty$ for all n, and considering the set $A \cap (A_n \times B_n)$ which increases to A as $n \to \infty$. ∎

***Corollary*.** *Under the conditions of theorem* 6.4, *if $A \in \mathscr{F} * \mathscr{G}$, $\lambda(A) = 0$ if and only if $\nu(A_x) = 0$ for almost all x, and if and only if $\mu(A^y) = 0$ for almost all y.*

This follows from the theorem using the fact that a non-negative measurable function can integrate to zero only if it is zero almost everywhere. ∎

***Theorem* 6.5.** *Given all the conditions of theorem 6.4, we write $\mathscr{H}$ for the product σ-field $\mathscr{F} * \mathscr{G}$.*

(i) *If $h: X \times Y \to \mathbf{R}^+$ is any non-negative $\mathscr{H}$-measurable function then*

$$\int h \, d\lambda = \int \left(\int h_x d\nu \right) d\mu = \int \left(\int h_y d\mu \right) d\nu.$$

(ii) *If $h: X \times Y \to \mathbf{R}^*$ is $\mathscr{H}$-measurable and λ-integrable, then $h_x: Y \to \mathbf{R}^*$ is ν-integrable for almost all x and $h_y: X \to \mathbf{R}^*$ is μ-integrable for almost all y. Further*

$$\int h d\lambda = \int f d\mu = \int g d\nu,$$

where $$f(x) = \int h_x d\nu \quad \textit{when } h_x \textit{ is } \nu\textit{-integrable},$$

$$g(y) = \int h_y d\mu \quad \textit{when } h_y \textit{ is } \mu\textit{-integrable}$$

and f, g are defined to be zero on the remaining null sets.

(iii) *If $f: X \times Y \to \mathbf{R}^*$ is $\mathscr{H}$-measurable and $\int (\int |f_x| \, d\nu) \, d\mu$ is finite, then*

$$\int f d\lambda = \int \left(\int f_y d\mu \right) d\nu = \int \left(\int f_x d\nu \right) d\mu.$$

Proof (i). If h is the indicator function of a set in $\mathscr{H}$ the result follows by theorem 6.4. Because of the linearity of the integration process it now follows for non-negative $\mathscr{H}$-simple functions (note that sections of an $\mathscr{H}$-simple function will be simple by theorem 6.1). If we now take a sequence $\{h^{(n)}\}$ of non-negative simple functions increasing to h, we will have the sections $\{h_x^{(n)}\}, \{h_y^{(n)}\}$ increasing to h_x, h_y respectively. Hence, as $n \to \infty$,

$$\int h^{(n)} d\lambda \to \int h \, d\lambda,$$

$$\int h_x^{(n)} d\nu \to \int h_x d\nu \quad \text{for all } x, \qquad \int h_y^{(n)} d\mu \to \int h_y d\mu \quad \text{for all } y,$$

and an application of the monotone convergence theorem (5.6) now suffices to complete the proof.

(ii) Since h is integrable, the positive and negative parts h^+, h^- are integrable. Apply (i) to each of these functions. Then

$$f^+(x) = \int h_x^+ \, d\nu$$

will always be defined, though it may take the value $+\infty$. Since $\int f^+(x)\,d\mu$ exists, we must have f^+ finite except for a set of zero μ-measure. Similarly, f^- is finite almost everywhere. If we put

$$f(x) = f^+(x) - f^-(x)$$

when both f^+, f^- are finite and $f(x) = 0$ otherwise, we see that

$$\begin{aligned}\int h\,d\lambda &= \int h^+ d\lambda - \int h^- d\lambda \\ &= \int f^+ d\mu - \int f^- d\mu = \int f\,d\mu.\end{aligned}$$

(iii) Again split f into positive and negative parts. Since

$$0 \leqslant f^+ \leqslant |f|,$$

we can apply (i) to each of the positive and negative parts to deduce that $\int f^+ d\lambda$ and $\int f^- d\lambda$ are both finite. The result now follows by (ii). ∎

We should remark that theorem 6.5 is one of the most useful tools in the theory of integration as we have developed it. This result again exhibits the power and neatness of the absolute integral.

We have been careful to define the product measure λ on the smallest σ-field $\mathscr{H}$ which contains $\mathscr{F} \times \mathscr{G}$. Some authors define product measure to be the completion of this λ obtained by the process of theorem 4.3. If one uses this definition then some of our statements have to be modified to exclude possible subsets of zero measure, though the essential content of the results remain valid. In particular, given a function $f(x, y)$ which is measurable with respect to the completed σ-field $\mathscr{H}$, one can only say that the section f_x is $\mathscr{G}$-measurable for almost all x. However, provided $\mathscr{F}$ and $\mathscr{G}$ are complete with respect to their respective measures, theorem 6.5 remains valid as stated.

We can use our definition of product measure to give an alternative definition of the integral of a non-negative measurable function.

***Theorem* 6.6.** *Suppose $(\Omega, \mathscr{F}, \mu)$ is a σ-finite measure space, $(\mathbf{R}, \mathscr{L}, \nu)$ denotes the real line with Lebesgue measure on it and τ is the product measure $\mu \times \nu$ defined on the product σ-field $\mathscr{H}$ in $\Omega \times \mathbf{R}$. Then if $E \in \mathscr{F}$ and $f: E \to \mathbf{R}^+$ is non-negative, f is $\mathscr{F}$-measurable over E if and only if $Q(E, f) \in \mathscr{H}$, and in this case,*

$$\int_E f\,d\mu = \tau(Q(E, f));$$

where $Q(E, f)$ is the ordinate set defined by

$$\{(x, y) \colon x \in E,\ y \in \mathbf{R},\ 0 \leqslant y < f(x)\}.$$

Proof. Suppose first that $Q(E,f) \in \mathscr{H}$. Then by theorem 6.1 all its sections are in $\mathscr{F}$. But the section of $Q(E,f)$ at $y = a$ is the set

$$\{x: f(x) > a\},$$

so that by definition, f is $\mathscr{F}$-measurable. Conversely, if f is $\mathscr{F}$-measurable then there is a sequence $\{f_n\}$ of $\mathscr{F}$-simple non-negative functions which increases to f. Now for any $\mathscr{F}$-simple function $Q(E,f_n)$ is a finite union of measurable rectangles and is therefore in $\mathscr{H}$. Also $Q(E,f_n)$ increases monotonely to $Q(E,f)$ so we must have $Q(E,f) \in \mathscr{H}$. Further if

$$f_n = \sum_{i=1}^{r_n} C_{n,i} \chi_{E_{n,i}},$$

with $E_{n,i}$ a disjoint partition of E,

$$\int_E f_n \, d\mu = \Sigma C_{n,i} \mu(E_{n,i}) = \Sigma \tau(E_{n,i} \times [0, C_{n,i})) = \tau(Q(E,f_n)).$$

If we now let $n \to \infty$ we obtain the desired result. ❚

Corollary. *If* $f: \mathbf{R} \to \mathbf{R}^+$ *is* $\mathscr{L}$*-measurable, then the ordinate set* $\{(x,y): a \leqslant x \leqslant b,\ 0 \leqslant y < f(x)\}$ *is* $\mathscr{L}^2$*-measurable and has planar Lebesgue measure* $\int_a^b f(x)\, dx$.

In many elementary accounts of integration the notion of 'area under the curve' is intuitively important. This last corollary makes this notion rigorous for the Lebesgue integral of non-negative functions.

It is possible to consider Euclidean k-dimensional space $\mathbf{R}^k$ as the Cartesian product of k distinct spaces $\mathbf{R}$. Since we have a natural measure $(\mathbf{R}, \mathscr{L}, \nu)$ on each of these spaces we could form the product measure defined on $\mathscr{F}^k$ the product σ-field in $\mathbf{R}^k$ by the process of theorem 6.2. How does this measure compare with Lebesgue measure in $\mathbf{R}^k$? Since all the extension processes used are unique, and the two measures clearly coincide on $\mathscr{P}^k = \mathscr{P} \times \mathscr{P} \times \ldots \times \mathscr{P}$, the half-open rectangles in $\mathbf{R}^k$, it is clear that the two measures coincide whenever both are defined. However, $\mathscr{L}^k$ is complete with respect to Lebesgue measure while $\mathscr{F}^k$ is not known to be so. To see that $\mathscr{F}^k$ is not complete it is sufficient to consider the product of a linear set which is not measurable in $\mathbf{R}$ with $(k-1)$ single point sets. This set cannot be in the product σ-field by exercise 6.1 (7), but it is a subset of a line in $\mathbf{R}^k$ and therefore it must be in $\mathscr{L}^k$. It follows that $\mathscr{L}^k$ is a larger σ-field than $\mathscr{F}^k$. Since $\mathscr{B}^k$, the class of Borel sets in $\mathbf{R}^k$ is the σ-field generated by $\mathscr{P}^k$, we also have $\mathscr{F}^k \supset \mathscr{B}^k$. If E is any set in $\mathscr{L}^1$ but not in $\mathscr{B}^1$ the Cartesian product of E with $(k-1)$ whole lines $\mathbf{R}$ will be in $\mathscr{F}^k$ but not in $\mathscr{B}^k$, so that $\mathscr{F}^k$ is a larger σ-field than $\mathscr{B}^k$.

If we consider the case $k = 2$, a function $f(x,y)$ which is $\mathscr{L}^2$-measurable need not be $\mathscr{F}^2$-measurable. Thus we can only say that the function $f_x(y) = f(x,y)$ considered as a function of y for fixed x is $\mathscr{L}$-measurable for almost all x. Thus in Theorem 6.5 (ii), if $f(x,y)$ is Lebesgue integrable we can deduce that $\phi(x) = \int f(x,y)\,dy$ exists and is finite except for an exceptional set of x of zero measure. As $\phi(x)$ is thus defined a.e. it can be integrated and

$$\int\int f(x,y)\,dx\,dy = \int \phi(x)\,dx.$$

Exercises 6.3

1. Suppose Ω is any set of cardinal greater than $\aleph_0$, and $\mathscr{F}$ is the σ-field of sets in Ω which are either countable or have a countable complement. For $E \in \mathscr{F}$, put $\mu(E) = 0$ if E is countable, $\mu(E) = 1$ if $(\Omega - E)$ is countable. Consider the Cartesian product of two copies of Ω and let E be a set in $\Omega \times \Omega$ which has countable x-sections for every x and y-sections whose complement is countable for every y. If h is the indicator function of E, then

$$\int h_y(x)\,\mu(dx) = 1, \quad \int h_x(y)\,\mu(dy) = 0.$$

Why does this not contradict theorem 6.4?

2. Suppose $(X, \mathscr{F}, \mu)\,(Y, \mathscr{G}, \nu)$ are σ-finite measure spaces and λ is the product measure on the product σ-field $\mathscr{H}$. Show that

(i) If $E, G \in \mathscr{H}$ are such that $\nu(E_x) = \nu(G_x)$ for almost all $x \in X$, then $\lambda(E) = \lambda(G)$.

(ii) If f, g are integrable functions on X, Y then $f(x)\,g(y)$ is integrable on $X \times Y$ and

$$\int f(x)\,g(y)\,d\lambda = \int f\,d\mu \int g\,d\nu.$$

3. $X = Y = [0,1]$ and $\mathscr{F}, \mathscr{G}$ are the Borel subsets. Let $\mu(E)$ be the Lebesgue measure of E, $\nu(E)$ the number of points in E. Form the product measure $\mu \times \nu$ on Borel subsets of the unit square. Then if D is the diagonal $\{(x,y);\ x = y\}$, D is measurable and

$$\int \nu(D_x)\,\mu(dx) = 1, \quad \int \mu(D^y)\,\nu(dy) = 0.$$

Why does this not contradict theorem 6.4?

4. If $f(x,y) = (x^2 - y^2)/(x^2 + y^2)^2$ show that

$$\int_0^1 \left\{\int_0^1 f(x,y)\,dy\right\} dx = \frac{\pi}{4},$$

$$\int_0^1 \left\{\int_0^1 f(x,y)\,dx\right\} dy = -\frac{\pi}{4},$$

where all the integrals are taken in the Lebesgue sense. Thus theorem 6.5 (iii) is not valid without the modulus sign. Similarly, show that

$$\int_0^1 \left\{ \int_1^\infty (e^{-xy} - 2e^{-2xy})\, dx \right\} dy \neq \int_1^\infty \left\{ \int_0^1 (e^{-xy} - 2e^{-2xy})\, dy \right\} dx.$$

5. If $f(x,y) = xy/(x^2+y^2)^2$, then

$$\int_{-1}^{+1} \left\{ \int_{-1}^{+1} f(x,y)\, dx \right\} dy = 0 = \int_{-1}^{+1} \left\{ \int_{-1}^{+1} f(x,y)\, dy \right\} dx$$

but the integral over the unit square in $\mathbf{R}^2$ does not exist.

6. Given a countable collection of probability spaces $(X_i, \mathscr{F}_i, \mu_i)$ and the product measure μ on the product σ-field, we can form the finite product measures $\tau_n = \mu_1 \times \mu_2 \times \ldots \times \mu_n$ and the product measure λ_n on the product space $\prod_{i=n+1}^{\infty} X_i$. Then, if $f(x_1, x_2, \ldots)$ is any μ-integrable function on $\prod_{i=1}^{\infty} X_i$ we have

$$\int f\, d\mu = \int \left\{ \int f(x_1, x_2, \ldots)\, d\lambda_n \right\} d\tau_n.$$

6.4 Radon–Nikodym theorem

We start with a definition.

Absolute continuity

Suppose $\mathscr{F}$ is a σ-ring of subsets of Ω and μ is a measure on $\mathscr{F}$. Then the set function $\nu: \mathscr{F} \to \mathbf{R}^*$ is said to be absolutely continuous with respect to μ if $\nu(E) = 0$ for every E in $\mathscr{F}$ with $\mu(E) = 0$. In this case we write $\nu \ll \mu$. If $(\Omega, \mathscr{F}, \mu)$ is a measure space and $f: \Omega \to \mathbf{R}^*$ is μ-integrable, then it is clear that

$$\nu(E) = \int_E f\, d\mu \quad (E \in \mathscr{F}),$$

defines a finite valued absolutely continuous set function ν. In fact, in § 5.4 we proved that ν was σ-additive, and that (corollary to theorem 5.6) given $\epsilon > 0$, there is a $\delta > 0$ such that for $E \in \mathscr{F}$,

$$\mu(E) < \delta \Rightarrow |\nu(E)| < \epsilon. \tag{6.4.1}$$

It is immediate that any set function ν which satisfies (6.4.1) is absolutely continuous with respect to μ. The conditions are equivalent for finite measures, but not in general (see exercise 6.4 (4)). There is a partial converse given by:

Lemma. *If $(\Omega, \mathscr{F}, \mu)$ is a measure space and $\nu: \mathscr{F} \to \mathbf{R}$ is finite valued, σ-additive and absolutely continuous with respect to μ, then ν satisfies condition* (6.4.1).

Proof. By the decomposition of § 3.2, any such ν is the difference of two finite measures, so it is sufficient to prove the result for a measure ν. Then if (6.4.1) is false, there is an $\epsilon > 0$ and a sequence $\{E_n\}$ of sets of $\mathscr{F}$ such that $\nu(E_n) > \epsilon$ and $\mu(E_n) < 2^{-n}$. Put $E = \limsup E_n$. Then

$$\mu(E) \leqslant \mu\left(\bigcup_{r=n+1}^{\infty} E_r\right) \leqslant \sum_{n+1}^{\infty} \mu(E_r) < 2^{-n},$$

so that $\mu(E) = 0$ while

$$\nu(E) = \lim \nu\left(\bigcup_{r=n+1}^{\infty} E_r\right) \geqslant \limsup \nu(E_r)$$

so that $\nu(E) \geqslant \epsilon$. This contradicts $\nu \ll \mu$. ∎

Thus we see that the indefinite integral of an integrable function defines an absolutely continuous set function. Our object in the present section is to obtain the converse of this statement under suitable conditions. It is convenient at the same time to consider a more general σ-additive set function and to decompose it into a maximal absolutely continuous component and a remainder which has to be concentrated on a μ-null set. It is convenient to give a further definition.

Singular set function

Given a measure space $(\Omega, \mathscr{F}, \mu)$ a set function ν: $\mathscr{F} \to \mathbf{R}^*$ is said to be singular with respect to μ if there is a set $E_0 \in \mathscr{F}$ for which $\mu(E_0) = 0$ and

$$\nu(E) = \nu(E \cap E_0), \quad \text{all } E \in \mathscr{F}. \tag{6.4.2}$$

This condition clearly means that the parts of Ω outside the null set E_0 make no contribution to ν. In fact if ν is also a measure we see that Ω can be dissected into two sets E_0, $E_1 \in \mathscr{F}$ such that

$$E_0 \cap E_1 = \varnothing, \; E_0 \cup E_1 = \Omega, \; \mu(E_0) = 0, \; \nu(E_1) = 0.$$

The symmetry of the relationship in this case is sometimes stressed by saying that μ and ν are *mutually singular*.

***Theorem* 6.7.** *Given a σ-finite measure space $(\Omega, \mathscr{F}, \mu)$ and a σ-additive, σ-finite set function ν, then there is a unique decomposition*

$$\nu = \nu_1 + \nu_2$$

into σ-additive set functions ν_i which are σ-finite and such that ν_1 is singular with respect to μ and $\nu_2 \ll \mu$. Further there is a finite valued measurable f: $\Omega \to \mathbf{R}$ such that

$$\nu_2(E) = \int_E f\,d\mu, \quad \textit{all } E \in \mathscr{F}.$$

The function f is unique in the sense that if we also have

$$\nu_2(E) = \int_E g\,d\mu$$

for all E in $\mathcal{F}$, then $f(x) = g(x)$ except in a set of zero μ-measure.

Corollary. *Under the conditions of the theorem if $\nu \ll \mu$ then there is a finite valued $f: \Omega \to \mathbf{R}$ such that*

$$\nu(E) = \int_E f\,d\mu \quad \text{for} \quad E \in \mathcal{F}.$$

Note. The decomposition of ν into absolutely continuous and singular components is often called the Lebesgue decomposition, while the integral representation is called the Radon–Nikodym theorem.

Proof. Since we can express Ω as a union of a countable set of disjoint sets on each of which both μ and ν are finite, there is no loss in generality in assuming that they are both finite on Ω. This applies to both the existence and uniqueness proofs. We first see that the decomposition is unique.

Let

$$\nu = \nu_1 + \nu_2 = \nu_3 + \nu_4,$$

where ν_1, ν_3 are singular and ν_2, ν_4 are absolutely continuous. Then

$$\nu_1 - \nu_3 = \nu_4 - \nu_2,$$

Taking the union of support sets of ν_1, ν_3 gives a set E_0 such that

$$(\nu_1 - \nu_3)(E) = (\nu_1 - \nu_3)(E \cap E_0), \quad \mu(E_0) = 0.$$

But $(\nu_4 - \nu_2)$ is absolutely continuous and therefore zero on any null set so that, for any $E \in \mathcal{F}$,

$$(\nu_4 - \nu_2)(E) = (\nu_1 - \nu_3)(E) = (\nu_1 - \nu_3)(E \cap E_0)$$
$$= (\nu_4 - \nu_2)(E \cap E_0) = 0.$$

Thus $\nu_1(E) = \nu_3(E)$, $\nu_2(E) = \nu_4(E)$ for all E. The uniqueness of the integral representation of ν_2 was proved in §5.4. Thus it is sufficient to find any decomposition and integral representation.

By theorem 3.3 we can decompose ν into the difference of two measures. It is therefore sufficient to prove the theorem when ν is a measure. Now let $\mathcal{H}$ be the class of non-negative measurable

$$f: \Omega \to \mathbf{R}^+$$

such that

$$\nu(E) \geq \int_E f\,d\mu \quad \text{for all } E \text{ in } \mathcal{F}$$

and put
$$\alpha = \sup\left\{\int f\,d\mu : f \in \mathscr{H}\right\}.$$

Let $\{f_n\}$ be a sequence of functions in $\mathscr{H}$ such that
$$\int f_n\,d\mu > \alpha - \frac{1}{n}.$$

Put $g_n(x) = \max\{f_1(x), f_2(x), \ldots, f_n(x)\}$. Then if $E \in \mathscr{F}$, and n is fixed we can decompose E into a disjoint union $E_1 \cup E_2 \cup \ldots \cup E_n$ of sets of $\mathscr{F}$ such that $g_n = f_j$ on E_j. Hence
$$\int_E g_n\,d\mu = \sum_{j=1}^{n} \int_{E_j} g_n\,d\mu = \sum_{j=1}^{n} \int_{E_j} f_j\,d\mu \leqslant \sum_{j=1}^{n} \nu(E_j) = \nu(E),$$
so that $g_n \in \mathscr{H}$ for all n. But $\{g_n\}$ is monotone increasing, and by the monotone convergence theorem, $f_0(x) = \lim\limits_{n\to\infty} g_n(x) \in \mathscr{H}$. Since
$$f_0(x) \geqslant f_n(x) \quad \text{for all } n,$$
we must have
$$\alpha = \int f_0(x)\,d\mu.$$
For each E in $\mathscr{F}$, put
$$\nu_2(E) = \int_E f_0\,d\mu, \quad \nu_1(E) = \nu(E) - \nu_2(E).$$

Then ν_2 is absolutely continuous with respect to μ, so it only remains to show that ν_1 is singular.

Consider the σ-additive set function
$$\lambda_n = \nu_1 - (1/n)\,\mu$$
and decompose Ω, using theorem 3.3. into positive and negative sets P_n, N_n such that $P_n \cup N_n = \Omega$, $P_n \cap N_n = \varnothing$, $E \subset P_n \Rightarrow \lambda_n(E) \geqslant 0$, $E \subset N_n \Rightarrow \lambda_n(E) \leqslant 0$. Then, for $E \subset P_n$,
$$\nu(E) = \nu_1(E) + \nu_2(E) \geqslant \nu_2(E) + \frac{1}{n}\mu(E) = \int_E \left(f_0 + \frac{1}{n}\right) d\mu.$$

This shows that the function equal to f_0 on N_n and $[f_0 + (1/n)]$ on P_n is in $\mathscr{H}$. This will give a larger integral than α unless $\mu(P_n) = 0$. If $P = \bigcup_{n=1}^{\infty} P_n$, then $\mu(P) = 0$. Further $\Omega - P \subset N_n$ for all n so that $\nu_1(\Omega - P) = 0$ and
$$\nu_1(E) = \nu_1(E \cap P) \quad \text{for all } E \text{ in } \mathscr{F},$$
that is, ν_1 is μ-singular.

In the case where $\nu \ll \mu$, by the uniqueness of the decomposition we must have $\nu = \nu_2$, and the integral representation of ν now follows. ▌

Remark. In the statement of theorem 6.7 we do not assert that the function f is integrable. A necessary and sufficient condition that f be integrable is that ν be finite. However, the use of the symbol $\int_E f d\mu$ asserts that either f_+ or f_- has a finite integral. This corresponds to the result of theorem 3.2 that ν cannot take both the values $\pm\infty$.

Derivative of a set function

If $(\Omega, \mathscr{F}, \mu)$ is a σ-finite measure space and

$$\nu(E) = \int_E f d\mu \quad \text{for} \quad E \text{ in } \mathscr{F},$$

then we write $f = d\nu/d\mu$ and call f the Radon–Nikodym derivative of ν with respect to μ.

One should emphasise that the derivative $d\nu/d\mu$ is not defined uniquely at any given point, it has to be considered as a function and then it becomes uniquely defined in the sense that any two functions representing the same derivative can differ only on a μ-null set.

Exercises 6.4

1. Show that if μ, ν are any two measures on a σ-ring $\mathscr{S}$, then $\nu \ll \mu + \nu$.

2. Suppose $F(x)$ is the Cantor function defined in §2.7 and ν is the Lebesgue–Stieltjes measure with respect to F. Show that ν is singular with respect to Lebesgue measure.

3. Suppose $(\Omega, \mathscr{F}, \mu)$ is a measure space with $\mu(\Omega) < \infty$ and ν is a measure, $\nu \ll \mu$. Show there is a set E such that $(\Omega - E)$ has σ-finite ν measure and for every measurable $F \subset E$, $\nu(F)$ is either 0 or ∞.

4. Let Ω be the set of positive integers,

$$\mu(E) = \sum_{n \in E} 2^{-n}, \quad \nu(E) = \sum_{n \in E} 2^n$$

then $\nu \ll \mu$, but (6.4.1) is not satisfied. This shows that (6.4.1) is a stronger condition than absolute continuity when ν is not finite.

5. Suppose Ω is an uncountable set, $\mathscr{F}$ is the class of sets which are either countable or have countable complements. For $E \in \mathscr{F}$, put $\mu(E) =$ the number of points in E, $\nu(E) = 0$ or 1 according as E is countable or not. Then clearly $\nu \ll \mu$, but no integral representation is possible. This shows that in the Radon–Nikodym theorem we cannot do without the condition that μ be σ-finite.

6. If λ, μ, ν are σ-finite measures on $\mathscr{F}$ and $\lambda \ll \mu$, $\mu \ll \nu$; show that $\lambda \ll \nu$ and

$$\frac{d\lambda}{d\nu} = \frac{d\lambda}{d\mu}\frac{d\mu}{d\nu}$$

except on a set of zero λ-measure.

7. λ, μ are σ-finite measures on $\mathscr{F}$ with $\mu \ll \lambda$. Then if f is μ-integrable

$$\int f\,d\mu = \int f \frac{d\mu}{d\lambda}\,d\lambda.$$

8. If λ, μ are σ-finite measures on $\mathscr{F}$ such that $\mu \ll \lambda$ and $\lambda \ll \mu$ then

$$\frac{d\mu}{d\lambda} = \left(\frac{d\lambda}{d\mu}\right)^{-1},$$

except for a set of zero λ-measure.

9. If μ, ν are σ-finite measures on $\mathscr{F}$ such that $\nu \ll \mu$, show that the set of points x at which $d\nu/d\mu$ is zero has zero ν-measure.

10. Suppose $\{\mu_i\}$ is a countable family of finite measures on a σ-field $\mathscr{F}$. Show that there exists a finite μ on $\mathscr{F}$ such that each of the μ_i is absolutely continuous with respect to μ.

11. Suppose

$$\bar{\mu}_n = \sum_{k=1}^{n} \mu_k \to \bar{\mu}, \quad \bar{\nu}_n = \sum_{k=1}^{n} \nu_k \to \bar{\nu},$$

where all the μ, ν with suffices are finite measures on a σ-field $\mathscr{F}$ and $\bar{\nu}_n$ is $\bar{\mu}_n$-continuous for all n. Show that

(i) $d\mu_1/d\bar{\mu}_n \to d\mu_1/d\bar{\mu}$ almost everywhere ($\bar{\mu}$).

(ii) If each μ_n is ν-continuous then $d\bar{\mu}_n/d\nu \to d\bar{\mu}/d\nu$ a.e. (ν).

(iii) $\bar{\nu}$ is $\bar{\mu}$-continuous and $d\bar{\nu}_n/d\bar{\mu}_n \to d\bar{\nu}/d\bar{\mu}$ a.e. ($\bar{\mu}$).

6.5 Mappings of measure spaces

In mathematical arguments one often needs to consider two spaces, X, Y with a mapping f: $X \to Y$. Such a mapping induces mappings on the classes of subsets of X and Y: if $E \subset X$, $f(E)$ denotes the set of y in Y with $y = f(x)$, $x \in E$, and if $F \subset Y$, $f^{-1}(F)$ denotes the set of x in X with $f(x) \in F$; further if $\mathscr{C}$ is a class of subsets of X, $f(\mathscr{C})$ denotes the class of sets $f(E)$ with $E \in \mathscr{C}$, and similarly for $f^{-1}(\mathscr{E})$ where $\mathscr{E}$ is a class of subsets of Y. We saw (§1.5) that f^{-1} preserves the structure of a class of subsets, so that if $\mathscr{S}$ is a σ-field in Y, $f^{-1}(\mathscr{S})$ is a σ-field in X. Sometimes the two spaces X, Y already have classes of subsets defined, and one can then examine the relationship of the mapping f to these.

Measurable transformation

Suppose f is a mapping from X into Y, $\mathscr{F}$ is a σ-field in X and $\mathscr{G}$ is a σ-field in Y, then we say that f is a measurable transformation from $(X, \mathscr{F})$ into $(Y, \mathscr{G})$ if $f^{-1}(E) \in \mathscr{F}$ for every E in $\mathscr{G}$. This condition can also be written $f^{-1}(\mathscr{G}) \subset \mathscr{F}$.

In Chapter 5 we discussed 'measurable functions'. In our new terminology these are measurable transformations from $(X, \mathscr{F})$ to $(\mathbf{R}^*, \mathscr{G})$ in which $\mathscr{G}$ is the σ-field of Borel sets in $\mathbf{R}^*$. Given mappings $f: X \to Y$, $g: Y \to Z$ we can consider the composition $g(f): X \to Z$ defined by $g(f)(x) = g(f(x))$. In particular if $g: Y \to \mathbf{R}^*$ is an extended real-valued function on Y, then $g(f)$ defines an extended real function on X.

Lemma. *If $f: X \to Y$ is a measurable transformation from $(X, \mathscr{F})$ into $(Y, \mathscr{G})$ and $g: Y \to \mathbf{R}^*$ is $\mathscr{G}$-measurable as a function with extended real values, then the composition $g(f)$ is $\mathscr{F}$-measurable.*

Proof. For any Borel set B in $\mathbf{R}^*$ we have

$$\begin{aligned}\{x: g(f)(x) \in B\} &= f^{-1}\{y: g(y) \in B\} \\ &= f^{-1}(E) \quad \text{for some } E \in \mathscr{G},\end{aligned}$$

and is therefore in $\mathscr{F}$. ∎

Remark. We obtained a special case of this lemma when we proved that a Borel measurable function of a measurable function is measurable (see §5.2).

If we start with a measure space $(X, \mathscr{F}, \mu)$ and f is a measurable transformation from $(X, \mathscr{F})$ into $(Y, \mathscr{G})$ it is natural to use f to define a measure ν on $\mathscr{G}$ by putting

$$\nu(E) = \mu(f^{-1}(E)) \quad \text{for} \quad E \in \mathscr{G}. \tag{6.5.1}$$

With this definition of ν it is immediate that $(Y, \mathscr{G}, \nu)$ is a measure space. If (6.5.1) holds we will write $\nu = \mu f^{-1}$. This allows us to carry out a 'change of variable' in an integral.

Theorem 6.8. *Suppose f is a measurable transformation from a measure space $(X, \mathscr{F}, \mu)$ to $(Y, \mathscr{G})$ and $g: Y \to \mathbf{R}^*$ is $\mathscr{G}$-measurable: then*

$$\int_Y g\, d(\mu f^{-1}) = \int_X g(f)\, d\mu$$

in the sense that if either integral exists so does the other and the two are equal

Proof. It is clearly sufficient to consider non-negative functions

$g: Y \to \mathbf{R}^+$. Suppose first that $g = \chi_E$, the indicator function of a set E in $\mathscr{G}$. Then

$$g(f)(x) = 1 \quad \text{if} \quad x \in f^{-1}(E),$$
$$= 0 \quad \text{if} \quad x \notin f^{-1}(E);$$

so that $g(f)$ is the indicator function of $f^{-1}(E)$, a set in $\mathscr{F}$. Thus, in this case, by (6.5.1)

$$\int g\, d(\mu f^{-1}) = \mu f^{-1}(E) = \mu(f^{-1}(E)) = \int g(f)\, d\mu.$$

By linearity, the result now follows for non-negative $\mathscr{G}$-simple functions g. If $\{g_n\}$ is an increasing sequence of non-negative simple functions converging to the measurable function g, then $g_n(f)$ will be an increasing sequence of simple functions converging to $g(f)$. The definition of the integral of a non-negative function now completes the proof.]

Sometimes in integration, when the variable is changed, one wants to integrate with respect to a new measure $\nu \neq \mu f^{-1}$. We can do this easily when μf^{-1} is absolutely continuous with respect to ν.

***Theorem* 6.9.** *Given σ-finite measure spaces $(X, \mathscr{F}, \mu)$ and $(Y, \mathscr{G}, \nu)$ and a measurable transformation f from $(X, \mathscr{F})$ into $(Y, \mathscr{G})$ such that μf^{-1} is absolutely continuous with respect to ν*

$$\int g(f)\, d\mu = \int g.\phi\, d\nu,$$

where ϕ is the Radon–Nikodym derivative $d(\mu f^{-1})/d\nu$, for every measurable $g: Y \to \mathbf{R}^*$ in the sense that, if either integral exists, so does the other and the two are equal.

Corollary. *If $q: \mathbf{R} \to \mathbf{R}^+$ is Lebesgue integrable,*

$$F(x) = \int_{-\infty}^{x} q(t)\, dt,$$

and μ_F is the Lebesgue–Stieltjes measure generated by F, then

$$\int_A^B g(x)\, dx = \int_a^b g(F(t))\, d\mu_F = \int_a^b g(F(t))\, q(t)\, dt$$

where $A = F(a)$, $B = F(b)$.

Proof. By theorem 6.8 we have

$$\int g(f)\, d\mu = \int g\, d(\mu f^{-1}).$$

Since μf^{-1} is absolutely continuous with respect to ν, there exists a measurable ϕ such that, for every $E \in \mathscr{G}$

$$\int_E \phi d\nu = (\mu f^{-1})(E).$$

If g is the indicator function of a measurable set E it now follows that

$$\int g d(\mu f^{-1}) = (\mu f^{-1})(E) = \int g.\phi d\nu$$

and the required result now follows by successive extension to functions g which are: (i) non-negative, simple, (ii) non-negative, measurable, (iii) measurable.

Under the conditions of the corollary we consider the mapping $F: \mathbf{R} \to \mathbf{R}$ given by the measure function F from the Lebesgue measure space $(\mathbf{R}, \mathscr{L}, \mu_F)$ to $(\mathbf{R}, \mathscr{B}, \mu)$. Theorem 6.8 then gives the first equality. If we define

$$\lambda(E) = \int_E q d\mu,$$

then $\lambda: \mathscr{L} \to \mathbf{R}$ is a measure which coincides with

$$\mu_F(E) = \int_E 1 d\mu_F$$

for intervals of $\mathscr{P}$ and therefore for all sets $E \in \mathscr{L}$. Hence the measure μ_F is absolutely continuous with respect to Lebesgue measure μ and q is a possible definition of the Radon–Nikodym derivative $d\mu_F/d\mu$. The second equality now follows from theorem 6.9. ▌

Remark. It is clear from the above that the function ϕ (or q in the corollary) plays the part of the Jacobian (or rather the absolute value of the Jacobian) in the theory of transformations of multiple integrals. In general it is not easy to obtain an explicit value for the Radon–Nikodym derivative $d(\mu f^{-1})/d\nu$, but in important special cases this can be done. In particular, if both spaces are $(\mathbf{R}^k, \mathscr{L}^k, \mu)$ with μ Lebesgue measure, and $f: \mathbf{R}^k \to \mathbf{R}^k$ is a linear transformation given by a non-singular matrix A so that $\mathbf{y} = A\mathbf{x}$ one can prove that

$$|f(E)| = \|A\| . |E|,$$

where $\|A\|$ denotes the absolute value of the determinant of A. (This can best be shown by expressing A as a product of elementary transformations, and proving the result for each elementary transformation.) This means that, in this case a possible Radon–Nikodym derivative is the constant function $\|A\|$.

Exercises 6.5

1. Show that the composition of two measurable transformations is measurable.

2. If f is a measurable transformation from $(X, \mathscr{F})$ into $(Y, \mathscr{S})$ and μ, ν are two measures on $\mathscr{F}$ such that $\mu \ll \nu$, show that $\mu f^{-1} \ll \nu f^{-1}$.

3. (*Integration by parts.*) If $F(x)$, $G(x)$ are non-negative continuous functions satisfying the conditions of §4.5 for a Stieltjes measure function and E is any Borel set, then

$$\int_E F(x)\, dG(x) + \int_E G(x)\, dF(x) = \mu_{FG}(E),$$

where μ_{FG} denotes the Lebesgue–Stieltjes measure generated by $F(x)\,G(x)$. In particular if

$$F(x) = \int_a^x f(t)\, dt, \quad G(x) = \int_a^x g(t)\, dt,$$

then
$$\int_a^b F(x)\, g(x)\, dx + \int_a^b f(x)\, G(x)\, dx = F(b)\, G(b) - F(a)\, G(a).$$

4. Suppose A is a non-singular $k \times k$-matrix defining a mapping from $\mathbf{R}^k$ to $\mathbf{R}^k$, then this is a measurable transformation from $(R^{(k)}, \mathscr{L}^{(k)})$ to itself. If μ_k denotes Lebesgue measure in $\mathbf{R}^k$ show that

$$\int f\, d\mu_k = \|A\| \int f(A)\, d\mu_k,$$

for any Lebesgue measurable f, where $f(A)$ denotes the composite map $f(A)(x) = f(Ax)$.

6.6* Measure in function space

We saw that points in the product space $\prod_{i \in I} X_i$ can be thought of as functions $f: I \to \bigcup_{i \in I} X_i$ in which $f(i) \in X_i$. In the particular case where X_i is the same space X for all i, the space $\prod_{i \in I} X_i$ reduces to the set of functions: $I \to X$. For this reason such a product space is often denoted by X^I. Since theorem 6.3 clearly extends to arbitrary product spaces we can produce a product measure in X^I starting from any measure μ on X with $\mu(X) = 1$. However, for non-countable I, such product measures are rarely of interest. In applications, the space X^I usually describes a stochastic process (see Chapter 15), and the product measure in X^I would correspond to complete independence (see Chapter 11) between the values in each of the coordinate spaces. Usually one wants to be able to define and use measures in X^I which are not product measures.

In our account we restrict X to be the real line $\mathbf{R}$ (it is easy to extend the theory to the case $X = \mathbf{C}$, but some restriction is needed for its validity), leaving the index set I completely arbitrary.

Borel sets in $\mathbf{R}^I$

If we assume the usual topology in $\mathbf{R}$, and denote the class of Borel sets in $\mathbf{R}$ by $\mathscr{B}$, then the class $\mathscr{C}$ of cylinder sets

$$\{f \in \mathbf{R}^I : f(i_k) \in B_k, \quad k = 1, 2, \ldots, n\}, \quad B_k \in \mathscr{B}$$

is a semi-ring of subsets in $\mathbf{R}^I$. The σ-field generated by $\mathscr{C}$ will be denoted by $\mathscr{B}^I$. If $\mathscr{B}^n$ denotes the class of Borel sets in $\mathbf{R}^n$, it is immediate that $\mathscr{B}^I$ can also be generated by the class of sets of the form

$$\{f \in \mathbf{R}^I : a_k < f(i_k) \leqslant b_k, \quad k = 1, 2, \ldots, n\}, \tag{6.6.1}$$

or of the form

$$\{f \in \mathbf{R}^I : (f(i_1), f(i_2), \ldots, f(i_n)) \in B^n\}, \quad B^n \in \mathscr{B}^n. \tag{6.6.2}$$

It is important to notice that no set in $\mathscr{B}^I$ can have restrictions on an uncountable set of coordinates. For, if E is a countable subset of I and $F = I - E$, a set of the form

$$\{f \in \mathbf{R}^I : f_E \in \mathscr{B}^E\}, \tag{6.6.3}$$

where f_E denotes the restriction of f to E, contains functions f which are not restricted on F. The class of subsets of $\mathbf{R}^I$ of the form (6.6.3) (for all possible countable sets $E \subset I$) is clearly a σ-field which contains the finite dimensional cylinder sets $\mathscr{C}$. Further, every set of the form (6.6.3) must be in $\mathscr{B}^I$, so that the Borel sets in $\mathbf{R}^I$ are precisely the sets of the form (6.6.3).

Our object will be to extend a measure which is already defined on sets of the form (6.6.1) to the σ-field $\mathscr{B}^I$. For a fixed finite set $i_1, i_2, \ldots, i_n \in I$, the sets of the form (6.6.1) clearly generate a σ-field containing those sets of $\mathbf{R}^I$ obtained by taking a Borel set in $\mathbf{R}_{i_1} \times \mathbf{R}_{i_2} \times \ldots \times \mathbf{R}_{i_n}$ and forming the cylinder with this set as base. If we are to have $\mu(\mathbf{R}^I) = 1$, then, for each fixed $i_1, i_2, \ldots, i_n$, our set function on sets of the form (6.6.2) must define a measure on the Borel sets of the Euclidean n-space $\mathbf{R}_{i_1} \times \ldots \times \mathbf{R}_{i_n}$ in which the whole space has measure 1.

It is clear that the measures given in the various Euclidean spaces of this type have to satisfy various *consistency relations*, if there is to be any hope of extending to a single measure on the whole of $\mathscr{B}^I$. For such a measure on $\mathscr{B}^I$ must yield the original system on restriction to sets of the form (6.6.2). These consistency conditions can be stated in terms of multidimensional distribution functions which generate the measures on sets (6.6.2), but we prefer to state them (equivalently) in terms of the measures.

We assume then that for each finite set of distinct indices $i_1, i_2, \ldots, i_n$ we have a measure $\mu i_1 i_2 \ldots i_n$ defined for the Borel sets in $\mathbf{R}^n$ such that

(I) $\mu i_1 \ldots i_n i_{n+1}(A \times \mathbf{R}) = \mu i_1 \ldots i_n(A)$, $A \in \mathscr{B}^n$.

(II) If π is a permutation of $(1, 2, \ldots, n)$ and ϕ: $\mathbf{R}^n \to \mathbf{R}^n$ is the mapping

$$\phi(x_1, \ldots, x_n) = (x_{\pi_1}, x_{\pi_2}, \ldots, x_{\pi_n}),$$

then

$$\mu i_{\pi_1} i_{\pi_2} \ldots i_{\pi_n} = \mu i_1 i_2 \ldots i_n \phi^{-1}.$$

The condition (I) says that putting on the additional condition $f(i_{n+1}) \in \mathbf{R}$ at a new index cannot effect the measure of the set since it imposes no restriction, and condition (II) makes precise the notion that the order in which the index set $i_1, i_2, \ldots, i_n$ is written should not have any effect on the measure of the (same) set. Both these consistency conditions are clearly necessary if there is to be any hope of extending the measures $\mu i_1 \ldots i_n$ to a single measure μ on $\mathbf{R}^I$. The fact that they are also sufficient was proved by Daniell in 1918 and rediscovered by Kolmogorov in 1933. We state it as

***Theorem* 6.10.** *If I is any infinite index set, and for each finite set $i_1, i_2, \ldots, i_n$ of different indices in I there is a measure $\mu i_1 i_2 \ldots i_n$ defined on the Borel subsets of $\mathbf{R}^n$ such that the family of all such measures satisfies the consistency conditions* (I) *and* (II), *there is a unique measure μ defined on $\mathscr{B}^I$ in $\mathbf{R}^I$ such that, for each $n \in \mathbf{Z}$, $B^n \in \mathscr{B}^n$,*

$$\mu\{f \in \mathbf{R}^I : (f(i_1), \ldots, f(i_n)) \in B^n\} = \mu i_1 i_2 \ldots i_n (B^n).$$

Proof. Let $\mathscr{S}$ denote the semi-ring of sets in $\mathbf{R}^I$ of the form (6.6.1) for some finite value of n. Let $\mathscr{R}$ denote the ring generated by $\mathscr{S}$ consisting of finite unions of disjoint sets in $\mathscr{S}$. Now $\mu i_1 i_2 \ldots i_n$ defines the measure of the set

$$\{f \in \mathbf{R}^I : a_k < f(i_k) \leqslant b_k, \quad k = 1, \ldots, n\} \tag{6.6.4}$$

and the consistency conditions (I) and (II) clearly ensure that the measure is uniquely defined and additive on $\mathscr{S}$ (for the sets of any finite class of sets in $\mathscr{S}$ can all be described by restrictions on the same finite set of coordinates, and therefore the measure can be given by a single measure of the family). It follows, by theorem 3.1, that there is a set function τ defined on the ring $\mathscr{R}$ which is additive and coincides with the measure $\mu i_1 \ldots i_n$ on a set of the form (6.6.1).

Further $\mathscr{B}^I$ is the σ-field generated by $\mathbf{R}$ and we can obtain the required measure μ on $\mathscr{B}^I$ by applying theorem 4.2 to the measure τ

—provided the conditions of that theorem are satisfied. It is immediate that $\mathbf{R}^I$ is σ-$\mathscr{R}$, for $\mathbf{R}^I \in \mathscr{S}$, and $\tau(\mathbf{R}^I) = 1$; so that the only condition which requires proof is that τ is a measure on $\mathscr{R}$. The proof of this fact is an extension of the method used in §§3.4, 4.5.

If τ is not a measure on $\mathscr{R}$, we can find a decreasing sequence $\{E_n\}$ of sets in $\mathscr{R}$ such that $\bigcap_{n=1}^{\infty} E_n = \varnothing$, but $\tau(E_n) \geqslant \delta > 0$ for all n. Now given any set C of the form (6.6.4), and $\epsilon > 0$, we can choose $\eta > 0$ such that

$$\tau(D) > \tau(C) - \epsilon$$

where
$$D = \{f \in \mathbf{R}^I : (f(i_1), f(i_2), \ldots, f(i_n)) \in P_\eta\}$$

and
$$P_\eta = \{a_k + \eta < x_k \leqslant b_k, \quad k = 1, 2, \ldots, n\},$$

since $\mu i_1 i_2 \ldots i_n$ is a measure. But now $\bar{P}_\eta \subset P_0$. This argument clearly extends to any non-empty subset in $\mathscr{R}$, and we can apply it by induction to the sequence $\{E_n\}$. Since in each of the sets E_n the value of f at only a finite set of indices is restricted, there is no loss of generality in assuming that in the sets $E_1, E_2 \ldots, E_n$ there is a restriction on f only at the first n of the indices in the sequence

$$i_1, i_2, \ldots, i_n, \ldots.$$

(If this condition is not satisfied one need only add additional sets in $\mathscr{R}$ to the sequence $\{E_n\}$ to obtain a new sequence of which the original is a subsequence.)

Thus we may assume that

$$E_n = \{f \in \mathbf{R}^I : (f(i_1), \ldots, f(i_n)) \in Q_n\},$$

where $Q_n \in \mathscr{E}^n$ the class of elementary figures in $\mathbf{R}^n$. The condition that E_n be a decreasing sequence now means that $Q_{n+1} \subset Q_n \times \mathbf{R}$. We apply the above procedure to each of the sets E_n to give a sequence $\{D_n\}$ of sets

$$D_n = \{f \in \mathbf{R}^I : (f(i_1), \ldots, f(i_n)) \in \bar{P}_n\}$$

such that $\bar{P}_n \subset Q_n$, $P_n \in \mathscr{E}^n$ and

$$\tau(D_n) > \tau(E_n) - \frac{\delta}{2^{n+1}}.$$

If we put
$$V_n = D_1 \cap D_2 \cap \ldots \cap D_n$$

then
$$\tau(V_n) = \tau(E_n) - \tau(E_n - V_n) \geqslant \tau(E_n) - \sum_{i=1}^{n} \tau(E_i - D_i) > \tfrac{1}{2}\delta$$

so that the sets $\{V_n\}$ form a monotone decreasing sequence of non-empty sets. In each V_n choose a point

$$f_n = \{f_n(i), \quad i \in I\}.$$

Now $$(f_{n+p}(i_1), f_{n+p}(i_2), \ldots, f_{n+p}(i_n)) \quad (p = 1, 2, \ldots)$$
defines a sequence of points in $\mathbf{R}^n$ which is a subset of the bounded closed set
$$(\bar{P}_1 \times \mathbf{R}^{n-1}) \cap (\bar{P}_2 \times \mathbf{R}^{n-2}) \cap \ldots \cap (\bar{P}_n) = F_n.$$
We can therefore find a subsequence of $\{f_{n+p}\}$ which, evaluated at the first n indices converges to a point of F_n. Since $\tau(V_n) > \frac{1}{2}\delta$, V_n is not empty and F_n is not empty since
$$V_n \subset \{f \in \mathbf{R}^I: (f(i_1), \ldots, f(i_n)) \in F_n\}.$$
Further $F_n \times \mathbf{R} \subset F_{n+1}$ $(n = 1, 2, \ldots)$, and we can now employ a standard diagonalisation argument to obtain a point in $\bigcap_{n=1}^{\infty} E_n$.

Obtain successively, by induction, infinite increasing sequences of positive integers
$$\nu_1 \supset \nu_2 \supset \ldots \supset \nu_k \supset \ldots$$
such that $\{f_n\}$ restricted to the sequence ν_k gives a sequence whose values at $i_1, i_2, \ldots, i_k$ converge to a point in F_k. Form the sequence ν obtained by taking the kth integer in the sequence ν_k. Then, for each k, ν is a subsequence of ν_k except for a finite number of terms at the beginning so that $\{(f_n(i_1), f_n(i_2), \ldots, f_n(i_k))\}$, $n \in \nu$ must converge to a point in $F_k \subset Q_k$. If we put $q_k = \lim_{n \in \nu} f_n(i_k)$ $(k = 1, 2, \ldots)$ the set
$$H = \{f \in \mathbf{R}^I: f(i_k) = q_k, k = 1, 2, \ldots\}$$
is non-empty, and $H \subset V_n \subset E_n$ for all n. This contradicts $\bigcap_{n=1}^{\infty} E_n = \varnothing$. ∎

Remark. For a finite index set I, theorem 6.10 is still true, but lacks any content as the measure $\mu i_1 \ldots i_n$ already is the required μ if
$$I = \{i_1, i_2, \ldots, i_n\}.$$

Brownian motion

We can set up a mathematical model for Brownian motion by applying theorem 6.10 to a particular family of finite dimensional distributions. Use the index set $T = \{t \in \mathbf{R}, t \geqslant 0\}$ which can be thought of as time and, for
$$0 < t_1 < \ldots < t_n,$$
define
$$\mu t_1 \ldots t_n \{f \in \mathbf{R}^I: a_i < f(t_i) \leqslant b_i, i = 1, \ldots, n\}$$
$$= \frac{1}{(2\pi)^{\frac{1}{2}n}} \int_{a_n}^{b_n} \exp\left[-\frac{(\xi_n - \xi_{n-1})^2}{2(t_n - t_{n-1})}\right] d\xi_n \int_{a_{n-1}}^{b_{n-1}} \exp\left[-\frac{(\xi_{n-1} - \xi_{n-2})^2}{2(t_{n-1} - t_{n-2})}\right] d\xi_{n-1}$$
$$\ldots \int_{a_2}^{b_2} \exp\left[-\frac{(\xi_2 - \xi_1)^2}{2(t_2 - t_1)}\right] d\xi_2 \int_{a_1}^{b_1} \exp\left(-\frac{\xi_1^2}{2t_1}\right) d\xi_1.$$

The fact that this defines a consistent family of measures on $\mathcal{S}$ which can be extended to all sets of the form (6.6.2) can be proved directly (it will follow from the discussion of the multinormal distribution in Chapter 14). Hence, we can apply theorem 6.10 to give a measure μ on $\mathcal{B}^T$ the class of Borel sets in $\mathbf{R}^T = \Omega$. This is called *Wiener measure* in the space of functions $f: T \to \mathbf{R}$, and is an example of a stochastic process which will be discussed more fully in Chapter 15.

However, let us use the example of Wiener measure to illustrate the inadequacy of theorem 6.10. This follows from the fact that the σ-field $\mathcal{B}^T$ is too small to contain interesting sets—for we have seen that it contains no set in which a non-countable set of time coordinates is restricted. Even if $\mathcal{B}^T$ is completed with respect to μ to give a probability measure, the completed σ-field is still too small. For if $A \subset \mathbf{R}^T$ is a set in which f is restricted at a non-countable set, the only set of $\mathcal{B}^T$ which is contained in A is the empty set. This means that A can only be measurable, if it has measure zero. But the same argument applies to the complement of A so that if both A and its complement involve restrictions on $f: T \to \mathbf{R}$ at a non-countable set of indices, then the outer measure of A must be 1, and the inner measure must be 0. In particular the set

$$\{f \in \mathbf{R}^T: a < f(t) \leqslant b \quad \text{for all} \quad t \in [t_1, t_2]\} \tag{6.6.4}$$

is not measurable, and if C is the set of functions $f: T \to \mathbf{R}$ which are continuous for all $t \in T$, C also has outer measure 1 (and inner measure 0). Various methods can be used to extend μ from $\mathcal{B}^T$ to a larger σ-field which includes C and (6.6.4) and other sets of interest. These have been studied in detail and the interested reader is advised to look in J. L. Doob, *Stochastic Processes* (Wiley, 1953).

6.7 Applications

In the second part of this book random variables will be defined as $\mathcal{F}$-measurable functions $f: \Omega \to \mathbf{R}^*$ where $(\Omega, \mathcal{F}, \mu)$ is a probability space. Although it is usual to work with general carrier spaces Ω, there is a sense in which the real line $\mathbf{R}$ has a structure sufficiently complicated to reproduce all the probability properties of the function f. In fact, in many treatises on probability theory, the carrier space Ω is barely mentioned. This attitude is partially justified by the following considerations.

Suppose $(\Omega, \mathcal{F}, \mu)$ is any finite measure space and $f: \Omega \to \mathbf{R}$ is $\mathcal{F}$-measurable. For all real x, define

$$F(x) = \mu\{y: f(y) \leqslant x\}. \tag{6.7.1}$$

Then $F(x) \to 0$ as $x \to -\infty$, $F(x) \to \mu(\Omega)$ as $x \to +\infty$, and $F: \mathbf{R} \to \mathbf{R}$ is continuous on the right. Thus we can define a Stieltjes measure μ_F using this particular F.

***Theorem* 6.11.** *Suppose $(\Omega, \mathscr{F}, \mu)$ is a finite measure space and $f: \Omega \to \mathbf{R}$ is $\mathscr{F}$-measurable, μ_F is the Lebesgue–Stieltjes measure in $\mathbf{R}$ given by (6.7.1) and $g: \mathbf{R} \to \mathbf{R}$ is Borel measurable, then $g(f)$ is $\mathscr{F}$-measurable and $\mu\{x: g(f)(x) \in B\}$ is determined by μ_F for every Borel set B. Further*

$$\int g(f)\, d\mu = \int g(x)\, dF(x)$$

in the sense that, if either side exists so does the other, and the two are equal.

Proof. $\{x: g(f)(x) \in B\} = \{x: f(x) \in g^{-1}(B)\}$ and $g^{-1}(B) = C$ is a Borel set so that $\{x: f(x) \in C\}$ is in $\mathscr{F}$, and $\mu\{x: f(x) \in C\}$ is uniquely determined by $\mu\{x: a < f(x) \leqslant b\} = F(b) - F(a)$ for all real a, b since $\mathscr{P}$, the class of half-open intervals generates the σ-field $\mathscr{B}$ of Borel sets in $\mathbf{R}$, and $F(b) - F(a) = \mu_F(a, b]$. Thus, for all B in $\mathscr{B}$,

$$\mu\{x: g(f)(x) \in B\} = \mu_F(g^{-1}(B)).$$

Now suppose g is an indicator function of a Borel set B. Then

$$\int g(f)\, d\mu = \mu\{x: f(x) \in B\}$$
$$= \mu_F(B) = \int g\, d\mu_F.$$

By linearity our result follows for non-negative simple functions and the monotone convergence theorem then gives it for non-negative Borel measurable g and then for all integrable g. ▌

Corollary. *In the notation of the theorem*

$$\int f d\mu = \int x dF(x).$$

Remark. There is an n-dimensional form of theorem 6.11 and corollary which links the behaviour of n measurable functions with a Lebesgue–Stieltjes distribution in $\mathbf{R}^n$—see Chapter 14.

Marginal distributions

Not all measures in product spaces are product measures. Suppose X, Y are spaces, then the projection $X \times Y \to X$ given by $p(x, y) = x$ defines a mapping. This will be a measurable transformation on $(X \times Y, \mathscr{F})$ into $(X, \mathscr{S})$ provided $E \times Y \in \mathscr{F}$ for every $E \in \mathscr{S}$. In this case, if μ is a finite measure on $\mathscr{F}$, μp^{-1} defines a measure on $\mathscr{S}$. In general it may not be a very interesting measure as there may be no sets of finite positive measure. However, if $(X \times Y, \mathscr{F}, \mu)$ is a finite measure space, then the measure μp^{-1} on $\mathscr{S}$ is called the marginal measure on X. The marginal measure on Y is similarly defined using a projection on Y. If μ is a probability measure these marginal measures are called marginal (probability) distributions.

If $F(x, y)$ is a distribution function in $\mathbf{R}^2$ (see §4.5) then

$$\lim_{y \to +\infty} F(x, y) \quad \text{and} \quad \lim_{x \to +\infty} F(x, y)$$

will again define 1-dimensional distribution functions, and it is immediate from theorem 4.8 that the corresponding Lebesgue–Stieltjes measures will be the marginal distributions of μ_F. If

$$F(x, y) = F_1(x) . F_2(y)$$

is the product of two 1-dimensional distributions, then μ_F will be the completion of the product measure $\mu_{F_1} \times \mu_{F_2}$ and F_1, F_2 will be the marginal distributions for F. Conversely, if μ_F is a product measure, then it must be the product of its marginal distributions so that $F(x, y) = F_1(x) F_2(y)$ is a necessary as well as a sufficient condition for μ_F to be a product of two probability measures.

Thick subsets

For any finite measure space $(\Omega, \mathscr{F}, \mu)$ we can generate the outer measure

$$\mu^*(E) = \inf \sum_{i=1}^{\infty} \mu(E_i) \tag{6.7.2}$$

the infimum being taken over all sequences of sets $\{E_i\}$ in $\mathscr{F}$ with $E \subset \bigcup_{i=1}^{\infty} E_i$. (Since μ is a measure on the σ-field $\mathscr{F}$, (6.7.2) is the same as $\mu^*(E) = \inf \mu(F)$ for $F \supset E, F \in \mathscr{F}$.) A subset E_0 of Ω is said to be *thick* in Ω if $\mu^*(E_0) = \mu(\Omega)$. Thus a subset E_0 is thick if and only if $(\Omega - E_0)$ contains no set in $\mathscr{F}$ of positive μ-measure. There is a sense in which the measure space can be projected onto any thick subset.

Theorem 6.12. *If E_0 is a thick subset of the finite measure space $(\Omega, \mathscr{F}, \mu)$, $\mathscr{F}_0 = \mathscr{F} \cap E_0$, and $\mu_0(E \cap E_0) = \mu(E)$ for any $E \in \mathscr{F}$, then $(E_0, \mathscr{F}_0, \mu_0)$ is a measure space.*

Proof. We first see that μ_0 is defined uniquely on $\mathscr{F}_0$. If $A_1, A_2 \in \mathscr{F}$ are such that $A_1 \cap E_0 = A_2 \cap E_0$, then we must have

$$(A_1 \triangle A_2) \cap E_0 = \varnothing,$$

so that $\mu(A_1 \triangle A_2) = 0$ and $\mu(A_1) = \mu(A_2)$.

Now suppose $\{B_n\}$ is a disjoint sequence of sets in $\mathscr{F}_0$ so that there is a sequence of sets $\{C_n\}$ in $\mathscr{F}$ with

$$B_n = C_n \cap E_0 \quad (n = 1, 2, \ldots).$$

Put
$$D_n = C_n - \bigcup_{i=1}^{n-1} C_i \quad (n = 1, 2, \ldots).$$

Then
$$D_n \cap E_0 = C_n \cap E_0,$$

so that $\mu(D_n \triangle C_n) = 0$. It follows that

$$\sum_{n=1}^{\infty} \mu_0(B_n) = \sum_{n=1}^{\infty} \mu(C_n) = \sum_{n=1}^{\infty} \mu(D_n) = \mu\left(\bigcup_{n=1}^{\infty} D_n\right) = \mu\left(\bigcup_{n=1}^{\infty} C_n\right) = \mu_0\left(\bigcup_{n=1}^{\infty} B_n\right)$$

so that μ_0 is a measure. ▌

Remark. This theorem shows that in a probability space $(\Omega, \mathscr{F}, P)$, the σ-field $\mathscr{F}$ can be extended to include any set E_0 not in it whose outer measure is 1. The effect of this extension is to discard all the points of Ω which are not in E_0. The device turns out to be useful in the theory of stochastic processes where, by a careful choice of E_0, one can obtain a probability on a useful class of subsets. In particular, for Wiener measure in $\mathbf{R}^T$ described in §6.6, it can be shown that the set C of continuous functions is thick and that the extension given by putting $E_0 = C$ is a useful one—see Chapter 15.

Exercises 6.7

1. Formulate and prove a theorem of the form of theorem 6.11 for n $\mathscr{F}$-measurable functions $f_i: \Omega \to \mathbf{R}$ $(i = 1, 2, \ldots, n)$.

2. Find the 2-dimensional distribution function $F(x, y)$ which generates the measure μ_F such that $\mu_F(R)$ is $1/\sqrt{2}$ (length of diagonal D in R) for any rectangle R, where D is the segment joining $(0, 0)$ to $(1, 1)$. Calculate the marginal distributions of μ_F, and show that μ_F is not a product measure.

3. If $(\Omega, \mathscr{F}, \mu)$ is a complete σ-finite measure space, and the outer measure μ^* is defined by (6.7.2) show that a set E is μ^*-measurable if and only if it is in $\mathscr{F}$.

4. Suppose $(\Omega, \mathscr{F}, \mu)$ is a finite measure space and E_0 is a subset of Ω such that, for $A_1, A_2 \in \mathscr{F}$,

$$A_1 \cap E_0 = A_2 \cap E_0 \Rightarrow \mu(A_1) - \mu(A_2).$$

Prove that E_0 is thick in Ω.

7

THE SPACE OF MEASURABLE FUNCTIONS

Throughout this chapter we will assume (unless stated otherwise) that $(\Omega, \mathscr{F}, \mu)$ is a σ-finite measure space, and that the σ-field $\mathscr{F}$ is complete with respect to μ. This implies that if $f: \Omega \to \mathbf{R}^*$, $g: \Omega \to \mathbf{R}^*$ are functions such that f is $\mathscr{F}$-measurable and $f = g$ a.e., then g is also $\mathscr{F}$-measurable. Thus, if M is the class of functions $f: \Omega \to \mathbf{R}^*$ which are $\mathscr{F}$-measurable, we say that f_1, f_2 in M are equivalent if $f_1 = f_2$ a.e. This clearly defines an equivalence relation in M and we can form the space $\mathscr{M}$ of equivalence classes with respect to this relation. When we think of a function f of M as an element of $\mathscr{M}$ we are really thinking of f as a representative of the class of $\mathscr{F}$-measurable functions which are equal to f a.e. As is usual we will use the same notation f for an element of M and $\mathscr{M}$. We can think of M or $\mathscr{M}$ as an abstract space, and the definition of convergence if given in terms of a metric will then impose a topological structure on the space. We will consider several such notions of convergence of which some, but not all, can be expressed in terms of a metric in $\mathscr{M}$. We will obtain the relationships between different notions of convergence, and in each case prove that the space is complete in the sense that for any Cauchy sequence there is a limit function to which the sequence converges. The main strategy used to prove completeness will be to find a suitable subsequence of the given sequence which clearly converges to a limit f and then show that f is a limit of the whole sequence. This extends the method used in §2.2 to show that $\mathbf{R}$ is complete.

7.1 Point-wise convergence

Given a sequence $\{f_n\}$ of functions where $f_n: E \to \mathbf{R}^*$ and a function $f: E \to \mathbf{R}^*$ $(E \subset \Omega)$, we say that f_n converges to f point-wise on E if, for each x in E, $f_n(x) \to f(x)$ as $n \to \infty$. This notion has a meaning if we restrict consideration to $\mathscr{M}$. If E is such that $\mu(\Omega - E) = 0$, and $f_n \to f$ point-wise on E, then we say that $f_n \to f$ a.e. For if $f_n \to f$ a.e., $f_n = g_n$ a.e. for each n, and $g_n \to g$ a.e., then

$$\{x: f(x) \neq g(x)\} \subset \{x: f_n(x) \not\to f(x)\}$$
$$\cup \{x: g_n(x) \not\to g(x)\} \cup \bigcup_{n=1}^{\infty} \{x: f_n(x) \neq g_n(x)\},$$

and each of these sets has zero measure so $f(x) = g(x)$ a.e. which means that $f = g$ in $\mathscr{M}$. $\{f_n\}$ is a Cauchy sequence (point-wise) on E if, given $x \in E$, $\epsilon > 0$, there is an integer N such that

$$|f_n(x) - f_m(x)| < \epsilon \quad \text{for} \quad n, m > N. \tag{7.1.1}$$

(This has meaning only if f_n: $E \to \mathbf{R}$ is finite valued.) Because $\mathbf{R}$ is complete it is clear that if $\{f_n\}$ is a Cauchy sequence on E, there must be an f: $E \to \mathbf{R}$ such that $f_n \to f$ point-wise on E.

Uniform convergence

If the sequence $\{f_n\}$ and the function f are finite valued functions on E to $\mathbf{R}$, we say that f_n converges uniformly to f on E if for each $\epsilon > 0$, there is an integer N such that

$$x \in E, \quad n \geqslant N \Rightarrow |f_n(x) - f(x)| < \epsilon.$$

Similarly, we say that the sequence is a Cauchy sequence uniformly on E if given $\epsilon > 0$, a single integer N can be chosen so that (7.1.1) is satisfied for all $x \in E$. Since a Cauchy sequence uniformly on E is certainly a Cauchy sequence on E and the existence of $\lim_{n\to\infty} f_m(x) = f(x)$ follows for each x, we can let $m \to \infty$ in (7.1.1) to deduce that a Cauchy sequence uniformly on E must have a limit function $f: E \to \mathbf{R}$ such that $f_n \to f$ uniformly on E.

If $\mu(\Omega - E) = 0$ and $f_n \to f$ uniformly on E, then we say that $f_n \to f$ uniformly a.e. All these notions have a meaning for functions which need not be measurable. However, the notion of convergence uniformly a.e. can be expressed in terms of a metric on the restricted class of measurable functions.

Essentially bounded functions

An $\mathscr{F}$-measurable function $f: \Omega \to \mathbf{R}^*$ is said to be essentially bounded if $\mu\{x: |f(x)| > a\} = 0$ for some real number a. In this case we define the essential supremum of f by

$$\text{ess sup}\,|f| = \inf\{a: \mu\{x: |f(x)| > a\} = 0\}.$$

Notice that, if $\text{ess sup}\,|f| = C$, then

$$E = \{x: |f(x)| > C\} = \bigcup_{k=1}^{\infty} \left\{x: |f(x)| > C + \frac{1}{k}\right\}$$

so that $\mu(E) = 0$ and $|f(x)| \leqslant C$ outside E. Thus $|f(x)| \leqslant C$ a.e., and if we define

$$f^*(x) = \begin{cases} f(x) & \text{if} \quad |f(x)| \leqslant C, \\ 0 & \text{if} \quad |f(x)| > C, \end{cases}$$

then $|f^*(x)| \leqslant C$ for all x and $f^* = f$ a.e. Further $\{x\colon |f^*(x)| > C-\epsilon\}$ has positive measure for all $\epsilon > 0$, so that it is non-empty and we must have $\sup|f^*| = C$. It is clear that, if $f = g$ a.e., then $\operatorname{ess\,sup} f = \operatorname{ess\,sup} g$, so that we can think of ess sup as a functional on the subset $\mathscr{L}_\infty$ of the essentially bounded functions of $\mathscr{M}$. If we define $(\alpha f+\beta g)$ by

$$\begin{aligned}(\alpha f+\beta g)(x) &= \alpha f(x)+\beta g(x) \quad \text{when} \quad f(x), g(x) \in \mathbf{R},\\ &= 0 \qquad\qquad\qquad\quad \text{otherwise;}\end{aligned}$$

it is clear that $(\alpha f+\beta g) \in \mathscr{L}_\infty$ if $f, g \in \mathscr{L}_\infty$ for any $\alpha, \beta \in \mathbf{R}$ so that $\mathscr{L}_\infty$ is a linear subspace of $\mathscr{M}$ (over the reals) Further

$$\rho_\infty(f,g) = \operatorname{ess\,sup}|f-g|$$

defines a metric in $\mathscr{L}_\infty$ for

(i) $\rho_\infty(f,g) = \rho_\infty(g,f)$;

(ii) $\rho_\infty(f,g) = 0$ if only if $f = g$ a.e.;

(iii) $\operatorname{ess\,sup}|f+g| \leqslant \operatorname{ess\,sup}|f| + \operatorname{ess\,sup}|g|$ so that

$$\rho_\infty(f,g) \leqslant \rho_\infty(f,h)+\rho_\infty(h,g).$$

Now it is clear that, if $\{f_n\}$ and f are functions in $\mathscr{L}_\infty$ such that $f_n \to f$ uniformly a.e., then $\rho_\infty(f_n, f) \to 0$ as $n \to \infty$. Conversely suppose $\rho_\infty(f_n, f) \to 0$, and let E_n be a set of $\mathscr{F}$ with $\mu(E_n) = 0$ and

$$\operatorname{ess\,sup}|f_n-f| = \sup_{x\in\Omega-E_n} |f_n(x)-f(x)|.$$

Put $E = \bigcup E_n$, then for $x \in \Omega - E$

$$|f_n(x)-f(x)| \leqslant \sup_{x\in\Omega-E_n} |f_n(x)-f(x)| = \operatorname{ess\,sup}|f_n-f|$$

so that $f_n \to f$ uniformly on $\Omega - E$ and $\mu(E) = 0$. A similar, but slightly more complicated argument shows that, in $\mathscr{L}_\infty$, a Cauchy sequence uniformly a.e. is the same as a Cauchy sequence in ρ_∞ norm.

Almost uniform convergence

Given functions $f_n\colon E \to \mathbf{R}^*$ $(n = 1, 2, \ldots)$ and $f\colon E \to \mathbf{R}^*$ each of which is finite a.e. on E we say that f_n converges almost uniformly to f on E if, for each $\epsilon > 0$, there is a set $F_\epsilon \subset E$, $F_\epsilon \in \mathscr{F}$, $\mu(F_\epsilon) < \epsilon$ such that $f_n \to f$ uniformly on $(E - F_\epsilon)$. The example $E = [0, 1] \subset \mathbf{R}$, $f_n(x) = x^n$ μ Lebesgue measure shows that it is possible for a sequence to converge almost uniformly on E while it does not converge uniformly a.e. on E. However, it is immediate from the definitions that convergence uniformly a.e. implies almost uniform convergence. What is more surprising is that, under suitable conditions, convergence a.e. implies almost uniform convergence.

***Theorem* 7.1.** *(Egoroff). Suppose $E \in \mathscr{F}$, $\mu(E) < \infty$, and $\{f_n\}$ is a sequence of measurable functions on $E \to \mathbf{R}^*$ which are finite a.e. and converge a.e. to a function $f: E \to \mathbf{R}^*$ which is also finite a.e. Then $f_n \to f$ almost uniformly in E.*

Proof. By omitting a subset of E of zero measure, we may assume that all the functions f_n and f are finite and that

$$f_n(x) \to f(x) \quad \text{for all} \quad x \in E.$$

For positive integers, m, n put

$$A_n^m = \bigcap_{i=n}^{\infty} \left\{x : |f_i(x) - f(x)| < \frac{1}{m}\right\}.$$

Then, for fixed m, $A_1^m, A_2^m, \ldots, A_n^m, \ldots$ is an increasing sequence of measurable sets converging to E. Since $\mu(E)$ is finite, by theorem 3.2 there is a positive integer $N_m = N_m(m)$ such that

$$\mu(E - A_i^m) < \epsilon/2^m \quad \text{for} \quad i \geqslant N_m.$$

If we put
$$F_\epsilon = \bigcup_{m=1}^{\infty} (E - A_{N_m}^m),$$

then $\mu(F_\epsilon) < \epsilon$. Further given $\delta > 0$ we can choose m so that $1/m < \delta$ and then

$$|f_i(x) - f(x)| < \delta \quad \text{for all} \quad i \geqslant N_m, \quad x \in (E - F_\epsilon),$$

so that $f_n \to f$ uniformly on $(E - F_\epsilon)$. ∎

Remark. The converse to theorem 7.1 is true and almost trivial. For if $\{f_n\}$, f are finite a.e. on E, measurable, and $f_n \to f$ almost uniformly, this means we can find sets F_n with $\mu(F_n) < 1/n$ such that $f_n \to f$ uniformly on $(E - F_n)$ and so $f_n \to f$ point-wise on $(E - F_n)$. Put

$$F = \bigcap_{n=1}^{\infty} F_n, \quad \text{then} \quad \mu(F) = 0$$

and $f_n \to f$ point-wise on $(E - F)$ so that $f_n \to f$ a.e. on E.

Exercises 7.1

1. Let X be the space of positive integers, $\mathscr{F}$ the class of all subsets of X, and $\mu(E)$ the number of integers in $E \subset X$. If $f_n(x)$ is the indicator function of $\{1, 2, \ldots, n\}$, then $f_n(x) \to 1$ for all x. However, f_n does not converge almost uniformly to 1, showing that theorem 7.1 is false without $\mu(E) < \infty$.

2. Suppose the conditions of theorem 7.1 are satisfied except that $\mu(E) = \infty$, show that given $P > 0$, there is a subset $F_P \subset E$ with $\mu(F_P) > P$ such that $f_n \to f$ uniformly on F_P but that there need not be a subset F with $\mu(F) = +\infty$ with $f_n \to f$ uniformly on F.

3. Suppose $E \in \mathscr{F}$, $\mu(E) < \infty$, $f_n: E \to \mathbf{R}^*$ $(n = 1, 2, \ldots)$ is a Cauchy sequence a.e. of measurable funtions each finite a.e. Prove there is a finite c and a measurable $F \subset E$ with $\mu(F) > 0$ such that, for every integer n, all $x \in F$, $|f_n(x)| \leqslant c$.

4. Suppose $E \in \mathscr{F}$, E has σ-finite measure, f_n $(n = 1, 2, \ldots)$ and f are finite a.e. on E and $f_n \to f$ a.e. on E. Show that there exists a sequence $\{E_i\}$ of sets in $\mathscr{F}$ such that

$$\mu\left(E - \bigcup_{i=1}^{\infty} E_i\right) = 0 \quad \text{and} \quad f_n \to f$$

uniformly on each E_i. By considering the measure of example 2, §3.1, and a suitable sequence of functions show that the condition that E has σ-finite measure is essential.

5. In §4.4 we produced a sequence of sets $\{Q_n\}$ each of which was not Lebesgue measurable. If we put $f_n(x) =$ indicator function of

$$[0, 1) - \bigcup_{i=1}^{n} Q_i, \quad \text{then} \quad f_n(x) \to 0 \quad \text{for all } x \text{ in } [0, 1].$$

Show that f_n does not converge almost uniformly so that theorem 7.1 fails if the functions are not measurable.

6. Suppose $f_h: E \to \mathbf{R}$, $h > 0$ is a continuous family of measurable functions, each finite valued, $\mu(E) < \infty$ and for each $x \in E, f_h(x) \to f(x)$ as $h \to 0$ where f is finite valued. Then if a continuous parameter version of Egoroff's theorem were valid we would have given $\epsilon > 0$, there exists $F_\epsilon \subset \mathscr{F}$, $F_\epsilon \subset E$, $\mu(F_\epsilon) < \epsilon$ such that $f_h(x) \to f(x)$ as $h \to 0$ uniformly on $(E - F_\epsilon)$. The following example shows that this extension is false. In Chapter 4, we saw that there is a non-measurable set $E \subset [0, 1)$ such that every point $x \in [0, 1)$ has a unique representation $x = y + q \pmod 1$, $y \in E$, q rational.

Prove that, if M is a measurable subset of $[0, 1]$ such that $M \cap E(r)$ is non-void for finitely many rationals r, then $|M| = 0$.

Arrange the rationals Q as a sequence $\{r_n\}$. For $x \in [0, 1)$ let $n(x)$ be the integer such that $x = y + r_{n(x)}$, $y \in E$. If $x/n(x) = \cdot\alpha_1\alpha_2\ldots$ (decimal representation not ending in 9 recurring), put $\phi(x) = \cdot\beta_1\beta_2\ldots$ where $\beta_{2k} = \alpha_k$ $(k = 1, 2, \ldots)$; and $\beta_{2k-1} = 1$ for $k = n(x)$, 0 otherwise. Put $f_h(x) = 1$, for $x = \phi(h)$, $f_h(x) = 0$ otherwise. Prove $f_h(x) \to 0$ as $h \to 0$ for each x. Show that if M any measurable set, $|M| > 0$, then $f_h(x) \not\to 0$ uniformly on M.

7. Suppose $\{f_n\}$ is a sequence of functions in $\mathscr{M}$, $f_n \to f$ a.e. and there is an integrable function g such that $|f_n| \leqslant g$ a.e. for all n. Show that $f_n \to f$ almost uniformly.

8. Define what is meant by saying that a sequence $\{f_n\}$ of a.e. finite valued functions is a Cauchy sequence almost uniformly, and show that this implies the existence of a limit function f such that $f_n \to f$ almost uniformly.

7.2 Convergence in measure

We now consider a different kind of 'nearness' in $\mathscr{M}$ in which the measure of the set where two functions differ by more than a fixed positive number is relevant. This time we make the definitions relative to the whole space Ω. Obvious changes give the corresponding concepts relative to a set E in $\mathscr{F}$. Given $\mathscr{F}$-measurable functions $f: \Omega \to \mathbf{R}^*$, $f_n: \Omega \to \mathbf{R}^*$ $(n = 1, 2, \ldots)$ we say that f_n converges in measure (μ) to f if, for each $\epsilon > 0$,

$$\lim_{n \to \infty} \mu\{x: |f_n(x) - f(x)| \geqslant \epsilon\} = 0.$$

Note that the definition only makes sense for functions in $\mathscr{M}$ which are finite a.e. We first see that the limit in measure is unique in $\mathscr{M}$. For suppose $f_n \to f$ in measure, $f_n \to g$ in measure; then if $\delta > 0$,

$$\{x: |f(x) - g(x)| > \delta\} \subset \{x: |f_n(x) - f(x)| > \tfrac{1}{2}\delta\} \cup \{x: |f_n(x) - g(x)| > \tfrac{1}{2}\delta\}$$

and both sets on the right can be made of arbitrarily small measure by choosing n large. This means that

$$\mu\{x: |f(x) - g(x)| > \delta\} = 0 \quad \text{for each} \quad \delta > 0,$$

and it follows that $f = g$ a.e. (by taking a sequence δ_n decreasing to zero).

We say that the sequence $\{f_n\}$ of functions in $\mathscr{M}$ is a Cauchy sequence in measure if, given $\epsilon > 0$, $\delta > 0$ there is an integer N such that

$$n \geqslant N, \quad m \geqslant N \Rightarrow \mu\{x: |f_n(x) - f_m(x)| > \epsilon\} < \delta.$$

The argument used to prove uniqueness of the limit also shows that

$$f_n \to f \text{ in measure} \Rightarrow \{f_n\} \text{ is a Cauchy sequence in measure.}$$

The converse is included in the following theorem.

***Theorem* 7.2.** *Suppose f and f_n $(n = 1, 2, \ldots)$ are functions in $\mathscr{M}$ which are finite a.e. Then*

(i) *$f_n \to f$ almost uniformly $\Rightarrow$ $f_n \to f$ in measure;*

(ii) *$\{f_n\}$ is a Cauchy sequence almost uniformly $\Rightarrow$ $\{f_n\}$ is a Cauchy sequence in measure;*

(iii) *$\{f_n\}$ is a Cauchy sequence in measure $\Rightarrow$ there is a subsequence $\{n_k\}$ such that $\{f_{n_k}\}$ is a Cauchy sequence almost uniformly;*

(iv) *$\{f_n\}$ is a Cauchy sequence in measure $\Rightarrow$ there is a function $g \in \mathscr{M}$ such that $f_n \to g$ in measure.*

Proof. (i) If $f_n \to f$ almost uniformly, for each $\epsilon > 0$, $\delta > 0$, we can find a set $E_\delta \in \mathscr{F}$ such that $\mu(E_\delta) < \delta$ and $f_n \to f$ uniformly on $(\Omega - E_\delta)$. Hence there is an N such that

$$|f_n(x) - f(x)| < \epsilon \quad \text{for} \quad n \geqslant N, \quad x \in (\Omega - E_\delta)$$

and then

$$\mu\{x: |f_n(x) - f(x)| \geqslant \epsilon\} \leqslant \mu(E_\delta) < \delta \quad \text{for} \quad n \geqslant N.$$

(ii) An argument similar to that in (i) will work.

(iii) Now suppose that f_n is a Cauchy sequence in measure. For each positive integer k, choose an integer m_k such that $m_k > m_{k-1}$ and

$$n \geqslant m_k, \quad m \geqslant m_k \Rightarrow \mu\{x: |f_n(x) - f_m(x)| \geqslant 2^{-k}\} \leqslant 2^{-k}.$$

Put
$$E_k = \{x: |f_{m_k}(x) - f_{m_{k+1}}(x)| \geqslant 2^{-k},$$

$$F_k = \bigcup_{i=k}^{\infty} E_i.$$

Then
$$\mu(F_k) \leqslant \sum_{i=k}^{\infty} \mu(E_i) \leqslant 2^{1-k}.$$

Given $\epsilon > 0$ we can choose k so that $\epsilon > 2^{1-k}$; then $\mu(F_k) < \epsilon$ and for all $x \in (\Omega - F_k)$ we have

$$|f_{m_i}(x) - f_{m_{i+1}}(x)| < 2^{-i} \quad \text{for all} \quad i \geqslant k.$$

Thus

$$j \geqslant i \geqslant k \Rightarrow |f_{m_i}(x) - f_{m_j}(x)| \leqslant \sum_{s=i}^{j-1} |f_{m_s}(x) - f_{m_{s+1}}(x)| < 2^{1-i}$$

so that the sequence f_{m_i} converges uniformly on $(\Omega - F_k)$; that is, since $\mu(F_k) \to 0$ as $k \to \infty$, it is a Cauchy sequence almost uniformly.

(iv) By (iii) we can obtain a subsequence f_{m_k} of the given sequence which is a Cauchy sequence almost uniformly. This means we can find a function $g \in \mathscr{M}$ such that $f_{m_k} \to g$ almost uniformly as $k \to \infty$. Now, for $\epsilon > 0$,

$$\{x: |f_n(x) - g(x)| \geqslant \epsilon\} \subset \{x: |f_n(x) - f_{m_k}(x)| \geqslant \tfrac{1}{2}\epsilon\} \cup \{x: |f_{m_k}(x) - g(x)| \geqslant \tfrac{1}{2}\epsilon\}.$$

Given $\delta > 0$ we can find a set $E_\delta \in \mathscr{F}$ and integers k_0, N such that $\mu(E_\delta) < \frac{1}{2}\delta$,

$$|f_{m_k}(x) - g(x)| < \tfrac{1}{2}\epsilon \quad \text{for} \quad k \geqslant k_0, \quad x \in \Omega - E_\delta,$$

and $\mu\{x: |f_n(x) - f_{m_k}(x)| \geqslant \frac{1}{2}\epsilon\} < \frac{1}{2}\delta$ for $n \geqslant N$, $m_k \geqslant N$.

It follows that

$$n \geqslant \max\{N, m_{k_0}\} \Rightarrow \mu\{x: |f_n(x) - g(x)| \geqslant \epsilon\} < \delta. \blacksquare$$

It is not difficult to see that convergence in measure does not necessarily imply convergence point-wise at any point, and so it certainly cannot imply almost uniform convergence of the whole sequence. For let

$$E_{r,k} = \left[\frac{r-1}{2^k}, \frac{r}{2^k}\right] \quad (r = 1, 2, \ldots, 2^k; \quad k = 1, 2, \ldots),$$

and arrange these intervals as a single sequence of sets $\{F_n\}$ by taking first those for which $k = 1$, then those with $k = 2$, etc. If μ denotes Lebesgue measure on $[0, 1]$, and $f_n(x)$ is the indicator function of F_n, then, for $0 < \epsilon < 1$,

$$\{x: |f_n(x)| \geqslant \epsilon\} = F_n$$

so that, for any $\epsilon > 0$, $\mu\{x: |f_n(x)| \geqslant \epsilon\} \leqslant \mu(F_n) \to 0$. This means that $f_n \to 0$ in measure in $[0, 1]$. However, at no point x in $[0, 1]$ does $f_n(x) \to 0$; in fact, since every x is in infinitely many of the sets F_n and infinitely many of the sets $(\Omega - F_n)$ we have

$$\liminf f_n(x) = 0, \quad \limsup f_n(x) = 1 \quad \text{for all} \quad x \in [0, 1].$$

Exercises 7.2

1. Suppose $\{f_n\}$ is a Cauchy sequence in measure, and f_{n_i}, f_{m_j} are two subsequences which converge to f, g, respectively. Prove that $f = g$ a.e.

2. Show that if $\{f_n\}$ is a Cauchy sequence in measure then every subsequence of $\{f_n\}$ is also a Cauchy sequence in measure.

3. If Ω is the set of positive integers and μ is the counting measure on the class $\mathscr{F}$ of all subsets, show that convergence in measure is equivalent to uniform convergence.

4. If $\mu(\Omega) = \infty$ can we say that convergence a.e. implies convergence in measure?

5. Suppose $\{A_n\}$ is a sequence of sets in $\mathscr{F}$, χ_n is the indicator function of A_n, and $d(A, B) = \mu(A \triangle B)$ for $A, B \in \mathscr{F}$. Show that $\{\chi_n\}$ is a Cauchy sequence in measure if and only if $d(A_n, A_m) \to 0$ as $n, m \to \infty$.

6. Suppose $\{f_n\}$ is a sequence of functions of M which are finite a.e. and $f_n \to f$ a.e. with f finite a.e. Show that, if (i) $\mu(\Omega) < \infty$, or (ii) $|f_n| \leqslant g_0$ for all n where g_0 is integrable; then $f_n \to f$ in measure.

7. Suppose $(\Omega, \mathscr{F}, \mu)$ is a finite measure space and $\{f_n\}$, $\{g_n\}$ are finite valued $\mathscr{F}$-measurable functions which converge in measure to f, g respectively. Show

(i) $|f_n|$ converges in measure to $|f|$;

(ii) for all real α, β the sequence $\{\alpha f_n + \beta g_n\}$ converges in measure to $(\alpha f + \beta g)$;

(iii) if $f = 0$ a.e., then f_n^2 converges in measure to f^2;

(iv) the sequence $\{f_n g\}$ converges in measure to fg;

(v) the sequence $\{f_n^2\}$ converges in measure to f^2;

(vi) the sequence $\{f_n g_n\}$ converges in measure to fg;

(vii) if $f_n \neq 0$ a.e. all n, $f \neq 0$ a.e., the sequence $\{1/f_n\}$ converges in measure to $1/f$.

Is the condition $\mu(\Omega) < \infty$ essential for all these results?

7.3 Convergence in pth mean

All the definitions of the present section can be made relative to an arbitrary E in $\mathscr{F}$. Since we could restrict μ to the σ-field $\mathscr{F} \cap E$ of subsets of E, there is no loss in generality in making our definitions in terms of Ω, the whole space. In Chapter 5 we saw that $f \in \mathscr{M}$ is μ-integrable (over Ω) if and only if $|f|$ is μ-integrable. Further we saw that the subset of L_1 of M consisting of μ-integrable functions is a linear space (here we define $(\alpha f + \beta g)(x)$ arbitrarily on the set of zero measure where it is not defined because it involves $+\infty + (-\infty)$). Further for $f, g \in L_1$,

$$\rho_1(f, g) = \int |f - g| \, d\mu$$

is finite. By theorem 5.6, $\rho(f, g) = 0$ if and only if $f = g$ a.e., so that if we take equivalence classes of functions equal a.e. to form the linear space $\mathscr{L}_1 \subset \mathscr{M}$ we see that

$$\rho_1(f, g) = \rho_1(g, f) \quad \text{for all} \quad f, g \in \mathscr{L}_1,$$

$$\rho_1(f, g) = 0 \quad \text{if and only if} \quad f = g \text{ in } \mathscr{L}_1.$$

The triangle inequality

$$\rho_1(f, h) \leqslant \rho_1(f, g) + \rho_1(g, h) \quad \text{for} \quad f, g, h \in \mathscr{L}_1$$

also follows by integrating

$$|f(x) - h(x)| \leqslant |f(x) - g(x)| + |g(x) - h(x)|,$$

so that ρ_1 defines a metric in the space $\mathscr{L}_1$.

Convergence in mean

A sequence $\{f_n\}$ of functions in L_1 (or in $\mathscr{L}_1$) is said to converge in mean to a function f in L_1 if $\rho_1(f_n, f) \to 0$ as $n \to \infty$. A sequence $\{f_n\}$ of functions in L_1 is a Cauchy sequence in mean if $\rho_1(f_n, f_m) \to 0$ as $n, m \to \infty$.

Convergence in mean is the special case $p = 1$ of convergence in pth mean. Since most of the proofs are the same for $p = 1$ and $p \geqslant 1$, it is convenient to consider this at the same time.

The class $\mathscr{L}_p$

For $p \geqslant 1$, a function f in M is said to be of class L_p if $|f|^p$ is μ-integrable. Since

$$|f(x)+g(x)| \leqslant \begin{cases} 2|f(x)|, & \text{if } \; |f(x)| \geqslant |g(x)|, \\ 2|g(x)|, & \text{if } \; |g(x)| \geqslant |f(x)|; \end{cases}$$

we have, for all x,

$$|f(x)+g(x)|^p \leqslant 2^p\{|f(x)|^p + |g(x)|^p\}. \tag{7.3.1}$$

Thus, if $f, g \in L_p$ we must have $(f \pm g) \in L_p$. With the usual convention about the set of zero measure where $(\alpha f + \beta g)$ may not be defined, it follows that L_p is a linear space. For $f, g \in L_p$ we define

$$\rho_p(f,g) = \left[\int |f-g|^p \, d\mu\right]^{1/p}$$

and notice again that $\rho_p(f,g) = 0$ if and only if $f = g$ a.e. so that in the space $\mathscr{L}_p \subset \mathscr{M}$ of equivalence classes we have

$$\rho_p(f,g) = \rho_p(g,f),$$
$$\rho_p(f,g) = 0 \quad \text{if and only if} \quad f = g \text{ in } \mathscr{L}_p.$$

We will prove in the next section that ρ_p satisfies the triangle inequality, which shows that it is a metric in $\mathscr{L}_p$. However, we can now define:

Convergence in pth mean

A sequence $\{f_n\}$ of functions in L_p (or in $\mathscr{L}_p$) is said to converge in pth mean to a function f in L_p if $\rho_p(f_n, f) \to 0$ as $n \to \infty$. A sequence $\{f_n\}$ of functions in L_p is a Cauchy sequence in pth mean if $\rho_p(f_n, f_m) \to 0$ as $n, m \to \infty$.

It is immediate, by (7.3.1) that convergence in pth mean to a function implies that we have a Cauchy sequence in pth mean. Completeness for this type of convergence can now be proved.

Theorem 7.3. *For $p \geqslant 1$, if $\{f_n\}$ is a sequence of functions in L_p which is a Cauchy sequence in pth mean, then there is an f in L_p such that $f_n \to f$ in pth mean.*

Proof. We again use the device of obtaining a subsequence which will converge a.e. to f. For any $\epsilon > 0$, let $N(\epsilon)$ denote an integer such that

$$\int |f_r - f_s|^p \, d\mu < \epsilon^{p+1} \quad \text{for} \quad r, s \geqslant N(\epsilon).$$

Put $N_k = N(\epsilon 2^{-k})$, and assume that $N_{k+1} > N_k$ for each integer k. Then

$$\mu(E(\epsilon, r, s)) < \epsilon \quad \text{for} \quad r, s \geq N(\epsilon),$$

where

$$E(\epsilon, r, s) = \{x : |f_r - f_s| \geq \epsilon\}.$$

If we put

$$E_k = E(\epsilon 2^{-k}, N_{k+1}, N_k),$$

$$F_k = \bigcup_{i=k}^{\infty} E_i,$$

we have $\mu(E_k) < 2^{-k}\epsilon$, $\mu(F_k) < 2^{1-k}\epsilon$, and if x is not in F_k,

$$|f_{N_{i+1}}(x) - f_{N_i}(x)| < \epsilon 2^{-i} \quad \text{for all} \quad i \geq k.$$

Hence the series $\sum_{i=1}^{\infty} (f_{N_{i+1}} - f_{N_i})$ converges outside $F = \bigcap_{k=1}^{\infty} F_k$ and $\mu(F) = 0$. Suppose then that $f_{N_i} \to f$ a.e. For a fixed integer r, if we put $g_i = |f_{N_i} - f_r|^p$, $g = |f - f_r|^p$ we obtain a sequence g_i of non-negative measurable functions with $\liminf g_i = \lim g_i = g$ a.e. By theorem 5.7 (Fatou) we have

$$\int g \, d\mu \leq \liminf_{i \to \infty} \int |f_{N_i} - f_r|^p \, d\mu < \epsilon \quad \text{if} \quad r > N(\epsilon).$$

Hence, g is integrable, so that $(f - f_r) \in L_p$ which implies that $f \in L_p$. We have also proved that

$$\int |f - f_r|^p \, d\mu < \epsilon \quad \text{if} \quad r > N(\epsilon)$$

so that $f_r \to f$ in pth mean. ▌

It is worth remarking at this stage that the theorem corresponding to theorem 7.3 for Riemann integrals over a finite interval is false. It is not difficult to construct an example of a sequence of functions whose pth powers are Riemann integrable and which Cauchy converges in pth mean, but for which the limit is necessarily discontinuous on a set of positive measure and so cannot be Riemann integrable by theorem 5.9 (see exercise 7.3 (10)). Thus theorem 7.3 exhibits another way in which the Lebesgue integral is a big improvement on the Riemann integral.

We now relate convergence in pth mean to convergence in measure.

***Theorem* 7.4.** *If $\{f_n\}$ is a sequence of functions of L_p ($p \geq 1$) which is a Cauchy sequence in pth mean then $\{f_n\}$ is a Cauchy sequence in measure. If $f_n \to f$ in pth mean, then $f_n \to f$ in measure.*

Proof. For any h in L_p, $\eta > 0$

$$\mu\{x : |h(x)| \geq \eta^{1/2p}\} > \eta^{\frac{1}{2}} \Rightarrow \int |h|^p \, d\mu \geq \eta.$$

If $\{f_n\}$ is not a Cauchy sequence in measure, then there is an $\epsilon > 0$, $\delta > 0$ for which

$$\mu\{x: |f_n(x) - f_m(x)| \geqslant \epsilon\} > \delta$$

for infinitely many n, m. If now $\eta > 0$ is small enough to ensure that $\epsilon \geqslant \eta^{1/2p}$, $\delta > \eta^{\frac{1}{2}}$ we have

$$\int |f_n(x) - f_m(x)|^p \, d\mu \geqslant \eta > 0$$

for infinitely many n, m so that $\{f_n\}$ is not a Cauchy sequence in pth mean. This proves the first statement: the second part of the theorem is proved similarly. ∎

Remark. The example after theorem 7.2 shows that $\{f_n\}$ may converge in pth mean but not converge a.e., though theorems 7.2, 7.3 together show that there must be a subsequence $\{f_{n_i}\}$ which converges a.e. If we consider Lebesgue measure in $\mathbf{R}$ and put

$$f_n(x) = \begin{cases} n^{-1/p} & \text{for } \quad x \text{ in } [0, n], \\ 0 & \text{otherwise,} \end{cases}$$

$$g_n(x) = \begin{cases} n^{1/p} & \text{for } \quad x \text{ in } [0, 1/n], \\ 0 & \text{otherwise.} \end{cases}$$

we see that $f_n \to 0$ uniformly (and therefore almost uniformly, a.e., and in measure) but not in pth mean. If $\Omega = [0, 1]$, then $g_n \to 0$ almost uniformly, a.e. and in measure, but not in pth mean so that even in a finite measure space we cannot deduce convergence in mean from other types of convergence without some additional condition, even if the functions concerned are all in $\mathscr{L}_p$. The next definition turns out to be appropriate:

Set functions equicontinuous at $\varnothing$

Suppose ν_i $(i \in I)$ is a family of set functions defined on a σ-field $\mathscr{F}$. The family is said to be equicontinuous at $\varnothing$ if, given $\epsilon > 0$ and any sequence $\{B_n\}$ of sets of $\mathscr{F}$ which decreases to $\varnothing$, there is an integer N such that

$$|\nu_i(B_n)| < \epsilon \quad \text{for all} \quad i \in I, \quad n \geqslant N.$$

In § 6.4 we saw that a set function ν was absolutely continuous with respect to μ if, given $\epsilon > 0$ there is a $\delta > 0$ such that, for

$$E \in \mathscr{F}, \quad \mu(E) < \delta \Rightarrow |\nu(E)| < \epsilon;$$

and that this condition was also necessary if ν was a finite valued measure. This makes the following definition reasonable:

Uniform absolute continuity

Any family ν_i $(i \in I)$ of set functions defined on $\mathscr{F}$ is said to be uniformly absolutely continuous with respect to μ if, given $\epsilon > 0$ there is a $\delta > 0$ such that, for $E \in \mathscr{F}$, $\mu(E) < \delta \Rightarrow |\nu_i(E)| < \epsilon$ for all i. To see what this condition means, suppose ν_i $(i \in I)$ is a family of measures each of which is absolutely continuous with respect to μ, but such that the family is *not* uniformly absolutely continuous. Then there is an $\epsilon > 0$, and a sequence $\{B_n\}$ of sets of $\mathscr{F}$ with indices $\{i_n\}$ such that

$$\mu(B_n) < 2^{-n}, \quad \nu_{i_n}(B_n) \geqslant \epsilon.$$

Put
$$A_k = \bigcup_{n=k}^{\infty} B_n, \quad C = \lim_{k \to \infty} A_k.$$

Then $\mu(C) = 0$ and $\lim (A_k - C) = \varnothing$. It follows that

$$\nu_{i_k}(A_k - C) = \nu_{i_k}(A_k) \geqslant \nu_{i_k}(B_k) \geqslant \epsilon > 0$$

so that, by considering the sequence $\{A_k - C\}$ which decreases to $\varnothing$, we see that the family ν_i $(i \in I)$ is not equicontinuous at $\varnothing$. Thus we have proved

Lemma. *Suppose ν_i $(i \in I)$ is a family of measures on $\mathscr{F}$ each of which is absolutely continuous w.r.t. μ. Then if the family is equicontinuous at $\varnothing$, it is uniformly absolutely continuous w.r.t. μ.*

***Theorem* 7.5.** *Suppose $\{f_n\}$ is a sequence of functions of L_p and*

$$\nu_n(E) = \int_E |f_n|^p \, d\mu, \quad (E \in \mathscr{F}, n = 1, 2, \ldots).$$

(i) *$\{f_n\}$ is a Cauchy sequence in pth mean if and only if $\{f_n\}$ is a Cauchy sequence in measure and the family $\{\nu_n\}$ of measures is equicontinuous at $\varnothing$.*

(ii) *The sequence $\{f_n\}$ converges to f in pth mean if f_n converges to f in measure and $\{\nu_n\}$ is equicontinuous at $\varnothing$.*

Proof. (i) Suppose first that $\{f_n\}$ is a Cauchy sequence in pth mean. Then by theorem 7.4 $\{f_n\}$ is a Cauchy sequence in measure. For each $\epsilon > 0$, there is an N such that

$$\int |f_n - f_N|^p \, d\mu < \frac{\epsilon}{2^{p+1}} \quad \text{for} \quad n \geqslant N.$$

Now suppose $\{B_k\}$ is a sequence of sets of $\mathscr{F}$ decreasing to $\varnothing$. Since ν_n is absolutely continuous for $n = 1, 2, \ldots, N$ we can find, by theorem 5.6 an integer k_0 such that

$$\int_{B_k} |f_n|^p < \frac{\epsilon}{2^{p+1}} \quad \text{for} \quad k \geqslant k_0 \quad (n = 1, 2, \ldots, N).$$

By (7.3.1) we obtain, for $n \geqslant N$, $k \geqslant k_0$,

$$\int_{B_k} |f_n|^p \, d\mu \leqslant 2^p \int_{B_k} |f_N|^p \, d\mu + 2^p \int_{B_k} |f_n - f_N|^p \, d\mu$$

$$< \frac{\epsilon}{2} + 2^p \int_{\Omega} |f_n - f_N|^p \, d\mu < \epsilon,$$

so that the sequence $\{\nu_n\}$ is equicontinuous at $\varnothing$.

In the other direction, since we assume that μ is σ-finite on Ω, there must be a sequence $\{E_n\}$ in $\mathscr{F}$ which decreases to $\varnothing$ and is such that $\mu(\Omega - E_n)$ is finite for all n. Given $\epsilon > 0$, the equicontinuity of ν_n now ensures that there is a set $E = E_N$ with $\mu(\Omega - E) < \infty$ and

$$\int_E |f_n|^p \, d\mu < \frac{\epsilon}{2^{p+2}} \quad \text{for all } n.$$

Thus, for all m, n, by (7.3.1)

$$\int_E |f_n - f_m|^p \, d\mu < \tfrac{1}{2}\epsilon. \tag{7.3.2}$$

Now put $\Omega - E = F$, $\mu(F) = \lambda$. By the lemma, the sequence $\{\nu_n\}$ of measures must be uniformly absolutely continuous. We can therefore find an $\eta > 0$ such that, for $B \in \mathscr{F}$, $\mu(B) < \eta$,

$$\nu_n(B) = \int_B |f_n|^p \, d\mu < \frac{\epsilon}{2^{p+3}}. \tag{7.3.3}$$

For each m, n put

$$C_{m,n} = \left\{x \colon |f_m(x) - f_n(x)| \geqslant \left(\frac{\epsilon}{4\lambda}\right)^{1/p}\right\}.$$

Then $$\int_{F - C_{m,n}} |f_m - f_n|^p \, d\mu \leqslant \frac{\epsilon}{4\lambda}\mu(F - C_{m,n}) \leqslant \frac{\epsilon}{4\lambda}\mu(F) = \tfrac{1}{4}\epsilon.$$

Since $\{f_n\}$ is a Cauchy sequence in pth mean we can find an n_0 such that $\mu(C_{m,n}) < \eta$ for $m \geqslant n_0$, $n \geqslant n_0$. This gives, by (7.3.3),

$$\int_{C_{m,n}} |f_m - f_n|^p \, d\mu \leqslant 2^p \int_{C_{m,n}} |f_m|^p \, d\mu + 2^p \int_{C_{m,n}} |f_n|^p \, d\mu < \tfrac{1}{4}\epsilon,$$

so that $$\int_F |f_m - f_n|^p \, d\mu < \tfrac{1}{2}\epsilon \quad \text{for} \quad m, n \geqslant n_0.$$

This, together with (7.3.2) gives

$$\int_\Omega |f_m - f_n|^p \, d\mu < \epsilon \quad \text{for} \quad m, n \geqslant n_0.$$

(ii) If $f_n \to f$ in measure, then $\{f_n\}$ is a Cauchy sequence in measure so that by (i) the condition that $\{\nu_n\}$ is equicontinuous implies that $\{f_n\}$ is a Cauchy sequence in pth mean. By theorem 7.3, there exists a $g \in L_p$ such that $f_n \to g$ in pth mean. By theorem 7.4 (i), $f_n \to g$ in measure so that we have $f = g$ a.e. and it follows that $f_n \to f$ in pth mean.]

We can now slightly strengthen the dominated convergence theorem (5.8).

***Theorem* 7.6.** *Suppose $p \geqslant 1$, and $\{f_n\}$ is a sequence of measurable functions with $|f_n|^p \leqslant h \in L_1$ for each n. If either $f_n \to f_0$ in measure or $f_n \to f_0$ a.e., then $f_n \to f_0$ in pth mean.*

Proof. We must have $\nu_n(E) = \int_E |f_n|^p \, d\mu \leqslant \int_E h d\mu$, so that the family $\{\nu_n\}$ is equicontinuous at $\varnothing$ by theorem 5.6. If $f_n \to f_0$ in measure we can apply theorem 7.5 (ii) to obtain the result. On the other hand, in the proof of theorem 7.5 (i) we only use convergence in measure on the subset F of Ω with $\mu(F)$ finite. On F, $f_n \to f_0$ a.e. implies $f_n \to f_0$ in measure by theorems 7.1, 7.2, so that the condition $f_n \to f_0$ a.e., together with equicontinuity at $\varnothing$ of $\{\nu_n\}$, implies convergence in pth mean of $\{f_n\}$.]

We have now defined convergence to a limit for sequences of functions in several different ways, and have proved completeness in each case. It may help to summarise the relationships by a number of diagrams (Figures 2 to 4). In each of these an arrow from A to B

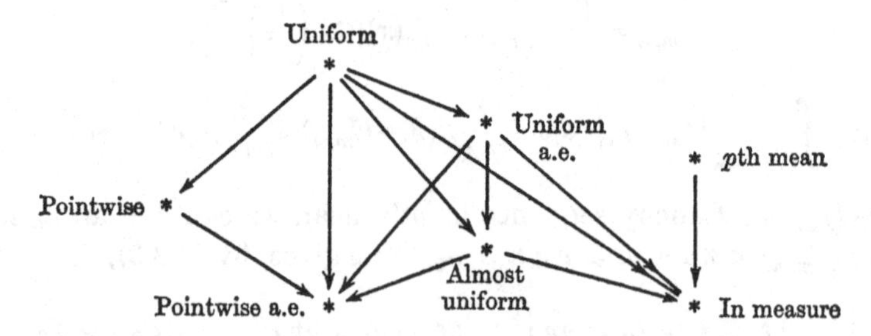

Fig. 2. No additional conditions.

indicates that convergence in sense A implies convergence in sense B. The lack of an arrow from A to B indicates that there is an example of a sequence satisfying the stated conditions which converges in sense A, but not in sense B. We assume throughout that we are considering functions in M which are a.e. finite.

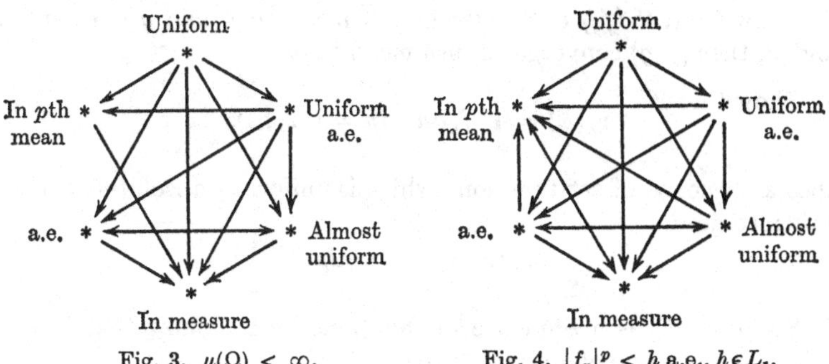

Fig. 3. $\mu(\Omega) < \infty$.

Fig. 4. $|f_n|^p < h$ a.e., $h \in L_1$.

Exercises 7.3

1. Check Figures 2, 3, 4, stating in each case the theorem or theorems which justifies $A \to B$, or the example which justifies the exclusion of $A \to B$.

2. Show that, if $\mu(\Omega) < \infty$, then the condition in theorem 7.5 that $\{\nu_n\}$ be equicontinuous at $\varnothing$ can be replaced by a condition that $\{\nu_n\}$ is uniformly absolutely continuous.

3. Show that if $\{f_n\}$ Cauchy converges uniformly a.e. and each f_n is integrable over E with $\mu(E)$ finite, then $f(x) = \lim f_n(x)$ a.e. is integrable over E and

$$\int_E |f_n - f| \, d\mu \to 0 \quad \text{as} \quad n \to \infty.$$

4. Suppose Ω is set of positive integers and μ is the counting measure. Then

(i) If

$$f_n(k) = \begin{cases} 1/n & \text{for} \quad 1 \leqslant k \leqslant n, \\ 0 & \text{for} \quad k > n, \end{cases}$$

show that $f_n(x) \to 0$ uniformly on Ω, each f_n is integrable but

$$\int f_n \, d\mu \nrightarrow \int \lim f_n \, d\mu.$$

This shows $\mu(E) < \infty$ is essential in question (3).

(ii) For the same sequence $\{f_n\}$ show that

$$\nu_n(E) = \int_E f_n \, d\mu$$

is uniformly absolutely continuous, but not equicontinuous at $\varnothing$. This shows that the condition $\mu(\Omega) < \infty$ is essential in question (2).

(iii) If

$$f_n(k) = \begin{cases} 1/k & \text{for} \quad 1 \leqslant k \leqslant n, \\ 0 & \text{for} \quad k > n, \end{cases}$$

show that $\{f_n\}$ is uniformly convergent on Ω, each f_n is integrable, but the limit is not.

5. Show that, if $\{f_n\}$ converges in pth mean to f, and g is essentially bounded, then $\{f_n g\}$ converges in pth mean to fg.

6. Show that if

$$\nu_n(E) = \int_E f_n d\mu \quad (n = 1, 2, \ldots),$$

defines a sequence of set functions which is uniformly absolutely continuous then so does

$$\lambda_n(E) = \int_E |f_n| \, d\mu.$$

7. Suppose $\{f_n\}$ is a sequence of functions in L_1. Show that $\{f_n\}$ is a Cauchy sequence in mean if and only if

$$\int_E f_n d\mu = x_n$$

is a Cauchy sequence of real numbers for every $E \in \mathscr{F}$, and $\{f_n\}$ is a Cauchy sequence in measure. Give an example of a sequence which does not converge in measure, for which

$$\lim_{n\to\infty} \int_E f_n d\mu = 0$$

for all E.

8. Suppose $\mu(\Omega) < \infty$, and for $f, g \in \mathscr{M}$, and a.e. finite;

$$\rho(f,g) = \int \frac{|f-g|}{1+|f-g|} d\mu.$$

Show that ρ defines a metric in the space of a.e. finite functions of $\mathscr{M}$, and that convergence in this metric is equivalent to convergence in measure.

9. Suppose $\mu(\Omega) < \infty \, (1 \leqslant q < p)$. Show that $\mathscr{L}_1 \supset \mathscr{L}_q \supset \mathscr{L}_p \supset \mathscr{L}_\infty$, and that $\rho_\infty(f,0) = \lim_{p\to\infty} \rho_p(f,0)$ for $f \in \mathscr{L}_\infty$. Show that the finite measure condition is essential.

By considering a suitable function on $[0,1]$ show that $\mathscr{L}_\infty \neq \bigcap_{p>1} \mathscr{L}_p$, but that if $f \in \bigcap_{p>1} \mathscr{L}_p - \mathscr{L}_\infty$, then $\rho_p(f,0) \to \infty$ as $p \to \infty$.

10. Suppose $\Omega = [0,1]$, μ is Lebesgue measure. Let K be a nowhere dense perfect set with positive measure and let $\{E_k\}$ be the set of disjoint open intervals such that $(0,1) - K = \bigcup_{i=1}^{\infty} E_i$. Let f_n be the indicator function of $F_n = \bigcup_{i=1}^{n} E_i$. Prove that $f_n \, (n = 1, 2, \ldots)$ is Riemann integrable and converges in mean to the indicator function f of $(0,1) - K$. By considering the construction of K, show that f is discontinuous a.e. on the set K of positive measure, and so cannot be Riemann integrable. This shows that the class of Riemann integrable functions is not complete with respect to convergence in mean.

7.4 Inequalities

We now obtain some inequalities which turn out to be important in several branches of analysis. We need to use the algebraic inequality

$$x^{\alpha}y^{\beta} \leqslant \alpha x + \beta y \tag{7.4.1}$$

for $x > 0, y > 0$, $\alpha > 0$, $\beta > 0$, $\alpha + \beta = 1$, which is strict unless $x = y$. This is most easily proved by taking logarithms to give

$$\alpha \log x + \beta \log y \leqslant \log(\alpha x + \beta y)$$

and using the fact that log: $\mathbf{R}^{+} \to \mathbf{R}$ is strictly concave since it has a negative second derivative.

Conjugate indices

If $p > 1$, $q > 1$ and $1/p + 1/q = 1$, we say that p and q are conjugate indices.

***Theorem* 7.7** (*Hölder*). *If p, q are conjugate indices, and $f \in L_p$, $g \in L_q$ then fg is integrable and, for each E in $\mathscr{F}$,*

$$\int_E |fg|\, d\mu \leqslant \left(\int_E |f|^p\, d\mu\right)^{1/p} \left(\int_E |g|^q\, d\mu\right)^{1/q}.$$

The inequality is strict unless there exist real numbers a, b such that $a|f|^p = b|g|^q$ a.e. on E.

Proof. If
$$\int_E |fg|\, d\mu = 0,$$

then the loose inequality is certainly satisfied and if the right-hand side is also zero then either $f = 0$ a.e. on E or $g = 0$ a.e. on E. In either case the condition $a|f|^p = b|g|^q$ a.e. is satisfied with $b = 0$ or $a = 0$, respectively. Hence we may assume that

$$\int_E |fg|\, d\mu > 0.$$

Put
$$E_1 = \{x: |g(x)| \leqslant |f(x)|^{p-1}\},$$
$$E_2 = \Omega - E_1.$$

Then
$$\text{for } x \in E_1, \quad |f(x)\,g(x)| \leqslant |f(x)|^p,$$
$$\text{for } x \in E_2, \quad |f(x)\,g(x)| \leqslant |g(x)|^q,$$

so that
$$|f(x)\,g(x)| \leqslant |f(x)|^p + |g(x)|^q, \quad \text{for all } x,$$

this implies that fg is integrable.

Given $E \in \mathscr{F}$, put $E_0 = E \cap \{x: f(x)\, g(x) = 0\}$. Then by our assumption $\mu(E-E_0) > 0$. For $x \in E-E_0$, we can apply (7.4.1) replacing

$$\alpha \text{ by } \frac{1}{p}, \quad \beta \text{ by } \frac{1}{q}, \quad x \text{ by } \left[\frac{|f(x)|^p}{\int_{E-E_0} |f|^p\, d\mu}\right] \quad \text{and} \quad y \text{ by } \left[\frac{|g(x)|^q}{\int_{E-E_0} |g|^q\, d\mu}\right].$$

This gives

$$\frac{|f(x)\, g(x)|}{\left(\int_{E-E_0} |f|^p\, d\mu\right)^{1/p} \left(\int_{E-E_0} |g|^q\, d\mu\right)^{1/q}} \leqslant \frac{|f(x)|^p}{p\int_{E-E_0} |f|^p\, d\mu} + \frac{|g(x)|^q}{q\int_{E-E_0} |g|^q\, d\mu}. \tag{7.4.2}$$

If we now integrate over $(E-E_0)$ and note that the right-hand side gives $1/p+1/q = 1$, we obtain the desired inequality for the integral over $(E-E_0)$. We can only obtain equality for the integrals if we have equality in (7.4.2) for almost all $x \in (E-E_0)$. The condition for equality in (7.4.1) now shows that we must have $a|f|^p = b|g|^q$ a.e. on $(E-E_0)$ where

$$\frac{1}{a} = \int_{E-E_0} |f|^p\, d\mu \quad \text{and} \quad \frac{1}{b} = \int_{E-E_0} |g|^q\, d\mu.$$

The inequality for the integrals over E now follows, and we can again only have equality if $f = g = 0$ a.e. on E_0, since otherwise the right-hand side is increased while the left-hand side remains the same on replacing $(E-E_0)$ by E. Thus equality can only occur if $a|f|^p = b|g|^q$ a.e. on E. ▌

Remark 1. The special case $p = q = 2$ of theorem 7.7 is called Schwartz's inequality. A simpler proof of this case is possible. See exercise 5.4 (13).

Remark 2. In the sense of theorem 7.7 the index conjugate to $p = 1$ is $q = \infty$. It is easy to prove directly that, if $f \in L_1$, $g \in L_\infty$ then

$$\int_E |fg|\, d\mu \leqslant \left(\int_E |f|\, d\mu\right) (\text{ess sup}\, |g\chi_E|).$$

***Theorem* 7.8** (*Minkowski*). *For $p \geqslant 1$, if $f, g \in L_p$ then $(f+g) \in L_p$ and for any $E \in \mathscr{F}$,*

$$\left\{\int_E |f+g|^p\, d\mu\right\}^{1/p} \leqslant \left(\int_E |f|^p\, d\mu\right)^{1/p} + \left(\int_E |g|^p\, d\mu\right)^{1/p}.$$

For $p > 1$, equality is strict unless there are real numbers $a > 0$, $b > 0$ such that $af = bg$ a.e. on E.

Proof. We already proved in §7.3 that L_p is a linear space. For $p = 1$, the result is immediate. For $p > 1$,

$$\int_E |f+g|^p \, d\mu = \int_E |f+g||f+g|^{p-1} \, d\mu$$

$$\leqslant \int_E |f||f+g|^{p-1} \, d\mu + \int_E |g||f+g|^{p-1} \, d\mu,$$

with equality if and only if f and g have the same sign a.e. in E. If we now apply theorem 7.7 to each of these integrals we obtain

$$\int_E |f+g|^p \, d\mu \leqslant \left(\int_E |f|^p \, d\mu\right)^{1/p} \left(\int_E |f+g|^p \, d\mu\right)^{1/q}$$

$$+ \left(\int_E |g|^p \, d\mu\right)^{1/p} \left(\int_E |f+g|^p\right)^{1/q}$$

with strict inequality unless there are numbers α, β, γ such that $\alpha|f|^p = \beta|f+g|^p = \gamma|g|^p$ a.e. We can only have equality in both inequalities if there is $a \geqslant 0$, $b \geqslant 0$ with $af = bg$ a.e. Provided it is not zero we now divide both sides by

$$\left(\int_E |f+g|^p \, d\mu\right)^{1/q}$$

to obtain the desired result. If

$$\int_E |f+g|^p \, d\mu = 0,$$

then the inequality is trivially satisfied, and equality is only possible if $f = g = 0$ a.e. ▌

The above theorem shows that

$$\rho_p(f, g) = \left(\int_E |f-g|^p \, d\mu\right)^{1/p}$$

defines a metric in $\mathscr{L}_p$.

We have proved the Hölder and Minkowski inequalities for general measure spaces. They are of course valid for Lebesgue measure in $\mathbf{R}$ either over a finite interval or over the whole line.

However, we can also apply these general theorems to the case where Ω is the set of positive integers $\mathbf{Z}$, and μ is the counting measure

$$\mu(E) = \text{number of integers in } E,$$

which makes all subsets $E \subset \mathbf{Z}$ measurable. Then functions $f: \mathbf{Z} \to \mathbf{R}$ and $g: \mathbf{Z} \to \mathbf{R}$ reduce to

$$f(i) = a_i, \quad g(i) = b_i \quad (i = 1, 2, \ldots),$$

where $\{a_i\}$ and $\{b_i\}$ are sequences of real numbers, and we can apply theorems 7.7, 7.8 to give:

Hölder. $$\sum_{i=1}^{\infty} |a_i b_i| \leqslant \left(\sum_{i=1}^{\infty} |a_i|^p\right)^{1/p} \left(\sum_{i=1}^{\infty} |b_i|^q\right)^{1/q}$$

in the sense that the convergence of both series on the right implies the convergence of the series on the left and the inequality. Further, equality is only possible if there is a constant k such that $|a_i|^p = k|b_i|^q$ for all i.

Minkowski.

$$\left(\sum_{i=1}^{\infty} |a_i + b_i|^p\right)^{1/p} \leqslant \left(\sum_{i=1}^{\infty} |a_i|^p\right)^{1/p} + \left(\sum_{i=1}^{\infty} |b_i|^p\right)^{1/p}$$

with equality if and only if there is a $k > 0$ such that $a_i = kb_i$ for all i.

It is interesting to see how these elementary inequalities (which can of course be proved directly) fall out of the general theorems by using a simple special measure.

Exercises 7.4

1. Suppose $\alpha > 0, \beta > 0, \gamma > 0, \alpha + \beta + \gamma = 1$ and $f \in L_\alpha$, $g \in L_\beta$, $h \in L_\gamma$. Show that

$$\int_E |fgh| \, d\mu \leqslant \left(\int_E |f|^{1/\alpha} d\mu\right)^{\alpha} \left(\int_E |g|^{1/\beta} d\mu\right)^{\beta} \left(\int_E |h|^{1/\gamma} d\mu\right)^{\gamma}.$$

Generalise to $n \geqslant 2$ functions.

2. If μ is Lebesgue measure on $I = (a, b)$ and $f \in L_p(I, \mu)$ show that there is a continuous g such that

$$\int_a^b |f(x) - g(x)|^p \, dx < \epsilon.$$

Deduce that $$\int_a^b |f(x+h) - f(x)|^p \, dx \to 0 \text{ as } h \to 0$$

(here f is given the value zero outside I). *Hint.* See theorem 5.10.

3. If $\mu(\Omega) < \infty$, $1 \leqslant p < q$ and $f \in L_q$ show that

$$[\rho_p(f, 0)] \leqslant k\rho_q(f, 0)$$

for a suitable constant k. Show that $\mu(\Omega) < \infty$ is essential.

4. Show by an example that theorem 7.8 is false for $p < 1$.

5. If p, q are conjugate indices, $f_n \to f_0$ in pth mean, $g_n \to g_0$ in qth mean, show that $f_n g_n \to f_0 g_0$ in mean.

6. By considering intervals of $\mathscr{P}^k$ whose coordinates are rational, and linear combinations of indicator functions of such intervals obtain a countable dense set in $\mathscr{L}_p$ for $\Omega = \mathbf{R}^k$, μ any Lebesgue–Stieltjes measure. Such a space $\mathscr{L}_p$ is therefore separable.

7.5* Measure preserving transformations from a space to itself

In §6.5 we discussed measurable transformations T from $(X, \mathscr{F}, \mu)$ to $(Y, \mathscr{S})$ and defined the measure μT^{-1} on $\mathscr{S}$ in terms of the transformation. A special case of this is obtained when $T: X \to X$ maps X into itself. We then say that T is *measure preserving* if, for every $E \in \mathscr{F}$, $\mu(T^{-1}(E)) = \mu(E)$. Given a mapping $T: X \to X$, we can define the iterates T^n obtained by composing T with itself n times. For convenience T^0 will denote the identity mapping, and T^{-n} will be defined as a set mapping

$$T^{-n}(E) = \{x : T^n(x) \in E\}$$

even if T^{-n} is not a point function.

If $\mathscr{F}$ is the σ-ring generated by a semi-ring $\mathscr{P}$, then it is immediate on applying the extension theorems of Chapters 3 and 4 that T is measure preserving if, and only if,

$$\mu(T^{-1}(E)) = \mu(E) \quad \text{for every } E \text{ in } \mathscr{P}.$$

If T is a (1, 1) transformation from X to itself, then it is said to be *invertible* and the condition for T to be measure preserving in this case can be written as $\mu(T(E)) = \mu(E)$ for all E in $\mathscr{F}$.

In §4.5 we considered the geometrical properties of Lebesgue measure and showed that the transformations of Euclidean space defined in terms of translations, rotations or reflexions are measure preserving. One can also prove that a matrix transformation of determinant 1 defines a measure preserving transformation in Euclidean space. All these are easily seen to be invertible.

If $\Omega = [0, 1)$ and $Tx = 2x$ reduced mod 1 then, for Lebesgue measure, T is seen to be measure preserving by considering the effect of T^{-1} on the dyadic intervals $[p/2^q, (p+1)/2^q)$ which form a semi-ring generating $\mathscr{B}$. If $x = .\alpha_1\alpha_2\ldots$ is the expansion of x as a binary 'decimal', then $Tx = .\alpha_2\alpha_3\ldots$. This T is not invertible.

It is worth remarking that the study of measure preserving transformations started with certain considerations in statistical mechanics. Suppose we have a system of k particles whose present state is described by a point in 'phase space' $\mathbf{R}^{6k}$ in which each particle determines

3 coordinates for position and 3 coördinates for momentum. Then the entire history of the system can be represented by a trajectory in phase space which is completely determined (assuming the laws of classical mechanics) by a single point on it. Thus for any (time) t we can define an invertible transformation T_t by saying that, for x in phase space, $T_t x$ denotes the state of a system which starts at x after a time t. One of the basic results in statistical mechanics (due to Liouville) states that, if the coordinates in phase space are correctly chosen, then the 'flow' in phase space leaves all volumes (i.e. Lebesgue measure in $\mathbf{R}^{6k}$) unchanged. This means that T_t becomes a measure preserving transformation in $(\mathbf{R}^{6k}, \mathscr{L}^{6k}, \mu)$. In practice k is enormous, and it is not possible to observe at any one moment all the particles of the system. Instead one asks questions like 'what is the probability that at time t the state of the system belongs to a given subset of phase space?' One then imposes conditions which ensure that this can be calculated by considering the 'average' behaviour of $T_t x$ as $t \to \infty$. To be more precise $T_{s+t} = T_s T_t$ so that $T_{nt} = (T_t)^n$ and one can consider a discrete model, count the proportion of times up to n that $T^i x \in E$ where $T = T_{t_0}$ and then let $n \to \infty$. In practice a set E in phase space is replaced by a function $f(x)$ (representing some physical measurement) and one considers the average behaviour in terms of the sequence

$$\frac{1}{n}\sum_{i=0}^{n-1} f(T^i x) \quad (n = 1, 2, \ldots).$$

This discussion of phase space provides a physical interpretation for the mathematical results which we now formulate precisely.

For the remainder of this section, T will denote a measure preserving transformation of Ω to itself, and $f: \Omega \to \mathbf{R}^*$ will denote an integrable function. We define

$$f_k(x) = f(T^k x) \quad (k = 0, 1, 2, \ldots).$$

Then f_k will be integrable and theorem 6.8 shows immediately that

$$\int f_k \, d\mu = \int f \, d\mu.$$

Before giving the proof (due to F. Riesz) of the point-wise ergodic theorem, we obtain a lemma which is an important step towards it.

Lemma. (*sometimes called the maximal ergodic theorem*). *Suppose E is the set of points $x \in \Omega$ such that*

$$\sum_{i=0}^{n} f_i(x) \geq 0$$

for at least one n: then

$$\int_E f d\mu \geqslant 0.$$

Proof. We first need a result about finite sequences of real numbers. Suppose $\alpha_1, \alpha_2 \ldots, \alpha_n \in \mathbf{R}$ and $m \leqslant n$. A term α_i of this sequence is called an *m-leader* if there is an integer p, $1 \leqslant p \leqslant m$, such that

$$\alpha_i + \alpha_{i+1} + \ldots + \alpha_{i+p-1} \geqslant 0.$$

For a fixed m, let α_k be the first m-leader, and let $(\alpha_k + \ldots + \alpha_{k+p-1})$ be the shortest non-negative sum that it leads. Then for every integer h with $k \leqslant h \leqslant k+p-1$, we must have $\alpha_h + \alpha_{h+1} + \ldots + \alpha_{k+p-1} \geqslant 0$, so that α_h is an m-leader. Now continue with the first m-leader in $\alpha_{k+p}, \ldots, \alpha_n$ and repeat the argument until all the m-leaders have been found. It follows that the sum of *all* the m-leaders of the original sequence must be non-negative, as it is the same as the sum of the non-negative shortest sums obtained by the above procedure.

We can now turn to the proof of the lemma and notice that, since f is integrable, we may assume that it is everywhere finite valued. If E_m denotes the set of x such that

$$\sum_{i=0}^{n} f_i(x) \geqslant 0$$

for at least one $n \leqslant m$, then E_m increases to E, so it is sufficient to prove

$$\int_{E_m} f d\mu \geqslant 0 \quad \text{for all } m.$$

For a positive integer n, let $s(x)$ be the sum of the m-leaders of the finite sequence $f_0(x), f_1(x), \ldots, f_{n+m-1}(x)$. Let A_k be the set of x for which $f_k(x)$ is an m-leader and let χ_k be the indicator function of A_k. Since A_k is measurable, and $s(x) = \sum_{k=0}^{n+m-1} \chi_k(x) f_k(x)$, s is measurable and integrable and $s(x) \geqslant 0$ so that

$$\sum_{k=0}^{n+m-1} \int_{A_k} f_k d\mu \geqslant 0. \tag{7.5.1}$$

Now notice that, for $k = 1, 2, \ldots, n-1$, $T(x) \in A_{k-1}$ if and only if $f_{k-1}(Tx) + \ldots + f_{k-1+p-1}(Tx) \geqslant 0$ for some $p \leqslant m$, which is equivalent to $f_k(x) + \ldots + f_{k+p-1}(x) \geqslant 0$ for some $p \leqslant m$ which in turn is the condition for $x \in A_k$. Thus $A_k = T^{-1}A_{k-1} = T^{-k}A_0$ for $k = 1, \ldots, n-1$. Hence by theorem 6.8,

$$\int_{A_k} f_k(x)\, d\mu = \int_{T^{-k}A_0} f(T^k x)\, d\mu = \int_{A_0} f(x)\, d\mu$$

so that the first n terms of (7.5.1) are all equal. Now $A_0 = E_m$, so that (7.5.1) implies

$$n\int_{E_m} f d\mu + m\int |f| d\mu \geqslant 0.$$

Divide by n, keep m fixed and let $n \to \infty$ to give

$$\int_{E_m} f d\mu \geqslant 0. ∎$$

***Theorem* 7.9** (*Birkhoff*). *Suppose T is a measure preserving transformation on a σ-finite measure space $(\Omega, \mathscr{F}, \mu)$ to itself and $f:\Omega \to \mathbf{R}^*$ is integrable. Then*

(i) $\frac{1}{n}\sum_{i=0}^{n-1} f(T^i x)$ *converges point-wise a.e.;*

(ii) *the limit function $f^*(x)$ is integrable and invariant under T (i.e. $f^*(Tx) = f^*(x)$ a.e.);*

(iii) *if $\mu(\Omega) < \infty$, then $\int f^* d\mu = \int f d\mu$.*

Proof (i). Suppose r, s are rational numbers $r < s$ and $B = B(r, s)$ is the set of points x for which

$$\liminf \left[\frac{1}{n}\sum_{i=0}^{n-1} f_i(x)\right] < r < s < \limsup \left[\frac{1}{n}\sum_{i=1}^{n-1} f_i(x)\right].$$

It is immediate that B is measurable and invariant under T. Our result will now follow if we can prove that $\mu(B(r,s)) = 0$ for all rational r, s. The first step in this direction is to show that $\mu(B) < \infty$.

We may assume without loss of generality that $s > 0$, for otherwise the argument can be carried out with f replaced by $-f$. Suppose $C \in \mathscr{F}$, $\mu(C) < \infty$, $C \subset B$, and χ is the indicator function of C. Apply the lemma to the function $(f - s\chi)$ to give

$$\int_E (f - s\chi) d\mu \geqslant 0,$$

where E is defined in the lemma. If $x \in B$, then at least one of the averages

$$\frac{1}{n}\sum_{i=0}^{n-1} f_i(x) > s > 0$$

so that at least one of the sums

$$\sum_{i=0}^{n-1} \{f(T^i x) - s\chi(T^i x)\} \geqslant 0,$$

and it follows that $x \in E$. Thus

$$\int_E f d\mu \geqslant \int_E s\chi d\mu \quad \text{so that} \quad \mu(C) \leqslant \frac{1}{s}\int_E |f| \, d\mu.$$

Since B has σ-finite measure and its subsets of finite measure have bounded measure, it follows that $\mu(B)$ is finite. Since B is invariant under T we can restrict our σ-field and measure to B and think of T as a measure preserving transformation on B. Apply the lemma again to the integrable function $(f-s)$, and note that in this case the set E of the lemma is the whole space B. This gives

$$\int_B (f-s)\,d\mu \geqslant 0.$$

Similarly, we can obtain $\displaystyle\int_B (r-f)\,d\mu \geqslant 0.$

Together these give $\displaystyle\int_B (r-s)\,d\mu \geqslant 0.$

Since $r < s$, we must have $\mu(B) = 0$.

(ii) Put

$$f^*(x) = \lim_{n\to\infty}\left\{\frac{1}{n}\sum_{i=0}^{n-1} f_i(x)\right\}$$

when the sequence converges. Then it is immediate that f^* is measurable and invariant. Further

$$\int\left|\frac{1}{n}\sum_{i=0}^{n-1} f_i(x)\right| d\mu \leqslant \frac{1}{n}\sum_{i=0}^{n-1}\int |f_i(x)|\,d\mu = \int |f(x)|\,d\mu$$

so that, by theorem 5.7 (Fatou) f^* is integrable, and

$$\int |f^*|\,d\mu \leqslant \int |f|\,d\mu.$$

(iii) For fixed n, put

$$D(k,n) = \left\{x: \frac{k}{2^n} \leqslant f^*(x) < \frac{k+1}{2^n}\right\}$$

and apply the lemma to the transformation T on the set $D(k,n)$ which can be assumed to be invariant. Then $f^*(x) \geqslant k/2^n$ in $D(k,n)$, so that at least one of the sums

$$\sum_{i=0}^{n-1}\left(f_i(x) - \frac{k}{2^n} + \epsilon\right) > 0$$

for each $\epsilon > 0$. Hence

$$\int_{D(k,n)} f\,d\mu \geqslant \left(\frac{k}{2^n} - \epsilon\right)\mu(D(k,n)),$$

and so we must have

$$\int_{D(k,n)} f\,d\mu \geqslant \frac{k}{2^n}\,\mu(D(k,n)).$$

Similarly
$$\int_{D(k,n)} f\,d\mu \leqslant \frac{k+1}{2^n}\mu(D(k,n))$$
and
$$\frac{k}{2^n}\mu(D(k,n)) \leqslant \int_{D(k,n)} f\,d\mu \leqslant \frac{k+1}{2^n}\mu(D(k,n)).$$
For each integer k, it follows that
$$\left|\int_{D(k,n)} f^*\,d\mu - \int_{D(k,n)} f\,d\mu\right| \leqslant \frac{1}{2^n}\mu(D(k,n));$$
and if we sum over k
$$\left|\int_\Omega f^*\,d\mu - \int_\Omega f\,d\mu\right| \leqslant \frac{1}{2^n}\mu(\Omega).$$
Since n is arbitrary we must have $\int f^*\,d\mu = \int f\,d\mu$. ∎

For applications to statistical mechanics one would expect the equilibrium value $f^*(x)$ to be independent of the point x, so that the limit function f^* of theorem 7.9 is a constant. Unfortunately this is not true without imposing an additional condition.

Ergodic transformation

A measure preserving transformation T is said to be ergodic (or metrically transitive or metrically invariant) if for all invariant sets E (sets for which $T^{-1}(E) = (E)$), $\mu(E) = 0$ or $\mu(\Omega - E) = 0$.

Lemma. *T is ergodic if and only if every measurable invariant function is constant a.e.*

Proof. Suppose g is measurable and invariant. Then $\{x: g(x) \geqslant a\}$ is invariant for all real a, and must either have zero measure or be the complement of a set of zero measure. Hence g = constant a.e., if T is ergodic. Conversely, if every measurable invariant function is constant a.e., since the indicator function of an invariant set is an invariant function, there cannot be any invariant sets other than null sets and complements of null sets. ∎

We can now apply this lemma to theorem 7.9 when T is ergodic. There are two cases:

(i) $\mu(\Omega) = +\infty$. Since the only constant which is integrable over a space of infinite measure is zero we deduce that
$$\frac{1}{n}\sum_{i=0}^{n-1} f_i(x) \to 0 \quad \text{a.e.}$$
(ii) $\mu(\Omega) < \infty$. We can integrate $f^* = c$ a.e. by (iii) to obtain
$$\frac{1}{n}\sum_{i=0}^{n-1} f_i(x) \to \frac{1}{\mu(\Omega)}\int f\,d\mu \quad \text{a.e.}$$

This last result ties up with our remarks about statistical mechanics, since it shows that the average value of f on the discrete trajectory approaches the average value of f in phase space for all starting points x except for a possible null set.

The reader who wishes to learn more about ergodic theory is advised to read P. R. Halmos, *Lectures on Ergodic Theory* (Chelsea, 1956).

Exercises 7.5

1. Suppose Ω is the real line, T is the translation $Tx = x+1$, f is the indicator function of $[0, 1]$. What is

$$f^*(x) = \lim \frac{1}{n} \sum_{i=0}^{n-1} f_i(x)$$

in this case? Show that theorem 7.9 (iii) is not satisfied without the condition $\mu(\Omega) < \infty$.

2. Suppose T is measure preserving and ergodic on $(\Omega, \mathscr{F}, \mu)$ and

$$\mu(\Omega) < \infty.$$

If f is non-negative measurable and

$$\frac{1}{n} \sum_{i=0}^{n-1} f(T^i x) \to c \in \mathbf{R} \text{ a.e.},$$

deduce that f is integrable.

3. Suppose Ω is five point space $\{a, b, c, d, e\}$, $\mathscr{F}$ is the set of all subsets, $\mu\{a\} = \mu\{b\} = \mu\{c\} = 1$ and $\mu\{d\} = \mu\{e\} = 2$, T is the permutation $(a, b, c)\,(d, e)$. Show that T is measure preserving but not ergodic. Find the f^* of theorem 7.9 if f is the indicator function of $\{a, b, e\}$.

4. Suppose $(\Omega, \mathscr{F}, \mathsf{P})$ is a probability measure. Form the doubly infinite Cartesian product of copies of $(\Omega, \mathscr{F}, \mathsf{P})$ labelled $(\ldots, -2, -1, 0, 1, \ldots, n, \ldots)$ and the product measure by the process of theorem 6.3. If a point of this product space is $w = (\ldots, w_{-1}, w_0, w_1, \ldots)$ and T is the shift $(Tw)_n = w_{n+1}$; show that T is measure preserving and ergodic.

5. If $\Omega = [0, 1)$, $Tx = 2x (\text{mod } 1)$, and μ is Lebesgue measure, show that T is ergodic. By applying the ergodic theorem to the indicator function of $[0, \frac{1}{2})$ deduce the Borel normal number theorem which states that in the binary expansion $0 \cdot \alpha_1 \alpha_2 \ldots \alpha_n \ldots$ of real numbers in $[0, 1)$, the density

$$\frac{1}{n} \sum_{i=1}^{n} \alpha_i \to \tfrac{1}{2} \quad \text{for almost all } x.$$

6. Suppose T is ergodic and measure preserving on a finite measure space $(X, \mathscr{F}, \mu)$ and f, g are the indicator functions of measurable sets F, G. Show that

$$\lim \left[\frac{1}{n} \sum_{i=0}^{n-1} \mu((T^{-i} F) \cap G)\right] = \frac{\mu(F)\,\mu(G)}{\mu(X)}.$$

8
LINEAR FUNCTIONALS

In this chapter all measure spaces $(\Omega, \mathscr{F}, \mu)$ will be σ-finite, and $\mathscr{F}$ will be complete with respect to μ, unless stated otherwise. In Chapter 7 we saw that $\mathscr{L}_p$ $(1 \leqslant p < \infty)$ with the metric

$$\rho_p(f,g) = \left\{\int |f-g|^p \, d\mu\right\}^{1/p},$$

and $\mathscr{L}_\infty$ with the metric

$$\rho_\infty(f,g) = \operatorname{ess\,sup} |f-g|,$$

were complete metric spaces. We also proved they were linear spaces (over the reals); and it is immediate that the metric defines a norm

$$\|f\|_p = \rho_p(f,0) \quad (1 \leqslant p \leqslant \infty),$$

for which the spaces are normed linear spaces. Thus

$$\|f\|_p > 0 \quad \text{if} \quad f \neq 0, \quad \|0\| = 0,$$

$$\|f+g\|_p \leqslant \|f\|_p + \|g\|_p,$$

$$\|\alpha f\|_p = |\alpha| \|f\|_p \quad \text{for} \quad \alpha \in \mathbf{R}.$$

We will omit the suffix p in $\|.\|_p$ when it is clear which $\mathscr{L}_p$ space is being considered.

It turns out that the space $\mathscr{L}_2$ has some special properties not shared by other $\mathscr{L}_p$ spaces. These can be postulated in terms of the difference between Hilbert space and Banach space, but we prefer to examine, in the first three sections of this chapter, the structure of $\mathscr{L}_2$ and then discuss later the analogous properties of more general normed linear spaces.

8.1 Dependence of $\mathscr{L}_2$ on the underlying $(\Omega, \mathscr{F}, \mu)$

In general, the structure of the space $\mathscr{L}_2$ depends on the underlying space $(\Omega, \mathscr{F}, \mu)$: when we want to emphasise this we use the notation $\mathscr{L}_2(\Omega, \mu)$. We first examine conditions on $(\Omega, \mathscr{F}, \mu)$ which will ensure that $\mathscr{L}_2(\Omega, \mu)$ is separable (in the topology of the norm). We later define (real) *Hilbert space* in terms of its abstract properties, and show that $\mathscr{L}_2(\Omega, \mu)$ is always a realisation of Hilbert space.

Countable basis for measure

In the measure space $(\Omega, \mathscr{F}, \mu)$ we can use the equivalence relation $A \sim B \Leftrightarrow \mu(A \triangle B) = 0$ to identify the subsets in $\mathscr{F}$ which differ only by a set of measure zero. If we denote the resulting quotient space by $\mathscr{F}_\mu$, it is clear that, when $\mu(\Omega) < \infty$, $\mathscr{F}_\mu$ is a metric space with the metric $\rho(A, B) = \mu(A \triangle B)$, and one can further show that $\mathscr{F}_\mu$ is complete. In this case we can define a dense subset by means of the topology of this metric. However, the notion of a dense subset in $\mathscr{F}_\mu$ can be extended to include the case $\mu(\Omega) = \infty$ by a device which makes sense provided μ is σ-finite on Ω. Thus we say that μ has a *countable basis* if there is a sequence $\{E_n\}$ of sets in $\mathscr{F}$ such that, given $\epsilon > 0$ and any $A \in \mathscr{F}$ with $\mu(A) < \infty$, there is a set E_k of the sequence for which

$$\mu(A \triangle E_k) < \epsilon.$$

In Chapters 3 and 4 we saw how measures could be obtained by extending a measure already defined on a semi-ring. If μ can be defined by extending a finite measure on a semi-ring $\mathscr{P}$ which contains only a countable number of sets, then μ has a countable basis. For the ring $\mathscr{R}$ generated by $\mathscr{P}$ is countable, and forms a basis for $\mathscr{F}$ by theorem 4.4. In particular, in the definition of Lebesgue measure, we could have used the countable semi-ring of $\frac{1}{2}$-open intervals, whose bounding coordinates are rationals, to generate the σ-field $\mathscr{B}^k$ of Borel sets in $\mathbf{R}^k$; so it follows that Lebesgue measure in $\mathbf{R}^k$ has a countable basis.

We first obtain a condition equivalent to the existence of a countable basis for μ.

Lemma. *A measure μ has a countable basis if and only if, for each $\epsilon > 0$, any collection $\mathscr{C} \subset \mathscr{F}$ of subsets of finite measure such that*

$$A, B \in \mathscr{C}, \quad A \neq B \Rightarrow \mu(A \triangle B) \geqslant \epsilon \tag{8.1.1}$$

is countable.

Proof. Suppose first that $\epsilon > 0$ is such that there is a non-countable $\mathscr{C}$ satisfying (8.1.1); and suppose if possible that μ has a countable basis $\mathscr{D}$. Then, for each $A \in \mathscr{C}$ we can find a set $E_A \in \mathscr{D}$ with

$$\mu(A \triangle E_A) < \tfrac{1}{3}\epsilon.$$

Then, if $A \neq B$,

$$\mu(E_A \triangle E_B) \geqslant \mu(A \triangle B) - \mu(A \triangle E_A) - \mu(B \triangle E_B) > \tfrac{1}{3}\epsilon > 0,$$

so that $E_A \neq E_B$. Thus if $\mathscr{D}$ is dense, it contains a non-countable subclass, which is a contradiction.

Conversely, suppose the condition is satisfied. Then for each positive integer n, the set Γ_n of those classes $\mathscr{C}_n$ which satisfy (8.1.1) with $\epsilon = 1/n$ can be partially ordered by inclusion. Clearly if $\Delta_n \subset \Gamma_n$ is a totally ordered set of classes, the union of all the classes in Δ_n is a class $\mathscr{C}_n$ which is a maximal element of Δ_n. By Zorn's lemma (see §1.6) it follows that we can obtain a maximal element in Γ_n with respect to this ordering. Thus we can find a class $\mathscr{C}_n^0 \subset \mathscr{F}$ satisfying (8.1.1), with $\epsilon = 1/n$, and such that, given $E \in \mathscr{F}$, there is at least one $A \in \mathscr{C}_n^0$ with $\mu(A \triangle E) < 1/n$, as otherwise E could be added to $\mathscr{C}_n^0$ to form a larger collection. But $\mathscr{C}_n^0$ is countable so $\mathscr{C} = \bigcup_{n=1}^{\infty} \mathscr{C}_n^0$ is countable and forms a basis for μ. ▌

***Theorem* 8.1.** *The space $\mathscr{L}_2(\Omega, \mu)$ of functions $f\colon \Omega \to \mathbf{R}^*$ which are square integrable is separable (in the norm topology) if and only if the measure μ has a countable basis.*

Proof. Suppose first that $\mathscr{L}_2$ is separable, so that there is a sequence $\{f_n\}$ in $\mathscr{L}_2$ such that for any $\epsilon > 0$, $f \in \mathscr{L}_2$ we can find an integer k with $\|f - f_k\| < \epsilon$. Let $\mathscr{C}$ be any collection of measurable sets of finite measure. Then for each $A \in \mathscr{C}$, the indicator function $\chi_A \in \mathscr{L}_2$, so there is an integer k_A such that

$$\|f_{k_A} - \chi_A\| < \tfrac{1}{3}\epsilon.$$

Then, if $\mathscr{C}$ satisfies (8.1.1), we must have

$$\|f_{k_A} - f_{k_B}\| > \tfrac{1}{3}\epsilon \quad \text{for} \quad A \neq B$$

so that $k_A \neq k_B$, and $\mathscr{C}$ must be countable. By the lemma this implies that μ has a countable basis.

Conversely suppose that μ has a countable basis $\mathscr{C}$. The set $\mathscr{S}$ of all simple functions

$$h = \sum_{i=1}^{n} r_i \chi_i$$

which are finite sums of rational multiples of indicator functions of sets of $\mathscr{C}$ is then countable. In order to prove $\mathscr{L}_2$ separable, it is sufficient to show that this set $\mathscr{S}$ is dense in $\mathscr{L}_2$.

From the definition of the integral, for any $f \in \mathscr{L}_2$, $\epsilon > 0$ we can find a set $E \in \mathscr{F}$ with $\mu(E) < \infty$ such that f is bounded on E and

$$\int_{\Omega - E} |f|^2 \, d\mu < \tfrac{1}{8}\epsilon^2.$$

On the set E, we can use the process of theorem 5.2 to approximate f uniformly by a simple function g taking only rational values

$$g = \sum_{k=1}^{n} r_k \chi_{E_k} \begin{cases} E_i \cap E_j = \varnothing \quad (i \neq j), \\ E = \bigcup_{i=1}^{n} E_i. \end{cases}$$

Using $\mu(E) < \infty$, this means that such a function g can be found with

$$\int_E |f-g|^2 \, d\mu < \tfrac{1}{8}\epsilon^2.$$

Then
$$\begin{aligned} \|f-g\|^2 &= \int_{\Omega-E} |f-g|^2 \, d\mu + \int_E |f-g|^2 \, d\mu \\ &= \int_{\Omega-E} |f|^2 \, d\mu + \int_E |f-g|^2 \, d\mu \\ &< \tfrac{1}{4}\epsilon^2; \end{aligned}$$

so that $\|f-g\| < \frac{1}{2}\epsilon$. If all the r_k in the representation of g are zero we are finished, so there is no loss in generality in assuming they are all non-zero. Since $\mathscr{C}$ is a basis for μ and $\mu(E_k) < \infty$, we can find sets $C_k \in \mathscr{C}$ such that

$$\mu(E_k \triangle C_k) < \left(\frac{\epsilon}{2r_k n}\right)^2 \quad (k = 1, 2, \ldots, n).$$

Then
$$\|r_k \chi_{E_k} - r_k \chi_{C_k}\|^2 = r_k^2 \mu(E_k \triangle C_k) < \left(\frac{\epsilon}{2n}\right)^2,$$

so that, if

$$h = \sum_{k=1}^{n} r_k \chi_{C_k}, \quad \text{we have} \quad \|g-h\| \leqslant \sum_{k=1}^{n} \|r_k \chi_{E_k} - r_k \chi_{C_k}\| < \tfrac{1}{2}\epsilon$$

and
$$\|f-h\| \leqslant \|f-g\| + \|g-h\| < \epsilon. ∎$$

Corollary. *If μ denotes Lebesgue measure in $\mathbf{R}^k$, then $\mathscr{L}_2(\mathbf{R}^k, \mu)$ is separable.*

To prove this we use the observation that Lebesgue measure in $\mathbf{R}^k$ has a countable basis. It is worth remarking that the classical method of proving this corollary is to approximate $f \in \mathscr{L}_2$ by a continuous function vanishing outside a finite interval, and then approximate this continuous function by a rational polynomial.

We end this section by making an important definition which is essential to a geometric understanding of linear spaces. We will see later than it is *not* possible to define an inner product in $\mathscr{L}_p$ for $p \neq 2$.

Inner product

For any normed linear space K over the reals a function (f,g) on $K \times K \to \mathbf{R}$ is called an inner product (or scalar product) if

(i) $(f,g) = (g,f)$;
(ii) $(f_1+f_2,g) = (f_1,g)+(f_2,g)$;
(iii) $(\lambda f,g) = \lambda(f,g)$, for $\lambda \in \mathbf{R}$;
(iv) $(f,f) = \|f\|^2$.

For the normed linear space $\mathscr{L}_2$ we can define

$$(f,g) = \int fg\,d\mu, \quad f,g \in \mathscr{L}_2,$$

since, by theorem 7.7, the product fg is integrable. It is a simple matter to check that, with this definition, (f,g) satisfies all the conditions (i)–(iv) for an inner product.

Exercises 8.1

1. For any normed linear space with an inner product, prove that

$$|(f,g)| \leqslant \|f\|\|g\|.$$

Hint. Consider $(f+\theta g, f+\theta g) \geqslant 0$ for all real θ.

2. If $(f,g) = 0$ in a normed linear space, show that

$$\|f+g\|^2 = \|f\|^2+\|g\|^2.$$

3. Suppose $(\Omega, \mathscr{F}, \mu)$ is a discrete measure space, i.e. there is a sequence $\{p_i\}$ of reals with $\Sigma|p_i| < \infty$ and a sequence $\{x_i\}$ in Ω such that $\mu(E) = \sum_{x_i \in E} p_i$. Prove that $\mathscr{L}_2(\Omega, \mu)$ is separable.

4. If $(X, \mathscr{F}, \mu)$, $(Y, \mathscr{G}, \nu)$ are σ-finite measure spaces each with countable bases, show that the product measure $\lambda = \mu \times \nu$ on $X \times Y$ also has a countable basis.

Hint. Consider finite unions of rectangles which are products of basic sets.

5. Generalise the above to countable products of spaces $(X_i, \mathscr{F}_i, \mu_i)$ with $\mu_i(X_i) = 1$. The example (8) below shows that it does not extend to arbitrary products.

6. Let Ω be any set and define the counting measure $\mu(E)$ = number of points in E when E is finite; $\mu(E) = +\infty$ otherwise. Show (i) if Ω is countable, the finite subsets of Ω form a countable basis; (ii) if Ω is uncountable, there is no countable basis for μ.

7. Show that any Lebesgue–Stieltjes measure $(\mathbf{R}^k, \mathscr{L}_F^k, \mu_F)$ has a countable basis.

8. Suppose I is a non-countable index set and for $\alpha \in I$, X_α denotes a 2-point space $\{0, 1\}$ with $\mu_\alpha\{0\} = \mu_\alpha\{1\} = \frac{1}{2}$. Form the product measure μ on the Cartesian product $\prod_{\alpha \in I} \{0, 1\} = \Omega$. Show that there is no countable basis for μ.

9. Show that, if μ has a countable basis, then $\mathscr{L}_p(\Omega, \mu)$, $1 \leqslant p < \infty$ is a separable space.

8.2 Orthogonal systems of functions

We now examine the part of the structure of $\mathscr{L}_2(\Omega, \mu)$ which is more intimately related to the inner product.

Linear dependence

In a linear space K, the finite set $\phi_1, \phi_2 \ldots, \phi_n$ is said to be linearly dependent if there are real numbers c_i, not all zero, such that

$$\sum_{i=1}^{n} c_i \phi_i = 0. \tag{8.2.1}$$

On the other hand, if (8.2.1) implies that all the c_i are zero, then we say that $\phi_1, \ldots, \phi_n$ are *linearly independent*. A set $E \subset K$ is said to be linearly independent if each of its finite subsets is linearly independent. Note that, when $K = \mathscr{L}_2$ the relation (8.2.1) becomes

$$\sum_{i=1}^{n} c_i \phi_i(x) = 0 \text{ a.e.}$$

Closed linear span

In a normed linear space K, given a family $\{\phi_\alpha\}$, $\alpha \in I$ of points of K the set $M\{\phi_\alpha\}$ of all finite linear combinations

$$\sum_{k=1}^{n} c_{\alpha_k} \phi_{\alpha_k}, \quad c_{\alpha_k} \in \mathbf{R} \tag{8.2.2}$$

is called the span of $\{\phi_\alpha\}$, and its closure (in the norm topology) is called the *closed linear span* of $\{\phi_\alpha\}$ and denoted by $\overline{M}\{\phi_\alpha\}$. Thus this set $\overline{M}$ consists of precisely those elements of K which can be approximated in norm by elements of the form (8.2.2).

Complete set

A family $\{\phi_\alpha\}$ $(\alpha \in I)$ in a normed linear space K is said to form a complete set if its closed linear span is the whole space.

Suppose now that a normed linear space K is separable, so that there is a sequence $x_1, x_2, \ldots, x_n, \ldots$ of points dense in K. By omitting, in turn, any point in the sequence which can be expressed as a linear combination of the previous ones we obtain a sequence $g_1, g_2, \ldots$ which is linearly independent, and has the same linear span as $\{x_n\}$. Since $\{x_n\}$ is dense, the closed linear span of $\{g_n\}$ must therefore be the whole space. Thus in any separable normed linear space there is a complete set of independent points which is either finite or enumerable. If there is a finite complete set $(g_1, g_2, \ldots, g_k)$ of independent points and K has an inner product, then we will see that K is isomorphic to Euclidean k-space. For $K = \mathscr{L}_2(\Omega, \mu)$, it is easy to see that K is finite-dimensional if μ is a discrete measure concentrated on a finite set of points, for then the indicator functions of these individual points will form a finite complete set. However, the interesting $\mathscr{L}_2$-spaces are infinite-dimensional. Any $(\Omega, \mathscr{F}, \mu)$ for which $\mathscr{F}$ contains infinitely many disjoint sets, each of finite positive measure, clearly generates an infinite-dimensional $\mathscr{L}_2(\Omega, \mu)$ since the indicator functions of this sequence form an independent set.

Orthogonal system

Two points x, y in a normed linear space K with an inner product are said to be orthogonal if $(x, y) = 0$. Any class $\{\phi_i\}$ $(i \in I)$ of points of K which are different from zero and pairwise orthogonal is called an *orthogonal system*. A non-zero element $x \in K$ is said to be normalized if $\|x\| = 1$, i.e. $(x, x) = 1$. An orthogonal system of normalized points is said to be an *orthonormal* system in K. Thus $\{\phi_i\}$ $(i \in I)$ is an orthonormal system if

$$(\phi_i, \phi_j) = \begin{cases} 1 & \text{for} \quad i = j \in I, \\ 0 & \text{for} \quad i \neq j. \end{cases}$$

Now any orthogonal system of points is certainly linearly independent for, if we take the inner product of (8.2.1) with ϕ_j we obtain $c_j(\phi_j, \phi_j) = 0$, so that $c_j = 0$. Further, if K is separable, any orthogonal system in K is countable. For any such system can be normalised to give an orthonormal system $\{\phi_i\}$ $(i \in I)$, and then

$$\|\phi_i - \phi_j\| = \sqrt{2} \quad \text{for} \quad i \neq j;$$

and, if $\{x_n\}$ is a countable dense set, we can find for each $i \in I$ an integer n_i such that $\|x_{n_i} - \phi_i\| < \frac{1}{2}$ and this gives

$$\|x_{n_i} - x_{n_j}\| > \sqrt{2} - 1 > 0 \quad \text{for} \quad i \neq j, \quad \text{so that} \quad n_i \neq n_j.$$

In the study of finite-dimensional normed linear spaces it is helpful to use a (finite) orthogonal normalized basis. In the general case, at least for K separable, it is also possible to find a complete orthonormal sequence for K. This can be done by first obtaining a complete independent sequence and then orthogonalising it by the process of the next theorem.

***Theorem* 8.2** (*Gram–Schmidt orthogonalisation process*). *If K is a normed linear space with an inner product and $x_1, x_2, \ldots, x_n, \ldots$ is a linearly independent sequence in K, then there is an orthonormal sequence $y_1, y_2, \ldots, y_n, \ldots$ such that*

(i) $y_n = a_{n1}x_1 + a_{n2}x_2 + \ldots + a_{nn}x_n$, $a_{nn} \neq 0$;

(ii) $x_n = b_{n1}y_1 + b_{n2}y_2 + \ldots + b_{nn}y_n$, $b_{nn} \neq 0$;

where a_{ij}, b_{ij} are real numbers. Further each y_i is uniquely determined (up to the sign) by these conditions.

Proof. If $y_1 = \alpha x_1$, then

$$(y_1, y_1) = \alpha^2(x_1, x_1) = 1$$

if α is suitably chosen. The conditions are therefore satisfied with $n = 1$ if $b_{11} = 1/a_{11} = \sqrt{(x_1, x_2)}$. (Note that the linear independence condition ensures that $\|x_1\| \neq 0$.) For $n > 1$, suppose $y_1, y_2, \ldots, y_{n-1}$ have been found to satisfy all the conditions. Then

$$x_n = b_{n1}y_1 + \ldots + b_{n,n-1}y_{n-1} + z_n,$$

where $b_{ni} = (x_n, y_i)$ $(i = 1, 2, \ldots, n-1)$ so that $(z_n, y_i) = 0$ for $i < n$. We must have $(z_n, z_n) > 0$, since otherwise $z_n = 0$ and $x_1, x_2, \ldots, x_n$ would be linearly dependent. If we put

$$y_n = \frac{z_n}{\sqrt{(z_n, z_n)}}, \quad b_{nn} = \sqrt{(z_n, z_n)},$$

then $(y_n, y_n) = 1$, $(y_n, y_i) = 0$ for $i < n$ and (ii) is satisfied. We can then deduce (i) since $b_{nn} \neq 0$. By induction the method of orthogonalisation is established.

The uniqueness of the process (apart from sign) follows since the values of the constants are all determined except for the $\pm$ sign in the square root which occurs at each stage. ∎

Corollary. *If $\mathscr{L}_2(\Omega, \mu)$ is separable, then there is a complete orthonormal sequence in $\mathscr{L}_2$.*

Proof. Start with a sequence $\{f_n\}$ which is dense in $\mathscr{L}_2$, and replace it first by a sequence $\{g_n\}$ of linearly independent functions with the same closed linear span. If this is an infinite sequence, use the process of theorem 8.2 to obtain the orthonormal sequence $\{h_n\}$. It is clear that this sequence has the same closed linear span as $\{g_n\}$ so that it is a complete set. On the other hand, if $\mathscr{L}_2$ is finite dimensional, we will obtain a finite set $\{g_1, g_2, \ldots, g_n\}$ whose linear span is $\mathscr{L}_2$. This finite set can be replaced by a finite orthonormal set using the process of theorem 8.2.]

In practice it is not always easy to prove that a given orthonormal sequence $\{\phi_1, \phi_2, \ldots\}$ is complete. Various methods for proving completeness will be given in the next section.

Exercises 8.2

1. Suppose $\Omega = [0,1)$, μ is Lebesgue measure, $f_0(x) \equiv 1$,

$$f_n(x) = \begin{cases} +1 & \text{if} \quad 2^{n-1}x \equiv y \in [0, \tfrac{1}{2}) \quad (\text{mod } 1), \\ -1 & \text{if} \quad 2^{n-1}x \equiv y \in [\tfrac{1}{2}, 1) \quad (\text{mod } 1). \end{cases}$$

The functions f_n: $[0, 1) \to \mathbf{R}$ are called the Rademacher functions. Show that they form an orthonormal sequence in $\mathscr{L}_2(\Omega, \mu)$.

2. For $\Omega = [-\pi, \pi]$, μ Lebesgue measure, show that the trigonometric functions

$$\frac{1}{\sqrt{2\pi}}, \quad \frac{1}{\sqrt{\pi}}\cos x, \quad \frac{1}{\sqrt{\pi}}\sin x, \ldots, \quad \frac{1}{\sqrt{\pi}}\cos nx, \frac{1}{\sqrt{\pi}}\sin nx, \ldots$$

form an orthonormal sequence in $\mathscr{L}_2(\Omega, \mu)$.

3. For $\Omega = [-1, 1]$, μ Lebesgue measure, show that the Legendre polynomials

$$P_n(x) = \frac{1}{2^n n!} \frac{d^n\{(x^2-1)^n\}}{dx^n} \quad (n = 0, 1, 2, \ldots),$$

are orthogonal in $\mathscr{L}_2(\Omega, \mu)$ and that the sequence $\sqrt{\{\tfrac{1}{2}(2n+1)\}}\, P_n(x)$ is orthonormal.

8.3 Riesz–Fischer theorem

This theorem is formulated and proved in Hilbert space. Since $\mathscr{L}_2$-spaces will be shown to be realisations of Hilbert space, we will deduce the classical theorem about the Fourier expansion as a trigonometric series of a function in $\mathscr{L}_2$ as a special case.

Hilbert space

Suppose H is a normed linear space with an inner product which is complete in the topology of the norm; then H is said to be a Hilbert space. (Note that some older text-books require separability in addition.) If H is finite-dimensional, then theorem 8.2 allows us to choose a finite orthogonal basis $e_1, e_2, \ldots, e_n$ for H. It is then clear that

$$x = \sum_{k=1}^{n} c_k e_k, \quad c_k = (x, e_k) \tag{8.3.1}$$

is the unique expansion of $x \in H$ in terms of this basis. For separable infinite-dimensional H, we have seen that there is always an orthonormal sequence $\{e_i\}$ which forms a complete set. The main objective of the present section is to obtain the extension of (8.3.1) to this infinite-dimensional case. However, in formulating the results we will not assume that H is separable. It will turn out that an expansion in the form (8.3.1) is still possible, and that at most countably many terms in any such expansion will be non-zero. Before leaving the finite-dimensional H, we should observe that any Hilbert space of dimension n is isomorphic to Euclidean n-space $\mathbf{R}^n$ with the usual scalar product. For (8.3.1) determines the point $(c_1, c_2, \ldots, c_n) \in \mathbf{R}^n$ and defines a (1, 1) mapping which then preserves the inner product.

Fourier coefficients

Given an orthonormal family (e_j), $j \in J$ in a Hilbert space H, and any point $x \in H$, the real numbers

$$c_j = (x, e_j) \quad (j \in J),$$

are called the *Fourier coefficients* of x on the orthonormal family, and the series

$$\sum_{j \in J} c_j e_j$$

is called the *Fourier series* of x. (Note we have not yet said in what sense (if any) this series converges.)

The choice of the Fourier coefficients c_j can be justified as follows. If I is a finite subset of the index set J, re-label the indices $1, 2, \ldots, n$ and consider the partial sum

$$s_n = \sum_{i=1}^{n} \alpha_i e_i \quad (n = 1, 2, \ldots).$$

Then
$$\|s_n - x\|^2 = \left(x - \sum_{i=1}^{n} \alpha_i e_i,\ x - \sum_{i=1}^{n} \alpha_i e_i\right),$$

$$= \|x\|^2 - 2\sum_{i=1}^{x} (x, \alpha_i e_i) + \sum_{i=1}^{n} \sum_{j=1}^{n} (\alpha_i e_i,\ \alpha_j e_j),$$

$$= \|x\|^2 - 2\sum_{i=1}^{n} \alpha_i c_i + \sum_{i=1}^{n} \alpha_i^2,$$

so that
$$\|s_n - x\|^2 = \|x\|^2 - \sum_{i=1}^{n} c_i^2 + \sum_{i=1}^{n} (\alpha_i - c_i)^2. \tag{8.3.2}$$

Thus $\|s_n - x\|$ will be a minimum when all the terms of the last series in (8.3.2) are zero, and $\sum_{i=1}^{n} \alpha_i e_i$ is the best approximation (in norm) to x when all the α_i are Fourier coefficients. This generalises the well-known geometrical theorem (for $\mathbf{R}^n$) which states that the length of the perpendicular from a point to a plane is smaller than the length of any other line joining the point to the plane: for $\left(x - \sum_{i=1}^{k} c_i e_i\right)$ is orthogonal to all linear combinations of the form $\sum_{i=1}^{k} \beta_i e_i$.

Bessel's inequality

We can make another deduction from (8.3.2) by noting that

$$\|s_n - x\|^2 \geqslant 0.$$

If we put $\alpha_i = c_i$ we obtain

$$\sum_{k=1}^{n} c_k^2 \leqslant \|x\|^2.$$

If we now define $\sum_{j \in J} c_j^2$ to be the supremum of $\sum_{j \in I} c_j^2$ for all finite subsets $I \subset J$ we find that

$$\sum_{j \in J} c_j^2 \leqslant \|x\|^2, \tag{8.3.3}$$

and this is known as *Bessel's inequality*. It follows as an immediate corollary that at most countably many Fourier coefficients of a point in H can be non-zero.

***Theorem* 8.3.** *If $\{e_j\}$ $(j \in J)$ is an orthonormal family in a Hilbert space, each of the following conditions is equivalent to $\{e_j\}$ being a complete family*

(i) $\sum_{j \in J} c_j^2 = \|x\|^2$ *for every* $x \in H$ *(Parseval),*

where $\{c_j\}$ are the Fourier coefficients of x;

(ii) *The finite partial sums* $s_I = \sum_{k \in I} c_k e_k$ *of the Fourier series of x converge to x in norm for all $x \in H$.*

Note. For any arbitrary index set J we say that $\sum_{j \in I} x_j$ converges in norm to x if, given $\epsilon > 0$, there is a finite set K such that if I is finite and $K \subset I \subset J$ then

$$\left\| \sum_{j \in I} x_j - x \right\| < \epsilon.$$

It is easy to check that, when J is countable and the x_j are real so that we have a real series this notion reduces to the usual definition of absolute convergence.

Proof. The conditions (i) and (ii) are clearly equivalent by (8.3.2). Now suppose that (ii) is satisfied. Then any x can be approximated in norm by a finite sum s_n which is a linear combination of $e_1, e_2, \ldots, e_n$.

Hence, each x is in the closed linear span of $\{e_j\}$ and the sequence must form a complete system.

Conversely suppose $\{e_j\}$ is a complete system. Then given $\epsilon > 0$, $x \in H$, there is a finite sum $y = \sum_{i=1}^{N} \alpha_i e_i$ for which $\|x-y\| < \epsilon$. But, if S_N is the corresponding partial Fourier sum, we know

$$\|x-y\|^2 \geqslant \|x-s_N\|^2,$$

so that, by (8.3.2)

$$\sum_{i=1}^{N} c_i^2 \geqslant \|x\|^2 - \epsilon.$$

Since ϵ is arbitrary, we can combine this with (8.3.3) to give

$$\sum_{j \in J} c_j^2 = \|x\|^2. \;\rbrack$$

From (8.3.3) we know that a given set $\{\beta_i\} (j \in J)$ of real numbers can only be the Fourier coefficients of a point in H if $\sum_{j \in J} \beta_i^2$ converges. It turns out that this condition is sufficient as well as necessary.

***Theorem* 8.4** (*Riesz–Fischer*). *Let $\{e_j\} (j \in J)$ be any orthonormal system (not necessarily complete) in a Hilbert space H, and let $\{\beta_j\}\, j \in J$ be any set of real numbers such that $\sum_{j \in J} \beta_j^2$ converges. Then there is a point $x \in H$ with Fourier coefficients $\beta_j = (x, e_j)$ such that the finite partial sums $s_I = \sum_{i \in I} \beta_i e_i$ converge to x in norm.*

Proof. Since $\sum_{j \in J} \beta_j^2$ converges the set of j for which $\beta_j \neq 0$ is countable and we may suppose then that these indices are renamed

$$\beta_1, \beta_2, \ldots, \beta_n, \ldots$$

(it may be only a finite set). Put

$$s_n = \sum_{i=1}^{n} \beta_i e_i.$$

Then

$$\|s_{n+p} - s_n\|^2 = \sum_{i=n+1}^{n+p} \beta_i^2.$$

Since $\Sigma \beta_i^2$ converges, it follows that $\{s_n\}$ is a Cauchy sequence in norm. Since H is complete, there must be an $x \in H$ such that $s_n \to x$ in norm. Further

$$(x, e_i) = (s_n, e_i) + (x - s_n, e_i)$$

$$= \beta_i + (x - s_n, e_i) \quad \text{for} \quad n \geqslant i$$

and, by exercise 8.1 (1)

$$|(x - s_n, e_i)| \leqslant \|e_i\| \,.\, \|x - s_n\| = \|x - s_n\| \to 0 \quad \text{as} \quad n \to \infty.$$

Since (x, e_i) is independent of n, we have $\beta_i = (x, e_i)$ for all i. $\rbrack$

Corollary. *An orthonormal system* $\{e_j\}$ $(j \in J)$ *in a Hilbert space* H *forms a complete system if and only if the only point* $x \in H$ *which is orthogonal to all the* $\{e_j\}$ *is the point* $x = 0$.

Proof. Suppose $\{e_j\}$ is a complete orthonormal set and $(x, e_j) = 0$ for all j. Then all the Fourier coefficients of x are zero and so

$$\|x\|^2 = \sum_{j \in J} c_j^2 = 0.$$

Conversely suppose $\{e_j\}$ is not a complete system. Then by theorem 8.3 (i), there is a point $x \in H$ with

$$\|x\|^2 > \sum_{j \in J} c_j^2, \quad \text{where} \quad c_j = (x, e_j).$$

By theorem 8.4, there is a $y \in H$ such that

$$c_j = (y, e_j), \quad \|y\|^2 = \sum_{j \in J} c_j^2.$$

Then the point $(x-y) \in H$ is orthogonal to all the e_j. But $\|x-y\| \neq 0$, since $\|x\| > \|y\|$. ∎

Remark. If the Hilbert space is separable we have already observed that any orthonormal system is countable. For a separable Hilbert space, therefore, it is natural to state and prove theorem 8.4 and corollary in terms of an arbitrary orthonormal sequence. No essential modifications to the proof are needed.

The space l_2

The class of all infinite sequences $c_1, c_2, \ldots, c_n, \ldots$ of real numbers for which $\sum_{k=1}^{\infty} c_k^2$ converges is called the space l_2. By using the discrete version of theorem 7.7 one can check that if $\{c_i\}, \{d_i\} \in l_2$,

$$(c, d) = \sum_{i=1}^{\infty} c_i d_i \tag{8.3.4}$$

converges and defines an inner product. Alternatively, if Ω is the set of positive integers, μ is the counting measure, $c_i = f(i)$, then $f \in \mathscr{L}_2(\Omega, \mu)$ if and only if $\sum_{k=1}^{\infty} c_k^2 < \infty$, and (8.3.4) is the usual inner product $(f, g) = \int fg \, d\mu$ in $\mathscr{L}_2$. Completeness and separability for l_2 can be proved directly, or they can be deduced from the corresponding properties of $\mathscr{L}_2(\Omega, \mu)$. Thus l_2 is a separable Hilbert space according to our definition—and historically l_2 was the space first considered in detail.

The justification for our abstract definition of Hilbert space is contained in the following theorem.

***Theorem* 8.5.** *An n-dimensional Hilbert space is isomorphic to* $\mathbf{R}^n$. *Any separable infinite-dimensional Hilbert space is isomorphic to* l_2.

Proof. The finite-dimensional case was considered earlier. If H is separable and infinite-dimensional we can select a complete orthonormal sequence $\{e_n\}$ and obtain the Fourier coefficients $\{c_n\}$ of a point $x \in H$. Then since $x \in H \Rightarrow \Sigma c_n^2 < \infty$, this defines a mapping from H to l_2. Conversely, every sequence in l_2 determines a point in H with these Fourier coefficients, by theorem 8.4. There is only one such function by the corollary to theorem 8.4. Thus to prove that we have an isomorphism it is sufficient to prove that the linear structure and inner product are preserved by the mapping. Suppose

$$x^{(1)}, x^{(2)} \in H$$

correspond to $\{c_i^{(1)}\}, \{c_i^{(2)}\} \in l_2$. Then it is immediate that

$$\alpha x^{(1)} \leftrightarrow \{\alpha c_i^{(1)}\} \quad (\alpha \in \mathbf{R});$$

$$x^{(1)} + x^{(2)} \leftrightarrow \{(c_i^{(1)} + c_i^{(2)})\};$$

$$\|x^{(1)}\|^2 = \Sigma(c_i^{(1)})^2, \quad \|x^{(2)}\|^2 = \Sigma(c_i^{(2)})^2;$$

$$\begin{aligned}\|x^{(1)} + x^{(2)}\|^2 &= \|x^{(1)}\|^2 + 2(x^{(1)}, x^{(2)}) + \|x^{(2)}\|^2 \\ &= \Sigma(c_i^{(1)})^2 + 2\Sigma c_i^{(1)} c_i^{(2)} + \Sigma(c_i^{(2)})^2,\end{aligned}$$

so that $(x^{(1)}, x^{(2)}) = \Sigma c_i^{(1)} c_i^{(2)}$. ∎

Corollary. *If* $(\Omega, \mathscr{F}, \mu)$ *is any measure space with a countable basis, then* $\mathscr{L}_2(\Omega, \mu)$ *is either finite-dimensional, when it is isomorphic to a Euclidean space* $\mathbf{R}^n$, *or it is infinite-dimensional when it is isomorphic to* l_2. *If* $(\Omega_1, \mathscr{F}_1, \mu_1)$, $(\Omega_2, \mathscr{F}_2, \mu_2)$ *are such that* $\mathscr{L}_2(\Omega_1, \mu_1)$ *and* $\mathscr{L}_2(\Omega_2, \mu_2)$ *are both infinite-dimensional and separable, then* $\mathscr{L}_2(\Omega_1, \mu_1)$ *and* $\mathscr{L}_2(\Omega_2, \mu_2)$ *are isomorphic.*

The theorems of this section were first obtained for trigonometric series of functions f in $\mathscr{L}_2([-\pi, \pi], \mu)$ where μ is Lebesgue measure. In order to obtain these special theorems one only has to prove that the functions

$$\frac{1}{\sqrt{2\pi}}, \quad \frac{1}{\sqrt{\pi}}\cos x, \quad \frac{1}{\sqrt{\pi}}\sin x, \quad \ldots, \quad \frac{1}{\sqrt{\pi}}\cos nx, \quad \frac{1}{\sqrt{\pi}}\sin nx, \ldots$$

form a complete orthonormal sequence in this $\mathscr{L}_2$-space. The steps necessary for this proof are contained in exercise 8.3 (2).

Exercises 8.3

1. Prove that a series $\sum_{n=1}^{\infty} a_n$ of real terms converges absolutely to s if and only if, for each $\epsilon > 0$ there is a finite set $I \subset \mathbf{Z}$ such that for every finite K with $I \subset K \subset \mathbf{Z}$ we have

$$|s - \sum_{n \in K} a_n| < \epsilon.$$

2. If $\Omega = [-\pi, \pi]$, μ is Lebesgue measure, $f \in \mathscr{L}_2(\Omega, \mu)$

$$a_m = \frac{1}{\pi}\int_{-\pi}^{\pi} f(x)\cos mx\,dx \quad (m = 0, 1, 2, \ldots);$$

$$b_m = \frac{1}{\pi}\int_{-\pi}^{\pi} f(x)\sin mx\,dx \quad (m = 1, 2, \ldots),$$

then the a_m, b_m are the classical Fourier coefficients of f. Prove:

(i) $\frac{1}{2}a_0^2 + \sum_{m=1}^{\infty}(a_m^2 + b_m^2) \leqslant \frac{1}{\pi}\int_{-\pi}^{\pi}\{f(x)\}^2\,dx;$

(ii) if $\{a_n\}$, $\{b_n\}$ are such that

$$\tfrac{1}{2}a_0^2 + \sum_{1}^{\infty}(a_m^2 + b_m^2) < \infty,$$

then there is a function $f \in \mathscr{L}_2$ for which these are the Fourier coefficients and such that

$$s_n(x) = \tfrac{1}{2}a_0 + \sum_{m=1}^{n}(a_m \cos mx + b_m \sin mx) \to f$$

in second mean;

(iii) if $\{a_n\}$, $\{b_n\}$ are the Fourier coefficients of f in the above sense

$$s_n(x) = \tfrac{1}{2}a_0 + \sum_{m=1}^{n}(a_m \cos mx + b_m \sin mx),$$

$$\sigma_n(x) = \frac{1}{n+1}[s_0(x) + s_1(x) + \ldots + s_n(x)],$$

then $\sigma_n(x) \to f(x)$ in $\mathscr{L}_2$ norm (and in fact $\sigma_n \to f$ a.e.);

(iv) $$\frac{1}{\pi}\int_{-\pi}^{\pi}\sigma_n^2(x)\,d\mu = \tfrac{1}{2}a_0^2 + \sum_{1}^{n}(a_k^2 + b_k^2)\left(1 - \frac{k}{n+1}\right)^2$$

$$\leqslant \tfrac{1}{2}a_0^2 + \sum_{1}^{n}(a_k^2 + b_k^2) \leqslant \tfrac{1}{2}a_0^2 + \sum_{1}^{\infty}(a_k^2 + b_k^2);$$

(v) since $\int_{-\pi}^{\pi}\sigma_n^2(x)\,d\mu \to \int_{-\pi}^{\pi} f^2(x)\,d\mu$ we have for $f \in \mathscr{L}_2$

$$\frac{1}{\pi}\int_{-\pi}^{\pi}\{f(x)\}^2\,d\mu = \tfrac{1}{2}a_0^2 + \sum_{1}^{\infty}(a_k^2 + b_k^2);$$

(vi) the trigonometric system of exercise 8.2 (2) is complete.

3. If $\{\phi_n\}$ is any orthonormal sequence in a Hilbert space H, then for $x \in H$, $(x, \phi_n) \to 0$ as $n \to \infty$.

4. If $f \in \mathscr{L}_2([-\pi, \pi], \mu)$ then, as $n \to \infty$,

$$\int_{-\pi}^{\pi} f(x) \sin nx \, dx \to 0, \quad \int_{-\pi}^{\pi} f(x) \cos nx \, dx \to 0.$$

5. If $\boldsymbol{l}_p$ is the space of real sequences $\{x_i\}$ for which

$$\sum_{i=1}^{\infty} |x_i|^p < \infty \quad \text{for} \quad 1 \leqslant p < \infty,$$

show that $\boldsymbol{l}_p$ is separable. $\boldsymbol{l}_\infty$ is the space of bounded real sequences with $\|x\| = \sup |x_i|$. By considering sequences of 0's and 1's show that $\boldsymbol{l}_\infty$ is not separable. Deduce that $\mathscr{L}_\infty$ is not separable if Ω has infinite but σ-finite measure.

8.4* Space of linear functionals

We start by defining a more general type of normed linear space.

Banach space

Any normed linear space over the reals which is complete in the topology determined by the norm is called a (real) Banach space.

We saw already that the $\mathscr{L}_p$ $(1 \leqslant p \leqslant +\infty)$ spaces are normed linear spaces and that each of them is complete in the norm topology so that each $\mathscr{L}_p$ is a Banach space. Euclidean n-space $\mathbf{R}^n$ with the usual metric provides a simpler example of a Banach space. $C[a, b]$, the space of continuous functions on a finite closed interval, with $\|f\| = \sup |f(x)|$, can also be easily seen to be a Banach space.

From our definition of Hilbert space, it follows that any Banach space will be a Hilbert space provided there is an inner product defined satisfying $\|f\|^2 = (f, f)$. The question immediately arises as to whether or not all Banach spaces are Hilbert spaces; or is it always possible to define an inner product in a Banach space? We can settle this as follows. If there is to be an inner product, then

$$\|f+g\|^2 = (f+g, f+g) = (f,f) + 2(f,g) + (g,g),$$
$$\|f-g\|^2 = (f-g, f-g) = (f,f) - 2(f,g) + (g,g),$$

so that on adding

$$\|f+g\|^2 + \|f-g\|^2 = 2\|f\|^2 + 2\|g\|^2. \tag{8.4.1}$$

Thus, a relation (8.4.1) for all f, g in the space is a necessary condition for the Banach space to have an inner product. One can also easily check that, if (8.4.1) is always satisfied, then

$$(f, g) = \tfrac{1}{2}\{\|f+g\|^2 - \|f\|^2 - \|g\|^2\} \tag{8.4.2}$$

satisfies all the conditions for an inner product, so that any Banach space which satisfies (8.4.1) is a Hilbert space if we define the inner product by (8.4.2). We can think of the condition (8.4.1) as a generalisation of the Euclidean theorem that in any parallelogram the sum of the squares on the diagonals is twice the sum of the squares on two adjacent sides. If this theorem is not valid in the Banach space K, then it is not possible to define an inner product on K. This allows us to show that $\mathscr{L}_p$ is not a Hilbert space for $p \neq 2$—see exercise 8.4 (2).

Linear functional

Given a linear space K over the reals a function $T: K \to \mathbf{R}$ is called a linear functional if, for all $x_1, x_2 \in K$, $\alpha, \beta \in \mathbf{R}$,

$$T(\alpha x_1 + \beta x_2) = \alpha T(x_1) + \beta T(x_2).$$

If K is a normed linear space, then T is continuous at $x_0 \in K$ if, given $\epsilon > 0$, there is a $\delta > 0$ with

$$\|x - x_0\| < \delta \Rightarrow |T(x) - T(x_0)| < \epsilon.$$

For a normed K the functional is said to be bounded if there is a real constant C such that

$$|T(x)| \leqslant C\|x\| \quad \text{for all} \quad x \in K.$$

It is immediate that a linear functional on a normed linear space is continuous everywhere if it is continuous at any one point. The connexion between continuity and boundedness is not quite so obvious.

Lemma. *A linear functional T on a normed linear space K is continuous if and only if it is bounded.*

Proof. Suppose first that T is bounded, then given $x_1 \in K$, $\epsilon > 0$, put $\delta = \epsilon . C^{-1}$ and we have

$$|T(x) - T(x_1)| = |T(x - x_1)| \leqslant C\|x - x_1\| < \epsilon$$

if $\|x - x_1\| < \delta$. Conversely, if T is continuous at $\mathbf{0} \in K$ we can choose $B > 0$ such that

$$|T(x)| \leqslant 1 \quad \text{for} \quad \|x\| \leqslant B.$$

Then for $x \in K$, $\left\| \dfrac{B}{\|x\|} x \right\| = B$, so that

$$|T(x)| = \frac{\|x\|}{B} T\left(\frac{Bx}{\|x\|}\right) \leqslant \frac{1}{B}\|x\|,$$

and T is bounded. ❚

Norm of a bounded functional

If T is a bounded linear functional on a normed linear space K, the smallest number C satisfying $|T(x)| \leqslant C\|x\|$ for all $x \in K$ is called the norm of T and we denote it by $\|T\|$. Because of linearity,

$$\|T\| = \sup_{x \neq 0} \frac{|T(x)|}{\|x\|} = \sup_{\|x\|=1} |T(x)|.$$

If T_1, T_2 are two linear functionals on a linear space K, $\alpha, \beta \in \mathbf{R}$ then

$$(\alpha T_1 + \beta T_2)(x) = \alpha T_1(x) + \beta T_2(x)$$

is also a linear functional on K; and the set of all linear functionals on K is a linear space. We can say more if K is normed.

Lemma. *If K^* denotes the set of bounded linear functionals on a normed linear space K, then K^* is a Banach space.*

Proof. $T \in K^*$, $\|T\| = 0$ implies that $T(x) = 0$ for all $x \in K$ which means that T is the null transformation,

$$\|T_1 + T_2\| = \sup_{\|x\|=1} |T_1(x) + T_2(x)| \leqslant \sup_{\|x\|=1} |T_1(x)| + \sup_{\|x\|=1} |T_2(x)|$$
$$= \|T_1\| + \|T_2\|,$$

and
$$\|\alpha T\| = \sup_{\|x\|=1} |\alpha T(x)| = |\alpha| \sup_{\|x\|=1} |T(x)| = |\alpha| . \|T\|.$$

This shows that K^* is a normed linear space with the norm

$$\|T\| = \sup_{\|x\|=1} |T(x)|.$$

It remains to show that K^* is complete. Suppose $\{T_n\}$ is a sequence in K^* such that

$$\|T_m - T_n\| \to 0 \quad \text{as} \quad m, n \to \infty.$$

Then, for each $x \in K$, $|T_m(x) - T_n(x)| \to 0$ as $m, n \to \infty$. The completeness of $\mathbf{R}$ now implies that there is a real number

$$y = T(x) = \lim_{n \to \infty} T_n(x).$$

Now T is clearly a linear transformation and since

$$|\|T_m\| - \|T_n\|| \leqslant \|T_m - T_n\|,$$

the real sequence $\{\|T_n\|\}$ must be bounded, say by C. Then

$$|T_n(x)| \leqslant C\|x\| \quad \text{for all} \quad x \in K$$

and all integers n, so that $|T(x)| \leqslant C\|x\|$ and T is a bounded linear functional, that is $T \in K^*$. Given $\epsilon > 0$, there is an integer $N = N(\epsilon)$ such that

$$|T_m(x) - T_n(x)| < \epsilon \quad \text{for} \quad \|x\| = 1, \quad x \in K, \quad m, n, \geqslant N.$$

If we let $m \to \infty$, then

$$|T(x) - T_n(x)| \leqslant \epsilon \quad \text{for} \quad \|x\| = 1 \quad (n \geqslant N),$$

so that $\|T - T_n\| \to 0$ as $n \to \infty$, and K^* is complete. ∎

Conjugate space

For any normed linear space K (in particular if K is a Banach space), the Banach space K^* of bounded linear functionals on K is called the conjugate space (or dual space) of K.

Linear subspace

A set H contained in a linear space K such that H is itself a linear space is called a linear subspace of K.

If K contains a point $x \neq 0$, it is clear that the set of all points αx, $\alpha \in R$ is a linear subspace H of K. Then, if we put,

$$T(\alpha x) = \alpha \quad \text{for all} \quad \alpha x \in H$$

it is immediate that $T \neq 0$ is a linear functional on H. However, it is not immediately obvious that the set of bounded linear functionals defined on the whole of K contains any $T \neq 0$. The existence of such a non-trivial T will follow if we can prove that linear functionals defined on a subspace can always be extended to the whole of K.

Theorem 8.6 (*Hahn–Banach extension*). *Suppose K is a linear subspace of a linear space H. Then any bounded linear functional on K can be extended to a bounded linear functional on H with the same norm.*

Proof. Suppose $f: K \to \mathbf{R}$ is the given functional and

$$a = \sup_{x \in K} |f(x)|/\|x\|.$$

Then $|f(x)| \leqslant a\|x\|$ for all x in K. Consider the class $\mathscr{C}$ of all linear functionals T defined on spaces J such that (i) $K \subset J \subset H$; (ii) $T(x) = f(x)$ for $x \in K$; (iii) $T(x) \leqslant a\|x\|$ for $x \in J$. We can partially order $\mathscr{C}$ by putting $g_1 < g_2$ if g_1 is defined on J_1, g_2 on J_2, $K \subset J_1 \subset J_2 \subset H$ and

$$g_1(x) = g_2(x) = f(x) \quad \text{for} \quad x \in K, \qquad g_1(x) = g_2(x) \quad \text{for} \quad x \in J_1.$$

By Zorn's lemma (§ 1.6) we can find a maximal element in this partial ordering. This must be an extension T of f defined on a subspace $J \subset H$ such that no further extension to a larger subspace is possible. It is clearly sufficient to show that, for this maximal $T \in \mathscr{C}$, we must have $J = H$.

Suppose not, then there is a point $z \in H - J$. We will obtain a contradiction by showing that T can be extended to the linear space J_z consisting of all points of the form $j + \alpha z$, $j \in J$, $\alpha \in \mathbf{R}$. Note first that, since $z \notin J$, the representation $(j+\alpha z)$ is unique. The extension to J_z will therefore be determined by its value at z. Now if $x, y \in J$,

$$T(x)-T(y) = T(x-y) \leqslant a\|x-y\| \leqslant a\|x+z\|+a\|-y-z\|$$

so that
$$-a\|-y-z\|-T(y) \leqslant a\|x+z\|-T(x).$$

Hence
$$\sup_{y\in J}[-a\|-y-z\|-T(y)] \leqslant \inf_{x\in J}[a\|x+z\|-T(x)].$$

Let t be any real number satisfying

$$\sup_{y\in J}[-a\|-y-z\|-T(y)] \leqslant t \leqslant \inf_{x\in J}[a\|x+z\|-T(x)], \qquad (8.4.3)$$

and put $\bar{T}(z) = t$: this implies

$$\bar{T}(k+\alpha z) = T(k)+\alpha t \quad \text{for} \quad k \in J.$$

Now put $y = x/\alpha$ in (8.4.3) and let $w = x+\alpha z$:

$$-a\left\|-\frac{w}{\alpha}\right\|-T\left(\frac{x}{\alpha}\right) \leqslant t \leqslant a\left\|\frac{w}{\alpha}\right\|-T\left(\frac{x}{\alpha}\right).$$

If $\alpha > 0$ multiply the right-hand inequality by α, while if $\alpha < 0$ multiply the left-hand inequality by α. Both cases give

$$a\|w\|-T(x) \geqslant \alpha t$$

so that
$$\bar{T}(w) \leqslant a\|w\|$$

and $\bar{T} \in \mathscr{C}$. Since J is a proper subset of J_z this establishes the existence of the extension. To see that the extension F has the same norm as f we need only note that $|F(x)| \leqslant a\|x\|$ for all x in H so that

$$\|F\| \leqslant a = \|f\|;$$

and
$$\|F\| = \sup_{\substack{x\in H\\ \|x\|=1}} |F(x)| \geqslant \sup_{\substack{x\in K\\ \|x\|=1}} |f(x)| = \|f\|. \blacksquare$$

Remark. In the above theorem, the only property of the norm which we used was that

$$\|x+y\| \leqslant \|x\|+\|y\| \quad \text{for all} \quad x, y \in H.$$

It is possible to state and prove the extension theorem in terms of any subadditive bounding functional. This gives

***Theorem* 8.6A** (*Hahn–Banach extension*). *Suppose K is a linear subspace of a linear space H, p is a subadditive functional on H such that $p(ax) = ap(x)$ for $a \geqslant 0$, $x \in H$; and f is a linear functional on K such that $f(x) \leqslant p(x)$ for all $x \in K$. Then there is a linear functional $\bar{f}$: $H \to \mathbf{R}$ such that*

$$\bar{f}(x) = f(x) \quad \text{for} \quad x \in K, \qquad \bar{f}(x) \leqslant p(x) \quad \text{for} \quad x \in H.$$

Exercises 8.4

1. If $(\Omega, \mathscr{F}, \mu)$ is such that there are two sets E_1, $E_2 \in \mathscr{F}$ with $\mu(E_1)$, $\mu(E_2)$ positive and finite, show by considering the indicator functions of E_1, E_2 that $\mathscr{L}_p(\Omega, \mu)$ does not satisfy (8.4.1) if $p \neq 2$; and therefore is not a Hilbert space.

2. Prove that m, the set of bounded real sequences $\{x_i\}$, is a Banach space with $\|x\| = \sup_i |x_i|$.

3. Suppose K is a Banach space, K^* is its dual, and K^{**} is the dual of K^*. Prove:

(i) if x is a fixed element in K, $X(f) = f(x)$ for $f \in K^*$ defines a linear functional on K^*;

(ii) for the above function $\|X\| = \|x\|$, so that $T(x) = X$ is a norm preserving map from K to K^{**};

(iii) this map T preserves the linear structure;

(iv) The set of elements X of K^{**} such that $X(f) = f(x)$ for some $x \in K$ forms a closed linear subspace of K^{**}.

4. If K is a linear subspace of a Banach space H show that a point y of H is in the closure of K if and only if $f(y) = 0$ for every linear functional $f \in H^*$ which vanishes on K.

5. In §4.4 we showed that it is not possible to define a *measure* on $[0, 1)$ which is defined for all subsets and invariant for translations (mod 1). The following steps will show that we can define such a *finitely additive* set function on all subsets of $[0, 1)$ with $\nu[0, 1) = 1$ and $\nu(E) = |E|$ when E is Lebesgue measurable.

(i) Let H be the set of all bounded functions f: $[0, 1) \to \mathbf{R}$ which are extended to be periodic in the whole of $\mathbf{R}$ by $f(x+1) = f(x)$. Prove H is a linear space.

(ii) Put
$$M(f; \alpha_1, \alpha_2, \ldots, \alpha_n) = \sup_{x \in \mathbf{R}} \frac{1}{n} \sum_{i=1}^{n} f(x+\alpha_i),$$
$p(f) = \inf M(f; \alpha_1, \ldots, \alpha_n)$ for all such finite sequences of real α_i. Prove that p is subadditive and $p(af) = ap(f)$ for $a \geqslant 0$.

(iii) If f: $[0, 1) \to \mathbf{R}$ is bounded and Lebesgue measurable, show that the Lebesgue integral $\mathscr{I}(f) \leqslant M(f; \alpha_1, \ldots, \alpha_n)$.

(iv) Show that the set of bounded measurable f: $[0, 1) \to R$ is a linear subspace of H, and $\mathscr{I}(f)$ is a bounded linear functional on this subspace.

(v) Use theorem 8.6A to extend $\mathscr{I}$ to a linear functional F defined on all of H.

(vi) Show that $F\{f(x+x_0)\} = F\{f(x)\}$ for all $x_0 \in \mathbf{R}$.

(vii) By considering indicator functions of subsets of $[0, 1)$, put $\nu(E) = F(\chi_E)$ for $E \subset [0, 1)$.

Prove $\nu(E_1 \cup E_2) = \nu(E_1) + \nu(E_2)$ if E_1, E_2 are disjoint,

$\nu(E) = |E|$ if E is Lebesgue measurable,

$\nu(E)$ is invariant under translations.

6. Similar arguments to those used in (5) above can be applied to the linear space V of bounded real functions $f: [0, +\infty) \to \mathbf{R}$. Show that there exists a bounded linear functional $\operatorname{Lim} f(x)$ on V such that, for $a, b \in \mathbf{R}$.

(i) $\operatorname{Lim} \{af(x) + bg(x)\} = a \operatorname{Lim} f(x) + b \operatorname{Lim} g(x)$;

(ii) $f(x) \geqslant 0$ in $[0, \infty) \Rightarrow \operatorname{Lim} f(x) \geqslant 0$;

(iii) $\operatorname{Lim} f(x + x_0) = \operatorname{Lim} f(x)$ for any $x_0 \geqslant 0$;

(iv) $\operatorname{Lim} f(x) = \lim_{x \to +\infty} f(x)$ if this exists.

Deduce a corresponding result for the space $\boldsymbol{m}$ of bounded real sequences.

8.5* The space conjugate to $\mathscr{L}_p$

We have seen that $\mathscr{L}_p$ $(1 \leqslant p \leqslant +\infty)$ is a Banach space and, *a fortiori*, a normed linear space. It follows from the lemma on p. 211 that the space of bounded linear functionals on $\mathscr{L}_p$ is also a Banach space. The object of the present section is to identify these conjugate spaces—at least up to an isomorphism.

***Theorem* 8.7.** *Suppose $(\Omega, \mathscr{F}, \mu)$ is a σ-finite measure space and $\mathscr{L}_p$, $1 \leqslant p < \infty$ is the linear space of $\mathscr{F}$-measurable functions $f: \Omega \to \mathbf{R}^*$ whose pth power is integrable, with the usual norm*

$$\|f\| = \left\{\int |f|^p \, d\mu\right\}^{1/p}.$$

Let $1/p + 1/q = 1$ (if $p = 1, q = \infty$). Then

(i) *for each $g \in \mathscr{L}_q$,*

$$F(f) = \int fg \, d\mu$$

defines a linear funtional on $\mathscr{L}_p$;

(ii) *given any bounded linear functional F on $\mathscr{L}_p$, there is a $g \in \mathscr{L}_q$ such that*

$$F(f) = \int fg \, d\mu,$$

and in this case

$$\|F\| = \left\{\int |g|^q \, d\mu\right\}^{1/q} \quad \text{if} \quad p > 1,$$

$$= \operatorname{ess\,sup} |g| \quad \text{if} \quad p = 1.$$

Proof. (i) This follows immediately from theorem 7.7 and the linearity properties of the integral; for

$$|F(f)| \leqslant \left\{\int |f|^p \, d\mu\right\}^{1/p} \left\{\int |g|^q \, d\mu\right\}^{1/q}.$$

(ii) Suppose first that $\mu(\Omega) < \infty$ and F is a linear functional on $\mathscr{L}_p$. For any measurable $E \subset \Omega$ put

$$\sigma(E) = F(\chi_E),$$

where χ_E is the indicator function of E. The linearity of F implies immediately that σ is finitely additive. Now suppose $E = \bigcup_{i=1}^{\infty} E_i$, E_i disjoint. Then $\mu\left(\bigcup_{i=1}^{N} E_i\right) \to \mu(E)$ as $N \to \infty$, so that in $\mathscr{L}_p$,

$$\|\chi_{Q_N} - \chi_E\| \to 0, \quad \text{where} \quad Q_N = \bigcup_{i=1}^{N} E_i.$$

Since F is continuous, we must have

$$\sum_{i=1}^{\infty} \sigma(E_i) = \lim_{N\to\infty} \sum_{i=1}^{N} \sigma(E_i) = \sigma(E)$$

so that σ is completely additive on $\mathscr{F}$. Further $\mu(E) = 0 \Rightarrow \sigma(E) = 0$, so that σ is absolutely continuous with respect to μ. By theorem 6.7 it now follows that there exists a function g which is integrable, such that

$$F(\chi_E) = \sigma(E) = \int_E g\,d\mu \quad \text{for all} \quad E \in \mathscr{F},$$

$$= \int \chi_E g\,d\mu.$$

This gives the required representation for F on the class of indicator functions of measurable sets. We must prove that $g \in \mathscr{L}_q$, and that the representation is valid on the whole of $\mathscr{L}_p$.

It is clear from linearity that the representation is valid for $\mathscr{F}$-simple functions. If $f_0 \in \mathscr{L}_p$, $f_0 \geqslant 0$, we can find a sequence f_n of simple functions which increases monotonely to f_0. Then by theorem 7.6, $f_n \to f_0$ in pth mean, and by the continuity of F

$$F(f_0) = \lim_{n\to\infty} F(f_n) = \lim \int f_n g\,d\mu = \int f_0 g\,d\mu$$

on applying the monotone convergence theorem to $\{f_n g_+\}$ and $\{f_n g_-\}$ separately. The restriction $f_0 \geqslant 0$ can be removed by considering f_+ and f_- separately, so that

$$F(f) = \int fg\,d\mu \quad \text{for all} \quad f \in \mathscr{L}_p.$$

Now suppose $p > 1$, and $g(t)$ has been obtained by the above process. Put

$$g_n(t) = \begin{cases} |g(t)|^{q-1} \operatorname{sign} g(t) & \text{if} \quad |g(t)|^{q-1} \leqslant n, \\ n \operatorname{sign} g(t) & \text{if} \quad |g(t)|^{q-1} > n. \end{cases}$$

Then each g_n is bounded and measurable, and is therefore in $\mathscr{L}_p$. Hence

$$|F(g_n)| = \left|\int g_n g\, d\mu\right| \leqslant \|F\|\|g_n\| = \|F\|\left(\int |g_n|^p\, d\mu\right)^{1/p}.$$

But
$$g_n g = |g_n|\,|g| \geqslant |g_n|\,|g_n|^{1/q-1} = |g_n|^p,$$

so that
$$\int |g_n|^p\, d\mu \leqslant \|F\|\left(\int |g_n|^p\, d\mu\right)^{1/p},$$

and
$$\left(\int |g_n|^p\, d\mu\right)^{1/q} \leqslant \|F\|.$$

But
$$|g_n|^p \to |g|^q \text{ a.e.}$$

so that, by theorem 5.5
$$\left(\int |g|^q\, d\mu\right)^{1/q} \leqslant \|F\|. \tag{8.5.1}$$

Before going on to prove equality in (8.5.1), let us now remove the restriction $\mu(\Omega) < \infty$. Suppose $\mu(\Omega) = \infty$, so that there is a sequence $\{Q_n\}$ of disjoint measurable sets with

$$\Omega = \bigcup_{i=1}^{\infty} Q_i, \quad \mu(Q_n) < \infty \quad \text{all } n.$$

We can apply the above argument to each of the spaces $(Q_n, \mathscr{F}_n, \mu)$ where $\mathscr{F}_n = \mathscr{F} \cap Q_n$. By the uniqueness of the derivative g in the Radon–Nikodym theorem, if $f \in \mathscr{L}_p(\Omega, \mu)$ and vanishes outside $R_N = \bigcup_{i=1}^{N} Q_i$,

$$F(f) = \int fg\, d\mu.$$

But if $f \geqslant 0$, we can apply the monotone convergence theorem to each of $\{f_n g_+\}$, $\{f_n g_-\}$ where $f_n = f\chi_{R_n}$ to obtain this representation by using the continuity of F. The final step is to use $f = f_+ - f_-$ so that the representation is valid on all of $\mathscr{L}_p$. Further (8.5.1) follows since it is true for the integral over each R_n.

Now by Hölder's inequality (theorem 7.7) we have

$$|F(f)| \leqslant \|f\|\left\{\int |g|^q\, d\mu\right\}^{1/q},$$

so that
$$\|F\| = \left\{\int |g|^q\, d\mu\right\}^{1/q} \quad \text{using (8.5.1).}$$

We need to modify the argument in the case $p = 1$, assuming that g has been defined as before as the Radon–Nikodym derivative of σ. For any $t > 0$, let E be a set such that $0 < \mu(E) < \infty$ and $|g(x)| > t$ for $x \in E$. Put $f(x) = \chi_E \operatorname{sign} g(x)$ and we have

$$F(f) \geqslant t\mu(E), \quad \|f\| = \mu(E)$$

so $\|F\| \geqslant t$. Since such a set E can be found for any $t < \text{ess sup}\,|g|$ we must have

$$\|F\| \geqslant \text{ess sup}\,|g|.$$

But
$$|F(f)| = \left|\int fg\,d\mu\right| \leqslant (\text{ess sup}\,|g|)\,\|f\|,$$

so that $\|F\| \leqslant \text{ess sup}\,|g|$.]

Corollary. *If H is a separable Hilbert space:*

(i) *for any fixed $h \in H$, the inner product $F(f) = (f, h)$ defines a bounded linear functional;*

(ii) *for any bounded linear functional F on H, there is an $h \in H$ such that $F(f) = (f, h)$ for all $f \in H$: further $\|F\| = \|h\|$.*

Proof. Choose a measure space $(\Omega, \mathscr{F}, \mu)$ such that $\mathscr{L}_2(\Omega, \mu)$ is a separable Hilbert space, and so is isomorphic to H. Now apply the theorem in the case $p = q = 2$.]

Note. One can also construct a direct proof of the Corollary without the restriction that H be separable; the case $p = q = 2$ of theorem 8.7 could then be deduced from this.

Reflexive Banach space

In exercise 8.4 (3) we proved that, for any Banach space H, $H^{**} \supset H$ in the sense that H is isomorphic to a Banach subspace of H^{**}. Those Banach spaces H for which $H = H^{**}$ are called *reflexive*. By our representation theorem 8.7, $\mathscr{L}_p$ is reflexive for $1 < p < \infty$. In general, $\mathscr{L}_1$ is not reflexive because $\mathscr{L}_\infty^*$ is bigger than $\mathscr{L}_1$: this will follow from exercises 8.5 (3,4). In fact very little is known about the structure of $\mathscr{L}_\infty^*$: the difficulty is that the axiom of choice, or something equivalent, is needed to construct $\mathscr{L}_\infty^*$ and this makes it impossible to get a hold on it.

Exercises 8.5

1. Suppose $1 \leqslant p \leqslant +\infty$, $1/p + 1/q = 1$ and $f_n \to f$ in $\mathscr{L}_p$ norm, $g_n \to g$ in $\mathscr{L}_q$ norm. Deduce that

$$\int f_n g_n\,d\mu \to \int fg\,d\mu.$$

2. If Ω is the set of positive integers and μ is counting measure, then $\mathscr{L}_p (1 \leqslant p < \infty)$ reduces to the set of sequences $\{x_i\}$ of real numbers such that $\sum_{i=1}^{\infty} |x_i|^p < \infty$; $\mathscr{L}_\infty$ reduces to the set m of bounded sequences.

3. Let $X = [-1, 1]$, μ Lebesgue measure. Show that the collection $\mathscr{C}$ of continuous functions $f: X \to \mathbf{R}$ is a closed linear subspace of $\mathscr{L}_\infty$ (provided any function f which is equal a.e. to a continuous function is identified with it). Hence, by theorem 8.6, extend the bounded linear functional $F(f) = f(0)$ from $\mathscr{C}$ to $\mathscr{L}_\infty$ without changing its norm. If possible, suppose there is an f_0 which is integrable and such that

$$F(f) = \int f f_0 d\mu \quad \text{for} \quad f \in \mathscr{L}_\infty.$$

Then, for the special sequence

$$f_n(x) = (1 - |x|^n),$$

we have $F(f_n) = 1$ for all n. Show that, for any $f_0 \in \mathscr{L}_\infty$,

$$\int f_n f_0 d\mu \to 0.$$

4. Extend example (3) to show that if Ω contains a disjoint sequence of measurable sets of finite positive measure, then $\mathscr{L}_1(\Omega, \mu)$ is a proper subspace of $\mathscr{L}_\infty^*$. Deduce that l_1 is not reflexive.

8.6* Mean ergodic theorem

In §7.6 we obtained the point-wise ergodic theorem for functions in $\mathscr{L}_1$. If the function is in $\mathscr{L}_2$ there is an alternative form of this theorem in which point-wise convergence is replaced by convergence in second mean. We saw that any $\mathscr{L}_2$ is a Hilbert space. A measure preserving transformation T on the underlying measure space then leads naturally to a mapping on the Hilbert space to itself which preserves the inner product (and norm). It is therefore possible to state the mean ergodic theorem in terms of the properties of such a mapping in Hilbert space, and deduce the $\mathscr{L}_2$ theorem by considering this as a realization of Hilbert space. However, we choose instead to state and prove it directly as a theorem about the structure of $\mathscr{L}_2(\Omega, \mu)$. It helps if we first show that bounded linear functionals on a Banach space can be used to separate a closed linear subspace K from a point not in K (see exercise 8.4 (4)).

***Theorem* 8.8.** *Suppose K is a linear subspace of a Banach space H, and $y \in H$ with $d(y, K) = \eta > 0$. Then there is a bounded linear functional F on H such that $\|F\| = 1$, $F(y) = \eta$, $F(x) = 0$ for all $x \in K$.*

Proof. Let J be the set of points of H of the form

$$x = z + ay, \quad z \in K, \quad a \in \mathbf{R}.$$

Then J is a linear subspace of H and the representation of points of J in this form is unique. Define a linear functional f on J by

$$f(z + ay) = a\eta.$$

Then f vanishes on K and, for $a \neq 0$,

$$\|z+ay\| = |a|\left\|\frac{z}{a}+y\right\| \geqslant |a|\,\eta = |f(z+ay)|,$$

so that $\|f\| \leqslant 1$. But if $\{z_n\}$ is a sequence in K for which $\|z_n - y\| \to \eta$ we have

$$\|f\|\|z_n - y\| \geqslant |f(z_n - y)| = |f(z_n) - f(y)| = |f(y)| = \eta$$

so that $\|f\| \geqslant 1$, on letting $n \to \infty$. Hence $\|f\| = 1$, and f has all the desired properties except that it is only defined on J, a linear subspace of H. Use theorem 8.6 to extend it to a linear functional F on the whole of H with $\|F\| = \|f\| = 1$.]

Corollary. *If $(\Omega, \mathscr{F}, \mu)$ is a σ-finite measure space, and K is a closed linear subspace of $\mathscr{L}_2(\Omega, \mu)$, and $y \in \mathscr{L}_2 - K$, then $y = z + x$ where $z \in K$ and $(x, w) = 0$ for all $w \in K$.*

Proof. $\mathscr{L}_2(\Omega, \mu)$ is a Banach space, and K is closed (in the metric ρ_2) so that $d(y, K) = \eta > 0$. Find the functional F satisfying the conditions of theorem 8.8 and represent it, by theorem 8.7, as

$$F(\mu) = (\mu, g) \quad \text{where} \quad g \in \mathscr{L}_2.$$

Now put $x = \eta g$, $z = y - x$ so that

$$(x, w) = \eta F(w) = 0 \quad \text{for all} \quad w \in K.$$

It only remains to show that $z \in K$. For $\epsilon > 0$ choose $k \in K$ such that

$$\|k-y\|^2 = (k-y, k-y) < \eta^2 + \epsilon.$$

Then

$$\begin{aligned} \|k-z\|^2 &= (k-y, k-y) + 2(x, k-y) + (x, x) \\ &= \|k-y\|^2 + 2\eta(g, k-y) + \eta^2\|g\|^2 \\ &= \|k-y\|^2 - 2\eta F(y) + \eta^2\|F\|^2 \\ &= \|k-y\|^2 - \eta^2 < \epsilon, \end{aligned}$$

so that there are points of K arbitrarily close to z, and we must have $z \in K$, since K is closed.]

Let us remind ourselves of the conditions under which we established theorem 7.9. $(\Omega, \mathscr{F}, \mu)$ is a σ-finite measure space, and T is a measure preserving transformation from Ω to itself. T^k is the result of repeating the transformation k times (T^0 is the identity map). For an $\mathscr{F}$-measurable function f which is finite a.e. we consider the sequence of means

$$g_n = \frac{1}{n}\sum_{i=0}^{n-1} f(T^i x). \tag{8.6.1}$$

***Theorem* 8.9.** *If $(\Omega, \mathscr{F}, \mu)$ and T satisfy the conditions in theorem 7.9, $f \in \mathscr{L}_2(\Omega, \mu)$, and g_n is defined by (8.6.1) then $\{g_n\}$ is a Cauchy sequence in second mean. Its limit (in second mean) f^* satisfies*

(i) *f^* is invariant under T, that is*

$$f^*(Tx) = f^*(x) \text{ a.e.};$$

(ii) $\|f^*\| \leqslant \|f\|$;

(iii) *for any function g in $\mathscr{L}_2$ which is invariant under T, $(g, f^*) = (g, f)$.*

Proof. (*a*) Suppose first that f is such that there is an $h \in \mathscr{L}_2$ such that

$$f(x) = h(Tx) - h(x) \text{ a.e.}$$

Then
$$g_n(x) = \frac{1}{n}\sum_{i=0}^{n-1} f(T^i x) = \frac{1}{n}[h(T^{n-1}x) - h(x)]$$

so that $\|g\| \leqslant 2\|h\|/n \to 0$ as $n \to \infty$.

(*b*) Now suppose f is the limit (in second mean) of a sequence $\{f_k\}$ such that, for each k, $f_k(x) = h_k(Tx) - h_k(x)$ with $h_k \in \mathscr{L}_2$. Then

$$\begin{aligned}\|g_n\| &\leqslant \frac{1}{n}\left\|\sum_{i=0}^{n-1}\{f(T^i x) - f_k(T^i x)\}\right\| + \frac{1}{n}\left\|\sum_{i=0}^{n-1} f_k(T^i x)\right\| \\ &\leqslant \frac{1}{n}\sum_{i=0}^{n-1}\|f(T^i x) - f_k(T^i x)\| + \frac{1}{n}\left\|\sum_{i=0}^{n-1} f_k T^i(x)\right\| \\ &\leqslant \|f - f_k\| + \frac{2}{n}\|h_k\|;\end{aligned}$$

so that we can make $\|g_n\| < \epsilon$ by first choosing f_k with $\|f - f_k\| < \frac{1}{2}\epsilon$ and then making n large.

The class of $f \in \mathscr{L}_2$ which satisfy either (*a*) or (*b*) is clearly a closed linear subspace K of $\mathscr{L}_2$. By the corollary to theorem 8.8, any $f \in \mathscr{L}_2$ can be written uniquely as

$$f = f_1 + f_2 \quad \text{where} \quad f_1 \in K,$$

and
$$(f_2, Tf - f) = 0 \quad \text{for all} \quad f \in \mathscr{L}_2.$$

Now
$$\begin{aligned}0 = (f_2, Tf - f) &= (f_2, Tf) - (f_2, f) \\ &= (T^{-1}f_2, f) - (f_2, f) = (T^{-1}f_2 - f_2, f) \quad \text{for all} \quad f \in \mathscr{L}_2,\end{aligned}$$

and in particular when f is the indicator function of a measurable set E of finite measure. Hence $T^{-1}f_2 = f_2$ a.e. so that f_2 is invariant under T. Hence

$$\frac{1}{n}\sum_{i=0}^{n-1} f_2(T^i x) = f_2(x) \text{ a.e.} \quad \text{for all} \quad n,$$

so that $f^* = f_2$ is the limit in second mean of $\{g_n\}$. Thus (i) and (ii) are proved. To prove (iii), suppose g is invariant under T; then

$$(T^i f, g) = (f, T^{-i} g) = (f, g)$$

so that $(g_n, g) = (f, g)$ for each n and the result follows on letting $n \to \infty$ since the inner product is continuous in the norm topology. ∎

Corollary. *Under the conditions of theorem* 8.9, *if T is ergodic, then the limit* (*in second mean*) $f^* = c$ *a.e. Also*

(i) *if* $\mu(\Omega) = \infty$, *then* $c = 0$,

(ii) *if* $\mu(\Omega) < \infty$, *then* $\int f^* \, d\mu = \int f \, d\mu$.

Proof. The only invariant functions are constants so (ii) of the theorem implies that $f^* = c$ a.e. Now if $\mu(\Omega) = \infty$, we have $\|f^*\|$ finite, so $c = 0$. If $\mu(\Omega) < \infty$, then the function $g(x) \equiv 1$ is in $\mathscr{L}_2$ and is invariant so that

$$(1, f^*) = \int f^* d\mu = (1, f) = \int f d\mu. \text{ ∎}$$

Exercises 8.6

1. If $\mu(\Omega) < \infty$, $f \in \mathscr{L}_p(\Omega, \mu)$, $1 < p < \infty$ and T is a measure preserving transformation, show that g_n, defined by (8.6.1), converges a.e. to a limit function $f^* \in \mathscr{L}_p$ such that $\rho_p(g_n, f^*) \to 0$. (This gives a simpler proof of theorem 8.9 for the case $\mu(\Omega) < \infty$).

Prove that (ii) of the corollary to theorem 8.9 is valid without the condition that T be ergodic.

2. Suppose X is an open subset of $\mathbf{R}^k$ of finite Lebesgue measure and $T: X \to X$ preserves Lebesgue measure and is ergodic. Show that, for almost all $x \in X$, the sequence $\{T^k x\}$ is dense in X.

9

STRUCTURE OF MEASURES IN SPECIAL SPACES

In the present book most of the theory of measure and integration has been developed in abstract spaces, and we have used the properties of special spaces only to illustrate the general theory. The present chapter, apart from § 9.4, is devoted to a discussion of properties which depend essentially on the structure of the space.

The first question considered is that of point-wise differentiation. In the Radon–Nikodym theorem 6.7 we defined the derivative $d\mu/d\nu$ of one measure with respect to another for suitable measures μ, ν: but the point function $d\mu/d\nu$ obtained is only determined in the sense that the equivalence class of functions equal almost everywhere is uniquely defined. This means that at *no* single point (except for those points which form sets of positive measure) is the derivative defined by the Radon–Nikodym theorem. In order to define $d\mu/d\nu$ at a point x, the local topological structure of the space near x has to be considered. It is possible to develop this local differentiation theory in fairly general spaces, but only at the cost of complicated and rather unnatural additional conditions: we have decided instead to give the detailed theory only in the space $\mathbf{R}$ of real numbers where the term derivative has a clear elementary meaning.

There are several ways of defining an integral with properties similar to those obtained in Chapter 5. So far in this book we have considered definitions which start from a given measure defined on a suitable class of sets. In § 9.4 we describe the Daniell integral and show that, under suitable conditions this can be obtained in terms of a measure. Then, for locally compact spaces, we discuss positive linear functionals on the space C_K of real-valued continuous functions which vanish outside a compact set, and show that these also correspond to integrals with respect to a suitable measure.

The final section of the chapter is devoted to the definition of Haar measure in topological spaces which have the algebraic structure of a group and in which the group operation is continuous. The details are given only for locally compact metric groups.

9.1 Differentiating a monotone function

We say that $f: I \to \mathbf{R}$ where I is an open interval in $\mathbf{R}$ (that is, a set of the form (a, b) with $a, b \in \mathbf{R}^*$), is *monotone increasing*, if

$$x_1, x_2 \in I, \quad x_1 < x_2 \Rightarrow f(x_1) \leqslant f(x_2).$$

At a given point x in I the function $f: I \to \mathbf{R}$ may not be differentiable but

$$D^+f(x) = \limsup_{h\to 0+} \frac{f(x+h)-f(x)}{h}, \quad D^-f(x) = \limsup_{h\to 0+} \frac{f(x)-f(x-h)}{h};$$

$$D_+f(x) = \liminf_{h\to 0+} \frac{f(x+h)-f(x)}{h}, \quad D_-f(x) = \liminf_{h\to 0+} \frac{f(x)-f(x-h)}{h}$$

are always uniquely determined in the extended real number system $\mathbf{R}^*$. These numbers are called the *derivates* of f at x. We say that f is differentiable at x if

$$D^+f(x) = D_+f(x) = D^-f(x) = D_-f(x) = Df(x) \neq \pm\infty.$$

It is clear that f is differentiable at x if and only if there is a real number $Df(x)$ such that, given $\epsilon > 0$ there is a $\delta > 0$ for which

$$\left|\frac{f(x+h)-f(x)}{h} - Df(x)\right| < \epsilon \quad \text{if} \quad 0 < |h| < \delta;$$

so that our definition is equivalent to that usually adopted in elementary texts on real analysis. When f is differentiable at x, we call $Df(x)$ the *derivative* of f at x.

If $f: I \to \mathbf{R}$ is continuous, but not monotone, it is possible that it is differentiable at no point x. However, a monotone $f: I \to \mathbf{R}$ must be continuous except at the points in a countable set, and the monotonicity further implies that there are some points x where the derivative exists. In fact we prove much more: the set of points x in I where f is not differentiable turns out to have zero measure. In order to prove this it is convenient first to obtain a new type of covering theorem. When in §2.2 we showed that a bounded closed interval K in $\mathbf{R}$ is compact we started with a covering of K by a family of open sets and we demanded that *all* of K be covered by a finite subfamily. However, in proving compactness we were not interested in economical covering, and the covering sets finally chosen could overlap. Clearly if we require that the covering sets must not overlap we can no longer require that all of K be covered. However, even if we are satisfied with a countable subcovering by disjoint sets of almost all of K (see exercise 9.1 (8)) additional conditions on the nature of the original covering are essential. A suitable form of these conditions now follows.

Vitali covering

For a subset $E \subset \mathbf{R}$, a class $\mathscr{J}$ of intervals is said to cover E in the Vitali sense if, given $x \in E$, $\epsilon > 0$ there is an interval $J \in \mathscr{J}$ with $x \in J$ and $0 < |J| < \epsilon$.

***Theorem* 9.1.** *Suppose $E \subset \mathbf{R}$ has finite Lebesgue outer measure and is covered in the Vitali sense by a class $\mathscr{J}$ of intervals. Then there is a countable disjoint subclass $\mathscr{J}_1 \subset \mathscr{J}$ such that*

$$|E - \bigcup\{J : J \in \mathscr{J}_1\}| = 0.$$

Proof. We use $|A|$ to denote the Lebesgue outer measure of A whether or not A is measurable. There is no harm in assuming that all the intervals J in $\mathscr{J}$ are closed since $|\bar{I}| = |I|$ for any interval I. We may further assume without loss of generality that there is an open set $G \supset E$ with $|G| < \infty$, and that all the intervals of $\mathscr{J}$ are contained in G.

We choose $\mathscr{J}_1$ by induction as follows. Let J_1 be any interval of $\mathscr{J}$. Suppose we have already chosen disjoint intervals $J_1, J_2, \ldots, J_m$ and let s_m be the supremum of the lengths of the intervals in $\mathscr{J}$ which do not intersect any of $J_1, J_2, \ldots, J_m$. Now $s_m < |G| < \infty$, and if E is not contained in $\bigcup_{i=1}^{m} J_i$, we must have $s_m > 0$. Thus if E is not already covered, we can choose J_{m+1} disjoint from $\bigcup_{i=1}^{m} J_i$ with $|J_{m+1}| > \frac{1}{2}s_m$. Now the theorem is proved if $E \subset \bigcup_{i=1}^{m} J_i$ for any finite m. Otherwise we obtain a sequence $\{J_n\}$ of disjoint sets so that

$$\sum_{i=1}^{\infty} |J_i| \leqslant |G| < \infty.$$

Now suppose, if possible, that

$$|E - \bigcup_{i=1}^{\infty} J_i| = \delta > 0.$$

We can choose N so that $\quad \sum_{i=N+1}^{\infty} |J_i| < \frac{1}{5}\delta,$

and put $\quad F = E - \bigcup_{i=1}^{N} J_i.$

F must be non-void and $\bigcup_{i=1}^{N} J_i$ is closed so we can find a point x in E

and an interval J of $\mathscr{J}$ containing x and short enough to be disjoint from $\bigcup_{i=1}^{N} J_i$. This implies $|J| \leqslant s_n < 2|J_{n+1}|$. Since

$$\lim_{n\to\infty} |J_n| = 0,$$

this J must meet at least one of the J_i for $i > n$. Let k be the smallest integer for which $J \cap J_k \neq \varnothing$. Then $|J| \leqslant s_{k-1} < 2|J_k|$, so the distance from x to the mid-point of J_k is at most $|J| + \frac{1}{2}|J_k| \leqslant \frac{5}{2}|J_k|$, and x must belong to the interval H_k which has the same centre as J_k and 5 times the length. Thus

$$F \subset \bigcup_{i=N+1}^{\infty} H_i$$

and
$$\delta = |F| \leqslant \sum_{i=N+1}^{\infty} |H_i| = 5 \sum_{i=N+1}^{\infty} |J_i| < \delta,$$

which establishes a contradiction. ▌

Corollary. *Under the conditions of theorem 9.1, for each $\epsilon > 0$ there is a finite set $J_1, J_2, \ldots, J_p$ of disjoint intervals of $\mathscr{J}$ such that*

$$\left| E - \bigcup_{i=1}^{p} J_i \right| < \epsilon.$$

Theorem 9.2. *Suppose $f: I \to \mathbf{R}$ is monotone increasing. Then the set E of points x in I for which f is differentiable at x satisfies $|I - E| = 0$. The derivative f' is Lebesgue measurable, and if $[a, b] \subset I$,*

$$\int_a^b f'(x)\,dx \leqslant f(b) - f(a).$$

Proof. It is clearly sufficient to prove the theorem for a finite closed interval $I = [c, d]$. The first step is to show that each of the subsets of I:

$$\{x: D^+f(x) > D_-f(x)\}, \quad \{x: D^-f(x) > D_+f(x)\},$$
$$\{x: D^+f(x) > D_+f(x)\}, \quad \{x: D^-f(x) > D_-f(x)\},$$

has zero Lebesgue measure. We give the details for the set

$$E = \{x: D^+f(x) > D_-f(x)\};$$

the proof for the others is similar. Now E is the (countable) union of sets

$$E_{u,v} = \{x: D^+f(x) > u > v > D_-f(x)\}$$

over rational pairs u, v. It is therefore sufficient to show that $|E_{u,v}| = 0$ for all pairs u, v with $u > v$.

Let $t = |E_{u,v}|$ and $\epsilon > 0$. Find an open set $G \supset E_{u,v}$ with

$$|G| < t+\epsilon.$$

For each $x \in E_{u,v}$, there is an arbitrarily small closed interval

$$[x-h, x] \subset G$$

with
$$f(x)-f(x-h) < vh.$$

By theorem 9.1, corollary we can find a finite disjoint collection $J_1, J_2, \ldots, J_N$ of such intervals whose interiors cover a subset F of $E_{u,v}$ with $|E_{u,v}-F| < \epsilon$. If we sum over these intervals

$$\sum_{n=1}^{N} \{f(x_n)-f(x_n-h_n)\} < v \sum_{n=1}^{N} h_n < v|G|$$
$$< v(t+\epsilon).$$

But each $y \in F$ is the left-hand end-point of an arbitrarily small interval $[y, y+k]$ which is contained in one of the J_i $(i = 1, 2, \ldots, N)$ and such that
$$f(y+k)-f(y) > uk.$$

Use theorem 9.1 again to find a disjoint collection $K_1, K_2, \ldots, K_P$ of such intervals which covers a subset H of F with $|H| > t-2\epsilon$. Summing over these intervals, since each K_i is contained in a J_n,

$$\sum_{i=1}^{N} \{f(x_n)-f(x_n-h_n)\} \geqslant \sum_{i=1}^{P} f(y_i+k_i)-f(y_i)$$
$$> u \sum_{i=1}^{P} k_i > u(t-2\epsilon)$$

so that
$$v(t+\epsilon) > u(t-2\epsilon).$$

Since $u > v$ and ϵ is arbitrary, we must have $t = 0$. Thus for almost all x in I,
$$g(x) = Df(x) = \lim_{h\to 0} \frac{f(x+h)-f(x)}{h}$$

exists as an element in $\mathbf{R}^*$ (we are thus allowing the value $\pm\infty$ for a limit). If we put

$$g_n(x) = n\left[f\left(x+\frac{1}{n}\right)-f(x)\right] \quad \text{for} \quad x \in [a, b]$$

where we re-define $f(x) = f(b)$ for $x \geqslant b$, then $g_n(x)$ is defined and measurable and $g_n(x) \to g(x)$ for almost all x in $[a, b]$ as $n \to \infty$ so that $g: I \to \mathbf{R}^*$ is Lebesgue measurable if we define it arbitrarily to be zero on the exceptional set where $Df(x)$ is not defined. By Fatou (theorem 5.7)

$$\begin{aligned}\int_a^b g(x)\,dx &\leqslant \liminf_{n\to\infty}\int_a^b g_n(x)\,dx\\ &= \liminf n\int_a^b\left\{f\left(x+\frac{1}{n}\right)-f(x)\right\}dx\\ &= \liminf\left[n\int_b^{b+(1/n)} f(x)\,dx - n\int_a^{a+(1/n)} f(x)\,dx\right]\\ &\leqslant f(b)-f(a).\end{aligned}$$

This shows that the function g is integrable and so finite almost everywhere. Thus f is differentiable a.e. in $[a,b]$. Since $[a,b]$ is an arbitrary subinterval of I, f is differentiable almost everywhere in I. ∎

Functions of bounded variation

A function $f\colon I\to\mathbf{R}$ is said to be of bounded variation on I if

$$\sum_{i=1}^{n}|f(x_i)-f(x_{i-1})|$$

is bounded above for all ordered finite sequences $x_0 < x_1 < \ldots < x_n$ in I. Clearly if $f\colon I\to\mathbf{R}$ is of bounded variation on I, it is also of bounded variation on each interval $J\subset I$. For an ordered sequence $\alpha = \{x_i\}$, $i = 0,1,\ldots,n$ put

$$p(\alpha) = \sum_{i=1}^{n}\max\,[0, f(x_i)-f(x_{i-1})],$$

$$n(\alpha) = -\sum_{i=1}^{n}\min\,[0, f(x_i)-f(x_{i-1})],$$

$$t(\alpha) = p(\alpha)+n(\alpha) = \sum_{i=1}^{n}|f(x_i)-f(x_{i-1})|.$$

If $f\colon[a,b]\to\mathbf{R}$ is of bounded variation on $[a,b]$, put

$$T_a^b = \sup_\alpha t(\alpha),\quad P_a^b = \sup_\alpha p(\alpha),\quad N_a^b = \sup_\alpha n(\alpha),$$

where each of the suprema is taken over all ordered finite sequences α in $[a,b]$. It is easy to check that, in this case

$$T_a^b = P_a^b+N_a^b,\quad f(b)-f(a) = P_a^b-N_a^b.$$

Now if $f\colon[a,b]\to\mathbf{R}$ is of bounded variation on $[a,b]$ we can put

$$g(x) = N_a^x,\ h(x) = P_a^x \quad \text{for all} \quad x\in[a,b]$$

so that $f(x)$ can be expressed as the difference of two non-decreasing functions of bounded variation.

Corollary (*Lebesgue*). *A function $f\colon I \to \mathbf{R}$ which is of bounded variation on each finite interval $[a,b] \subset I$ must be differentiable at x for almost all x in I.*

Proof. In each finite $[a,b]$ we can express f as the difference of two monotone increasing functions g and h. Each of these is differentiable almost everywhere in $[a,b]$ by theorem 9.2. Hence the difference f is differentiable almost everywhere in $[a,b]$. ▌

Exercises 9.1

1. Show that, if $g\colon I \to \mathbf{R}$, $h\colon I \to \mathbf{R}$ are each monotone increasing, then $f = g - h$ is of bounded variation on each $[a,b] \subset I$.

2. If $f\colon I \to \mathbf{R}$ is of bounded variation on each $[a,b] \subset I$, show that the limits $f(x+0)$, $f(x-0)$ exist at each interior point of I.

3. If c is an interior point of I and $f\colon I \to \mathbf{R}$ has a (local) maximum at c, show that $D^+f(c) \leqslant 0$, $D_-f(c) \geqslant 0$.

4. If $f\colon [a,b] \to \mathbf{R}$ is continuous and $D^+f(x) \geqslant 0$ for all x in $[a,b)$, show that $f(b) \geqslant f(a)$.

5. Define

$$\begin{aligned} f(0) &= 0, \\ f(x) &= x^2 \sin x^{-2} \quad \text{for} \quad x \neq 0; \\ g(0) &= 0, \\ g(x) &= x^2 \sin x^{-1} \quad \text{for} \quad x \neq 0. \end{aligned}$$

Which of the functions f, g is of bounded variation on $[-1,1]$?

6. Give an example of a function for which all the four derivates are different at $x = 0$.

7. For any Lebesgue measurable $f\colon I \to \mathbf{R}$, prove that $D^+f(x)$ is Lebesgue measurable.

8. Show that theorem 9.1 as stated in $\mathbf{R}$ is false in $\mathbf{R}^n$ for $n \geqslant 2$.

Hint. Take a Vitali covering of $[0,1]$ and for each J of covering consider $J \times [0, \frac{2}{3}]$ and $J \times [\frac{1}{3}, 1]$. This will give a covering in the sense of our definition of the unit square $[0,1] \times [0,1]$. Show theorem 9.1 is not satisfied.

(In fact a more complicated construction shows that theorem 9.1 fails even if we require each point of the set to be covered by an interval J of arbitrarily small diameter.)

9. Show that theorem 9.1 is true in $\mathbf{R}^n$ for all n if we restrict the covering to cubes. (In fact it can be shown that it is true if there is a constant K such that the ratio of the lengths of longest and shortest sides is bounded for the intervals in $\mathscr{J}$.)

10. For the Cantor ternary function $g\colon [0,1] \to [0,1]$ show that $g'(x) = 0$ for all $x \in [0,1] - C$.

(This shows that we cannot hope, in general, for equality in

$$\int_a^b f'(x)\,dx \leqslant f(b)-f(a).)$$

11. Prove that a convergent series of non-decreasing real functions can be differentiated term by term a.e.

9.2 Differentiating the indefinite integral

The 'fundamental theorem of the integral calculus' states that, if $f:[a,b]\to\mathbf{R}$ is a continuous function and

$$F(x) = \int_a^x f(t)\,dt$$

then F: $[a,b]\to\mathbf{R}$ is differentiable in (a,b) with $F'(x)=f(x)$. The object of this section is to obtain the analogous theorem for the Lebesgue integral, where it is not appropriate to assume that f is continuous. (Of course, if $f:[a,b]\to\mathbf{R}$ is continuous on $[a,b]$, we know that $F'(x)=f(x)$ for all x in (a,b) since the Lebesgue integral coincides with the Riemann integral in this case.) The first thing to note is that, even for a monotonic function F, we cannot claim that, in general,

$$\int_a^b F'(x)\,dx = F(b)-F(a), \tag{9.2.1}$$

see exercise 9.1 (10). We will, however, obtain necessary and sufficient conditions for the truth of (9.2.1).

Lemma. *If $f:[a,b]\to\mathbf{R}^*$ is Lebesgue integrable on $[a,b]$ and*

$$\int_a^x f(t)\,dt = 0 \quad \textit{for all} \quad x \textit{ in } [a,b],$$

then $f(t)=0$ for almost all t in $[a,b]$.

Note. This strengthens the result of theorem 5.5 (vii).

Proof. If the lemma is false then at least one of the sets

$$\{t: f(t)<0\}, \quad \{t: f(t)>0\}$$

has positive measure. If $|\{t:f(t)>0\}|>0$ then we can find a $\delta>0$ for which $|E|>0$, where $E=\{t:f(t)>\delta\}$. Now choose a closed set $F\subset E$ with $|F|>0$, and consider the open set $G=(a,b)-F$. Then

$$0=\int_a^b f\,dm = \int_F f\,dm + \int_G f\,dm.$$

But G is the disjoint union of a countable collection of open intervals

$$(a_n,b_n) \quad\text{and}\quad \int_{a_n}^{b_n} f\,dm = 0$$

for each n. Since the integral defines a σ-additive set function we must have

$$\int_G f\,dm = 0 \quad \text{so that} \quad \int_F f\,dm = 0$$

and this contradicts $\int_F f\,dm \geqslant \delta|F| > 0.$ ∎

Let us now consider the properties of any function F which is an indefinite integral, that is

$$F(x) = \int_a^x f(t)\,dt$$

for a function $f\colon [a,b] \to \mathbf{R}^*$ which is Lebesgue integrable. It is immediate from theorem 5.6 that F is continuous on $[a,b]$, but more can be said: since it is the difference of the indefinite integrals of f_+ and f_- it must be the difference of two monotone functions and therefore it is of bounded variation. In fact, we saw in theorem 5.6 that the set function

$$\nu(E) = \int_E f\,dm\colon E \text{ measurable}, \quad E \subset [a,b]$$

is absolutely continuous; that is that $\nu(E) \to 0$ as $m(E) \to 0$. This means in particular that given $\epsilon > 0$, there is a $\delta > 0$ such that if $E = \bigcup_{k=1}^{n} I_k$ is a finite disjoint union of intervals in $[a,b]$ for which

$$\sum_{k=1}^{n} m(I_k) < \delta, \quad \text{then} \quad |\nu(E)| = \left|\sum_{k=1}^{n} \nu(I_k)\right| < \epsilon.$$

In fact, by considering separately the intervals I_k for which ν is positive and negative we can find $\delta > 0$ such that

$$\sum_{k=1}^{n} m(I_k) < \delta \Rightarrow \sum_{k=1}^{n} |\nu(I_k)| < \epsilon.$$

In terms of the indefinite integral F this means that the function $F\colon [a,b] \to \mathbf{R}$ is such that, for each $\epsilon > 0$ there is a $\delta > 0$ for which

$$\sum_{i=1}^{n} (b_i - a_i) < \delta \Rightarrow \sum_{i=1}^{n} |F(b_i) - F(a_i)| < \epsilon \tag{9.2.2}$$

for any finite class of disjoint intervals $(a_i, b_i) \subset (a,b)$. Any function $F\colon I \to \mathbf{R}$ which satisfies this condition on every finite interval $(a,b) \subset I$ is said to be *absolutely continuous* on I.

It is immediate that any function $F\colon I \to \mathbf{R}$ which is absolutely continuous is of bounded variation on each finite interval $[a,b] \subset I$. For if we put $\epsilon = 1$ in (9.2.2) and choose $\delta > 0$, then any finite dissection of $[a,b]$ can be split into K sets of intervals (by inserting extra division points if necessary) each of total length less than δ, where $K = [(b-a)/\delta]+1$; and it follows that, for any dissection of $[a,b]$

$$\sum_{r=1}^{n} |F(x_r)-F(x_{r-1})| \leqslant K.$$

By the corollary to theorem 9.2 we now see that any function F which is absolutely continuous is differentiable except on a set of zero measure.

***Theorem* 9.3.** *Suppose $f: [a,b] \to \mathbf{R}^*$ is Lebesgue integrable on $[a,b]$ and $F: [a,b] \to \mathbf{R}$ satisfies*

$$F(x) = F(a) + \int_a^x f(t)\,dt.$$

Then F is differentiable with $F'(x) = f(x)$ for almost all x in $[a,b]$.

Proof. Assume first that f is bounded on $[a,b]$ so that for a suitable M in R, $|f(x)| \leqslant M$, for all x in $[a,b]$. Now we know that F is absolutely continuous and therefore differentiable almost everywhere. Put

$$f_n(x) = n\left[F\left(x+\frac{1}{n}\right) - F(x)\right].$$

Then $|f_n| \leqslant M$ and $f_n(x) \to F'(x)$ almost everywhere; so, by theorem 5.8 for $a \leqslant c \leqslant b$,

$$\int_a^c F'(x)\,dx = \lim \int_a^c f_n(x)\,dx = \lim n \int_a^c \left[F\left(x+\frac{1}{n}\right) - F(x)\right] dx$$

$$= \lim \left[n \int_c^{c+(1/n)} F(x)\,dx - n \int_a^{a+(1/n)} F(x)\,dx\right]$$

$$= F(c) - F(a) = \int_a^c f(x)\,dx$$

since F is continuous. Hence

$$\int_a^c \{F'(x) - f(x)\}\,dx = 0$$

for all c in $[a,b]$ so that $F'(x) = f(x)$ almost everywhere.

Now suppose that $f: [a,b] \to \mathbf{R}^*$ is integrable but not bounded. From the definition of the integral it is sufficient to prove the theorem when $f \geqslant 0$. Put

$$g_n(x) = \min\,[n, f(x)]$$

and

$$G_n(x) = \int_a^x [f(t) - g_n(t)]\,dt.$$

Since $f-f_n \geqslant 0$, G_n is monotone increasing and so has a non-negative derivative almost everywhere. Since f_n is bounded (by n) we know that

$$\frac{d}{dx}\left\{\int_a^x f_n(t)\,dt\right\} = f_n(x) \text{ a.e.},$$

so that the derivative

$$F'(x) = G_n'(x) + \frac{d}{dx}\left\{\int_a^x f_n(t)\,dt\right\} \geqslant f_n(x),$$

and exists almost everywhere. Since this is true for each integer n,

$$F'(x) \geqslant f(x) \text{ a.e.} \tag{9.2.3}$$

Hence
$$\int_a^b F'(x)\,dx \geqslant \int_a^b f(x)\,dx = F(b)-F(a),$$

and by theorem 9.2 we must have

$$\int_a^b F'(x)\,dx = F(b)-F(a) = \int_a^b f(x)\,dx,$$

and
$$\int_a^b \{F'(x)-f(x)\}\,dx = 0.$$

This with (9.2.3) implies that $F'(x) = f(x)$ a.e. ∎

Lemma. *If* $F\colon [a,b] \to \mathbf{R}$ *is absolutely continuous on* $[a,b]$ *and*

$$F'(x) = 0 \text{ a.e.},$$

then F *is constant.*

Proof. Suppose $a \leqslant c \leqslant b$, and $E = \{x \in [a,c];\ F'(x) = 0\}$. For a fixed $\epsilon > 0$, there are arbitrarily small intervals $[x, x+h]$ for each $x \in E$ such that

$$|F(x+h)-F(x)| < \epsilon h.$$

Choose $\delta > 0$ to satisfy (9.2.2) in the definition of absolute continuity and use theorem 9.1 to obtain a finite collection $[x_k, y_k]$ of intervals with

$$|F(y_k)-F(x_k)| < \epsilon(y_k - x_k)$$

which cover all of E except for a subset of measure less than δ. Order these intervals so that

$$y_0 = a \leqslant x_1 < y_1 \leqslant x_2 < \ldots \leqslant y_n \leqslant c = x_{n+1},$$

and
$$\sum_{i=0}^{n} |x_{i+1}-y_i| < \delta.$$

By (9.2.2) this implies $\displaystyle\sum_{i=0}^{n} |F(x_{i+1})-F(y_i)| < \epsilon$

and, from the choice of the covering family

$$\sum_{i=0}^{n} |F(y_i) - F(x_i)| < \epsilon(c-a)$$

so that

$$|F(c)-F(a)| = \left| \sum_{i=0}^{n} \{F(x_{i+1}) - F(y_i)\} + \sum_{i=0}^{n} \{F(y_i) - F(x_i)\} \right|$$
$$< \epsilon(c-a+1).$$

Since ϵ is arbitrary, we have $F(c) = F(a)$. ▮

***Theorem* 9.4.** *A function $F: I \to \mathbf{R}$ is an indefinite integral, that is there is a measurable $f: I \to \mathbf{R}^*$ such that*

$$F(b) - F(a) = \int_a^b f(x)\, dx$$

for all $[a,b] \subset I$, if and only if F is absolutely continuous on I.

Proof. We have already seen that any indefinite integral is absolutely continuous. Conversely suppose $F: I \to \mathbf{R}$ is absolutely continuous. Then F is differentiable almost everywhere in $[a,b]$ and

$$|F'(x)| \leqslant F_1'(x) + F_2'(x) \text{ a.e.,}$$

where $F = F_1 - F_2$ expresses F as the difference of two monotone functions. By theorem 9.2, F' in integrable on $[a,b]$. Put

$$G(x) = \int_a^x F'(t)\, dt.$$

Then G is absolutely continuous and so is $H = F - G$. But, by theorem 9.3,

$$H' = F' - G' = F' - F' = 0 \text{ a.e.}$$

so that H is constant by the lemma. Hence

$$F(x) = \int_a^x F'(t)\, dt + F(a). ▮$$

Corollary. *Every absolutely continuous function $F: I \to \mathbf{R}$ is the indefinite integral of its derivative.*

Density

Given a set $A \subset \mathbf{R}$, $x \in \mathbf{R}$ consider the ratio

$$\frac{|I \cap A|}{|I|}$$

for all intervals I containing x where $|E|$ denotes the Lebesgue outer measure of E. If this ratio converges to a limit as $|I| \to 0$, then this limit is called the density of A at x and denoted $\tau(x, A)$. The point x is called a point of density for A if $\tau(x, A) = 1$, and a point of dispersion for A if $\tau(x, A) = 0$. We can obtain the following as a corollary of theorem 9.4.

Lemma (*Lebesgue*). *If* $A \subset \mathbf{R}$, *A is Lebesgue measurable, then*

$$\tau(x, A) = 1 \quad \textit{for almost all} \quad x \in A,$$

$$\tau(x, A) = 0 \quad \textit{for almost all} \quad x \in \mathbf{R} - A.$$

Proof. Suppose $a < x < b$. Then the indicator function χ_A is Lebesgue integrable over $[a, b]$. Hence

$$F(x) = \int_a^x \chi_A \, dx$$

is differentiable almost everywhere and

$$F'(x) = 1 \quad \text{for almost all } x \text{ in } [a, b] \cap A,$$

$$F'(x) = 0 \quad \text{for almost all } x \text{ in } [a, b] \cap (\mathbf{R} - A).$$

But if x is such that $F'(x) = 1$, there is for each $\epsilon > 0$ a $\delta > 0$ such that

(i) $1 \geqslant \dfrac{|[x, x+h] \cap A|}{h} > 1 - \epsilon \quad \text{for} \quad 0 < h < \delta,$

(ii) $1 \geqslant \dfrac{|[x-k, x] \cap A|}{k} > 1 - \epsilon \quad \text{for} \quad 0 < k < \delta;$

and so $\quad 1 \geqslant \dfrac{|[x-k, x+h] \cap A|}{h+k} > 1 - \epsilon \quad \text{for} \quad 0 < h, k < \delta,$

which is precisely the condition for $\tau(x, A) = 1$. A similar proof shows that, at points x where $F'(x) = 0$ we have $\tau(x, A) = 0$. ∎

Exercises 9.2

1. If $F: I \to \mathbf{R}$ is absolutely continuous, show that F^p is absolutely continuous for each $p \geqslant 1$, but not, in general, for $p < 1$.

2. If $F: [a, b] \to \mathbf{R}$ is such that F' exists everywhere in (a, b) and is bounded show that

$$\int_a^b F'(x) \, dx = F(b) - F(a).$$

For $F(x) = x^2 \sin 1/x^2$ $(x \neq 0)$, $F(0) = 0$ show that $F'(x)$ exists for all x but is not Lebesgue integrable over $[-1, 1]$. (This shows that even the Lebesgue integral is not strong enough to integrate all derivatives.)

3. Construct a subset $A \subset \mathbf{R}$ for which $\tau(0, A) = \frac{1}{2}$.

4. Extend the density result to non-measurable sets A by showing that for any $A \subset \mathbf{R}$, $\tau(x, A) = 1$ for all x in A except a subset of zero measure.

Hint. Assume A is contained in a finite interval, and take a measurable set $B \supset A$ with $|B| = |A|$.

Deduce that a set $A \subset \mathbf{R}$ is measurable if and only if $\tau(x, A) = 0$ for almost all x in $(\mathbf{R} - A)$.

5. Prove that the Cantor function $g: [0, 1] \to [0, 1]$ defined in §2.7 is monotone increasing and continuous but not absolutely continuous.

6. The function $f: [0, 1] \to \mathbf{R}$ is absolutely continuous on $[\epsilon, 1]$ for each $\epsilon > 0$. Can one deduce that f is absolutely continuous on $[0, 1]$? Does the additional condition that f is of bounded variation on $[0, 1]$ help?

9.3 Point-wise differentiation of measures

In theorem 4.8 we proved that all measures μ in $\mathbf{R}$ defined for Borel sets and finite on bounded sets are Lebesgue–Stieltjes measures: that is, there is a monotone increasing function $F: \mathbf{R} \to \mathbf{R}$ which is continuous on the right such that $\mu = \mu_F$ on $\mathscr{B}$. Because of this correspondence we can obtain properties of such Borel measures in terms of the corresponding properties of F.

***Lemma* 1.** *Suppose μ_F is the Lebesgue–Stieltjes measure with respect to the function $F: \mathbf{R} \to \mathbf{R}$ which is continuous on the right. Then μ_F is absolutely continuous with respect to Lebesgue measure m if and only if F is absolutely continuous.*

Proof. Suppose first that F is absolutely continuous. Then, by theorem 9.4

$$\mu_F(a, b] = F(b) - F(a) = \int_a^b F'(t)\, dt$$

so that, for $E \in \mathscr{P}$, μ_F coincides with the set function

$$\nu(E) = \int_E F'\, dm.$$

But the extension of a measure from $\mathscr{P}$ to $\mathscr{B}$ is unique, so that $\mu_F = \nu$ on $\mathscr{B}$, and μ_F must therefore be absolutely continuous with respect to m.

Conversely, if μ_F is absolutely continuous with respect to m, by the Radon–Nikodym theorem there is an $f \geq 0$ such that

$$\mu_F = \int_E f dm \quad \text{for} \quad E \in \mathscr{B}.$$

Hence $$\mu_F(0,x] = F(x)-F(0) = \int_0^x f(t)\,dt \quad \text{for} \quad x \geq 0,$$

$$\mu_F(x,0] = F(0)-F(x) = -\int_0^x f(t)\,dt \quad \text{for} \quad x < 0,$$

so that $F:\mathbf{R}\to\mathbf{R}$ is an indefinite integral and must therefore be absolutely continuous. ▌

Atom

Given any measure space $(X,\mathscr{F},\mu)$ in which $\mathscr{F}$ contains all single point sets the point $x \in X$ is said to be an atom for the measure μ if $\mu\{x\} > 0$. A measure μ with no atoms is said to be *non-atomic*. Now if μ is σ-finite, the set of atoms of μ is countable. In this case if we put

$$\nu(E) = \sum_{\substack{x\in E \\ \mu\{x\}\neq 0}} \mu\{x\}$$

we obtain a new measure ν defined on all subsets of X, and ν is a *discrete measure* as defined in §3·1. Further, the set function

$$\tau = \mu - \nu$$

defined on $\mathscr{F}$ is clearly non-atomic and so

$$\mu = \nu + \tau$$

is a decomposition of a σ-finite measure μ into the sum of a discrete measure and a non-atomic measure. This decomposition is clearly unique. Thus we have proved

***Lemma* 2.** *Given a σ-finite measure space $(X,\mathscr{F},\mu)$ in which $\mathscr{F}$ contains all single point sets there is a unique decomposition of μ,*

$$\mu = \nu + \tau$$

for which ν is a discrete measure on X and τ is a non-atomic measure on $\mathscr{F}$.

***Lemma* 3.** *A measure μ on $\mathscr{B}$ (the Borel sets of* $\mathbf{R}$*) which is finite on bounded intervals is a discrete measure if and only if $\mu = \mu_F$ where F is a jump function, that is,*

$$\left.\begin{aligned} F(x) &= \sum_{0<x_i\leq x} p_i \quad \textit{for} \quad x \geq 0, \\ -F(x) &= \sum_{x<x_i\leq 0} p_i \quad \textit{for} \quad x < 0, \end{aligned}\right\} \tag{9.3.1}$$

where the measure μ has atoms x_i of weight (or measure) p_i.

Proof. It is clear that if $F: \mathbf{R} \to \mathbf{R}$ satisfies (9.3.1) then

$$\mu_F(a,b] = \sum_{a<x_i\leqslant b} p_i = \sum_{x_i\in(a,b]} p_i$$

so that μ_F coincides with the discrete measure

$$\nu(E) = \sum_{x_i\in E} p_i$$

for $E \in \mathscr{P}$. By uniqueness of extension μ_F must be a discrete measure.

Conversely, if μ is a discrete measure with atoms x_i of weight p_i, an application of the theorem 4.8 shows that $\mu = \mu_F$ with F a jump function. ∎

Lemma 4. *A measure μ defined on $\mathscr{B}$ which is finite on bounded intervals is non-atomic if and only if $\mu = \mu_F$ for a continuous $F: \mathbf{R} \to \mathbf{R}$.*

Proof. If F is continuous, then

$$0 \leqslant \mu_F\{x\} < \mu_F(x-h, x] = F(x) - F(x-h)$$

for all $h > 0$, so that $\mu_F\{x\} = 0$.

Conversely if F is not continuous at x_0, then

$$\mu_F\{x_0\} = \lim_{n\to\infty} \mu_F\left(x_0 - \frac{1}{n}, x_0\right] = F(x_0) - F(x_0 - 0)$$

so x_0 is an atom. ∎

Singular monotone function

Any function $F: I \to \mathbf{R}$ which is continuous and monotone increasing, such that $F'(x) = 0$ for all x in I except for a set of zero Lebesgue measure, is said to be *singular*. The function g defined in §2.7 clearly satisfies these conditions without being constant.

Lemma 5. *A function $F: \mathbf{R} \to \mathbf{R}$ is singular if and only if the Lebesgue–Stieltjes measure μ_F is non-atomic and singular with respect to Lebesgue measure.*

Proof. The continuity of F is equivalent to the condition that μ_F be non-atomic by lemma 4. Now a measure ν is singular with respect to Lebesgue measure if and only if any absolutely continuous τ satisfying $\tau(E) \leqslant \nu(E)$ for all E in $\mathscr{B}$ must be zero. Now if $F'(x) > 0$ on a set of positive measure, the set function

$$\tau(E) = \int_E F'(x)\,dx$$

is not always zero and $\tau \leqslant \mu_F$ by theorem 9.2 so μ_F is not singular with respect to Lebesgue measure. Conversely, if μ_F is not singular a non-null absolutely continuous measure $\tau \leqslant \mu_F$ can be found, and this corresponds to a function G, that is

$$G(b)-G(a) = \int_a^b G'(x)\,dx.$$

But $F'(x) \geqslant G'(x)$ when both are defined, so $F'(x) > 0$ on a set of positive measure. ∎

***Theorem* 9.5** (*Lebesgue*). *Given any function $F: \mathbf{R} \to \mathbf{R}$ which is monotone increasing and continuous on the right, there is a decomposition of F*

$$F = F_1 + F_2 + F_3$$

where F_1 is a jump function,

F_2 is singular,

F_3 is absolutely continuous.

This decomposition is unique if we insist that $F_1(0) = F_2(0) = 0$.

Proof. Use the function F to define a Lebesgue–Stieltjes measure μ_F on $\mathscr{B}$. Decompose μ_F with respect to Lebesgue measure m by theorem 6.7 so that

$$\mu_F = \nu_1 + \nu_3$$

with $\nu_3 \ll m$ and ν_1 singular with respect to m. Decompose ν_1 by lemma 2,

$$\nu_1 = \lambda_1 + \lambda_2,$$

where λ_1 is discrete and λ_2 is non-atomic.

Let F_1, F_2 be the monotone functions (with $F_1(0) = F_2(0) = 0$) obtained by theorem 4.8 for which $\lambda_1 = \mu_{F_1}$, $\lambda_2 = \mu_{F_2}$ on $\mathscr{B}$. Then by lemmas 3 and 5, F_1 is a jump function, and F_2 is a singular function. If one applies theorem 4.8 to ν_3 one obtains an absolutely continuous G_3 for which $\nu_3 = \mu_{G_3}$. Finally, put $F_3(x) = G_3(x) + F(0)$ and we still have F_3 absolutely continuous, and $\nu_3 = \mu_{F_3}$. Now

$$F(x)-F(0) = F_1(x)-F_1(0)+F_2(x)-F_2(0)+F_3(x)-F_3(0)$$

for all x so that $\qquad F(x) = F_1(x)+F_2(x)+F_3(x).$

The uniqueness follows from the uniqueness of the decomposition $\mu_F = \lambda_1 + \lambda_2 + \nu_3$, and theorem 4.8. ∎

In $\mathbf{R}$ we can also use the connexion between μ_F and F to define differentiation. Thus if $F: I \to \mathbf{R}$ is differentiable at x_0, this means that

$$\frac{F(x_0+h)-F(x_0-k)}{h+k} \to F'(x_0) \quad \text{as} \quad h, k \to 0$$

with $h > 0, k > 0$, and

$$\frac{\mu_F(x_0-k, x_0+h]}{|(x_0-k, x_0+h]|} \to F'(x_0).$$

This can be written

$$\frac{\mu_F(J)}{|J|} \to F'(x_0) \quad \text{as} \quad |J| \to 0$$

for intervals J containing x_0, and we can write $d\mu_F/dm\,(x_0)$ for the value of this limit. More generally, if μ, ν are two measures in $\mathbf{R}$ which are finite for bounded sets then

$$\lim_{\substack{|J|\to 0\\ x\in J}} \left[\frac{\mu(J)}{\nu(J)}\right],$$

when it exists, is called the derivative of μ with respect to ν at the point x.

In $\mathbf{R}^n$ we can consider the values of the ratio

$$\frac{\mu(J)}{\nu(J)} \tag{9.3.2}$$

for rectangles J (in $\mathscr{P}^n$) containing a fixed point x and ask whether or not this ratio approaches a limit as diam $(J) \to 0$. The existence of this limit for all x except for a set of zero ν measure can be proved when ν is Lebesgue measure: the limit in this case is called the *strong derivate* of μ at x. This result is harder to prove than the result in § 9.1 because theorem 9.1 is not valid without some restriction on the ratio of the sides of the covering class $\mathscr{J}$. Essentially similar methods to those of § 9.1 will work if only cubes J are considered. On the other hand if in (9.3.2) one considers rectangles with arbitrary orientation an example can be given for which the limit exists nowhere.

Differentiation point-wise in abstract spaces can also be defined in terms of suitable 'nets', and the theorems of this chapter can be obtained if sufficient conditions are imposed. Since the results are not often used in practice, we will not state them in detail.

Exercises 9.3

1. Enumerate the rationals as a sequence $\{r_i\}$. By considering the discrete measure with mass $1/i^2$ at r_i $(i = 1, 2, \ldots)$ define a jump function which is constant in no interval.

2. Give an example of a singular function which is constant in no interval.

3. If F, G are two monotone real functions differentiable at x_0 with $G'(x_0) \neq 0$, show that

$$\frac{d\mu_F}{d\mu_G}(x_0) = \lim_{\substack{x_0\in J\\ |J|\to 0}} \frac{\mu_F(J)}{\mu_G(J)} \quad \text{exists and equals} \quad \frac{F'(x_0)}{G'(x_0)}.$$

9.4* The Daniell integral

Our approach in this book has been to regard measure as the primitive concept, and to define the integration process in terms of a given measure. One important alternative is to start with an 'integral' defined on a suitable class of functions, extend its definition to a larger domain with desirable properties and then obtain measure as a by-product at a later stage. In the present section we describe this alternative approach: it is convenient to use it in the following section to obtain the integral representation of an important class of linear functionals.

For an arbitrary space X, we consider a family L of functions f: $X \to \mathbf{R}$ satisfying

(i) L is a linear space over the reals;

(ii) for each $f \in L$, the function $f_+ \in L$, where

$$f_+(x) = \max(0, f(x)).$$

Now if we define, for each $f, g \in L$, $x \in X$

$$(f \vee g)(x) = \max(f(x), g(x)),$$

$$(f \wedge g)(x) = \min(f(x), g(x)),$$

the relations

$$f_+ = f \vee 0, \quad f \vee g = (f-g) \vee 0 + g, \quad f \wedge g = f + g - (f \vee g);$$

show that

(iii) if $f, g \in L$, then $f \vee g, f \wedge g \in L$.

Any family L satisfying conditions (i) and (ii) (and therefore (iii)) is called a *vector lattice* of functions. Suppose $\mathscr{I}$ is a linear functional on L (considered as a real linear space), then we say $\mathscr{I}$ is *positive* if

$$f \in L, \quad f \geqslant 0 \Rightarrow \mathscr{I}(f) \geqslant 0.$$

A positive linear functional $\mathscr{I}$ on L is said to be a *Daniell functional* if, for every increasing sequence $\{f_n\}$ of functions of L

$$\mathscr{I}(g) \leqslant \lim_{n \to \infty} \mathscr{I}(f_n) \tag{9.4.1}$$

for each $g \in L$ satisfying $g(x) \leqslant \lim_{n \to \infty} f_n(x)$ for all $x \in X$. (Note that $\lim_{n \to \infty} f_n(x)$ will be $+\infty$ if the sequence $\{f_n(x)\}$ is unbounded, and even if $\lim f_n$ exists as a function with finite values we do not assume that it is in L.)

In particular, this implies that, if $\mathscr{I}$ is a Daniell functional, $\{f_n\}$ a monotone sequence in L such that $f(x) = \lim_{n \to \infty} f_n(x)$, $x \in X$ defines a function in L then $\mathscr{I}(f) = \lim_{n\to\infty} \mathscr{I}(f_n)$. For if $\{f_n\}$ is increasing then $f \geqslant f_n$ for all n, so $\mathscr{I}(f) \geqslant \mathscr{I}(f_n)$ since $\mathscr{I}$ is positive, which with (9.4.1) gives the required equality. Thus a Daniell functional is continuous in the sense that for any sequence $\{f_n\}$ in L which decreases monotonically to the zero function we must have $\mathscr{I}(f_n) \to 0$. Any Daniell functional is therefore an 'integral' in the sense discussed in §5.1. However, for the integral to be useful we want the domain L to be as large as possible: if $\{f_n\}$ is an increasing sequence in L which is bounded above by an element of L we would certainly want $\lim f_n$ to be in L. The Daniell integral is essentially the result of extending a Daniell functional $\mathscr{I}$ from L to a class $L_1 \supset L$: it turns out that this extension can be carried out in two stages.

Suppose $\mathscr{I}$ is a Daniell functional on a vector lattice L. Denote by L^+ the set of functions f: $X \to \mathbf{R}^*$ which are limits of monotone increasing functions of L. L^+ is not a linear space but

$$\alpha, \beta \geqslant 0 \quad f, g \in L^+ \Rightarrow \alpha f + \beta g \in L^+.$$

Then if $\{f_n\}$ is an increasing sequence in L, $\{\mathscr{I}(f_n)\}$ is an increasing sequence in $\mathbf{R}$ which has a unique limit in $\mathbf{R} \cup \{+\infty\}$. We can define $\mathscr{I}$ in L^+ by

$$\mathscr{I}(\lim f_n) = \lim \mathscr{I}(f_n).$$

This definition is proper because if $\{f_n\}, \{g_n\}$ are two monotone sequences each converging to h in L^+, condition (9.4.1) gives, for fixed n,

$$f_n \leqslant h = \lim g_n \Rightarrow \mathscr{I}(f_n) \leqslant \lim \mathscr{I}(g_n)$$

so that $\lim \mathscr{I}(f_n) \leqslant \lim \mathscr{I}(g_n)$ and the opposite inequality can be similarly obtained. It is clear that $\mathscr{I}$ is linear on L^+ in the sense that

$$\alpha \geqslant 0, \quad \beta \geqslant 0; \quad f, g \in L^+ \Rightarrow \mathscr{I}(\alpha f + \beta g) = \alpha \mathscr{I}(f) + \beta \mathscr{I}(g).$$

For an arbitrary function $f: X \to \mathbf{R}^*$ we define the *upper integral* $\mathscr{I}^*(f)$ by

$$\mathscr{I}^*(f) = \inf_{\substack{g \geqslant f \\ g \in L^+}} \mathscr{I}(g),$$

where we adopt the (usual) convention that the infimum of the empty set is $+\infty$. Similarly, the *lower integral* $\mathscr{I}_*(f)$ is defined by

$$\mathscr{I}_*(f) = -\mathscr{I}^*(-f),$$

and we say that a function $f: X \to \mathbf{R}^*$ is *integrable* (with respect to $\mathscr{I}$) if $\mathscr{I}^*(f) = \mathscr{I}_*(f)$ and is finite. The class of integrable functions will be denoted by $L_1 = L_1(\mathscr{I}, L)$. For $f \in L_1$ we call the common value of $\mathscr{I}^*(f), \mathscr{I}_*(f)$ the integral of f and denote it by $\mathscr{I}(f)$. We now show that this functional $\mathscr{I}$ on L_1 is a Daniell functional which extends $\mathscr{I}$, and that L_1 has the closure properties desired. It is convenient to obtain a number of preliminary results before stating the theorem.

***Lemma* 1.** *If $\{g_n\}$ is a sequence of non-negative functions in L^+, then*

$$g = \sum_{n=1}^{\infty} g_n \quad \text{is in } L^+ \quad \text{and} \quad \mathscr{I}(g) = \sum_{n=1}^{\infty} \mathscr{I}(g_n).$$

Proof. It is clear that a non-negative function $f: X \to \mathbf{R}^+$ belongs to L^+ if and only if there is a sequence $\{f_n\}$ of non-negative functions in L with $f = \sum_{n=1}^{\infty} f_n$. By definition, in this case

$$\mathscr{I}(f) = \sum_{n=1}^{\infty} \mathscr{I}(f_n).$$

Hence, each function g_n can be expressed as a sum

$$g_n = \sum_{v=1}^{\infty} f_{n,v} \quad \text{with} \quad f_{n,v}: X \to \mathbf{R}^+, \quad f_{n,v} \in L.$$

It follows that

$$g = \sum_n \sum_v f_{n,v}$$

is a countable sum of non-negative functions of L and so must be in L^+. Further since all the terms are non-negative, the order of summation is immaterial and

$$\mathscr{I}(g) = \sum_{n=1}^{\infty} \left(\sum_{v=1}^{\infty} \mathscr{I}(f_{n,v}) \right). \blacksquare$$

***Lemma* 2.** *For arbitrary functions $f: X \to \mathbf{R}^*$, $g: X \to \mathbf{R}^*$:*

(i) $\mathscr{I}^*(f+g) \leqslant \mathscr{I}^*(f) + \mathscr{I}^*(g)$;

(ii) *if* $c \geqslant 0$, $\mathscr{I}^*(cf) = c\mathscr{I}^*(f)$;

(iii) *if* $f \leqslant g$, $\mathscr{I}^*(f) \leqslant \mathscr{I}^*(g)$, $\mathscr{I}_*(f) \leqslant \mathscr{I}_*(g)$;

(iv) $\mathscr{I}_*(f) \leqslant \mathscr{I}^*(f)$;

(v) *if* $f \in L^+$, $\mathscr{I}_*(f) = \mathscr{I}^*(f) = \mathscr{I}(f)$.

Proof. (i), (ii) and (iii) follow immediately from the definitions. It is worth noting in (i), that we can put $(f+g)(x) = +\infty$ at those points x for which one of $f(x)$ is $+\infty$ and the other is $-\infty$ so that (i) is true whatever the value in $\mathbf{R}^*$ chosen for $(f+g)(x)$ at such points x.

(iv) Since $0 = \mathscr{I}(0) = \mathscr{I}(f-f) \leqslant \mathscr{I}^*(f) + \mathscr{I}^*(-f)$ by (i), it follows that $\mathscr{I}_*(f) = -\mathscr{I}_*(-f) \leqslant \mathscr{I}^*(f)$.

(v) If $f \in L^+$, then by definition $\mathscr{I}^*(f) = \mathscr{I}(f)$. Now if $g \in L$, then

$-g \in L \subset L^+$ so that $\mathscr{I}_*(g) = \mathscr{I}(g)$. But each f in L^+ is the limit of an increasing sequence $\{g_n\}$ in L. Thus $f \geqslant g_n$ so $\mathscr{I}_*(f) \geqslant \mathscr{I}_*(g_n) = \mathscr{I}(g_n)$ and $\mathscr{I}_*(f) \geqslant \lim \mathscr{I}(g_n) = \mathscr{I}(f)$.]

***Lemma* 3.** *If $\{g_n\}$ is a sequence of functions on X to $\mathbf{R}^+$, and*

$$g = \sum_{n=1}^{\infty} g_n, \quad \textit{then} \quad \mathscr{I}^*(g) \leqslant \sum_{n=1}^{\infty} \mathscr{I}^*(g_n).$$

Proof. If $\mathscr{I}^*(g_n) = +\infty$ for some n, or if the series $\Sigma \mathscr{I}^*(g_n)$ diverges there is nothing to prove. Otherwise, given $\epsilon > 0$, for each integer n choose $h_n \geqslant g_n, h_n \in L^+$ such that $\mathscr{I}^*(g_n) > \mathscr{I}(h_n) - \epsilon 2^{-n}$. Then $h = \Sigma h_n \in L^+$ by lemma 1, $h \geqslant g$ and

$$\mathscr{I}^*(g) \leqslant \mathscr{I}(h) = \Sigma \mathscr{I}(h_n) < \epsilon + \sum_{n=1}^{\infty} \mathscr{I}^*(g_n).$$

Since ϵ is arbitrary the result is proved.]

***Theorem* 9.6.** *Given a Daniell functional $\mathscr{I}$ on a vector lattice L of functions on X to $\mathbf{R}$, the process defining a functional $\mathscr{J}$ on the set L_1 determines a Daniell functional on a lattice L_1 which extends $\mathscr{I}$. Further, if $\{f_n\}$ is an increasing sequence of functions in L_1 and $f = \lim f_n$, then $f \in L_1$ if and only if $\lim \mathscr{J}(f_n)$ is finite in which case $\mathscr{J}(f) = \lim \mathscr{J}(f_n)$.*

Proof. Lemma 2(v) shows that $L_1 \supset L$ and that $\mathscr{J}$ is an extension of $\mathscr{I}$. Now if $g \in L_1$ so does cg for c in $\mathbf{R}$ since

$$c \geqslant 0 \Rightarrow \mathscr{I}^*(cf) = c\mathscr{I}^*(f) = c\mathscr{I}_*(f) = \mathscr{I}_*(cf),$$

$$c < 0 \Rightarrow \mathscr{I}^*(cf) = c\mathscr{I}_*(f) = c\mathscr{I}^*(f) = \mathscr{I}_*(cf).$$

Further, if f and g are both in L_1, using lemma 2(i),†

$$-\mathscr{I}_*(f+g) = \mathscr{I}^*(-f-g) \leqslant -\mathscr{J}(f) - \mathscr{J}(g)$$

so

$$\mathscr{I}_*(f+g) \geqslant \mathscr{J}(f) + \mathscr{J}(g) \geqslant \mathscr{I}^*(f+g);$$

and, by lemma 2(iv), $f+g \in L_1$ and

$$\mathscr{J}(f+g) = \mathscr{J}(f) + \mathscr{J}(g).$$

Thus L_1 is a real linear space, and $\mathscr{J}$ is a linear functional on L_1.

To prove that L_1 is a lattice it is sufficient to prove that

$$f \in L_1 \Rightarrow f_+ \in L_1.$$

† As pointed out in the proof of lemma 2(i) the inequality is valid, whatever value in $\mathbf{R}^*$ is chosen for $(f+g)(x)$ at points x where $f(x) = +\infty$, $g(x) = -\infty$. The proof given then shows that, for $f, g \in L_1$, $(f+g) \in L_1$ whatever values are assumed at such points.

For a fixed f in L_1 and each $\epsilon > 0$, choose functions g, h in L^+ such that $-h \leqslant f \leqslant g$ and

$$\mathscr{I}(g) < \mathscr{I}(f)+\epsilon < \infty, \quad \mathscr{I}(h) \leqslant -\mathscr{I}(f)+\epsilon < \infty.$$

Now $g = (g \vee 0)+(g \wedge 0)$ and $g \wedge 0 \in L^+$; so $\mathscr{I}(g \vee 0) \leqslant \mathscr{I}(g)-\mathscr{I}(g \wedge 0) < \infty$. Thus $g_+ = g \vee 0 \in L^+$ and $\mathscr{I}(g_+) < \infty$. Similarly, $-h_- = h \wedge 0 \in L^+$ and $h_- \leqslant f_+ \leqslant g_+$. But $(g+h) \geqslant 0$; and separate consideration of each possible pair of signs for g, h shows that $g_+ - h_- \leqslant g+h$. Hence

$$\mathscr{I}(g_+)+\mathscr{I}(-h_-) \leqslant \mathscr{I}(g)+\mathscr{I}(h) < 2\epsilon.$$

But
$$-\mathscr{I}(-h_-) \leqslant \mathscr{I}_*(f_+) \leqslant \mathscr{I}^*(f_+) \leqslant \mathscr{I}(g_+)$$

so that $\mathscr{I}^*(f_+)-\mathscr{I}_*(f_+) < 2\epsilon$. Since ϵ is arbitrary and $\mathscr{I}(g_+)$ is finite we have $f_+ \in L_1$ as required.

Now suppose $\{f_n\}$ is an increasing sequence of functions in L_1 and $f = \lim f_n$. Then if $\lim \mathscr{I}(f_n) = +\infty$, and $g \leqslant f, g \in L_1$ it is clear that $\mathscr{I}(g) \leqslant \lim \mathscr{I}(f_n)$ since $\mathscr{I}(g)$ is finite. On the other hand if $\lim \mathscr{I}(f_n)$ is finite, put $h = f - f_1$. Then $h \geqslant 0$ and

$$h = \sum_{n=1}^{\infty} (f_{n+1} - f_n).$$

By lemma 3,
$$\mathscr{I}^*(h) \leqslant \sum_{n=1}^{\infty} \{\mathscr{I}(f_{n+1}) - \mathscr{I}(f_n)\}$$

$$= \lim \mathscr{I}(f_n) - \mathscr{I}(f_1)$$

so that
$$\mathscr{I}^*(f) = \mathscr{I}^*(f_1+h) \leqslant \mathscr{I}^*(f_1)+\mathscr{I}^*(h) \leqslant \lim \mathscr{I}(f_n).$$

But $f_n \leqslant f$ so that $\mathscr{I}_*(f) \geqslant \lim \mathscr{I}(f_n)$, and we must have

$$\mathscr{I}^*(f) = \mathscr{I}_*(f) = \lim \mathscr{I}(f_n).$$

This means that, if $\lim \mathscr{I}(f_n)$ is finite, then f is in L_1, and

$$\mathscr{I}(f) = \lim \mathscr{I}(f_n).$$

The positive functional $\mathscr{I}$ therefore satisfies (9.4.1) and must be a Daniell functional on L_1. ∎

Remark. There may be some functions $f: X \to \mathbf{R}^*$ which take the values $\pm\infty$ at some points but are still in L_1. In the course of the proof we saw that it made no difference to the linear functional $\mathscr{I}$ what value was assigned to $(bf+cg)(x)$ at points x where the usual calculation leads to $+\infty + (-\infty)$. It is in this sense that $\mathscr{I}$ is a linear functional on the real linear space L_1. However, we will shortly see that all functions $f: X \to \mathbf{R}^*$ in L_1 must take finite values at 'almost all' points, so that the set where $(bf+cg)$ is not determined by the laws of algebra is always small (relative to $\mathscr{I}$).

Now if one starts with a Daniell functional $\mathscr{I}$ on a vector lattice L which is already closed for monotone limits, i.e. if $\{f_n\}$ is a monotone sequence in L and $\lim \mathscr{I}(f_n)$ is finite, then $f = \lim f_n$ is in L; the extension process defined will lead to nothing new as the part of L^+ on which $\mathscr{I}$ is finite is in L and this will give $L = L_1$.

Daniell integral

Any Daniell functional $\mathscr{I}$ on a vector lattice L_1 of functions on X to $\mathbf{R}^*$ such that the limit f of a monotone sequence $\{f_n\}$ of functions in L_1 is in L_1 provided $\lim \mathscr{I}(f_n)$ is finite is called a Daniell integral.

We now see how one can obtain a theory of measure if one starts with a linear operator $\mathscr{I}$ satisfying these conditions. The definitions are made so that the integral (in the sense of Chapter 5) with respect to the measure recovers the operator $\mathscr{I}$. Starting with a Daniell integral $\mathscr{I}$ we say that a non-negative function $f: X \to \mathbf{R}^+$ is *measurable* (with respect to $\mathscr{I}$) if $g \in L_1 \Rightarrow f \wedge g \in L_1$. We say that a set $A \subset X$ is *measurable* (with respect to $\mathscr{I}$) if the indicator function χ_A is measurable; while the set A is *integrable* if $\chi_A \in L_1$. In order to ensure that the class of measurable functions and sets has useful properties we will further assume that the space X is measurable, that is, that the constant function $f(x) \equiv 1$ is measurable.

Lemma 4. *If X is measurable, then the class $\mathscr{A}$ of sets measurable with respect to $\mathscr{I}$ is a σ-field. If $f: X \to \mathbf{R}^+$ is any non-negative integrable function, the set $E_\alpha = \{x: f(x) > \alpha\} \in \mathscr{A}$ for all $\alpha \in \mathbf{R}$.*

Proof. Given f, g non-negative measurable functions, the lattice properties of L_1 immediately give that $f \vee g$ and $f \wedge g$ are measurable. But

$$\chi_{A \cap B} = \chi_A \wedge \chi_B, \quad \chi_{A \cup B} = \chi_A \vee \chi_B$$

so that $A, B \in \mathscr{A} \Rightarrow A \cap B$ and $A \cup B \in \mathscr{A}$. Further for any set E,

$$g \wedge \chi_E = (g \vee 0 + g \wedge 0) \wedge \chi_E = (g \vee 0) \wedge \chi_E + g \wedge 0$$

so that if $g \in L_1$, $g \vee 0$ and $g \wedge 0 \in L_1$ and

$$(g \vee 0) \wedge \chi_{A-B} = (g \vee 0) \wedge \chi_A - (g \vee 0) \wedge \chi_{A \cap B} \quad \in L_1,$$

$$(g \wedge 0) \wedge \chi_{A-B} = g \wedge 0 \in L_1,$$

so that $g \wedge \chi_{A-B} \in L_1$. Thus $\mathscr{A}$ is a ring, and since $X \in \mathscr{A}$, we have proved that $\mathscr{A}$ is a field. To show that $\mathscr{A}$ is a σ-field one need only use the fact that L_1 is closed for monotone limits which are bounded, since $E_n = \bigcup_{i=1}^{n} A_i$ is monotone and so is χ_{E_n}.

Now if $f: X \to \mathbf{R}^+$ is non-negative and is in L_1, $E_\alpha = X$ for $\alpha \leqslant 0$. If $\alpha = 0$ put $h = f$; while if $\alpha > 0$ put $h = [\alpha^{-1}f - (\alpha^{-1}f) \wedge 1]$. Then $h \in L_1$, and in either case $h(x) > 0$ for $x \in E_\alpha$ and $h(x) = 0$ for $x \in X - E_\alpha$. For each integer n, put $f_n = 1 \wedge (nh)$. Then $f_n \in L_1$ and the sequence $\{f_n\}$ increases monotonely to χ_{E_α}. Hence χ_{E_α} is measurable, so E_α is measurable. ∎

***Theorem* 9.7** (*Stone*). *Suppose $\mathcal{J}$ is a Daniell integral on the class L_1 of functions $f: X \to R^*$, and X is a measurable set with respect to $\mathcal{J}$. Then*

$$\mu(E) = \mathcal{J}(\chi_E) \text{ when } E \text{ is integrable,}$$

$$\mu(E) = +\infty \text{ otherwise}$$

defines a measure μ on the σ-field $\mathcal{A}$ of measurable sets. A function $f: X \to \mathbf{R}^*$ is in L_1 if and only if it is integrable with respect to this measure μ, and

$$\mathcal{J}(f) = \int f \, d\mu \quad \text{for all} \quad f \in L_1.$$

Proof. It is immediate that $\mu(\varnothing) = 0$. If B is integrable and A is measurable with $A \subset B$, the definitions ensure that A is integrable and $0 \leqslant \mu(A) \leqslant \mu(B)$. This inequality is trivially satisfied when B is measurable but not integrable, so μ is monotone on $\mathcal{A}$.

Now let $\{E_n\}$ be a disjoint sequence in $\mathcal{A}$ and $E = \bigcup_{n=1}^{\infty} E_n$. If at least one of the E_n fails to be integrable, then E is not integrable and

$$\mu(E) = +\infty = \Sigma\mu(E_n). \tag{9.4.2}$$

If each of the sets E_n is integrable, then E will be integrable if and only if $\Sigma\mu(E_n) < \infty$ by theorem 9.6, since $\chi_E = \Sigma\chi_{E_n}$ and in this case

$$\mu(E) = \Sigma\mu(E_n) < \infty.$$

It is clear from the statement of theorem 9.6 that (9.4.2) will be satisfied if $\Sigma\mu(E_n) = +\infty$. Thus in all cases, μ is σ-additive on $\mathcal{A}$.

Now lemma 4 ensures that $\mathcal{A}$ is a σ-field, and that any non-negative g-integrable functions is $\mathcal{A}$-measurable. Since each g-integrable function is the difference of two non-negative g-integrable functions it follows that any f in L_1 is $\mathcal{A}$-measurable.

Consider a non-negative $f: X \to \mathbf{R}^+$ in L_1. For each pair (r, s) of positive integers put

$$E_{r,s} = \{x: f(x) > r/s\}.$$

Now $E_{r,s} \in \mathscr{A}$ and $\chi_{E_{r,s}} \in L_1$ (that is, $\mu(E_{r,s}) < \infty$) since

$$\chi_{E_{r,s}} = \chi_{E_{r,s}} \wedge \left(\frac{r}{s} f\right).$$

Put
$$f_n = \frac{1}{s} \sum_{r=1}^{s^2} \chi_{E_{r,s}}, \quad s = 2^n,$$

and note that $\{f_n\}$ is a monotone sequence in L_1 which converges to f. Hence $\mathscr{I}(f) = \lim \mathscr{I}(f_n)$. But

$$\mathscr{I}(f_n) = \frac{1}{s} \sum_{r=1}^{s^2} \mathscr{I}(\chi_{E_{r,s}}) = \frac{1}{s} \sum_{r=1}^{s^2} \mu(E_{r,s}) = \int f_n \, d\mu,$$

and from the definition of the integral of a non-negative $\mathscr{A}$-measurable function we have

$$\mathscr{I}(f) = \lim \int f_n \, d\mu = \int f \, d\mu.$$

Conversely, if $f: X \to \mathbf{R}^+$ is non-negative and integrable with respect to μ, then each of the sets $E_{r,s}$ is in $\mathscr{A}$ and has finite μ-measure. Hence $\chi_{E_{r,s}}$ and therefore f_n are in L_1. Since

$$\int f \, d\mu = \lim \int f_n \, d\mu = \lim \mathscr{I}(f_n) < \infty,$$

by theorem 9.6, $f = \lim f_n$ is in L_1. This completes the representation theorem for non-negative functions. But for both the functional $\mathscr{I}$, and the integral with respect to μ we have a decomposition $f = f_+ - f_-$ of any integrable $f: X \to \mathbf{R}^*$ as the difference of two non-negative integrable functions, so we can deduce the representation for arbitrary integrable functions. ∎

An obvious question arising is that of uniqueness for the measure μ in theorem 9.7. This cannot always be obtained, but we give an outline of the uniqueness proof under suitable conditions in exercises 9.4 (8, 9).

Exercises 9.4

1. Show that the condition (9.4.1) for a positive linear functional is equivalent to saying that, if $\{u_n\}$ is a sequence of non-negative functions in L and $\phi \in L$ satisfies $\phi \leqslant \Sigma u_n$, then $\mathscr{I}(\phi) \leqslant \Sigma \mathscr{I}(u_n)$.

2. If $(\Omega, \mathscr{F}, \mu)$ is a σ-finite measure space, L is the class of μ-integrable functions and $\mathscr{I}(f) = \int f d\mu$, show that $\mathscr{I}$ is a Daniell functional on L.

3. Let J be the class of continuous functions on $\mathbf{R}$ to $\mathbf{R}$ which are zero outside $[-K, K]$ for some K and put

$$\mathscr{I}(f) = \int_{-\infty}^{\infty} f(x)\,dx \text{ in the Riemann sense.}$$

Show that $\mathscr{I}$ is a Daniell functional on J.

4. If $\mathscr{J}$ is a Daniell integral defined on the class L_1 prove that

$$f \in L_1 \Rightarrow |f| \in L_1.$$

5. (Fatou for Daniell integral.) Suppose $\{f_n\}$ is a sequence of non-negative functions in L_1. Prove that $\liminf f_n$ is in L_1 if $\liminf \mathscr{I}(f_n) < \infty$ and in this case

$$\mathscr{J}(\liminf f_n) \leqslant \liminf \mathscr{J}(f_n).$$

6. (Dominated convergence.) Suppose $\{f_n\}$ is a convergent sequence in L_1 such that $|f_n| \leqslant g$ for all n where $g \in L_1$. Then if $f = \lim f_n$, $f \in L_1$ and

$$\mathscr{J}(f) = \lim \mathscr{J}(f_n).$$

7. Suppose μ is a measure on a field $\mathscr{A}$ of subsets of X, and L is the family of finite linear combinations of indicator functions of sets of $\mathscr{A}$ with finite measure. Show that L is a vector lattice and if $\mathscr{I}$ is defined on L to be integration with respect to μ, then $\mathscr{I}$ is a Daniell functional. Discuss its extension $\mathscr{J}$ to a Daniell integral.

8. Suppose $\mathscr{I}$ is a Daniell functional on a vector lattice L, and $\mathscr{I}'$ is an extension of $\mathscr{I}$ to a Daniell functional on a vector lattice $L' \supset L$. If $\mathscr{I}$ and $\mathscr{I}'$ are extended to give Daniell integrals over L_1 and L_1' show that $L_1' \supset L_1$ and $\mathscr{J}'$ is an extension of $\mathscr{J}$.

9. Suppose L is a fixed vector lattice containing the constant function 1 and $\mathscr{B}$ is the smallest σ-field of subsets of X such that each function in L is measurable $\mathscr{B}$. Prove that for each Daniell integral $\mathscr{J}$ on L_1 there is a unique measure μ on $\mathscr{B}$ such that

$$\mathscr{I}(f) = \int f d\mu \quad \text{for all} \quad f \in L.$$

Hint. If $\mathscr{A}$ is σ-field of sets measurable w.r.t. $\mathscr{J}$, $\mathscr{A} \supset \mathscr{B}$. Existence of μ follows from theorem 9.7. To prove uniqueness it is sufficient to show that for any such μ,

$$\mu(B) = \mathscr{J}(\chi_B) \quad \text{for all} \quad B \in \mathscr{B}.$$

Use questions 8 and 7 above to extend the two Daniell functionals—one given and the other defined in terms of the integrals with respect to μ.

9.5* Representation of linear functionals

In this section we restrict our attention to topological spaces X which are locally compact and *Hausdorff*. A topological space is Hausdorff if given two distinct points $x, y \in X$, there are open sets G, H with $x \in G$, $y \in H$, $G \cap H = \varnothing$. The family of functions $f: X \to \mathbf{R}$ which are continuous on X and vanish outside a compact subset of X is called $C_0(X)$. If we define the *support* of a function $f: X \to \mathbf{R}$ to be the closure of the set $\{x: f(x) \neq 0\}$, then $C_0(X)$ is the family of those continuous functions $f: X \to \mathbf{R}$ which have compact support.

Baire sets and measure

The class of Baire sets is the smallest σ-field $\mathscr{C}$ of subsets of X such that each function f in $C_0(X)$ is $\mathscr{C}$-measurable. Thus $\mathscr{C}$ is the σ-field generated by the sets of the form $\{x: f(x) > \alpha\}$, $f \in C_0(X)$, $\alpha \in \mathbf{R}$. A measure μ is called a Baire measure on X if μ is defined on the σ-field $\mathscr{C}$ of Baire subsets, and $\mu(K)$ is finite for each compact set K in $\mathscr{C}$.

Clearly $C_0(X)$ is a normed linear space if we put

$$\|f\| = \sup_{x \in X} |f(x)|,$$

and we will also use the fact that $C_0(x)$ is a vector lattice. This allows us to identify the positive linear functionals on $C_0(X)$.

***Theorem* 9.8** (*Riesz*). *Suppose X is locally compact Hausdorff, and $\mathscr{I}$ is a positive linear functional on the space $C_0(X)$ of continuous functions $f: X \to \mathbf{R}$ with compact support. Then there is a Baire measure μ on X such that*

$$\mathscr{I}(f) = \int f\, d\mu \quad \textit{for all} \quad f \in C_0(X).$$

Proof. The first step is to show that $\mathscr{I}$ must be a Daniell functional on $C_0(X)$. Suppose $f \in C_0(X)$, $\{f_n\}$ is an increasing sequence in $C_0(X)$ and $f \leqslant \lim f_n$. In order to prove that $\mathscr{I}(f) \leqslant \lim \mathscr{I}(f_n)$ it is sufficient to show that $\mathscr{I}(f) = \lim \mathscr{I}(g_n)$ where $g_n = f \wedge f_n$ so that

$$f = \lim g_n \leqslant \lim f_n.$$

But then, if we put $h_n = f - g_n$ we obtain a decreasing sequence of functions of $C_0(X)$ whose limit is zero. Let K be the support of h_1, then there is a function ϕ in $C_0(X)$ which is non-negative and satisfies $\phi(x) = 1$ for $x \in K$.† For each $x \in K$, $\epsilon > 0$ there is an n_x such that $h_{n_x}(x) < \frac{1}{2}\epsilon$ and, since h_{n_x} is continuous, there is an open set G_x for which $x \in G_x$ and $h_{n_x}(t) < \epsilon$ for $t \in G_x$.

† This uses a separation property of X; see, for example page 146 of J. L. Kelley *General Topology*, Van Nostrand (1955).

Since K is compact there is a finite subcovering $G_{x_1}, \ldots, G_{x_s}$ of K. If $N = \max[n_{x_1}, \ldots, n_{x_s}]$ we have $h_n(x) < \epsilon$ for all x in K, $n \geq N$. Thus

$$0 \leq h_n < \epsilon\phi$$

so that

$$0 \leq \mathscr{I}(h_n) < \epsilon\mathscr{I}(\phi).$$

Since ϵ is arbitrary we must have $\lim \mathscr{I}(h_n) = 0$ which implies condition (9.4.1).

We can now apply theorem 9.7 to the extension $\mathscr{J}$ of $\mathscr{I}$ to

$$L_1 \supset C_0(X)$$

to obtain a measure μ on the σ-field $\mathscr{A}$ which contains the Baire sets and such that, for $f \in C_0(X)$,

$$\mathscr{I}(f) = \mathscr{J}(f) = \int f d\mu.$$

By considering the above function ϕ which is in $C_0(X)$ and takes the value 1 on the compact K, we see that

$$\mu(K) = \mathscr{J}(\chi_K) \leq \mathscr{J}(\phi) = \int \phi d\mu < \infty,$$

so that the measure μ we have obtained is finite on compact sets. ▌

When X is compact, $C_0(X)$ is the same as $C(X)$ the space of continuous $f: X \to \mathbf{R}$, so that in this case the positive linear functionals on $C(X)$ correspond to finite Baire measures. Further, because of exercise 9.5 (9) there is uniqueness. This gives

Corollary. *If X is a compact topological space and $C(X)$ is the set of continuous functions $f: X \to \mathbf{R}$, then there is a (1, 1) correspondence between positive linear functionals $\mathscr{J}$ on $C(X)$ and finite Baire measures μ on X given by*

$$\mathscr{J}(f) = \int f d\mu.$$

If we want to consider more general linear functionals on $C(X)$, it is convenient to express these as the difference of two positive linear functionals so that theorem 9.8 can be applied. This can be done for bounded linear functionals.

If L is a vector lattice of bounded functions $f: X \to \mathbf{R}$, then L is a normed linear space with $\|f\| = \sup |f(x)|$. A bounded linear functional F has a norm

$$\|F\| = \sup_{\|f\| \leq 1} |F(f)|.$$

Theorem 9.9. *Suppose L is a vector lattice of bounded functions $f: X \to \mathbf{R}$ which contains the constant function 1. Then for each bounded linear functional F on L, there are two positive linear functionals F^+ and F^- such that $F = F^+ - F^-$ and $\|F\| = F^+(1) + F^-(1)$.*

Proof. For each $f \geqslant 0$ in L put

$$F^+(f) = \sup_{\substack{0 \leqslant g \leqslant f \\ g \in L}} F(g).$$

Since $F(0) = 0$, $F^+(f) \geqslant 0$ and $F^+(f) \geqslant F(f)$. Further

$$F^+(cf) = cF^+(f) \quad \text{for} \quad c \geqslant 0.$$

If f, g are two non-negative functions in L, such that $0 \leqslant \phi \leqslant f$, $0 \leqslant \chi \leqslant g$, then $0 \leqslant \phi + \chi \leqslant f + g$, so that $F^+(f+g) \geqslant F(\phi) + F(\chi)$. Taking suprema over all such ϕ, χ in L gives

$$F^+(f+g) \geqslant F^+(f) + F^+(g).$$

To obtain the reverse inequality consider $\chi \in L$ such that

$$0 \leqslant \chi \leqslant f+g: \quad \text{then} \quad 0 \leqslant \chi \wedge f \leqslant f \quad \text{and} \quad 0 \leqslant \chi - (\chi \wedge f) \leqslant g$$

so that
$$F(\chi) = F(\chi \wedge f) + F[\chi - (\chi \wedge f)]$$
$$\leqslant F^+(f) + F^+(g)$$

and taking the supremum over such χ gives

$$F^+(f+g) \leqslant F^+(f) + F^+(g).$$

For an arbitrary $f \in L$, let p, q be two constants such that $(f+p)$ and $(f+q)$ are both non-negative. Then

$$F^+(f+p+q) = F^+(f+p) + F^+(q) = F^+(f+q) + F^+(p)$$

so that
$$F^+(f+p) - F^+(p) = F^+(f+q) - F^+(q).$$

This means that the value of $[F^+(f+p) - F^+(p)]$ is independent of p and we can define $F^+(f)$ to be this value. Thus F^+ is now defined on L,

$$F^+(f+g) = F^+(f) + F^+(g) \quad \text{for all} \quad f, g \in L,$$

and
$$F^+(cf) = cF^+(f) \quad \text{for} \quad c \geqslant 0, \quad f \in L.$$

But $F^+(-f) + F^+(f) = F^+(0) = 0$ so we have $F^+(-f) = -F^+(f)$ and F^+ is a positive linear functional on L.

But $F^+(f) \geqslant F(f)$ so that $F^- = F^+ - F$ is also a positive linear functional on L.

Now
$$\|F\| \leqslant \|F^+\| + \|F^-\| = F^+(1) + F^-(1).$$

To establish the opposite inequality consider functions $f \in L$ for which $0 \leqslant f \leqslant 1$. Since $|2f - 1| \leqslant 1$

$$\|F\| \geqslant F(2f-1) = 2F(f) - F(1).$$

Taking the supremum over such f gives

$$\|F\| \geq 2F^+(1)-F(1) = F^+(1)+F^-(1).\blacksquare$$

Corollary. *Let X be a compact Hausdorff space and $C(X)$ the set of continuous functions $f: X \to \mathbf{R}$. Then there is a $(1,1)$ correspondence between finite signed Baire measures ν on X and the dual space to $C(X)$ given by*

$$F(f) = \int f d\nu.$$

Moreover, $\|F\| = |\nu|(X)$.

Proof. If one starts with a finite signed Baire measure ν, then by theorem 3.3, there is a decomposition $\nu = \nu_+ - \nu_-$ into the difference of two finite Baire measures. Clearly

$$F(f) = \int f d\nu_+ - \int f d\nu_-$$

then defines a bounded linear functional on $C(X)$ since each function f in $C(X)$ is bounded and measurable with respect to the class of Baire sets.

Conversely given a bounded linear functional F on $C(X)$, this can be decomposed by theorem 9.9 into the difference $F = F^+ - F^-$ of two positive linear functionals. Apply theorem 9.8 and corollary to find finite Baire measures μ_1, μ_2 with

$$F^+(f) = \int f d\mu_1, \quad F^-(f) = \int f d\mu_2.$$

If we put $\nu = \mu_1 - \mu_2$, then ν is a finite Baire measure and

$$F(f) = \int f d\nu.$$

Now
$$|F(f)| \leq \int |f|\, d|\nu|$$
$$\leq \|f\| . |\nu|(X)$$

so that $\|F\| \leq |\nu|(X)$. Further

$$|\nu|(X) \leq \mu_1(X) + \mu_2(X) = F^+(1) + F^-(1) = \|F\|$$

so we have $\|F\| = |\nu|(X)$.

To prove that ν is uniquely determined by F, suppose there are two signed measures ν_1, ν_2 with

$$\int f d\nu_1 = \int f d\nu_2 \quad \text{for each} \quad f \in C(X).$$

Decompose $\lambda = \nu_1 - \nu_2$ by theorem 3.2 to give $\lambda = \lambda_+ - \lambda_-$. Then

$$\int f d\lambda_+ = \int f d\lambda_- \quad \text{for all} \quad f \in C(X),$$

so that by the uniqueness proved in exercise 9.4 (9), $\lambda_+ = \lambda_-$ on the Baire sets. Hence $\nu_1 = \nu_2$. ▌

Exercises 9.5

1. Show that in a locally compact separable metric space the class of Baire sets is the same as the class of Borel sets.

2. Suppose μ is a Baire measure on a locally compact space X. Let H be the union of all open Baire sets G for which $\mu(G) = 0$. The complement $F = X - H$ is closed and called the *support* of μ. Prove

(i) if G is an open Baire set and $G \cap F \neq \varnothing$ then $\mu(G) > 0$;
(ii) if K is a compact Baire set with $K \cap F = \varnothing$, then $\mu(K) = 0$;
(iii) if $f \in C_0(X)$ and $f \geqslant 0$, $\int f d\mu = 0$ if and only if $f \equiv 0$.

3. The corollary to theorem 9.9 is not valid on $C_0(X)$ for X locally compact Hausdorff. A *Radon measure* ϕ on a locally compact space is defined to be a linear functional on $C_0(X)$ which is continuous in the sense that, for each compact K, $\epsilon > 0$ there is a $\delta > 0$ such that $|f(x)| < \delta$ for all x, with the support of f contained in K, implies that $|\phi(f)| < \epsilon$. Prove every positive linear functional is a Radon measure.

For $\mathbf{R}$ and the usual topology define

$$\phi(f) = \sum_{r=-\infty}^{\infty} (-1)^r f(r) \quad \text{for} \quad f \in C_0(\mathbf{R}).$$

Show that ϕ is a Radon measure, but that ϕ does not correspond to any signed Baire measure.

9.6* Haar measure

There is a general method of defining a measure on an important class of topological spaces which have the algebraic structure of a group. For notational purposes we will represent the group operation in the set X by multiplication. We do not assume that the group operation is commutative. For subsets A, B of X and an element $x \in X$ we define

$$xA = \{xy : y \in A\},$$
$$AB = \{xy : x \in A, y \in B\},$$
$$A^{-1} = \{x : x^{-1} \in A\},$$

and call xA and Ax respectively the left translation and right translation of A by x. We also require the algebraic operations to be continuous in the topology of X. The theory of Haar measure can be developed for any such topological group which is locally compact and Hausdorff, but in this section we will make the additional (unnecessary) assumption that the topology is determined by a metric ρ.

A set X is a *metric group* if X is a group and there is a metric ρ such that in (X,ρ), the group operation is continuous. In particular

$$\left.\begin{array}{l}\lim_{n\to\infty} x_n = x_0\\ \lim_{n\to\infty} y_n = y_0\end{array}\right\} \Rightarrow \left\{\begin{array}{l}\lim_{n\to\infty} x_n y_n = x_0 y_0,\\ \lim_{n\to\infty} x_n^{-1} = x_0^{-1}.\end{array}\right.$$

We will, for the remainder of the section, assume that X is a metric group which is locally compact in the topology of the metric.

We are interested in measures for which the translation of A by any element x leaves the measure invariant. For example, the space $\mathbf{R}$ of real numbers is clearly a metric group with ordinary addition for the group operation. Given a set $E \subset \mathbf{R}$, and a point $x \in \mathbf{R}$, xE denotes the set of real numbers of the form $x+y$ with $y \in E$. We showed in §4.5 that Lebesgue measure in $\mathbf{R}$ is invariant under translations in the sense that, for measurable E, $|E| = |xE|$. The notation of an invariant measure in a topological group should be thought of as a generalisation of this property of Lebesgue measure in $\mathbf{R}$. To be precise, a measure μ defined on the class $\mathscr{B}$ of Borel subsets of X is called a *left Haar measure* if

(i) μ is invariant under left translations; that is for every $E \in \mathscr{B}$, $x \in X$, $\mu(xE) = \mu(E)$;

(ii) for every compact set C, $\mu(C) < \infty$;

(iii) for every non-void open set G, $\mu(G) > 0$.

Conditions (ii) and (iii) eliminate such trivial measures as the zero measure, and the measure which is $+\infty$ except on the null set. A *right Haar measure* is one for which left translation invariance is replaced by invariance for right translations. We give the details of construction for a left Haar measure: obvious modifications would give the right Haar measure.

Let $\mathscr{C}_0$ be the class of non-empty open subsets of X whose closures are compact. The important consequence of local compactness is that every compact K in X can be covered by a finite number of sets of $\mathscr{C}_0$. The sets $\varnothing$, X added to $\mathscr{C}_0$ form the class $\mathscr{C}$. The first step is to define a suitable set function λ on $\mathscr{C}$.

Suppose $H \in \mathscr{C}_0$, and G is any non-empty open set. Then

$$\mathscr{G} = \{xG \colon x \in \bar{H}G^{-1}\}$$

is a class of open sets covering $\bar{H}$ since, if $y \in \bar{H}$, $g \in G$, $x = yg^{-1}$, $y = xg \in xG$. But $\bar{H}$ is compact so there is a finite subclass of $\mathscr{G}$, which covers $\bar{H}$. Let the smallest number of sets of $\mathscr{G}$ which cover $\bar{H}$ be denoted

$$(H:G).$$

This is a measure of the relative sizes of H and G. It is immediate that, for $A, B, C \in \mathscr{C}_0$

$$1 \leqslant (A:C) \leqslant (A:B)(B:C).$$

Now compare all sets with some fixed $H_0 \in \mathscr{C}_0$, and put, for each non-empty open set $G, H \in \mathscr{C}_0$,

$$\lambda_G(H) = \frac{(H:G)}{(H_0:G)}.$$

Now, for fixed H, $\lambda_G(H)$ is a bounded function of G since

$$0 < \frac{1}{(H_0:H)} \leqslant \lambda_G(H) \leqslant (H:H_0). \tag{9.6.1}$$

If e is the identity element of the group X and

$$S_n = S\left(e, \frac{1}{n}\right) \quad (n = 1, 2, \ldots)$$

is the open sphere centre e radius $1/n$, then for each fixed $H \in \mathscr{C}_0$

$$\lambda_{S_n}(H) \quad (n = 1, 2, \ldots)$$

is a bounded sequence of real numbers. Put

$$\lambda(H) = \operatorname{Lim} \lambda_{S_n}(H)$$

where Lim is the generalized limit defined for all sequences in $\boldsymbol{m}$ using the Hahn–Banach theorem to extend the definition from $\boldsymbol{c}$ to $\boldsymbol{m}$ while preserving the norm (see exercise 8.4.(7)). Finally, put $\lambda(\varnothing) = 0$, $\lambda(X) = +\infty$ if X is not compact (and so not in $\mathscr{C}_0$).

Lemma. *The set function λ defined on $\mathscr{C}$ has the following properties:*

(i) $0 < \lambda(H) < \infty$ *for every* $H \in \mathscr{C}_0$;

(ii) *if* $H_1, H_2 \in \mathscr{C}_0$, $d(H_1, H_2) > 0$ *then*

$$\lambda(H_1 \cup H_2) = \lambda(H_1) + \lambda(H_2);$$

(iii) *for any* $H_1, H_2 \in \mathscr{C}_0$,

$$\lambda(H_1 \cup H_2) \leqslant \lambda(H_1) + \lambda(H_2);$$

(iv) *if* $H_1, H_2 \in \mathscr{C}_0, H_1 \subset H_2$ *then*

$$\lambda(H_1) \leqslant \lambda(H_2);$$

(v) *for any* $x \in X$, $H \in \mathscr{C}_0$, $\lambda(xH) = \lambda(H)$.

Proof. By (9.6.1), $\lambda_{S_n}(H)$ is bounded below and above so that

$$0 < \frac{1}{(H_0 : H)} \leqslant \lambda(H) \leqslant (H : H_0) < \infty.$$

This establishes (i). Further, for each $H \in \mathscr{C}_0$, G open, the covering ratios $(xH : G) = (H : G)$ for all $x \in X$; so that the sequence $\lambda_{S_n}(H)$ is invariant under left translations: therefore λ is also and (v) is proved.

If $H_1, H_2 \in \mathscr{C}_0$ and $d(H_1, H_2) = \eta > 0$, then for $1/n < \eta$ we must have

$$(H_1 \cup H_2 : S_n) = (H_1 : S_n) + (H_2 : S_n),$$

$$\lambda_{S_n}(H_1 \cup H_2) = \lambda_{S_n}(H_1) + \lambda_{S_n}(H_2),$$

and (ii) is now established by taking generalised limits.

Now for any open G, and $H_1, H_2 \in \mathscr{C}_0$

$$(H_1 \cup H_2 : G) \leqslant (H_1 : G) + (H_2 : G),$$

so

$$\lambda_{S_n}(H_1 \cup H_2) \leqslant \lambda_{S_n}(H_1) + \lambda_{S_n}(H_2);$$

this implies (iii) and a similar argument gives (iv). ∎

We now define a set function μ^* for all subsets of X by

$$\mu^*(E) = \inf \sum_{i=1}^{\infty} \lambda(H_i), \tag{9.6.2}$$

where the infimum is taken over all coverings $\{H_i\}$ of E by sets in $\mathscr{C}$.

***Theorem* 9.10.** *In a locally compact metric group, the set function μ^* given by* (9.6.2) *is a metric outer measure. The restriction μ of μ^* to the class $\mathscr{B}$ of Borel sets is a left Haar measure.*

Proof. In the definition of outer measure given in §3.1, condition (i) is obvious, (ii) follows from (iv) of the lemma, and subadditivity (iii) follows from (9.6.2) as in the proof of theorem 4.2. Thus μ^* is an outer measure. Now suppose $E_1, E_2 \subset X$ with $d(E_1, E_2) > 0$. If $E_1 \cup E_2$ cannot be covered by a sequence from $\mathscr{C}_0$, then at least one of the sets E_1, E_2 cannot be covered by such a sequence and

$$\mu^*(E_1 \cup E_2) = \mu^*(E_1) + \mu^*(E_2) \tag{9.6.3}$$

since both sides are $+\infty$. If E_1, E_2 can be covered by sequences from $\mathscr{C}_0$, first choose open sets $G_1 \supset E_1$, $G_2 \supset E_2$ for which $d(G_1, G_2) > 0$ and let $\{H_i\}$ be a sequence of sets from $\mathscr{C}_0$ covering $E_1 \cup E_2$ with

$$\Sigma\lambda(H_i) \leqslant \mu^*(E_1 \cup E_2) + \epsilon.$$

For each i, put $\quad H_i^1 = G_1 \cap H_i, \quad H_i^2 = G_2 \cap H_i.$

Then by (ii) and (iv) of the lemma, for each integer i,

$$\lambda(H_i) \geqslant \lambda(H_i^1 \cup H_i^2) = \lambda(H_i^1) + \lambda(H_i^2)$$

and so $$\mu^*(E_1) + \mu^*(E_2) \leqslant \Sigma\lambda(H_i) \leqslant \mu^*(E_1 \cup E_2) + \epsilon.$$

Since this is true for each $\epsilon > 0$, and μ^* is subadditive we have established (7.6.3) so that μ^* is a metric outer measure.

Now apply theorem 4.1 to μ^* to obtain a measure μ on a class $\mathscr{M}$ of μ^*-measurable sets. Since μ^* is a metric outer measure, this class $\mathscr{M}$ includes the open sets and therefore the Borel sets $\mathscr{B}$ (see exercise 4.3 (4)); so that the restriction of μ^* to $\mathscr{B}$ defines a measure on $\mathscr{B}$. If we now examine the conditions for μ to be left Haar measure we see that (v) of the lemma implies that μ^* is left translation invariant.

If K is any compact set in X, there is a finite subclass of $\mathscr{C}_0$ which covers K so that

$$\mu^*(K) \leqslant \sum_{i=1}^{n} \lambda(H_i) < \infty$$

so that condition (ii) for a Haar measure is satisfied. Now suppose G is any non-void open set in X. If $x \in G$, pick $\epsilon > 0$ such that $S(x, \epsilon) \subset G$ and put $E = S(x, \frac{1}{2}\epsilon)$ so that $\bar{E} \subset G$. Since X is locally compact we may assume ϵ is small enough to make $\bar{E}$ compact so that $E \in \mathscr{C}_0$. If $\mu(G) = \infty$ then $\mu(G) > 0$; so we may suppose $\mu(G) < \infty$. For each $\eta > 0$ there is a sequence $\{H_i\}$ from $\mathscr{C}_0$ such that

$$\bigcup_{i=1}^{\infty} H_i \supset G \supset \bar{E}, \quad \Sigma\lambda(H_i) \leqslant \mu^*(G) + \eta.$$

But $\bar{E}$ is compact so a finite number of the H_i must cover $\bar{E}$. Then if

$$\bigcup_{i=1}^{m} H_i \supset \bar{E},$$

$$\lambda(E) \leqslant \lambda\left(\bigcup_{i=1}^{m} H_i\right) \leqslant \sum_{i=1}^{m} \lambda(H_i) \leqslant \mu^*(G) + \eta,$$

and since η is arbitrary we have

$$\mu(G) = \mu^*(G) \geqslant \lambda(E) > 0$$

so that μ satisfies condition (iii) for a Haar measure.]

Corollary. *For any compact metric group X there is a left Haar measure P defined on a σ-field $\mathscr{F}$ which includes the Borel sets such that $(X, \mathscr{F}, \mathsf{P})$ is a probability space.*

Proof. If X is compact, the above construction gives a left Haar measure in which

$$0 < \mu(X) < \infty$$

with μ defined on a σ-field $\mathscr{F}$ which is complete with respect to μ. If we put

$$\mathsf{P}(E) = \frac{\mu(E)}{\mu(X)} \quad \text{for} \quad E \in \mathscr{F}$$

it is clear that $(X, \mathscr{F}, \mathsf{P})$ is a probability space. ▌

Exercises 9.6

1. Suppose Ω is the set of positive real numbers with the usual metric and multiplication for the group operation. If $(1, e)$ is the reference set H_0 used in the definition of Haar measure μ in Ω, show that

$$\mu(a, b) = \log b/a \quad \text{for each interval } (a, b) \subset \Omega.$$

(Here e is the base for Napierian logarithms.)

2. With $X = \mathbf{R}$ and addition for the group operation define Haar measure μ with $(0, 1)$ taken as the reference set H_0. Show that μ is Lebesgue measure in $\mathbf{R}$.

3. Let X be the set of 2×2 matrices of the form

$$\begin{pmatrix} x & y \\ 0 & x \end{pmatrix}$$

with $x > 0$ and multiplication for the group operation. Define a metric in X by using the Euclidean metric in $\mathbf{R}^2$. Show that in the topology of this metric, X is a locally compact metric group.

Define

$$F\begin{pmatrix} x & y \\ 0 & x \end{pmatrix} = -\frac{y}{x}.$$

Map the Lebesgue–Stieltjes measure μ_F in the right half-plane $x > 0$ of $\mathbf{R}^2$ onto the set X, and show that the result is both a left and a right Haar measure.

4. X is the set of 2×2 matrices of the form

$$\begin{pmatrix} x & y \\ 0 & 1 \end{pmatrix} \quad (x > 0)$$

metrised by the Euclidean metric in $\mathbf{R}^2$. As in question 3, obtain a measure in X by mapping the Lebesgue–Stieltjes measure μ_F of question 3 into X. Show that this is a left Haar measure but not a right Haar measure.

5. If μ is a left Haar measure on X and ν is defined by $\nu(E) = \mu(E^{-1})$, show ν is a right Haar measure.

6. The left Haar measure of theorem 9.10 is regular in the sense that $\mu^*(E) = \inf\{\mu(G): G \supset E, G \text{ open}\}$.

7. Haar measure is obviously not unique since for any Haar measure $\mu, c > 0$ the measure $c\mu$ is also a Haar measure. However, on a compact metric group, with the condition $\mu(X) = 1$ it can be proved that the Haar measure is essentially unique.

8. If A, B are two compact sets with $\mu(A) = \mu(B) = 0$, does it follow that $\mu(AB) = 0$?

9. If μ is a Haar measure in X, then X has a discrete topology if and only if $\mu\{x\} \neq 0$ for at least one $x \in X$.

10. If a Haar measure μ on X is finite prove that X is compact.

11. In a locally compact metric group X show that a Haar measure μ on X is σ-finite if and only if X is σ-compact.

10

WHAT IS PROBABILITY?

10.1 Probability statements

The mathematical theory of probability is concerned with the mathematical problems which arise in connexion with the manipulation of probability statements. We must therefore begin by taking a look at the way in which such statements can occur, and the relations between them.

The most primitive kind of probability statement first arose in the study of games of chance, and is of the following type:

(A) *If a fair coin is tossed, the probability of its falling heads is* $\frac{1}{2}$

The point of this statement is that it is impossible to predict which of the two outcomes, heads or tails, will occur. Moreover, there is symmetry between them (this being the burden of the adjective 'fair') so that, in some sense, neither is more likely than the other. Thus we assign to each outcome one half of the total probability, which by convention is 1.

Another statement of an entirely similar kind is

(B) *If a die is thrown, the probability of a* 5 *being shown is* $\frac{1}{6}$

Here there are six outcomes, supposed by symmetry to be equally likely, so that the probability $\frac{1}{6}$ is assigned to each From simple statements like (B) can be built up more complex ones, such as

(C) *If a die is thrown, the probability of either a* 2 *or a* 5 *being shown is* $\frac{1}{6}+\frac{1}{6}=\frac{1}{3}$

More generally, if E is an *event* made up of exactly r of the six possible outcomes, it is assigned a probability, denoted by $\mathsf{P}(E)$, of $\frac{1}{6}r$.

Thus probability statements can be made in the context of a general situation in which an *experiment* (such as tossing a coin or throwing a die) has a finite number of possible outcomes, which may be denoted by $\omega_1, \omega_2, \ldots, \omega_N$. It is supposed that there are reasons of symmetry which allow the assumption that no outcome is more likely than any other, and each outcome is therefore assigned the probability $1/N$. An event E is a subset of the set $\Omega = \{\omega_1, \omega_2, \ldots, \omega_N\}$ of outcomes, and, if it contains M elements, it is assigned the probability M/N. In other words

$$\mathsf{P}(E) = \text{probability of } E = \frac{\text{number of outcomes for which } E \text{ occurs}}{\text{total number of outcomes}}. \tag{10.1.1}$$

Another, more complicated, example occurs in connexion with certain card games. If a pack of 52 cards is well shuffled before being dealt to the players, there are $N = 52!$ different orders in which the cards may appear. The object of shuffling the cards is presumably to ensure that no order is more likely than any other, so that we have exactly the situation described in the last paragraph, where the outcomes $\omega_1, \omega_2, \ldots, \omega_{52!}$ are the possible orders of the cards in the pack. A typical event might consist of all those ω_n for which a certain player received all four aces, and the probability of this event may be obtained from (10.1.1) (in principle by direct enumeration, although of course much shorter methods exist). This sort of probability statement can be found in any book on the game of Bridge.

As a final example of this type of probability statement, consider the problem of tossing a fair coin n times. Any possible outcome may be described in the obvious way by a sequence ω like $HHTH\ldots T$ of H's and T's, n in all. Thus the experiment has $N = 2^n$ possible outcomes, which might plausibly be supposed to be equally likely, so that each would be assigned the probability $1/N = 2^{-n}$. An example of an event defined in this situation is that consisting of those sequences ω containing exactly r heads (where r is an integer in $0 \leqslant r \leqslant n$). The number of such sequences is well known to be given by the binomial coefficient

$$\binom{n}{r} = \frac{n!}{r!\,(n-r)!},$$

and it follows that the probability of the event is

$$\mathsf{P}(r \text{ heads in } n \text{ tosses}) = \binom{n}{r} 2^{-n}. \tag{10.1.2}$$

(This is a special case of the *binomial distribution* which will be studied in more detail in the next chapter.)

The idea of probability illustrated above is open to a number of rather serious objections, of which one is that it will only work when there is some notion of symmetry to which one can appeal in order to justify the assignment of equal probabilities to different outcomes. Take, for example, the statement (A). No coin is entirely symmetrical (indeed, if it were, we could not distinguish heads from tails), and we must therefore take into account the possibility of bias, of heads being more or less likely to occur than tails, which (A) implicitly excludes. In other words, we should want to allow the probability that the coin will fall heads to be different from $\frac{1}{2}$.

In order to see how this can be done, suppose that we take a particular coin and toss it repeatedly. Suppose that in n tosses we observe $r(n)$ heads and $n-r(n)$ tails, so that the *frequency* of heads in the first n tosses is $f(n) = r(n)/n$. It is a matter of observation in experiments of this type that, if the conditions under which the experiment is conducted do not alter as time goes on, the numbers $f(n)$ seem to settle down, as n becomes large, to a definite value p. This phenomenon, which is often called the *stability of statistical ratios*, is observed, not merely in coin tossings, but in many other situations in which an experiment may be repeated a large number of times under substantially constant conditions. It should be stressed that, since it is physically impossible to carry out more than a finite number of tosses, there is no way of deciding experimentally whether $f(n)$ actually converges to p as n tends to infinity, so that p cannot be defined as the mathematical limit of the sequence $\{f(n)\}$. Nevertheless, if we find that $f(n)$ seems to be approaching a number p, then we can express this fact by the statement

(A′) *The probability of throwing a head (with this particular coin) is p*

Conversely, a statement such as (A′) can be interpreted as meaning that if the coin is tossed a large number of times, then we expect the frequency with which heads are shown to be nearly p. This would be called the *frequency interpretation* of the statement (A′).

The relation between this interpretation and the symmetry interpretation of (A) is that, if we have a coin which appears to exhibit symmetry between heads and tails, then we expect heads and tails to turn up approximately equally often, so that in (A′) we should have $p = \frac{1}{2}$. This is in accordance with our intuitive feelings in the matter, since if an apparently symmetrical coin turned up heads substantially more often than tails, we would strongly suspect some unobserved asymmetry, either in the constitution of the coin or in the mechanism of tossing.

An exactly similar situation arises in the case of the throwing of a die. If, in a long sequence of trials, the frequency with which a 5 is thrown settles down to a number p_5, this fact can be expressed by saying that:

(B′) *The probability of a 5 being thrown is* p_5

Similarly, (C) would be replaced by an assertion of the form:

(C′) *The probability of either a 2 or a 5 being thrown is* $p_2 + p_5$

Probability statements are not of course confined to games of chance. For instance, one can make statements about the sex of an unborn child. The crudest statement here would be one based on an assumption of symmetry, that the probability of a child being a boy is $\frac{1}{2}$. This, however, is clearly unsatisfactory, because there is a manifest lack of symmetry in the problem, and indeed it is an observed fact that the proportion of male births is significantly different from $\frac{1}{2}$. A statement of the frequency type would be that the probability of the child being a boy is p, where p is the proportion of male births observed in the population. More refined statements can be made by taking into account special characteristics of the family in question.

All the situations considered so far have involved an 'experiment' having a finite number of possible outcomes $\omega_1, \omega_2, \ldots, \omega_N$, of which one and only one can occur. Sometimes there are reasons of symmetry which allow us to assign to each outcome the probability $1/N$. More commonly, however, we associate with ω_i a probability $p_i = \mathsf{P}\{\omega_i\}$ which represents the proportion of times in which ω_i occurs in a long series of repetitions of the experiment. An event E is a collection of outcomes, i.e. a subset of the set $\Omega = \{\omega_1, \omega_2, \ldots, \omega_N\}$, and we assign to it the sum of the probabilities of the outcomes which belong to it:

$$\mathsf{P}(E) = \sum_{\{i;\, \omega_i \in E\}} p_i. \qquad (10.1.3)$$

Not all situations, however, are quite as simple as this. Suppose, for example, that instead of the sex of an unborn child we were interested in conjecturing its weight at birth. Then the outcomes ω of the experiment would correspond to real numbers, so that the set Ω would be infinite, and indeed uncountable. We would still want to talk about the probability of, for instance, the child's weight exceeding eight pounds, but this would no longer be given by the simple formula (10.1.3). It is this sort of situation which necessitates the use of measure theory in formulating and studying appropriate models on which probability statements can be based.

We have now looked at a few simple examples of two types of probability statement, the 'symmetry' type and the 'frequency' type. There are other kinds of probability statement which are used. For instance, there are statements which express a 'degree of belief' in the occurrence of an event or the truth of an assertion.† It is found, however, that such statements are manipulated according to the same rules as those of a frequency type, and they therefore share the same mathematical theory. The exact nature of these rules of manipulation will be the concern of the rest of this chapter.

Exercises 10.1

1. If a fair die is thrown twice, what is the probability that an even number appears each time? What is the probability that the sum of the two scores is 5?

2. The numbers $1, 2, 3, \ldots, n$ are arranged in random order, all possible orders being equally likely. Find the probability that 1, 2, 3 appear as neighbours in that order.

3. Suppose that a room contains n people. Show that (if leap years are neglected) the probability that no two of them have the same birthday is

$$\left(1-\frac{1}{365}\right)\left(1-\frac{2}{365}\right)\ldots\left(1-\frac{n-1}{365}\right).$$

4. A lake contains N fish. Of these m are caught, marked, and thrown back. After some time a catch of n of the N fish is made, in such a way that all possible selections of n out of the N are equally likely. Show that the probability that exactly r of these n fish are marked is

$$p(r,m,n,N) = \binom{m}{r}\binom{N-m}{n-r}\Big/\binom{N}{n}.$$

Very often, although r, m, n are known, N is not, and we would like to use the observed value of r to estimate N. A common method (called *maximum likelihood estimation*) is to estimate N by the integer $\hat{N}$ which maximises $p(r,m,n,N)$. By considering $p(r,m,n,N)/p(m,n,r,N-1)$ prove that $\hat{N}$ is the integer part of mn/r.

10.2 The algebra of events

Each of the probability statements of the last section is of the form: 'The probability $\mathsf{P}(E)$ of an event E is a number p'. In this section we study the notion of an event and the relations between events.

The setting for a probability statement is, as has been indicated, a random experiment. An event can be regarded as a conjecture about the result of such an experiment. In the simplest cases, the experiment has only a finite number of possible outcomes $\omega_1, \omega_2, \ldots, \omega_N$,

† The reader will find a development of the theory of such statements in D. V. Lindley, *Introduction to Probability and Statistics from a Bayesian Viewpoint* (Cambridge 1964).

and then an event E is just a subset of the set Ω of outcomes. In more complicated situations it is still possible to regard any event as made up of a number of outcomes (or *elementary events*, as they are often called), but the set Ω of outcomes will in general be infinite.

Thus suppose that the result of an experiment is completely specified by the outcome ω. Then by an event we mean a subset E of the set Ω of all outcomes. There are two special events which are of importance. One is the event $E = \Omega$, which is the event that always occurs, and the other is the event $E = \varnothing$, which never occurs. In other words, every outcome implies Ω, and none implies $\varnothing$.

If A and B are two events, they have as subsets of Ω, an intersection $A \cap B$. An outcome belongs to $A \cap B$ if and only if it belongs to both A and B, so that $A \cap B$ is the event '*both A and B occur*'. Similarly $A \cup B$ is the event '*either A or B (or both) occurs*'. These interpretations obviously extend to general unions and intersections: if $\mathscr{A}$ is any collection of events, then

$$\bigcap_{A \in \mathscr{A}} A \equiv \text{'every } A \in \mathscr{A} \text{ occurs'},$$

and

$$\bigcup_{A \in \mathscr{A}} A \equiv \text{'at least one } A \in \mathscr{A} \text{ occurs.'}$$

If E is any event, then it has a complement $(\Omega - E)$. This event contains an outcome ω if and only if E does not contain ω and therefore $(\Omega - E)$ is the event '*not E*' which occurs if and only if E does not occur. Similarly, if A and B are events, $(A - B)$ is the event '*A occurs but B does not*'.

Thus all the notions of set theory can be interpreted in terms of obvious intuitive relations between events. Set-theoretic identities also have an interpretation. For example, the identity

$$\Omega - (A \cap B) = (\Omega - A) \cup (\Omega - B)$$

means that, if it is not true that both A and B occur, then either A does not occur or B does not occur, and conversely.

When Ω is finite, we usually consider every subset of Ω as an event whose probability is defined. For infinite Ω, however, this may not be possible, for just the same reason that Lebesgue measure cannot be defined for *all* subsets of $\mathbf{R}$. This is no real restriction in practice; we are not, for instance, likely to be interested in the probability that the weight of an unborn child lies in a non-Lebesgue-measurable subset of $\mathbf{R}$. It does mean, however, that we have to confine our attention to a class $\mathscr{C}$ of subsets of Ω, which does not necessarily contain every subset.

If attention is to be confined to events belonging to $\mathscr{C}$, it is clear that $\mathscr{C}$ ought to contain Ω and $\varnothing$. Moreover, if A belongs to $\mathscr{C}$ then so must $(\Omega - A)$, so that we can talk about the non-occurrence of an event. Finally, if A and B are two events in $\mathscr{C}$, it should be possible to talk about 'A and B' and 'A or B', so that $\mathscr{C}$ should contain $A \cap B$ and $A \cup B$. In other words, $\mathscr{C}$ should be a field of subsets of Ω. (In fact, for reasons to be explained in the next section, it is usual to take $\mathscr{C}$ to be a σ-field.)

So far, then, we have an experiment whose result is completely described by an element ω of some set Ω. We also have a field $\mathscr{C}$ of subsets of Ω, whose members are events or conjectures about the experiment. The operations of union, intersection and complementation have natural interpretations in terms of relations between events.

Exercises 10.2

1. Prove the identity

$$(A \cup B) \cap (B \cup C) \cap (C \cup A) = (A \cap B) \cup (B \cap C) \cup (C \cap A)$$

and give its interpretation when A, B, C are events.

2. If $\mathscr{C}$ is a σ-field and $E_1, E_2, \ldots$ are events in $\mathscr{C}$, show that

$$\limsup E_n = \bigcap_{N=1}^{\infty} \bigcup_{n=N}^{\infty} E_n$$

is the event 'E_n occurs for infinitely many n', and

$$\liminf E_n = \bigcup_{N=1}^{\infty} \bigcap_{n=N}^{\infty} E_n$$

is the event 'E_n occurs for all sufficiently large n'.

3. Suppose that a coin is tossed repeatedly and that ω_n is equal to 1 if the nth toss is a head and 0 if it is a tail, so that the outcome of the experiment is the sequence $\omega = (\omega_1, \omega_2, \ldots,)$. Write

$$\xi(\omega) = \sum_{n=1}^{\infty} \omega_n 2^{-n}.$$

If $\lambda_1, \lambda_2, \ldots, \lambda_n$ are all either 0 or 1, show that the image under ξ of the set of ω for which $\omega_1 = \lambda_1, \omega_2 = \lambda_2, \ldots, \omega_N = \lambda_N$ is a Borel subset of $[0, 1]$ with Lebesgue measure 2^{-N}.

10.3 Probability as measure

Let a random experiment have outcomes $\omega \in \Omega$, and let $\mathscr{C}$ be a field of subsets of Ω. Suppose that, to each member E of $\mathscr{C}$ there has been assigned a real number $\mathsf{P}(E)$ representing the probability of E. Then

$$\mathsf{P}: \mathscr{C} \to \mathbf{R}$$

is a set function on $(\Omega, \mathscr{C})$. It is, however, a set function having very particular properties. To see this we adopt the frequency viewpoint, regarding $\mathsf{P}(E)$ as the proportion of times E occurs in a long series of repetitions of the experiment.

The first thing to be said is that $\mathsf{P}(E)$, being a proportion, must satisfy

$$0 \leqslant \mathsf{P}(E) \leqslant 1 \quad (E \in \mathscr{C}). \tag{10.3.1}$$

Secondly, since $\varnothing$ never occurs and Ω always does,

$$\mathsf{P}(\varnothing) = 0, \quad \mathsf{P}(\Omega) = 1. \tag{10.3.2}$$

Thirdly, let A and B be disjoint members of $\mathscr{C}$, so that $A \cap B = \varnothing$. This means that A and B never occur together, so that the proportion of times that $A \cup B$ occurs is just the sum of the proportions for A and B separately. Therefore

$$\mathsf{P}(A \cup B) = \mathsf{P}(A) + \mathsf{P}(B), \quad (A, B \in \mathscr{C}, \; A \cap B = \varnothing). \tag{10.3.3}$$

Thus, if P is to have a sensible frequency interpretation, it must satisfy (10.3.1)–(10.3.3). The reader will readily verify that the same is rather trivially true of the symmetry interpretation.

In terms of the concepts introduced in Chapter 3, the conditions (10.3.1)–(10.3.3) state exactly that P is a non-negative finitely additive set function on $(\Omega, \mathscr{C})$ with the additional property that $\mathsf{P}(\Omega) = 1$.

Since $\mathscr{C}$ is a ring, it follows from theorem 3.2 that P will be a measure on $\mathscr{C}$ if (and only if) it is continuous at $\varnothing$. Thus, if P is not a measure, there must be a decreasing sequence of sets $A_n \in \mathscr{C}$ with $\bigcap A_n = \varnothing$ such that $\mathsf{P}(A_n)$ does not tend to zero as $n \to \infty$. Since $\mathsf{P}(A_n)$ decreases with n, there will then exist $\delta > 0$ with $\mathsf{P}(A_n) > \delta$ for all n. Now, if ω is any outcome

$$\omega \notin \bigcap_{n=1}^{\infty} A_n,$$

so that ω belongs to at most finitely many A_n. Let $\nu(\omega)$ be the largest integer for which $\omega \in A_n$. Then

$$\mathsf{P}\{\omega;\, \nu(\omega) \geqslant n\} = \mathsf{P}(A_n) \geqslant \delta.$$

Hence, with any outcome ω of the experiment, we can associate a positive integer $\nu(\omega)$ in such a way that, however large we choose n, the probability that $\nu(\omega) \geqslant n$ is at least δ. This seems to be a possibility which it is safe, as well as mathematically convenient, to exclude. It will therefore be assumed that P is a *measure* (i.e. a non-negative σ-additive set function) on $(\Omega, \mathscr{C})$.

If we recall the definition of §4.5 of a *probability measure* as a measure satisfying $\mathsf{P}(\Omega) = 1$, we see that P must be a probability measure on $(\Omega, \mathscr{C})$. Now $\mathscr{C}$ is a ring, and hence, from theorem 4.2, P can be uniquely extended to give a measure on the σ-ring generated by $\mathscr{C}$. Since Ω belongs to $\mathscr{C}$, it belongs to the generated σ-ring, which is thus a σ-field. Thus P can be uniquely extended to a probability measure on a σ-field $\mathscr{F}_0$ containing $\mathscr{C}$, and we can therefore enlarge the class of subsets which we are prepared to consider as events to include all the members of $\mathscr{F}_0$.

We have now arrived at the notion of probability as a probability measure on a σ-field $\mathscr{F}_0$ of subsets of the set Ω of outcomes. Now suppose that E is an element of $\mathscr{F}_0$ such that $\mathsf{P}(E) = 0$. How should we interpret this statement? In terms of a frequency interpretation, it asserts that $r(n)/n$ converges to zero as $n \to \infty$, where $r(n)$ is the number of times E occurs in n repetitions of the experiment. Thus E is not necessarily the impossible event $\varnothing$. For instance, it is possible that the weight of a baby at birth might be *exactly* eight pounds, but it is not an event which can be expected to occur in a significant proportion of births. Indeed, it is a characteristic of experiments with uncountably many outcomes that some at least of these outcomes have zero probability without being impossible.

However, whatever one's interpretation of a probability model, it is clear that an event E with zero probability must, in some sense, be negligible. If this is so, then of course every subset of E should be in the same position, so that we should want every subset of E to be an event with zero probability. For this reason it is usual to complete $\mathscr{F}_0$ for the measure P, to obtain a σ-field $\mathscr{F}$. Then P has (by theorem 4.3) a unique extension to $\mathscr{F}$, and it is reasonable to consider as events all members of $\mathscr{F}$.

Thus finally we arrive at the following mathematical model for a random experiment, called a probability space.

Probability space

A probability space is a triple $(\Omega, \mathscr{F}, \mathsf{P})$, where Ω is a set of points ω, $\mathscr{F}$ a σ-field of subsets of Ω, and P a complete probability measure on $(\Omega, \mathscr{F})$.

The points ω are the possible outcomes of the experiment, the elements of $\mathscr{F}$ are the events determined by the experiment, and P assigns to each event its probability.

If $E \in \mathscr{F}$ has $\mathsf{P}(E) = 0$, and $E_1 \subset E$, the completeness of P implies that $E_1 \in \mathscr{F}$ and $\mathsf{P}(E_1) = 0$. Dually, if $E \in \mathscr{F}$, $\mathsf{P}(E) = 1$, and $E_1 \supset E$, then $E_1 \in \mathscr{F}$ and $\mathsf{P}(E_1) = 1$. An event E with $\mathsf{P}(E) = 1$ is said to occur *almost surely*, or *almost certainly*, the word 'almost' being a reminder that $\mathsf{P}(E) = 1$ does not necessarily imply that $E = \Omega$.

The definition of a probability space makes it clear that probability theory will be based on measure theory. However, the problems considered in the former are often of a different type from those of the latter, being motivated by the practical uses of probability statements. In particular probability theory is concerned largely with the concept of *independence*, which has no place in measure theory. Before going on to discuss this concept in the next section, we consider some very simple examples of probability spaces.

Example 1. Let $\Omega = \{\omega_1, \omega_2, \ldots, \omega_N\}$ be a finite set, and let $p_1, p_2, \ldots, p_N$ satisfy

$$p_i \geqslant 0, \quad \sum_{i=1}^{n} p_i = 1. \tag{10.3.4}$$

Let $\mathscr{F}$ consist of all subsets E of Ω, and define

$$\mathsf{P}(E) = \sum_{\{i;\, \omega_i \in E\}} p_i.$$

Then it is trivial to verify that $(\Omega, \mathscr{F}, \mathsf{P})$ is a probability space, and it is one which serves as a model for any random experiment with only finitely many outcomes. If $p_i = N^{-1}$ for all i we have the situation envisaged by the symmetry interpretation, and then

$$\mathsf{P}(E) = (\text{number of elements in } E)/N.$$

Example 2. Let $\Omega = \{\omega_1, \omega_2, \ldots\}$ be a countable set, and let $p_i\ (i = 1, 2, \ldots)$ satisfy

$$p_i \geqslant 0, \quad \sum_{i=1}^{\infty} p_i = 1. \tag{10.3.5}$$

Then, if $\mathscr{F}$ consists of all subsets of Ω, and if, for $E \in \mathscr{F}$,

$$\mathsf{P}(E) = \sum_{\{i;\, \omega_i \in E\}} p_i,$$

$(\Omega, \mathscr{F}, \mathsf{P})$ is a probability space. This would be a model for an experiment with a countable infinity of possible outcomes. Notice that a symmetry interpretation is now impossible, since (10.3.5) cannot be satisfied if all the p_i are equal.

Example 3. Let F be any distribution function on $\mathbf{R}$, i.e. any non-decreasing function, continuous from the right, with

$$\lim_{x\to-\infty} F(x) = 0, \quad \lim_{x\to+\infty} F(x) = 1.$$

Take $\Omega = \mathbf{R}$, $\mathscr{F} = \mathscr{L}_F$, and P the Stieltjes measure defined on $\mathscr{L}_F$ by F. Then $(\Omega, \mathscr{F}, \mathsf{P})$ is a probability space, which would serve as a model for an experiment whose result is completely described by a single real number ω. The distribution function F satisfies

$$F(x) = \mathsf{P}\{\omega;\ \omega \leqslant x\}, \tag{10.3.6}$$

and its exact form will, of course, depend on the particular situation which the model is supposed to represent.

Other more complicated probability spaces will appear in later chapters.

Exercises 10.3

Let $(\Omega, \mathscr{F}, \mathsf{P})$ be a probability space, and let $E_1, E_2 \ldots$ belong to $\mathscr{F}$.

1. Prove that

$$\mathsf{P}(E_1 \cup E_2) + \mathsf{P}(E_1 \cap E_2) = \mathsf{P}(E_1) + \mathsf{P}(E_2).$$

2. Prove that

$$\mathsf{P}\{\Omega - (E_1 \cup E_2 \cup E_3)\} = 1 - \mathsf{P}(E_1) - \mathsf{P}(E_2) - \mathsf{P}(E_3) + \mathsf{P}(E_1 \cap E_2) + \mathsf{P}(E_2 \cap E_3) + \mathsf{P}(E_3 \cap E_1) - \mathsf{P}(E_1 \cap E_2 \cap E_3),$$

and derive a similar formula for

$$\mathsf{P}\{\Omega - (E_1 \cup E_2 \cup \ldots \cup E_n)\}.$$

3. Prove that $$\mathsf{P}\left(\bigcup_{i=1}^{\infty} E_i\right) \leqslant \sum_{i=1}^{\infty} \mathsf{P}(E_i).$$

4. Prove that

$$\mathsf{P}(E_1 \cap E_2 \cap \ldots \cap E_n) \geqslant 1 - \sum_{i=1}^{n} \mathsf{P}(\Omega - E_i).$$

5. Prove that $$\mathsf{P}(\liminf E_n) \leqslant \liminf \mathsf{P}(E_n),$$
$$\mathsf{P}(\limsup E_n) \geqslant \limsup \mathsf{P}(E_n),$$

and give examples to show that these inequalities can be strict.

6. Show that the probability that exactly r of the events $E_1, E_2, \ldots, E_n$ occur is

$$\sum_{j=0}^{n-r} (-1)^j \binom{r+j}{j} S_{r+j},$$

where $$S_m = \sum_{1 \leqslant i_1 < i_2 < \ldots < i_m \leqslant n} \mathsf{P}(E_{i_1} \cap E_{i_2} \cap \ldots \cap E_{i_m}).$$

10.4 Conditional probability

In the preceding sections we have seen how probability statements can be set within a mathematical model in which probability is regarded as a measure on a σ-field of subsets of the set of all possible outcomes of a random experiment. Thus any situation involving ideas of probability is formulated as a mathematical model consisting of a suitable probability space $(\Omega, \mathscr{F}, \mathsf{P})$.

If E is any event (i.e. any $\mathscr{F}$-measurable subset of Ω), it has been seen that it is possible to interpret $\mathsf{P}(E)$ in the following way: if in a succession of n repetitions of the experiment E occurs $r_E(n)$ times, then we expect the frequency $r_E(n)/n$ to approximate, for large n, to $\mathsf{P}(E)$.

Now let A, B be two events. Consider a series of n repetitions of the experiment, but restrict attention to those ($r_B(n)$ in number) in which B occurs. Of these, A occurs in $r_{A \cap B}(n)$, so that the frequency of occurrence of A, in those repetitions in which B occurs, is

$$\begin{aligned} f(n) &= r_{A \cap B}(n)/r_B(n) \\ &= \{r_{A \cap B}(n)/n\} \div \{r_B(n)/n\}. \end{aligned}$$

Therefore, when n is large, $f(n)$ approximates to

$$\mathsf{P}(A \cap B)/\mathsf{P}(B),$$

so long as $\mathsf{P}(B) > 0$. This quantity therefore represents the probability that A occurs, when it is given that B occurs.

Conditional probability

If A and B are two events, and $\mathsf{P}(B) > 0$, then the conditional probability of A given B (or the probability of A conditional on B), denoted by $\mathsf{P}(A|B)$, is defined by

$$\mathsf{P}(A|B) = \mathsf{P}(A \cap B)/\mathsf{P}(B). \tag{10.4.1}$$

Example. If a symmetrical die is thrown, the probability that a 3 or a 6 is thrown conditional on an even number being thrown is

$$\begin{aligned} \mathsf{P}(\{3,6\}|\{2,4,6\}) &= \mathsf{P}(\{3,6\} \cap \{2,4,6\})/\mathsf{P}\{2,4,6\} \\ &= \mathsf{P}\{6\}/\mathsf{P}\{2,4,6\} \\ &= \tfrac{1}{6}/\tfrac{1}{2} = \tfrac{1}{3}. \end{aligned}$$

An important property of conditional probability is that, as a function of A, $\mathsf{P}(A|B)$ is a probability measure on $(\Omega, \mathscr{F})$ for which the set B has probability one.

***Theorem* 10.1.** *Let $(\Omega, \mathscr{F}, \mathsf{P})$ be a probability space, and let $B \in \mathscr{F}$ have $\mathsf{P}(B) > 0$. Then the function $\mathsf{P}^B : \mathscr{F} \to \mathbf{R}$ given by $\mathsf{P}^B(A) = \mathsf{P}(A|B)$ is a probability measure on $(\Omega, \mathscr{F})$ with $\mathsf{P}^B(B) = 1$.*

Proof. From (10.4.1) it is immediate that $\mathsf{P}^B(A) \geqslant 0$ and that $\mathsf{P}^B(\varnothing) = 0$, so that P^B is a non-negative set function. Moreover, if $A_1, A_2, \ldots$ are disjoint members of $\mathscr{F}$ with union A, then

$$\begin{aligned}\mathsf{P}^B(A) &= \mathsf{P}(A \cap B)/\mathsf{P}(B) = \mathsf{P}\{\bigcup_n (A_n \cap B)\}/\mathsf{P}(B) \\ &= \{\sum_n \mathsf{P}(A_n \cap B)\}/\mathsf{P}(B) = \sum_n \mathsf{P}^B(A_n),\end{aligned}$$

so that P^B is a measure. Finally

$$\mathsf{P}^B(\Omega) = \mathsf{P}(B)/\mathsf{P}(B) = 1,$$

and

$$\mathsf{P}^B(B) = \mathsf{P}(B)/\mathsf{P}(B) = 1. \blacksquare$$

The measure P^B is not in general complete, since $\mathsf{P}^B(\Omega - B) = 0$, and $\Omega - B$ will in general have subsets not belonging to $\mathscr{F}$. However, if $\mathscr{F}_B = \{A \in \mathscr{F};\ A \subset B\}$, and if P_B is the restriction of P^B to $\mathscr{F}_B$, it is easy to verify that P_B is a complete measure on $(B, \mathscr{F}_B)$, so that $(B, \mathscr{F}_B, \mathsf{P}_B)$ is a probability space.

The function $\mathsf{P}(\cdot|\cdot)$ has a number of simple but useful properties, some of which are given in the following theorem.

***Theorem* 10.2.** *Let $(\Omega, \mathscr{F}, \mathsf{P})$ be a probability space.*

(i) *If $A, B \in \mathscr{F}$ and $\mathsf{P}(B) > 0$, then*

$$A \subset B \Rightarrow \mathsf{P}(A|B) = \mathsf{P}(A)/\mathsf{P}(B),$$

and

$$A \supset B \Rightarrow \mathsf{P}(A|B) = 1.$$

(ii) *If $A, B, C \in \mathscr{F}$ and $\mathsf{P}(B \cap C) > 0$, then*

$$\mathsf{P}(A|B \cap C) = \mathsf{P}(A \cap B|C)/\mathsf{P}(B|C).$$

(iii) *If $\{D_1, D_2, \ldots\}$ is an $\mathscr{F}$-dissection of Ω with $\mathsf{P}(D_n) > 0$, then, for any $A \in \mathscr{F}$,*

$$\mathsf{P}(A) = \sum_{n=1}^{\infty} \mathsf{P}(A|D_n)\mathsf{P}(D_n), \tag{10.4.2}$$

and if $\mathsf{P}(A) > 0$,

$$\mathsf{P}(D_n|A) = \mathsf{P}(A|D_n)\mathsf{P}(D_n)/\mathsf{P}(A). \tag{10.4.3}$$

(iv) *If $\{C_1, C_2, \ldots\}$ is a monotonic sequence in $\mathscr{F}$, whose limit C has $\mathsf{P}(C) > 0$, then, for any $A \in \mathscr{F}$,*

$$\mathsf{P}(A|C) = \lim_{n\to\infty} \mathsf{P}(A|C_n).$$

The proofs are easy consequences of the definition (10.4.1) and are left to the reader. Notice that it is a corollary of (iii) that, if $\{D_n\}$ is an $\mathscr{F}$-dissection of Ω, if $A \in \mathscr{F}$, and if $\mathsf{P}(A) > 0$ and $\mathsf{P}(D_n) > 0$, then

$$\mathsf{P}(D_n|A) = \frac{\mathsf{P}(A|D_n)\,\mathsf{P}(D_n)}{\sum_{r=1}^{\infty} \mathsf{P}(A|D_r)\,\mathsf{P}(D_r)}. \tag{10.4.4}$$

This formula is a celebrated result known as *Bayes's theorem*, which finds a number of important applications, and plays a central role in 'Bayesian inference'. It is worth noting that (10.4.4) continues to hold even if some of the D_r have zero probability, so long as we then interpret $\mathsf{P}(A|D_r)\,\mathsf{P}(D_r)$ as zero.

Now let A and B be two events, and suppose that $0 < \mathsf{P}(B) < 1$. Then the two conditional probabilities

$$\alpha = \mathsf{P}(A|B), \quad \beta = \mathsf{P}(A|\Omega - B)$$

are defined, α representing the probability of A given that B occurs, and β the corresponding probability when B does not occur. If it should happen that $\alpha > \beta$, then A is more likely to happen if B does than if B does not. Hence the occurrence of B influences the probability of A. On the other hand, if $\alpha < \beta$, B influences A in the other direction, A being less likely to occur if B does than otherwise. It may be, however, that we wish to assume that A is independent of B in the sense that the occurrence or non-occurrence of B in no way influences A, and in that case we would require that α and β should be equal.

If $\alpha = \beta$, then

$$\begin{aligned} \mathsf{P}(A) &= \alpha\mathsf{P}(B) + \beta\mathsf{P}(\Omega - B) \quad \text{from (10.4.2)} \\ &= \alpha[\mathsf{P}(B) + \mathsf{P}(\Omega - B)] \\ &= \alpha, \end{aligned}$$

so that
$$\mathsf{P}(A) = \mathsf{P}(A|B) = \mathsf{P}(A \cap B)/\mathsf{P}(B),$$

or
$$\mathsf{P}(A \cap B) = \mathsf{P}(A)\,\mathsf{P}(B).$$

This motivates the following definition.

Statistical independence

Two events A and B are said to be statistically independent if

$$\mathsf{P}(A \cap B) = \mathsf{P}(A)\,\mathsf{P}(B). \tag{10.4.5}$$

The discussion preceding the definition shows that we would expect events independent in some causal sense to be statistically independent, but it should never be forgotten that causal and statistical independence are two entirely separate concepts. In the rest of this book the adjective 'statistical' will be dropped, and the term 'independent' will always be used in the sense of the above definition. Notice that although in the discussion preceding the definition we have assumed that $0 < \mathsf{P}(B) < 1$, the definition makes sense even if $\mathsf{P}(B)$ is 0 or 1. The definition is also evidently symmetrical between A and B.

Equation (10.4.5) bears an interesting resemblance to (10.3.3). Thus:

$$\mathsf{P}(A \cup B) = \mathsf{P}(A) + \mathsf{P}(B) \quad \text{if} \quad A \text{ and } B \text{ are disjoint,}$$

$$\mathsf{P}(A \cap B) = \mathsf{P}(A)\,\mathsf{P}(B) \quad \text{if} \quad A \text{ and } B \text{ are independent.}$$

These are often called respectively the addition and multiplication laws of probability. It is important, however, to observe the difference between them. The first is a postulated property of probability, and the condition of disjointness is a purely set-theoretic notion, defined without reference to P. The second, however, is just the definition of independence, and this is a concept which depends on the probability measure P.

It will become clear in later chapters that, from a mathematical point of view, it is just this concept of independence which enriches the measure-theoretic structure of $(\Omega, \mathscr{F}, \mathsf{P})$ and yields the distinctive problems and results of the theory of probability.

To conclude this section we consider in detail an example of a type which has some practical application.

Example. We take a simple example from genetics (without claiming that it represents exactly any particular genetic situation). Suppose we have a large human population each member of which can be of one of three types, denoted by AA, Aa and aa. Let the respective proportions of these types in the population be α, β, γ. Then the probability that a person drawn at random from the population is of type AA is α, and so on.

It often happens that it is impossible to decide of an individual whether he is of type AA or Aa, although type aa is readily recognisable. Call a person of type AA or Aa normal, one of type aa abnormal. Then, if we choose an individual from the population and find him normal, we may ask for the probability that he is of type AA. The answer is clearly given by conditional probabilities; symbolically

$$\mathsf{P}(AA|\text{normal}) = \mathsf{P}(AA)/\mathsf{P}(\text{normal}) = \alpha/(\alpha+\beta) = p \text{ (say)},$$

$$\mathsf{P}(Aa|\text{normal}) = \mathsf{P}(Aa)/\mathsf{P}(\text{normal}) = \beta/(\alpha+\beta) = q = 1-p,$$

$$\mathsf{P}(aa|\text{normal}) = 0.$$

Now suppose that we take at random a normal male and a normal female from the population, and that they marry. We assume that the probability that a person is of a given type is independent of sex, and that the types of the male and female are statistically independent. Then the probability that they are both of type AA is the product of the probabilities for each separately, i.e.

$$\mathsf{P}(AA \,\&\, AA) = p^2.$$

Similarly

$$\mathsf{P}(AA \,\&\, Aa) = pq,$$

$$\mathsf{P}(Aa \,\&\, AA) = pq,$$

$$\mathsf{P}(Aa \,\&\, Aa) = q^2.$$

Now suppose that they produce a child. What is the probability that he is abnormal? It is a consequence of the laws of genetics that this can happen only if both parents are of type Aa, and that it then happens with probability $\frac{1}{4}$. Thus

$$\mathsf{P}(\text{abnormal child} \mid Aa \,\&\, Aa) = \tfrac{1}{4}.$$

It follows from (10.4.2) that

$$\begin{aligned}\mathsf{P}(\text{abnormal child}) &= \mathsf{P}(\text{abnormal child} \mid Aa \,\&\, Aa)\,\mathsf{P}(Aa \,\&\, Aa)\\ &= \tfrac{1}{4}q^2.\end{aligned}$$

A rather more complex problem is the following. Suppose that the two normal parents have produced children $C_1, C_2, \ldots, C_n$, all normal. What is the probability that the next, C_{n+1}, is abnormal? We have

$$\mathsf{P}(C_{n+1} \text{ abnormal} \mid C_1, C_2, \ldots, C_n \text{ normal}) = \frac{\mathsf{P}(C_1, C_2, \ldots, C_n \text{ normal}, C_{n+1} \text{ abnormal})}{\mathsf{P}(C_1, C_2, \ldots, C_n \text{ normal})}.$$

Now it is usual to assume that, conditional on the types of the parents, the types of their children are independent of one another. Hence

$$\begin{aligned}\mathsf{P}(C_1, C_2, \ldots, C_n \text{ normal}) &= \mathsf{P}(C_1, C_2, \ldots, C_n \text{ normal} \mid AA \,\&\, AA)\,\mathsf{P}(AA \,\&\, AA)\\ &\quad + \mathsf{P}(C_1, C_2, \ldots C_n \text{ normal} \mid AA \,\&\, Aa)\,\mathsf{P}(AA \,\&\, Aa)\\ &\quad + \mathsf{P}(C_1, C_2, \ldots, C_n \text{ normal} \mid Aa \,\&\, AA)\,\mathsf{P}(Aa \,\&\, AA)\\ &\quad + \mathsf{P}(C_1, C_2, \ldots, C_n \text{ normal} \mid Aa \,\&\, Aa)\,\mathsf{P}(Aa \,\&\, Aa)\\ &= 1.p^2 + 1.qp + 1.pq + \prod_{j=1}^{n} \mathsf{P}(C_j \text{ normal} \mid Aa \,\&\, Aa)\,q^2\\ &= p^2 + 2pq + (\tfrac{3}{4})^n q^2\\ &= 1 - q^2\{1 - (\tfrac{3}{4})^n\}.\end{aligned}$$

Similarly

$$\begin{aligned}\mathsf{P}(C_1, C_2, \ldots, C_n \text{ normal}, C_{n+1} \text{ abnormal}) \\ &= \mathsf{P}(C_1, C_2, \ldots, C_n \text{ normal}, C_{n+1} \text{ abnormal} \,|\, Aa \,\&\, Aa)\, q^2 \\ &= (\tfrac{3}{4})^n \tfrac{1}{4} q^2.\end{aligned}$$

Thus finally

$$\mathsf{P}(C_{n+1} \text{ abnormal} | C_1, C_2, \ldots, C_n \text{ normal}) = \tfrac{1}{4} q^2 \frac{(\frac{3}{4})^n}{1 - q^2\{1 - (\frac{3}{4})^n\}}.$$

As n increases, this decreases rapidly, converging exponentially fast to zero.

Exercises 10.4

1. Two dice are thrown. Find the probability that one of the dice shows a 6, conditional on the sum of the two scores being 8.

2. Find the probability that a bridge hand contains the king of spades, given that it contains all four aces.

3. Define a probability space $(\Omega, \mathscr{F}, \mathsf{P})$ by taking $\Omega = (0, 1)$, $\mathscr{F} = \mathscr{L}$, the σ-field of Lebesgue measurable subsets of Ω, and P to be Lebesgue measure. Let the binary expansion of $\omega \in \Omega$ be

$$\omega = \sum_{n=1}^{\infty} \omega_n 2^{-n},$$

where $\omega_n = 0$ or 1. If m and n are distinct integers, and a and b are equal to 0 or 1, show that the events $\{\omega;\ \omega_m = a\}$ and $\{\omega;\ \omega_n = b\}$ are independent.

4. If $(\Omega, \mathscr{F}, \mathsf{P})$ is a probability space, and if $\mathscr{F}_1$ and $\mathscr{F}_2$ are sub-σ-fields of $\mathscr{F}$, we say that $\mathscr{F}_1$ and $\mathscr{F}_2$, are independent if, for any $E_1 \in \mathscr{F}_1$, $E_2 \in \mathscr{F}_2$,

$$\mathsf{P}(E_1 \cap E_2) = \mathsf{P}(E_1)\, \mathsf{P}(E_2).$$

Show that events A and B are independent if and only if the σ-fields $\mathscr{F}(A)$ and $\mathscr{F}(B)$ which they generate are independent.

5. The example discussed at the end of this section was analysed without specifying the probability space $(\Omega, \mathscr{F}, \mathsf{P})$ on which it was defined. This is very common practice; it is both tedious and unnecessary always to give a detailed specification. Suggest a suitable probability space for the simple problem in which there is only one child.

10.5 Independent trials

So far we have been concerned with a random experiment described by a probability space $(\Omega, \mathscr{F}, \mathsf{P})$. However, in discussing the frequency interpretation of probability statements given by P, we have talked of continued repetitions of the experiment. The object of this section is to show how this notion of continued repetition is introduced into the mathematical theory. Before doing this, however, we must extend a little the concept of independence introduced in the last section.

Let $(\Omega, \mathscr{F}, \mathsf{P})$ be a probability space, and let A, B, C be events (i.e. members of $\mathscr{F}$). Now suppose that A is independent of B, B of C, and C of A, so that

$$\mathsf{P}(A \cap B) = \mathsf{P}(A)\,\mathsf{P}(B), \quad \mathsf{P}(B \cap C) = \mathsf{P}(B)\,\mathsf{P}(C),$$
$$\mathsf{P}(A \cap C) = \mathsf{P}(A)\,\mathsf{P}(C). \quad (10.5.1)$$

If these conditions are satisfied A, B, C are said to be *pairwise independent*. Pairwise independence is often, however, too weak a requirement. It can happen that A, B, C are pairwise independent, so that A is independent of B and C separately, but that A is not independent of $B \cap C$. The independence of A and $B \cap C$ is equivalent to the condition

$$\mathsf{P}(A \cap B \cap C) = \mathsf{P}(A)\,\mathsf{P}(B \cap C),$$

which in virtue of (10.5.1) is equivalent to

$$\mathsf{P}(A \cap B \cap C) = \mathsf{P}(A)\,\mathsf{P}(B)\,\mathsf{P}(C). \quad (10.5.2)$$

Thus we may often wish to add to (10.5.1) the additional requirement (10.5.2).

This idea can be extended to any collection of events, finite or infinite, as in the following definition.

Independence

If $(\Omega, \mathscr{F}, \mathsf{P})$ is a probability space, a collection of events $\mathscr{C} \subset \mathscr{F}$ is said to be independent if, for any finite subcollection

$$\{E_1, E_2, \ldots, E_n\} \subset \mathscr{C},$$
$$\mathsf{P}\left\{\bigcap_{j=1}^{n} E_j\right\} = \prod_{j=1}^{n} \mathsf{P}(E_j). \quad (10.5.3)$$

If (10.5.3) holds only for subcollections of two events, $\mathscr{C}$ is said to be *pairwise independent*. Thus independence implies pairwise independence, but not conversely (see exercise 10.5.5).

As an example of a random experiment which can be continually repeated, consider the tossing of a coin. Here the probability space has only two elements, which we may denote by H (heads) and T (tails), and the probability measure P must have the form

$$\mathsf{P}(H) = p, \quad \mathsf{P}(T) = 1-p,$$

where $0 \leqslant p \leqslant 1$.

Now suppose that the coin is tossed n times. This compound experiment can clearly itself be regarded as a random experiment. The different outcomes are described by sequences ω of H's and T's, n in all, so that the set of all outcomes is the set of all such sequences, which is exactly the Cartesian product

$$\Omega^n = \Omega \times \Omega \times \ldots \times \Omega$$

of n copies of Ω. Since Ω^n is finite, we take every subset of Ω^n to define an event, so that we consider the σ-field $\mathscr{F}^n$ of all subsets of Ω^n. Thus the experiment of tossing a coin n times will be described by a probability space $(\Omega^n, \mathscr{F}^n, \mathsf{P}^n)$, where P^n is some probability measure on the Cartesian product Ω^n.

How then is the measure P^n to be chosen? The answer comes by observing that it is implicit in the idea of a series of tosses that they are carried out in such a way that the result of one does not influence the others. Thus the successive tosses should be assumed to be (statistically) independent. Suppose for example that $n = 3$, and that we require the probability of the outcome HHT. This event is the intersection of the three events.

$$E_1 = \{H \text{ occurs at first toss}\},$$

$$E_2 = \{H \text{ occurs at second toss}\},$$

$$E_3 = \{T \text{ occurs at third toss}\}.$$

Because these are events defined on different tosses, they are assumed to be statistically independent, and therefore

$$\begin{aligned} \mathsf{P}^3(HHT) &= \mathsf{P}^3(E_1 \cap E_2 \cap E_3) \\ &= \mathsf{P}^3(E_1)\, \mathsf{P}^3(E_2)\, \mathsf{P}^3(E_3) \\ &= \mathsf{P}(H)\, \mathsf{P}(H)\, \mathsf{P}(T) \\ &= p^2(1-p). \end{aligned}$$

An exactly similar argument carries through for any value of n, and leads to the conclusion that, if ω is a sequence containing $r(\omega)$ heads and $n - r(\omega)$ tails, then

$$\mathsf{P}^n(\omega) = p^{r(\omega)}(1-p)^{n-r(\omega)}. \tag{10.5.4}$$

This determines the probability of any outcome $\omega \in \Omega^n$, and thus defines the probability measure P^n by

$$\mathsf{P}^n(E) = \sum_{\omega \in E} p^{r(\omega)}(1-p)^{n-r(\omega)} \quad (E \subset \Omega^n).$$

We say that this probability space $(\Omega^n, \mathscr{F}^n, \mathsf{P}^n)$ describes the experiment consisting of n *independent trials* (or *repetitions*) of the experiment described by $(\Omega, \mathscr{F}, \mathsf{P})$. Of course, it may be that we want to set up a model in which different trials are, for some reason, not independent. In this case we would still use Ω^n, but a different probability measure would be appropriate.

The reader will by now have observed that $(\Omega^n, \mathscr{F}^n, \mathsf{P}^n)$ is exactly the product, as defined in Chapter 6, of n copies of $(\Omega, \mathscr{F}, \mathsf{P})$. This shows how to proceed in more general circumstances. Let a random experiment described by a probability space $(\Omega, \mathscr{F}, \mathsf{P})$ be repeated n times. Then any outcome of the compound experiment is a sequence $(\omega_1, \omega_2, \ldots, \omega_n)$ of n elements of Ω, and therefore the set of outcomes is the Cartesian product

$$\Omega^n = \Omega \times \Omega \times \ldots \times \Omega$$

of n copies of Ω. If $E = E_1 \times E_2 \times \ldots \times E_n$ $(E_j \in \mathscr{F})$ is a member of the semi-ring

$$\mathscr{F} \times \mathscr{F} \times \ldots \times \mathscr{F},$$

then E has probability

$$\mathsf{P}^n(E) = \mathsf{P}(E_1)\,\mathsf{P}(E_2)\ldots\mathsf{P}(E_n).$$

As in § 6.2, P^n extends to a probability measure on the σ-field

$$\mathscr{F}^{*n} = \mathscr{F} * \mathscr{F} * \ldots * \mathscr{F}$$

generated by $\mathscr{F} \times \mathscr{F} \times \ldots \times \mathscr{F}$. If $\mathscr{F}^n$ denotes the completion of $\mathscr{F}^{*n}$ with respect to P^n, the probability space $(\Omega^n, \mathscr{F}^n, \mathsf{P}^n)$ is the appropriate one to describe the compound experiment consisting of the n repetitions of the experiment $(\Omega, \mathscr{F}, \mathsf{P})$.

In some problems it is convenient to think of the experiment being repeated an infinite number of times. In this case the appropriate probability space is the product $(\Omega^\infty, \mathscr{F}^\infty, \mathsf{P}^\infty)$ of countably many copies of $(\Omega, \mathscr{F}, \mathsf{P})$. These ideas will be dealt with in more detail in Chapter 13.

Let us now return to the coin-tossing problem. Equation (10.5.4) gives the probability of any outcome of the n tosses of the coin. Now let E_r be the event that, of these n tosses, exactly r are heads, where r is an integer in $0 \leqslant r \leqslant n$. Then

$$\begin{aligned} \mathsf{P}^n(E_r) &= \sum_{\{\omega;\, r(\omega)=r\}} \mathsf{P}^n(\omega) \\ &= \sum_{\{\omega;\, r(\omega)=r\}} p^{r(\omega)}(1-p)^{n-r(\omega)} \\ &= p^r(1-p)^r \sum_{\{\omega;\, r(\omega)=r\}} 1. \end{aligned}$$

The number of sequences ω with $r(\omega) = r$ is given by the binomial coefficient $\binom{n}{r}$, and so

$$\mathsf{P}^n(E_r) = \binom{n}{r} p^r(1-p)^{n-r} \quad (r = 0, 1, 2, \ldots, n) \qquad (10.5.5)$$

$$= b(n, p;\, r) \quad \text{(say)}.$$

Thus the probability of throwing r heads in n tosses is given by the expression $b(n, p;\, r)$. Notice that $\{E_0, E_1, \ldots, E_n\}$ is a dissection of Ω^n so that

$$\sum_{r=0}^{n} b(n, p;\, r) = 1,$$

which is, of course, a special case of the binomial theorem. If n and p are fixed, and r is allowed to vary, $b(n, p;\, r)$ measures the 'distribution of probability' over the sets E_r. The set of numbers

$$\{b(n, p;\, r);\quad r = 0, 1, 2, \ldots, n\}$$

is called the *binomial distribution* with parameters n, p.

When n is large the probabilities $b(n, p;\, r)$ become difficult to calculate, but there is an interesting approximation due to de Moivre, Stirling and Laplace. We recall Stirling's formula for $m!$, which can be written

$$\log m! = k + (m + \tfrac{1}{2}) \log m - m + \theta_m,$$

where $k = \frac{1}{2} \log (2\pi)$ and $\theta_m = O(m^{-1})$ as $m \to \infty$. Applying this to the factorials in $\binom{n}{r}$,

$$-\log b(n, p; r) = k + \tfrac{1}{2} \log \frac{r(n-r)}{n} + r \log \frac{r}{np} + (n-r) \log \frac{n-r}{n(1-p)}$$
$$- \theta_n + \theta_r + \theta_{n-r}.$$

Now write $\sigma = [p(1-p)]^{\frac{1}{2}}$, $x = x(r) = (r - np)/\sigma n^{\frac{1}{2}}$, and restrict x to lie in some finite interval (a, b). Then a little manipulation yields $-\log b(n, p; r) = k + \log \sigma n^{\frac{1}{2}} + \frac{1}{2}x^2 + O(n^{-\frac{1}{2}})$ as $n \to \infty$ uniformly in $a \leqslant x \leqslant b$. Thus, uniformly in $a \leqslant x \leqslant b$,

$$b(n, p; r) = \sigma^{-1}(2\pi n)^{-\frac{1}{2}} e^{-\frac{1}{2}x^2}\{1 + O(n^{-\frac{1}{2}})\}. \qquad (10.5.6)$$

Now observe that

$$\sum_{\{r;\, a \leqslant x(r) \leqslant b\}} \sigma^{-1} n^{-\frac{1}{2}} e^{-\frac{1}{2}x^2(r)}$$

is an approximating Riemann sum to the integral

$$\int_a^b e^{-\frac{1}{2}x^2}\, dx.$$

Hence, it follows from (10.5.6) that

$$\lim_{n\to\infty} \sum_{np+a\sigma\sqrt{n}\leqslant r\leqslant np+b\sigma\sqrt{n}} b(n,p;\,r) = (2\pi)^{-\frac{1}{2}}\int_a^b e^{-\frac{1}{2}x^2}\,dx. \tag{10.5.7}$$

This result is a special case of the *central limit theorem*, which will be proved in much greater generality in Chapter 13. It gives an approximation, for large n, to the probability that the number of heads in n tosses will be between $np+a\sigma\sqrt{n}$ and $np+b\sigma\sqrt{n}$.

It is well known (and proved at the end of §11.7) that

$$\int_{-\infty}^{\infty} e^{-\frac{1}{2}x^2}\,dx = (2\pi)^{\frac{1}{2}}.$$

Hence, for any $\epsilon > 0$, we can choose c so large that

$$(2\pi)^{-\frac{1}{2}}\int_{-c}^{c} e^{-\frac{1}{2}x^2}\,dx > 1-\tfrac{1}{2}\epsilon.$$

It then follows from (10.5.7) that, for all n greater than some n_0,

$$\sum_{np-c\sigma\sqrt{n}\leqslant r\leqslant np+c\sigma\sqrt{n}} b(n,p;\,r) > 1-\epsilon.$$

Thus, if we toss the coin $n(> n_0)$ times, there is probability at least $1-\epsilon$ that the number r of heads will satisfy

$$\left|\frac{r}{n}-p\right| \leqslant \frac{c\sigma}{\sqrt{n}}. \tag{10.5.8}$$

If we remember that, in the frequency interpretation, p is regarded as the value to which the proportion r/n of heads settles down as n becomes large, we see that (10.5.8) expresses a very reassuring consistency property of the probability model. It shows that the model attaches high probability to an event which implies that r/n is near to p. In Chapter 13 we shall prove much more precise and general theorems of this type.

Exercises 10.5

1. If $A_1, A_2, \ldots, A_n$ are independent events with $p_k = \mathsf{P}(A_k)$, find the probability that none of the events occurs, and show that it is always less than

$$\exp\left(-\sum_{k=1}^{n} p_k\right).$$

2. A die is such that the probability that it shows a 1 is p_1, a 2 is p_2, and so on. If it is thrown n times show that the probability that it shows r_1 1's, r_2 2's, ..., r_6 6's, where r_k are non-negative integers with

$$r_1+r_2+\ldots+r_6 = n,$$

is
$$\frac{n!}{r_1!\,r_2!\ldots r_6!}\,p_1^{r_1}p_2^{r_2}\ldots p_k^{r_k}.$$

(This is an example of the *multinomial distribution*; see §14.5.)

3. A coin with probability p of showing heads is tossed repeatedly. Find the probability p_n that exactly n tosses will have elapsed before k heads have been observed, and show that p_n is the coefficient of z^n in the expansion of
$$\left\{\frac{pz}{1-(1-p)z}\right\}^k.$$

(This is an example of the *negative binomial* distribution. Notice that $\sum_{n=1}^{\infty} p_n = 1$, which can be interpreted using the probability space
$$(\Omega^\infty, \mathscr{F}^\infty, \mathsf{P}^\infty)$$
as meaning that it is almost certain that k heads will appear if the coin is tossed for a sufficiently long time.)

4. Show that, for fixed n, p, the probability $b(n,p;\,r)$ attains its largest value $\hat{b}(n,p)$ when r is the integer part of $(n+1)p$. Show that, for fixed p,
$$\hat{b}(n,p) = O(n^{-\frac{1}{2}})$$
as $n \to \infty$.

5. Let Ω consist of the integers $1, 2, \ldots, 9$, each being assigned probability $\frac{1}{9}$. Show that the events
$$\{1,2,3\},\quad \{1,4,5\},\quad \{2,4,6\}$$
are pairwise independent but not independent.

6. Construct an example of a probability space with three events A, B, C which satisfy
$$\mathsf{P}(A \cap B \cap C) = \mathsf{P}(A)\,\mathsf{P}(B)\,\mathsf{P}(C)$$
but are not independent.

7. Let $\mathscr{A}$ be an independent set of events on a probability space $(\Omega, \mathscr{F}, \mathsf{P})$. Let $\mathscr{A}_1$ and $\mathscr{A}_2$ be disjoint subsets of $\mathscr{A}$, and let $\mathscr{F}_i$ $(i = 1, 2)$ be the σ-field generated by $\mathscr{A}_i$. Show that $\mathscr{F}_1$ and $\mathscr{F}_2$ are independent in the sense of exercise 10.4 (4). Deduce that any set A in $\mathscr{F}_1 \cap \mathscr{F}_2$ satisfies $\mathsf{P}(A) = 0$ or $\mathsf{P}(A) = 1$.

8. If $\mathscr{E}$ is a random experiment, denote by $\mathscr{E}^n$ the experiment consisting of n independent repetitions of $\mathscr{E}$. Are the probability spaces for $(\mathscr{E}^m)^n$ and $\mathscr{E}^{mn}$ the same? Do they give the same probability statements?

9. Show that, if a coin with probability p of falling heads is tossed n times, the probability that less than np of these tosses results in a head tends to $\frac{1}{2}$ as $n \to \infty$.

11

RANDOM VARIABLES

11.1 Random variables as measurable functions

In the last chapter there was set up the basic mathematical model for a random situation, a probability space $(\Omega, \mathscr{F}, \mathsf{P})$. The points ω of the set Ω represent the 'outcomes' of a random 'experiment'. The elements of the σ-field $\mathscr{F}$ represent events about which probability statements can be made, the probability of $A \in \mathscr{F}$ being $\mathsf{P}(A)$, and P being a complete probability measure on $(\Omega, \mathscr{F})$.

To take a very simple example, suppose that a fair coin is tossed three times. This is an experiment which has eight outcomes:

$$\omega_1 = HHH, \quad \omega_2 = HHT, \quad \omega_3 = HTH, \quad \omega_4 = HTT,$$

$$\omega_5 = THH, \quad \omega_6 = THT, \quad \omega_7 = TTH, \quad \omega_8 = TTT.$$

As in § 10.5, each of these outcomes has probability $\frac{1}{8}$. Thus

$$\Omega = \{\omega_1, \omega_2, \ldots, \omega_8\},$$

$\mathscr{F}$ is the class of all subsets of Ω, and P assigns to a set A containing r elements the probability $\frac{1}{8}r$.

It may be that we are interested not in the entire outcome of the experiment, but just in, say, the number X of heads. This is a quantity which can take any of the values, 0, 1, 2, 3, according to the outcome of the experiment; it is an example of what is called a *random variable*. This is, however, a somewhat misleading term, because, as a little thought will show, X is not a variable but a function, assigning to each possible outcome ω a number $X(\omega)$, the number of H's in ω. More specifically, X is a function from Ω into $\mathbf{R}$, given by

$$X(\omega_1) = 3,$$

$$X(\omega_2) = X(\omega_3) = X(\omega_5) = 2,$$

$$X(\omega_4) = X(\omega_6) = X(\omega_7) = 1,$$

$$X(\omega_8) = 0.$$

An important property of random variables is that probability statements can be made about them. For instance, in the above example

$$\mathsf{P}\{\omega;\ X(\omega) = 2\} = \mathsf{P}\{\omega_2, \omega_3, \omega_5\} = \tfrac{3}{8},$$

a statement which is usually abbreviated to

$$\mathsf{P}(X = 2) = \tfrac{3}{8};$$

the probability of getting 2 heads in 3 tosses is $\frac{3}{8}$.

A more complicated example is that of the birth of a baby. To specify the outcome ω, it would be necessary to describe the baby in complete detail, so that Ω (which might be regarded as the class of all conceivable babies) would be very large. We may, however, be interested in certain quantities such as the weight X of the baby, which is a quantity subject to random variation. Clearly X can be regarded as a function from Ω into $\mathbf{R}$, which associates with each baby ω its weight $X(\omega)$.

More generally, if $(\Omega, \mathscr{F}, \mathsf{P})$ is a probability space, we consider functions $X: \Omega \to \mathbf{R}^*$ which assign, to each outcome ω, a real number $X(\omega)$. Such functions cannot, however, be allowed to be entirely arbitrary. This is because we want to be able to talk about such things as 'the probability that X lies in the interval (a, b)', or symbolically

$$\mathsf{P}\{\omega; a < X(\omega) < b\}.$$

Thus we demand that the sets of the form

$$\{\omega;\ a < X(\omega) < b\}$$

should belong to $\mathscr{F}$, i.e. that $X: \Omega \to \mathbf{R}^*$ should be measurable. This leads to the following definition.

Random variable

A random variable on a probability space $(\Omega, \mathscr{F}, \mathsf{P})$ is a measurable function from $(\Omega, \mathscr{F})$ into $\mathbf{R}^*$.

It is very often convenient to drop the symbol ω from probability statements involving random variables, so that, for instance, the statement

$$\mathsf{P}\{\omega;\ a < X(\omega) < b\} = p$$

might be written more succinctly as

$$\mathsf{P}(a < X < b) = p.$$

In fact, we shall see that many manipulations involving random variables can be (and very often are) carried out without explicit reference to the underlying space Ω.

Because random variables are functions, there are a number of operations which can be carried out on them. For example, if X and Y are random variables, not taking more than one of the values $+\infty$ and $-\infty$, their sum $X + Y$ is defined by

$$(X + Y)(\omega) = X(\omega) + Y(\omega).$$

Moreover, if $f: \mathbf{R}^* \to \mathbf{R}^*$ is Borel measurable, the function

$$Z: \Omega \to \mathbf{R}^*$$

defined by

$$Z(\omega) = f\{X(\omega)\}$$

is measurable by the lemma of §6.5, and so defines a random variable Z which will be written as $f(X)$. If c is any real number, there is a trivial random variable $\hat{c}$ defined by

$$\hat{c}(\omega) = c \quad (\omega \in \Omega).$$

In what follows we shall not distinguish between $\hat{c}$ and c; for instance, if in the coin-tossing example just given Y is the number of tails, then

$$X + Y = 3,$$

and 3 here denotes the trivial random variable $\hat{3}$ which takes the value 3 regardless of the outcome of the experiment.

It is sometimes convenient to consider 'random variables' taking values in spaces other than $\mathbf{R}^*$. For instance, if $\mathbf{C}$ is the complex plane, we shall call a measurable function from Ω into $\mathbf{C}$ a *complex random variable*. Some of our results remain true for complex random variables; the details are left to the reader.

Exercises 11.1

1. Let Ω consist of the positive integers, with

$$\mathsf{P}\{\omega\} = 2^{-\omega},$$

and let $X(\omega)$ be the residue of ω modulo k. Show that X is a random variable, and find

$$\mathsf{P}(X = r) \quad (r = 0, 1, 2, \ldots, k-1).$$

2. In the situation described in exercise 10.2 (3), show that ξ is a random variable and that, for any Borel subset B of $[0, 1]$, $\mathsf{P}(\xi \in B)$ is the Lebesgue measure of B.

3. If X is a random variable, show that $|X|$ is. Give an example to show that the converse is false.

11.2 Expectations

Consider a probability space $(\Omega, \mathscr{F}, \mathsf{P})$ for which Ω has only a finite number N of elements (and $\mathscr{F}$ contains all subsets of Ω). Then we can write $\Omega = \{\omega_1, \omega_2, \ldots, \omega_N\}$, and P is determined by the numbers

$$p_j = \mathsf{P}\{\omega_j\} \quad (j = 1, 2, \ldots, N).$$

Let X be a random variable, and with

$$X(\omega_j) = \xi_j.$$

We might look on this experiment as one arising in a casino, the ω_j being the outcomes of a game, and ξ_j being the amount (possibly negative) that a particular gambler wins if the outcome is ω_j.

Now suppose that the gambler plays the game a large number M of times. According to the frequency interpretation of the probabilities p_j, the outcome ω_j will occur about Mp_j times, and so he will win $Mp_j \xi_j$ on this outcome. Hence his total winnings will approximate to

$$\sum_{j=1}^{N} Mp_j \xi_j,$$

and so his average winning per play will be approximately

$$\sum_{j=1}^{N} p_j \xi_j.$$

This quantity is called the *expectation* of the random variable X, and is denoted by $\mathsf{E}(X)$, so that

$$\mathsf{E}(X) = \sum_{j=1}^{N} \mathsf{P}\{\omega_j\} X(\omega_j). \tag{11.2.1}$$

The reader will recognise (11.2.1) as the integral of the function X with respect to the measure P. This suggests the following definition of the expectation of a random variable defined on a quite general probability space, not necessarily finite.

Expectation

If X is a random variable on a probability space $(\Omega, \mathscr{F}, \mathsf{P})$, the expectation of X is defined by

$$\mathsf{E}(X) = \int_{\Omega} X \, d\mathsf{P}, \tag{11.2.2}$$

provided that the integral exists.

(The term 'mean value', or more shortly 'mean', is sometimes used in place of 'expectation'.)

The properties of the integral given in Chapter 5 can be translated at once into results about the operator E. In particular, the following corollary of theorem 5.5 will be used a great deal.

***Theorem* 11.1.** (i) *If X and Y are random variables on the same probability space, if $\mathsf{E}(X)$ and $\mathsf{E}(Y)$ exist and if $\mathsf{E}(X)+\mathsf{E}(Y)$ is defined, then $\mathsf{E}(X+Y)$ exists and*

$$\mathsf{E}(X+Y) = \mathsf{E}(X)+\mathsf{E}(Y). \tag{11.2.3}$$

(ii) *If X is a random variable whose expectation exists, and c is a finite real number, then $\mathsf{E}(cX)$ exists and*

$$\mathsf{E}(cX) = c\mathsf{E}(X). \tag{11.2.4}$$

In other words, subject to finiteness conditions, the operator E is *linear* on the space of random variables defined on a given probability space.

If X is a random variable, and $f: \mathbf{R}^* \to \mathbf{R}^*$ is Borel measurable, then $f(X)$ is a random variable, and we may look at its expectation. By definition, this is

$$\mathsf{E}f(X) = \int_\Omega f(X)\, d\mathsf{P},$$

provided that the integral exists. Particular cases which will be of importance later are

$$\mathsf{E}(X^n) \quad (n = 1, 2, \ldots, \text{the } \textit{moments} \text{ of } X)$$

and $$\mathsf{E}(e^{tX}) \quad (t \in \mathbf{R}; \text{the } \textit{moment generating function} \text{ of } X).$$

Let X be a random variable for which $\mu = \mathsf{E}(X)$ exists and is finite. Then the integral

$$\operatorname{var}(X) = \mathsf{E}\{(X-\mu)^2\} = \int_\Omega (X-\mu)^2\, d\mathsf{P} \tag{11.2.5}$$

exists (but may equal $+\infty$); it is called the *variance* of X. Its importance is that it measures the 'dispersion' of X, in a sense illustrated by the following theorem.

***Theorem* 11.2** (*Tchebychev's inequality*). *Let X be a random variable for which $\mu = \mathsf{E}(X)$ and $\sigma^2 = \operatorname{var}(X)$ are finite ($\sigma > 0$). Then, for any $\lambda \geq 1$,*

$$\mathsf{P}\{|X-\mu| \geq \lambda\sigma\} \leq \lambda^{-2}. \tag{11.2.6}$$

Proof. Let E be the event $\{|X-\mu| \geq \lambda\sigma\}$, i.e.

$$E = \{\omega; |X(\omega)-\mu| \geq \lambda\sigma\}.$$

Then
$$\begin{aligned}\lambda^2\sigma^2\mathsf{P}(E) &= \int_E \lambda^2\sigma^2\, d\mathsf{P} \\ &\leq \int_E (X-\mu)^2\, d\mathsf{P} \\ &\leq \int_\Omega (X-\mu)^2\, d\mathsf{P} = \sigma^2,\end{aligned}$$

so that, since $\sigma > 0$, $$\mathsf{P}(E) \leq \lambda^{-2}. ∎$$

The point about this inequality is that, when λ is large, the probability that X lies outside the interval $(\mu - \lambda\sigma, \mu + \lambda\sigma)$ is small, being at most λ^{-2}. In particular, if σ is small, X is near its expectation μ with high probability.

Exercises 11.2

1. Let m, n be positive integers with $m < n$. Show that $z^m \leqslant 1+z^n$ $(z \geqslant 0)$, and deduce that, for any random variable X,

$$\mathsf{E}|X|^n < \infty \Rightarrow \mathsf{E}|X|^m < \infty.$$

2. Prove that, if $\mathsf{E}(e^{tX}) < \infty$ for some $t > 0$ and for some $t < 0$, then $\mathsf{E}(X^n)$ exists and is finite for all n. Is the converse true?

3. Let X be a random variable with expectation μ and variance σ^2 $(0 < \sigma < \infty)$. Show that

$$(X-\mu)/\sigma$$

is a random variable with expectation 0 and variance 1.

4. By considering a random variable which takes only the values $-\lambda\sigma, 0, \lambda\sigma$, show that Tchebychev's inequality is best possible, in the sense that equality can be obtained for a suitable choice of X.

5. Prove the following generalisation of Tchebychev's inequality. If X is a random variable, $g: \mathbf{R}^* \to \mathbf{R}^+$ Borel measurable, and B a Borel subset of $\mathbf{R}^*$ with the property that $g(x) \geqslant a > 0$ for all $x \in B$, then

$$\mathsf{P}(X \in B) \leqslant a^{-1}\mathsf{E}\{g(X)\}.$$

(Tchebychev's inequality corresponds to the special case $g(x) = (x-\mu)^2$.)

6. If X has $\mathsf{E}(X) = 0$ and $\operatorname{var}(X) = 1$, and if $|X| \leqslant M$, show that, for $\lambda < 1$,

$$\mathsf{P}(|X| \geqslant \lambda) \geqslant \frac{1-\lambda^2}{M^2-\lambda^2}.$$

7. $X_1, X_2, \ldots, X_n$ are non-negative random variables with

$$\mu_j = \mathsf{E}(X_j) < \infty.$$

Show that, for any $c > 0$,

$$\mathsf{P}(X_1+X_2+\ldots+X_n \geqslant c) \leqslant c^{-1}(\mu_1+\mu_2+\ldots+\mu_n).$$

8. Let X be a random variable with $\mathsf{E}(X^2) < \infty$. Show that the function

$$s(a) = \mathsf{E}(X-a)^2$$

attains its minimum when $a = \mathsf{E}(X)$.

9. Show that, if $\mathsf{E}(X^2) < \infty$,

$$\operatorname{var}(X) = \mathsf{E}(X^2) - \{\mathsf{E}(X)\}^2.$$

10. Let X have $\mathsf{E}(X) = 0$, $\operatorname{var}(X) = 1$. Tchebychev's inequality gives

$$\mathsf{P}(|X| \geqslant \lambda) \leqslant \lambda^{-2}.$$

Show that

$$\mathsf{P}(|X| \geqslant \lambda) = o(\lambda^{-2})$$

as $\lambda \to \infty$.

11.3 Distributions of random variables

A random variable X on a probability space $(\Omega, \mathscr{F}, \mathsf{P})$ is, by definition, a measurable mapping from $(\Omega, \mathscr{F})$ into $(\mathbf{R}^*, \mathscr{B}^*)$. It therefore follows from the results of § 6.5 that X induces a measure P' on $(\mathbf{R}^*, \mathscr{B}^*)$ by

$$\mathsf{P}'(B) = \mathsf{P}(X \in B) = \mathsf{P}\{\omega;\ X(\omega) \in B\} \quad (B \in \mathscr{B}^*).$$

The measure P' satisfies $\mathsf{P}'(\mathbf{R}^*) = 1$, and is thus a probability measure, but it is not in general complete. If $\mathscr{F}'$ denotes the completion of $\mathscr{B}^*$ with respect to P', and if P' is extended to $\mathscr{F}'$, then $(\mathbf{R}^*, \mathscr{F}', \mathsf{P}')$ is a probability space.

In many problems, we are interested in probability statements about X, such as

$$\mathsf{P}(3 < X < 7) = p.$$

These are made in the context of the original space $(\Omega, \mathscr{F}, \mathsf{P})$, but they can be rephrased in terms of $(\mathbf{R}^*, \mathscr{F}', \mathsf{P}')$; the above statement becomes

$$\mathsf{P}'(3, 7) = p.$$

Thus, if we are interested in only one random variable, we can work wholly in terms of the induced probability space $(\mathbf{R}^*, \mathscr{F}', \mathsf{P}')$. This is useful because, in general, $\mathbf{R}^*$ will be much simpler than Ω, the set of all possible outcomes.

The measure P' is called the *distribution* of the random variable X, the idea being that it expresses the way in which probability is distributed over the possible values of X.

Now suppose that X is almost certainly finite, so that

$$\mathsf{P}(X \in \mathbf{R}) = 1.$$

Then $\mathsf{P}'(\mathbf{R}) = 1$, so that P' is concentrated on $\mathbf{R}$. It thus induces a probability measure P'' on a σ-field $\mathscr{F}''$ of subsets of $\mathbf{R}$, and $\mathscr{F}''$ contains all Borel subsets of $\mathbf{R}$. It now follows from theorem 4.8 that there exists a distribution function F (i.e. a non-decreasing, right-continuous function $F: \mathbf{R} \to [0, 1]$ with

$$\lim_{x \to -\infty} F(x) = 0, \quad \lim_{x \to \infty} F(x) = 1),$$

such that

$$\mathscr{L}_F \subset \mathscr{F}'',$$

and such that P'' coincides on $\mathscr{L}_F$ with the Lebesgue–Stieltjes measure μ_F determined by F. Moreover,

$$\begin{aligned} F(x) &= \mu_F(-\infty, x] \\ &= \mathsf{P}''(-\infty, x] \\ &= \mathsf{P}'(-\infty, x] \\ &= \mathsf{P}(X \leqslant x), \end{aligned}$$

and, if $B \in \mathscr{B}$,

$$\begin{aligned} \mathsf{P}(X \in B) &= \mathsf{P}'(B) \\ &= \mathsf{P}''(B) \\ &= \mu_F(B) \\ &= \int_B dF(x). \end{aligned}$$

Hence we have proved the following result.

***Theorem* 11.3.** *If the random variable X on $(\Omega, \mathscr{F}, \mathsf{P})$ is finite with probability one, then*

$$F(x) = \mathsf{P}\{\omega;\ X(\omega) \leqslant x\} \tag{11.3.1}$$

determines a function $F: \mathbf{R} \to [0,1]$ which is non-decreasing, right-continuous, and satisfies

$$\lim_{x \to -\infty} F(x) = 0, \quad \lim_{x \to \infty} F(x) = 1.$$

For any Borel set $B \subset \mathbf{R}$,

$$\mathsf{P}\{\omega;\ X(\omega) \in B\} = \int_B dF(x). \tag{11.3.2}$$

The function $F(x)$ is called the *distribution function* of the random variable X. The theorem shows that, from a knowledge of the distribution function of a random variable, there can be deduced the probability that the variable lies in any Borel subset of $\mathbf{R}$. One can also deduce the value of the expectation $\mathsf{E}(X)$, and indeed the expectation of any Borel function of X. This is the content of the following theorem, which is simply a restatement of theorem 6.11.

***Theorem* 11.4.** *If the random variable X is finite with probability one, and if $g: \mathbf{R} \to \mathbf{R}^*$ is Borel measurable, then*

$$\mathsf{E}\{g(X)\} = \int_R g(x)\, dF(x), \tag{11.3.3}$$

where F is defined by (11.3.1), *and the equality is to be interpreted in the sense that, if either side exists, then so does the other and they are then equal.*

If the finite random variable X has distribution function F, and if $g: \mathbf{R} \to \mathbf{R}$ is Borel measurable, then

$$Y = g(X)$$

is a finite random variable, and one may ask for its distribution function G. This is easily found since, by definition,

$$\begin{aligned} G(y) &= \mathsf{P}(Y \leqslant y) \\ &= \mathsf{P}\{g(X) \leqslant y\} \\ &= \mathsf{P}\{X \in g^{-1}((-\infty, y])\} \\ &= \int_{g^{-1}\{(-\infty, y]\}} dF(x) \end{aligned}$$

by (11.3.2). This result takes a particularly simple form if g is continuous and non-decreasing, for then, if

$$\gamma(y) = \sup\{x;\, g(x) \leqslant y\},$$

we have
$$g^{-1}\{(-\infty, y]\} = (-\infty, \gamma(y)],$$

and so
$$G(y) = F[\gamma(y)].$$

Exercises 11.3

1. If X has distribution function F, define a random variable Y by

$$Y = F(X).$$

Show that, if F is continuous, Y has the distribution function

$$\begin{aligned} G(y) &= 0 \quad (y < 0) \\ &= y \quad (0 \leqslant y \leqslant 1) \\ &= 1 \quad (y > 1). \end{aligned}$$

What happens if F is not continuous?

2. If X has distribution function F, find the distribution function of X^2.

3. If X is a random variable which is not almost certainly finite, we can still define a 'distribution function' F by (11.3.1). Show that F has all the properties of a distribution function except that

$$\lim_{x \to -\infty} F(x) = \mathsf{P}(X = -\infty) \geqslant 0,$$

$$\lim_{x \to \infty} F(x) = 1 - \mathsf{P}(X = +\infty) \leqslant 1.$$

State and prove analogues of theorems 11.3 and 11.4 for this case.

11.4 Types of distribution function

The results of the last section show that the properties of a single finite random variable can be deduced from those of its distribution function. The properties of distribution functions were studied in Chapter 9, and we may now apply the results of that chapter.

The simplest type of random variable X is one which takes only a finite number of possible values $\xi_1, \xi_2, \ldots, \xi_N$. If

$$p_j = \mathsf{P}(X = \xi_j),$$

then
$$p_j \geqslant 0, \quad \sum_{j=1}^{N} p_j = 1,$$

and it is clear that the distribution function of X is given by

$$F(x) = \sum_{\{j;\, \xi_j \leqslant x\}} p_j.$$

A slight generalisation is to consider a random variable which takes a countable number of possible values $\xi_1, \xi_2, \ldots$. Then

$$p_j = \mathsf{P}(X = \xi_j)$$

satisfies
$$p_j \geqslant 0, \quad \sum_{j=1}^{\infty} p_j = 1, \tag{11.4.1}$$

and the corresponding distribution function is

$$F(x) = \sum_{\{j;\, \xi_j \leqslant x\}} p_j. \tag{11.4.2}$$

A random variable which takes on at most countably many values is called a *discrete* random variable (and the adjective is also attached to its distribution). Thus a variable is discrete if and only if its distribution function is a jump function. In most applications, the numbers ξ_j have no finite limit point, but this need not be so; the numbers $\{\xi_j\}$ might, for instance, be the rationals. Some examples of discrete distributions will be given in § 11.6.

A second type of distribution function which arises quite commonly is constructed as follows. Let f be a measurable function on $\mathbf{R}$ which satisfies

$$f(x) \geqslant 0, \quad \int_{-\infty}^{\infty} f(x)\, dx = 1. \tag{11.4.3}$$

Then it is clear that the function

$$F(x) = \int_{-\infty}^{x} f(\xi)\, d\xi \tag{11.4.4}$$

is a distribution function.

Any function satisfying (11.4.3) is called a *probability density*. If a random variable has distribution function given by (11.4.4) it is said to have the probability density f, and any random variable having a probability density is called a *continuous random variable* (and is said to have a continuous distribution). Thus, by theorem 9.4, a random variable is continuous if and only if its distribution function is *absolutely continuous*, which in turn occurs if and only if its probability distribution is absolutely continuous with respect to Lebesgue measure.

Notice that, by theorem 9.3, if f_1 and f_2 are probability densities of the same random variable, then $f_1 = f_2$ almost everywhere. Moreover, if a continuous random variable X has distribution function F and probability density f, then

$$f(x) = F'(x) \quad \text{almost everywhere.} \tag{11.4.5}$$

For this case, theorems 11.3 and 11.4 can be written in the useful alternative forms

$$\mathsf{P}(X \in B) = \int_B f(x)\,dx, \tag{11.4.6}$$

$$\mathsf{E}\{g(X)\} = \int_{\mathbf{R}} g(x) f(x)\,dx. \tag{11.4.7}$$

Some important examples of probability densities will be given in §11.7.

Most of the random variables which arise in applications are either discrete or continuous. It is certainly not true, however, that every random variable is of one or other of these types. In fact, Lebesgue's decomposition (theorem 9.5) shows that every distribution function is uniquely expressible in the form

$$F = F_1 + F_2 + F_3,$$

where

(i) F_1 is a jump function,

(ii) F_2 is absolutely continuous,

(iii) F_3 is singular, i.e. continuous with $F_3' = 0$ almost everywhere.

A random variable with distribution function F is discrete if and only if $F_2 = F_3 = 0$, and continuous if and only if $F_1 = F_3 = 0$. The singular component F_3 is generally (but not invariably) absent in practice.

Exercises 11.4

1. Show that $F(x) = \sum_{n=1}^{\infty} 2^{-2n+1}[2^{n-1}x + \frac{1}{2}], \quad 0 \leqslant x \leqslant 1,$

where $[z]$ is the integer part of z, is a discrete distribution function, the set $\{\xi_j\}$ consisting of the numbers of the form $r2^{-s}$, where r and s are integers.

2. Show that the distribution function of a random variable X is continuous if and only if

$$\mathsf{P}(X = x) = 0$$

for all x.

3. Show that Cantor's function (denoted by f in §2.7) is a singular distribution function, and that a random variable X having this distribution function satisfies

$$\mathsf{P}(X \in C) = 1,$$

where C is the Cantor set.

11.5 Independent random variables

In Chapter 10 the concept of (statistical) independence of events are defined. This idea can be extended to random variables. Suppose that $(\Omega, \mathscr{F}, \mathsf{P})$ is a probability space, and that X and Y are two random variables defined on this space, i.e. two measurable functions from Ω into $\mathbf{R}^*$. For any numbers $x, y \in \mathbf{R}^*$, consider the two events

$$\{\omega;\ X(\omega) \leqslant x\}, \quad \{\omega;\ Y(\omega) \leqslant y\}.$$

If, for every choice of x, y, these two events are independent, then X and Y are said to be independent. More generally, we make the following definition.

Independence

A collection of random variables $\{X_i; i \in I\}$ on a probability space $(\Omega, \mathscr{F}, \mathsf{P})$ is said to be independent if, for any $x_i \in \mathbf{R}^*$ $(i \in I)$, the collection of events

$$\{\omega;\ X_i(\omega) \leqslant x_i\} \quad (i \in I),$$

is independent.

If E is any event, then the indicator function χ_E defined by

$$\begin{aligned} \chi_E(\omega) &= 1 \quad (\omega \in E) \\ &= 0 \quad (\omega \notin E), \end{aligned}$$

is clearly a random variable. Moreover, two events A and B are independent if and only if χ_A and χ_B are independent random variables. It is sometimes convenient to talk of a random variable X and an event E being independent; this means simply that X and χ_E are independent random variables.

An important example of independent random variables arises in the situation of independent trials described in §10.5. If $(\Omega^n, \mathscr{F}^n, \mathsf{P}^n)$ is the product of n copies of $(\Omega, \mathscr{F}, \mathsf{P})$, so that the elementary events are n-tuples $(\omega_1, \omega_2, \ldots, \omega_n)$ of elements of Ω, and if $X_1, X_2, \ldots, X_n$ are random variables on Ω^n with the property that X_j is a function only of ω_j, then it is easily verified that $X_1, X_2, \ldots, X_n$ are independent.

There is a special technique for dealing with independent random variables. Let $X_1, X_2, \ldots, X_n$ be independent finite random variables defined on a probability space $(\Omega, \mathscr{F}, \mathsf{P})$, and let F_j be the distribution function of X_j. Denote by X the measurable function from $(\Omega, \mathscr{F})$ into $(\mathbf{R}^n, \mathscr{B}^n)$ defined by

$$X(\omega) = (X_1(\omega), X_2(\omega), \ldots, X_n(\omega)).$$

Then X induces a probability measure P' on $(\mathbf{R}^n, \mathscr{B}^n)$ by

$$\mathsf{P}'(B) = \mathsf{P}\{\omega;\ X(\omega) \in B\} \quad (B \in \mathscr{B}^n).$$

For any real numbers $x_1, x_2, \ldots, x_n$, take for B the set

$$B = \{y = (y_1, y_2, \ldots, y_n);\quad y_1 \leqslant x_1, y_2 \leqslant x_2, \ldots, y_n \leqslant x_n\}.$$

Then, because $X_1, X_2, \ldots, X_n$ are independent,

$$\begin{aligned} \mathsf{P}'(B) &= \mathsf{P}(X_j \leqslant x_j;\ j = 1, 2, \ldots, n) \\ &= \prod_{j=1}^{n} \mathsf{P}(X_j \leqslant x_j) \\ &= \prod_{j=1}^{n} F_j(x_j). \end{aligned}$$

The results of §4.5 show that P' is therefore just the Cartesian product of the Stieltjes measures

$$\mu_{F_1}, \mu_{F_2}, \ldots, \mu_{F_n};$$

in other words

$$\mathsf{P}'(B) = \int \cdots \int_{(x_1, x_2, \ldots, x_n) \in B} dF_1(x_1)\, dF_2(x_2) \ldots dF_n(x_n).$$

Let g be any Borel measurable function from $\mathbf{R}^n$ into $\mathbf{R}^*$. Then, from theorem 6.8,

$$\begin{aligned} \mathsf{E}\{g(X_1, X_2, \ldots, X_n)\} &= \int_{\mathbf{R}^n} g(x_1, x_2, \ldots, x_n)\, d\mathsf{P}' \\ &= \int_{\mathbf{R}} \cdots \int_{\mathbf{R}} g(x_1, x_2, \ldots, x_n)\, dF_1(x_1)\, dF_2(x_2) \ldots dF_n(x_n). \end{aligned}$$

Thus we have proved the following result.

***Theorem* 11.5.** *Let $X_1, X_2, \ldots, X_n$ be independent random variables on $(\Omega, \mathscr{F}, \mathsf{P})$ which are almost certainly finite, and let F_j be the distribution function of X_j. Then, for any $B \in \mathscr{B}^n$,*

$$\mathsf{P}\{(X_1, X_2, \ldots, X_n) \in B\} = \int \cdots \int_{(x_1, x_2, \ldots, x_n) \in B} dF_1(x_1)\, dF_2(x_2) \ldots dF_n(x_n). \tag{11.5.1}$$

If $g\colon \mathbf{R}^n \to \mathbf{R}^*$ is Borel measurable, then

$$\mathsf{E}\{g(X_1, X_2, \ldots, X_n)\} = \int_{\mathbf{R}} \cdots \int_{\mathbf{R}} g(x_1, x_2, \ldots, x_n)\, dF_1(x_1)\, dF_2(x_2) \ldots dF_n(x_n), \tag{11.5.2}$$

in the sense that, if either side exists, then so does the other, and they are then equal.

The theorem shows that, if we have a finite number of random variables with known distributions, and if we know them to be independent, then we can calculate the probability of any event, and the expectation of any random variable, which is defined in terms of these random variables. As we shall see in Chapter 14, this is only true because of the assumption of independence.

As an important example of the use of theorem 11.5, let X and Y be independent finite random variables with distribution functions F and G respectively, and consider the random variable

$$Z = X + Y.$$

What is its distribution function H? By definition, for any real z,

$$\begin{aligned} H(z) &= \mathsf{P}(Z \leqslant z) \\ &= \mathsf{P}(X + Y \leqslant z) \\ &= \int_{\{(x,y);\, x+y \leqslant z\}} dF(x)\, dG(y), \quad \text{by} \quad (11.5.1), \\ &= \int_{\mathbf{R}} dG(y) \int_{(-\infty, z-y]} dF(x), \quad \text{by theorem 6.5}, \\ &= \int_{\mathbf{R}} F(z-y)\, dG(y). \end{aligned}$$

Hence

$$H(z) = \int_{\mathbf{R}} F(z-y)\, dG(y). \tag{11.5.3}$$

The right-hand side of this equation is called the *Stieltjes convolution* of F and G, and is sometimes denoted by $F * G$.

Equation (11.5.3) takes a special form if the variables X and Y are continuous. Thus suppose that X has a probability density f and Y has a probability density g. Then

$$\begin{aligned} H(z) &= \int_{-\infty}^{\infty} F(z-y)\, g(y)\, dy \\ &= \int_{-\infty}^{\infty} \left\{ \int_{-\infty}^{z} f(\zeta - y)\, d\zeta \right\} g(y)\, dy \\ &= \int_{-\infty}^{z} d\zeta \int_{-\infty}^{\infty} f(\zeta - y)\, g(y)\, dy. \end{aligned}$$

Hence, by Fubini's theorem,

$$h(z) = \int_{-\infty}^{\infty} f(z-y)\,g(y)\,dy \tag{11.5.4}$$

exists for almost all z, and

$$H(z) = \int_{-\infty}^{z} h(\zeta)\,d\zeta.$$

Therefore Z is a continuous random variable, having a probability density h given by (11.5.4).

It is possible in an exactly similar way to compute the distributions of differences, products and quotients of independent random variables in terms of their distributions. Some examples will be found in the exercises at the end of this section, but there is one consequence of theorem 11.5 which is sufficiently important to have the status of a theorem.

***Theorem* 11.6.** *If X and Y are independent random variables with finite means, then*

$$\mathsf{E}(XY) = \mathsf{E}(X)\,\mathsf{E}(Y). \tag{11.5.5}$$

Proof. From (11.5.2), if X and Y have distribution functions F and G, respectively,

$$\begin{aligned}\mathsf{E}(XY) &= \int_{-\infty}^{\infty}\int_{-\infty}^{\infty} xy\,dF(x)\,dG(y)\\ &= \int_{-\infty}^{\infty} x\,dF(x)\int_{-\infty}^{x} y\,dG(y)\\ &= \mathsf{E}(X)\,\mathsf{E}(Y).\,\rbrack\end{aligned}$$

This theorem is important in itself, and it also leads to another result of great importance.

***Theorem* 11.7.** *If X and Y are independent random variables with $\mathsf{E}(X^2) < \infty$, $\mathsf{E}(Y^2) < \infty$, then*

$$\operatorname{var}(X+Y) = \operatorname{var}(X) + \operatorname{var}(Y). \tag{11.5.6}$$

Proof. Let $\xi = \mathsf{E}(X), \eta = \mathsf{E}(Y)$, $X' = X-\xi$, $Y' = Y-\eta$. Then X' and Y' are independent, and since $\mathsf{E}(X+Y) = \xi+\eta$,

$$\begin{aligned}\operatorname{var}(X+Y) &= \mathsf{E}\{(X+Y-\xi-\eta)^2\}\\ &= \mathsf{E}\{(X'+Y')^2\}\\ &= \mathsf{E}\{X'^2+Y'^2+2X'Y'\}\\ &= \mathsf{E}(X'^2)+\mathsf{E}(Y'^2)+2\mathsf{E}(X'Y').\end{aligned}$$

But $$\mathsf{E}(X'Y') = \mathsf{E}(X')\,\mathsf{E}(Y') = 0,$$

and so $$\operatorname{var}(X+Y) = \mathsf{E}\{(X-\xi)^2\}+\mathsf{E}\{(Y-\eta)^2\}$$
$$= \operatorname{var}(X)+\operatorname{var}(Y).]$$

Equation (11.5.6) is very reminiscent of the formula (11.2.3)

$$\mathsf{E}(X+Y) = \mathsf{E}(X)+\mathsf{E}(Y).$$

It is therefore important to bear in mind the crucial difference between them; (11.2.3) holds for any pair of random variables (with finite mean), but (11.5.6) follows from the strong assumption of independence.

Exercises 11.5

1. Let X and Y be independent random variables whose values are non-negative integers, and write

$$a_j = \mathsf{P}(X=j), \quad b_j = \mathsf{P}(Y=j).$$

If $Z = X+Y$, show that $\mathsf{P}(Z=n) = \sum_{j=0}^{n} a_j b_{n-j}$.

Verify (11.5.5) directly for this case.

2. Let X and Y be independent with

$$\left.\begin{aligned} \mathsf{P}(X=j) &= e^{-\alpha}\alpha^j/j! \\ \text{and} \quad \mathsf{P}(Y=j) &= e^{-\beta}\beta^j/j! \end{aligned}\right\} \quad (j=0,1,2,\ldots;\alpha,\beta>0).$$

Show that $$\mathsf{P}(X+Y=j) = e^{-\gamma}\gamma^j/j!,$$
where $\gamma = \alpha+\beta$.

3. X and Y are independent random variables with probability densities f and g, respectively. Show that $Z = XY$ has a probability density

$$h(z) = \int_{-\infty}^{\infty} f(y^{-1}z)\,g(y)\,|y|^{-1}\,dy.$$

What is the corresponding result for X/Y?

4. If $X_1, X_2, \ldots, X_n$ are independent, and X_j has distribution function F_j, show that

$$X^* = \max(X_1, X_2, \ldots, X_n)$$

has distribution function $$F^*(x) = \prod_{j=1}^{n} F_j(x),$$

and that $$X_* = \min(X_1, X_2, \ldots, X_n)$$

has distribution function

$$F_*(x) = 1-\prod_{j=1}^{n}[1-F_j(x)].$$

5. If, in the last example,

$$F_j(x) = 0 \quad (x < 0)$$
$$= x \quad (0 \leqslant x \leqslant 1)$$
$$= 1 \quad (x > 1)$$

for all j, show that the *range* $R = X^* - X_*$ has probability density

$$f(r) = n(n-1)\, r^{n-2}(1-r) \quad (0 \leqslant r \leqslant 1)$$
$$= 0 \quad \text{otherwise.}$$

Deduce the values of $\mathsf{E}(R)$ and var (R).

6. X is a random variable with $0 < \text{var}(X) < \infty$, a and b are numbers with $a \neq 0$, and $Y = aX + b$. Show that

$$\mathsf{E}(XY) \neq \mathsf{E}(X)\,\mathsf{E}(Y).$$

7. $X_1, X_2, \ldots, X_n$ are independent, and, for each j,

$$\mathsf{E}(X_j) = \mu, \quad \text{var}(X_j) = \sigma^2,$$

where $0 < \sigma < \infty$. Show that

$$Z = \frac{X_1 + X_2 + \ldots + X_n - n\mu}{\sigma n^{\frac{1}{2}}}$$

has

$$\mathsf{E}(Z) = 0, \quad \text{var}(Z) = 1.$$

8. Let X and Y be independent with distribution functions F and G, and let $X + Y$ have distribution function H. Show that

(i) F continuous $\Rightarrow$ H continuous,

(ii) F absolutely continuous $\Rightarrow$ H absolutely continuous. Show that neither of these results need hold if independence is not assumed.

11.6 Discrete distributions

A discrete random variable X has been defined as one which (almost surely) takes values in a countable set $\{\xi_1, \xi_2, \ldots\}$. Its distribution function

$$F(x) = \sum_{\{j;\, \xi_j \leqslant x\}} p_j$$

is determined by the set $\{\xi_j\}$ and by the numbers

$$p_j = \mathsf{P}(X = \xi_j),$$

which must satisfy $$p_j \geqslant 0, \quad \sum_{j=1}^{\infty} p_j = 1.$$

An example of a discrete random variable appears in § 10.5. Suppose that we have n independent trials (such as tosses of a coin), each one of which results in a 'success' (a head, say) with probability p, and a 'failure' (tail) with probability $1-p$. Consider the random variable X defined as the number of successes in the n trials. Clearly X takes on values from the finite set $\{0, 1, 2, \ldots, n\}$. The probabilities of these values are given by (10.5.5).

$$\mathsf{P}(X = r) = \binom{n}{r} p^r (1-p)^{n-r} = b(n, p; r).$$

The distribution function F of X is

$$F(x) = \sum_{r \leqslant x} \binom{n}{r} p^r (1-p)^{n-r}. \tag{11.6.1}$$

A random variable having this distribution function is said to have the *binomial distribution* $B(n, p)$ with parameters n and p, or more concisely to be a $B(n, p)$ random variable.

We have seen that the expectation of a random variable is determined by its distribution function. Let X have the binomial distribution $B(n, p)$. Then, from theorem 11.4,

$$\begin{aligned} \mathsf{E}(X) &= \int_{\mathbf{R}} x\, dF(x) \\ &= \sum_{r=0}^{n} r \binom{n}{r} p^r (1-p)^{n-r} \\ &= np \sum_{r=1}^{n} \binom{n-1}{r-1} p^{r-1} (1-p)^{n-r} \\ &= np, \end{aligned}$$

by the binomial theorem.

Thus the expectation of any $B(n,p)$ random variable is np. The usual way of expressing this fact is to say that *the mean of the distribution $B(n,p)$ is np*, the word 'mean' being approximately synonymous with 'expectation'.

Another way of computing the mean of the binomial distribution which is instructive is as follows. Consider the n trials, and, for each j, define a random variable Y_j to be equal to 1 if the jth trial is a success and equal to 0 if it is a failure. Then

$$\mathsf{P}(Y_j = 0) = 1-p, \quad \mathsf{P}(Y_j = 1) = p.$$

The total number X of successes is clearly

$$X = Y_1 + Y_2 + \ldots + Y_n.$$

Now
$$\mathsf{E}(Y_j) = (1-p).0 + p.1 = p,$$

and so
$$\mathsf{E}(X) = \mathsf{E}(Y_1) + \mathsf{E}(Y_2) + \ldots + \mathsf{E}(Y_n) = np.$$

This method also yields the variance of X, because the Y_j, being defined on different trials, are independent. It follows that

$$\operatorname{var}(X) = \operatorname{var}(Y_1) + \operatorname{var}(Y_2) + \ldots + \operatorname{var}(Y_n),$$

but, since $Y_j^2 = Y_j$,

$$\begin{aligned}\operatorname{var}(Y_j) &= \mathsf{E}(Y_j^2) - (\mathsf{E}Y_j)^2 \quad \text{(exercise 11.2.(9))}\\ &= \mathsf{E}(Y_j) - (\mathsf{E}Y_j)^2\\ &= p - p^2 = p(1-p),\end{aligned}$$

we have
$$\operatorname{var}(X) = np(1-p).$$

Thus we can make the following assertion.

The binomial distribution $B(n,p)$ has mean np and variance $np(1-p)$

A number of interesting distributions arise from the binomial distribution. Of these the most important is the Poisson distribution, which can be derived in the following way. Suppose that we have a large number n of independent trials, but the probability p of success is very small, in such a way that the expectation $\mu = np$ of the number of successes is moderate. In other words, keep $\mu = np$ fixed and let n tend to infinity. Then the probability of r successes is

$$\begin{aligned}b(n,p;\, r) &= b(n, \mu/n;\, r)\\ &= \binom{n}{r}\left(\frac{\mu}{n}\right)^r\left(1-\frac{\mu}{n}\right)^{n-r}\\ &= \frac{\mu^r}{r!}\,\frac{n(n-1)\ldots(n-r+1)}{(n-\mu)^r}\left(1-\frac{\mu}{n}\right)^n.\end{aligned}$$

Hence, as $n \to \infty$, $\qquad b(n, p: r) \to \mu^r e^{-\mu}/r!.$

The numbers $\quad p(\mu; r) = \mu^r e^{-\mu}/r! \quad (r = 0, 1, 2, \ldots) \qquad (11.6.2)$

form a probability distribution over the non-negative integers so long as $\mu \geqslant 0$, since

$$p(\mu; r) \geqslant 0$$

and

$$\sum_{r=0}^{\infty} p(\mu; r) = e^{-\mu} \sum_{r=0}^{\infty} \frac{\mu^r}{r!} = e^{-\mu} . e^{\mu} = 1.$$

A random variable with the corresponding distribution function

$$F(x) = \sum_{r \leqslant x} \mu^r e^{-\mu}/r!$$

is said to have the *Poisson distribution* $P(\mu)$. The limiting result above shows that, for large n, the distribution $B(n, \mu/n)$ approximates to $P(\mu)$. We shall see later, however, that the Poission distribution is important in its own right, and not merely as an approximation to the binomial distribution.

If X is a $P(\mu)$ random variable, then

$$\begin{aligned} \mathsf{E}(X) &= \sum_{r=0}^{\infty} r p(\mu; r) \\ &= e^{-\mu} \sum_{r=1}^{\infty} \frac{\mu^r}{(r-1)!} \\ &= e^{-\mu} . \mu e^{\mu} \\ &= \mu, \end{aligned}$$

so that $P(\mu)$ has mean μ. Moreover,

$$\begin{aligned} \mathsf{E}\{X(X-1)\} &= e^{-\mu} \sum_{r=2}^{\infty} r(r-1) \frac{\mu^r}{r!} \\ &= e^{-\mu} \sum_{r=2}^{\infty} \frac{\mu^r}{(r-2)!} \\ &= e^{-\mu} \mu^2 e^{\mu} \\ &= \mu^2, \end{aligned}$$

so that

$$\begin{aligned} \operatorname{var}(X) &= \mathsf{E}\{X(X-1) + X\} - (\mathsf{E}X)^2 \\ &= \mu^2 + \mu - \mu^2 \\ &= \mu. \end{aligned}$$

We can therefore assert that

The Poisson distribution $P(\mu)$ has mean μ and variance μ

Now consider an infinite sequence of independent trials (as defined in §10.5) with constant probability $p(0 < p < 1)$ of success. Let X_n be the number of successes in the first n trials. Then, as above, X_n is a $B(n, p)$ random variable. Suppose that we fix an integer k, and continue taking trials until exactly k successes have been recorded, and let Z_k be the number of trials which this requires. In other words,

$$Z_k = \min\{n;\ X_n = k\}.$$

Then Z_k is a random variable, since

$$\{Z_k \leqslant n\} = \{X_n \geqslant k\}.$$

It takes on the values $k, k+1, k+2, \ldots, +\infty$, the last value corresponding to the possibility that $X_n < k$ for all n. For $r = 0, 1, 2, \ldots$,

$$\begin{aligned} \mathsf{P}(Z_k = k+r) &= \mathsf{P}(X_{k+r-1} = k-1,\ Y_{k+r} = 1) \\ &= \mathsf{P}(X_{k+r-1} = k-1)\,\mathsf{P}(Y_{k+r} = 1) \\ &= b(k+r-1, p;\ k-1)\,p \\ &= \binom{k+r-1}{k-1} p^k (1-p)^r. \end{aligned}$$

Hence $$\mathsf{P}(Z_k = k+r) = \binom{k+r-1}{k-1} p^k (1-p)^r \quad (r = 0, 1, 2, \ldots). \tag{11.6.3}$$

Notice that the right-hand side of (11.6.3) is equal to

$$\frac{(k+r-1)(k+r-2)\ldots k}{r!} p^k (1-p)^r = \binom{-k}{r} p^k (p-1)^r;$$

because of the occurrence of $\binom{-k}{r}$ Z_k is said to have a *negative binomial distribution*. Notice also that

$$\begin{aligned} \sum_{r=0}^{\infty} \mathsf{P}(Z_k = k+r) &= p^k \sum_{r=0}^{\infty} \binom{-k}{r} (p-1)^r \\ &= p^k (1+p-1)^{-k} = 1, \end{aligned}$$

so that $$\mathsf{P}(Z_k = \infty) = 0.$$

Thus the number of trials needed for k successes is almost certainly finite.

A simple consequence of this last fact is that, for any k,

$$\mathsf{P}(X_n < k \quad \text{for all } n) = 0.$$

Hence, since X_n increases with n,

$$\mathsf{P}(X_n \to \infty \quad \text{as} \quad n \to \infty) = 1.$$

Much more precise results concerning the behaviour of X_n as $n \to \infty$ will be proved later (§ 13.3).

An important special case of the negative binomial distribution occurs when $k = 1$. We then have

$$\mathsf{P}(Z_1 = 1+r) = p(1-p)^r = (1-q)q^r,$$

if $q = 1-p$. A random variable W with

$$\mathsf{P}(W = r) = (1-q)q^r \quad (r = 0, 1, 2, \ldots) \tag{11.6.4}$$

(where $0 \leqslant q \leqslant 1$) is said to have the *geometric distribution* $G(q)$.

Finally, a trivial distribution which nevertheless deserves a name; if X is a random variable such that, for some x,

$$\mathsf{P}(X = x) = 1,$$

it is said to have the degenerate distribution $D(x)$.

More details of these and other discrete distributions, and of the continuous distributions to be studied in the next section, may be found in M. G. Kendall and A. Stuart, *The Advanced Theory of Statistics*, Vol. I (Griffin, 1958).

Exercises 11.6

1. Obtain the variance of $B(n,p)$ directly from (11.6.1).

2. If $F_{n,p}$ denotes the distribution function of $B(n,p)$, and x is an integer show that

$$\frac{\partial F_{n,p}(x)}{\partial p} = -n\binom{n-1}{x} p^x(1-p)^{n-1-x}.$$

Deduce that, if x is an integer between 0 and n,

$$F_{n,p}(x) = 1 - \int_0^p n\binom{n-1}{x} \alpha^x(1-\alpha)^{n-1-x}\,d\alpha.$$

3. Show that the distribution function F_μ of $P(\mu)$ satisfies

$$F_\mu(x) = 1 - \frac{1}{x!}\int_0^\mu \beta^x e^{-\beta}\,d\beta,$$

when x is a non-negative integer.

4. Let X, Y be independent, with distributions $P(\alpha)$, $P(\beta)$ respectively $(\alpha, \beta > 0)$. Show that, if r, n are integers with $0 \leqslant r \leqslant n$,

$$\mathsf{P}(X = r \mid X + Y = n) = b(n, p; r),$$

where $p = \alpha/(\alpha+\beta)$.

5. Show that the distribution $G(q)$ has mean $q/(1-q)$ and variance $q/(1-q)^2$. Find the mean and variance of the negative binomial distribution. (The quickest way of doing the second part is to deduce the result from that of the first part—if you can.)

11.7 Continuous distributions

The distribution function of a continuous random variable X is determined by its probability density f, which must satisfy

$$f \geqslant 0, \quad \int_{-\infty}^{\infty} f\,dx = 1.$$

Conversely any function satisfying these two conditions defines a distribution function by

$$F(x) = \int_{-\infty}^{x} f(y)\,dy.$$

By far the most important probability density is that given by

$$f(x) = (2\pi)^{-\frac{1}{2}}\, e^{-\frac{1}{2}x^2}. \tag{11.7.1}$$

That this is a probability density is a consequence of its obvious positivity and of the well known formula

$$\int_{-\infty}^{\infty} e^{-\frac{1}{2}x^2}\,dx = (2\pi)^{\frac{1}{2}}, \tag{11.7.2}$$

a proof of which is sketched at the end of this section. A random variable with probability density (11.7.1) is said to have the *standard normal distribution* (sometimes called the 'Gaussian' distribution), which, for a reason which will become clear later, is denoted by $N(0,1)$.

The probability density (11.7.1) first arose as an approximation to the binomial distribution. Let X_n have the distribution $B(n,p)$, for some p in $0 < p < 1$ which will be kept fixed (unlike the case of the Poisson approximation, in which p converged to zero). Then

$$\mathsf{E}(X_n) = np, \quad \operatorname{var}(X_n) = np(1-p).$$

It follows that, if we write

$$\sigma(n) = \{np(1-p)\}^{\frac{1}{2}},$$

the random variable $\quad X_n^* = (X_n - np)/\sigma(n)$

has $\quad \mathsf{E}(X_n^*) = 0, \quad \operatorname{var}(X_n^*) = 1.$

X_n^* is a discrete random variable, taking on the values

$$(r - np)/\sigma(n) \quad (r = 0, 1, 2, \ldots, n).$$

Its distribution function is

$$F_n(x) = \mathsf{P}(X_n^* \leqslant x)$$
$$= \mathsf{P}(X_n \leqslant np + \sigma(n)\, x)$$
$$= \sum_{r \leqslant np+\sigma(n)x} b(n, p;\ r).$$

When n is at all large, this sum is difficult to compute, but we saw in §10.5 that it could be approximated for large values of n. In fact, equation (10.5.7) shows that, if $a < b$,

$$F_n(b) - F_n(a) = \sum_{np+\sigma(n)\, a < r \leqslant np+\sigma(n)\, b} b(n, p;\ r)$$
$$\to (2\pi)^{-\frac{1}{2}} \int_a^b e^{-\frac{1}{2}x^2}\, dx$$

as $n \to \infty$. Hence, if $F(x) = (2\pi)^{-\frac{1}{2}} \int_{-\infty}^{x} e^{-\frac{1}{2}y^2}\, dy$

denotes the distribution function of $N(0, 1)$, we have

$$\lim_{n\to\infty} [F_n(b) - F_n(a)] = F(b) - F(a).$$

Now write $$\hat{F}(x) = \limsup_{n\to\infty} F_n(x).$$

Then $$\hat{F}(b) = \limsup_{n\to\infty} \{F_n(a) + [F_n(b) - F_n(a)]\}$$
$$= \hat{F}(a) + [F(b) - F(a)],$$

so that, for some c, and all x,

$$\hat{F}(x) = c + F(x).$$

Letting $x \to -\infty$ in $$0 \leqslant \hat{F}(x) = c + F(x)$$

gives $$0 \leqslant c.$$

Letting $x \to +\infty$ in $$1 \geqslant \hat{F}(x) = c + F(x)$$

gives $$1 \geqslant c + 1,$$

so that $c = 0$. Thus $$\limsup_{n\to\infty} F_n(x) = F(x).$$

Similarly $$\liminf_{n\to\infty} F_n(x) = F(x),$$

so that we have proved that

$$\lim_{n\to\infty} F_n(x) = (2\pi)^{-\frac{1}{2}} \int_{-\infty}^{x} e^{-\frac{1}{2}y^2}\, dy. \tag{11.7.3}$$

We express (11.7.3) by saying that, as $n \to \infty$, X_n^* has asymptotically the distribution $N(0,1)$. Such a statement will be written symbolically as

$$X_n^* \rightsquigarrow N(0,1) \quad (n \to \infty). \tag{11.7.4}$$

At first sight it may seem surprising that the distribution function of the discrete random variable X_n^* can be approximated by an absolutely continuous function. However, the possible values of X_n^* (i.e. the jumps of F_n) extend from

$$-n^{\frac{1}{2}}\left(\frac{p}{1-p}\right)^{\frac{1}{2}} \quad \text{to} \quad n^{\frac{1}{2}}\left(\frac{1-p}{p}\right)^{\frac{1}{2}}$$

and are evenly spaced a distance

$$n^{-\frac{1}{2}}[p(1-p)]^{\frac{1}{2}}$$

apart, so that when n is large they are fairly densely scattered. Thus F_n has a large number of small jumps, and as n becomes large approximates to the smooth curve F.

Convergence of distribution functions will be studied in more detail in Chapter 12. In Chapter 13 we shall see that (11.7.4) is but a special case of much more general theorems, which indicate that many types of random variable might be expected to approximate in distribution to the normal form $N(0,1)$.

If X is a random variable with distribution $N(0,1)$, and if k is a non-negative integer, then

$$\mathsf{E}(X^k) = (2\pi)^{-\frac{1}{2}} \int_{-\infty}^{\infty} x^k e^{-\frac{1}{2}x^2}\,dx$$

exists. By symmetry $\mathsf{E}(X^k) = 0$ for k odd, while for k even $\mathsf{E}(X^k)$ can be turned, by means of the substitution $y = \frac{1}{2}x^2$, into a gamma-function integral. Alternatively, one can proceed by integration by parts. For example, with $k = 2$,

$$\begin{aligned}
\mathsf{E}(X^2) &= (2\pi)^{-\frac{1}{2}} \int_{-\infty}^{\infty} x\,.\,x e^{-\frac{1}{2}x^2}\,dx \\
&= (2\pi)^{-\frac{1}{2}}\left\{[-x e^{-\frac{1}{2}x^2}]_{-\infty}^{\infty} + \int_{-\infty}^{\infty} e^{-\frac{1}{2}x^2}\,dx\right\} \\
&= (2\pi)^{-\frac{1}{2}}\{0 + (2\pi)^{\frac{1}{2}}\} = 1.
\end{aligned}$$

It follows that $$\mathsf{E}(X) = 0, \quad \operatorname{var}(X) = 1;$$

The standard normal distribution $N(0,1)$ has mean 0 and variance 1

This result is of course to be expected from (11.7.4), since X_n^* has mean 0 and variance 1.

Let X be an $N(0,1)$ random variable, and let μ, σ be real numbers with $\sigma > 0$. Then

$$Y = \mu + \sigma X$$

is a random variable with

$$\mathsf{E}(Y) = \mu, \quad \operatorname{var}(X) = \sigma^2.$$

The distribution function of Y is easily calculated;

$$\begin{aligned} \mathsf{P}(Y \leqslant y) &= \mathsf{P}(\mu + \sigma X \leqslant y) \\ &= \mathsf{P}(X \leqslant (y-\mu)/\sigma) \\ &= (2\pi)^{-\frac{1}{2}} \int_{-\infty}^{(y-\mu)/\sigma} e^{-\frac{1}{2}x^2}\, dx. \end{aligned}$$

Making the substitution $\eta = \mu + \sigma x$,

$$\mathsf{P}(Y \leqslant y) = \sigma^{-1}(2\pi)^{-\frac{1}{2}} \int_{-\infty}^{y} \exp\left\{-\frac{(\eta-\mu)^2}{2\sigma^2}\right\} d\eta.$$

Hence Y has the probability density

$$\sigma^{-1}(2\pi)^{-\frac{1}{2}} \exp\left\{-\frac{(y-\mu)^2}{2\sigma^2}\right\}. \qquad (11.7.5)$$

Any random variable having this probability density is said to have the *normal distribution $N(\mu, \sigma^2)$ with mean μ and variance σ^2.*

Although the normal distribution is the most important continuous distribution, it is by no means the only one. Indeed, any non-negative function f with integral 1 will determine such a distribution. We list below some examples which frequently occur.

(i) *The uniform distribution $U(a, b)$*

For real numbers $a < b$, this has density

$$\left.\begin{aligned} f(x) &= (b-a)^{-1} \quad \text{if} \quad a \leqslant x \leqslant b, \\ &= 0 \qquad\qquad \text{otherwise.} \end{aligned}\right\} \qquad (11.7.6)$$

Notice that a random variable X with this distribution satisfies

$$\mathsf{P}(a < X < b) = 1.$$

(ii) *The exponential distribution* $E(\lambda)$

For $\lambda > 0$ this has density

$$\left.\begin{aligned} f(x) &= \lambda e^{-\lambda x} \quad (x \geqslant 0) \\ &= 0 \qquad\quad (x < 0). \end{aligned}\right\} \tag{11.7.7}$$

(iii) *The Cauchy distribution*

This has density $$f(x) = [\pi(1+x^2)]^{-1}, \tag{11.7.8}$$

and its main use is in providing counter-examples.

Proof of (11.7.2): consider the integral

$$I = \int_{-\infty}^{\infty}\int_{-\infty}^{\infty} e^{-\frac{1}{2}(x^2+y^2)}\,dx\,dy.$$

If we make the substitution $x = r\cos\theta$, $y = r\sin\theta$ we get

$$\begin{aligned} I &= \int_0^{\infty}\int_0^{2\pi} e^{-\frac{1}{2}r^2}\,r\,dr\,d\theta \\ &= 2\pi\int_0^{\infty} r e^{-\frac{1}{2}r^2}\,dr \\ &= 2\pi[-e^{-\frac{1}{2}r^2}]_0^{\infty} = 2\pi, \end{aligned}$$

so that $$\begin{aligned} 2\pi &= \int_{-\infty}^{\infty} e^{-\frac{1}{2}x^2}\,dx \int_{-\infty}^{\infty} e^{-\frac{1}{2}y^2}\,dy \\ &= \left\{\int_{-\infty}^{\infty} e^{-\frac{1}{2}x^2}\,dx\right\}^2. \end{aligned}$$

Hence $$\int_{-\infty}^{\infty} e^{-\frac{1}{2}x^2}\,dx = (2\pi)^{\frac{1}{2}}.$$

Exercises 11.7

1. The random variable X_μ has distribution $P(\mu)$. Show that

$$\mu^{-\frac{1}{2}}(X_\mu - \mu) \rightsquigarrow N(0,1)$$

as $\mu \to \infty$.

2. Use integration by parts to derive a recurrence relation for the moments $\mathsf{E}(X^{2k})$ of an $N(0,1)$ random variable X, and deduce that

$$\mathsf{E}(X^n) = (n-1)(n-3)\ldots 3.1$$

whenever n is an even integer.

3. Find the means and variances of $U(a,b)$ and $E(\lambda)$. Show that the mean of the Cauchy distribution does not exist.

4. X and Y are independent $N(0,1)$ random variables. Show that X/Y has a Cauchy distribution.

5. X has distribution $U(0, 2\pi)$. Show that $Y = \tan X$ has a Cauchy distribution.

6. X and Y are independent, with distributions $N(\mu_1, \sigma_1^2)$ and $N(\mu_2, \sigma_2^2)$ respectively. Show that $X+Y$ has distribution $N(\mu_1+\mu_2, \sigma_1^2+\sigma_2^2)$.

7. A distribution is said to be *stable with exponent* α if, whenever X and Y are independent and have this distribution, $\beta(X+Y)$ has the same distribution, where $\beta = 2^{-1/\alpha}$. Prove that

(i) $N(0,1)$ is stable with $\alpha = 2$,

(ii) the Cauchy distribution is stable with $\alpha = 1$.

8. Show that, if n is a non-negative integer,

$$\begin{aligned} f_n(x) &= x^n e^{-x}/n! \quad (x \geqslant 0) \\ &= 0 \quad\quad\quad\;\; (x < 0) \end{aligned}$$

is a probability density. Show that the corresponding distribution (called the gamma distribution) has mean $(n+1)$ and variance $(n+1)$.

9. Let Φ denote the distribution function of $N(0,1)$. Show that, if

$$I_k = \int_x^\infty y^{-2k} e^{-\frac{1}{2}y^2}\, dy \quad (x > 0),$$

then
$$I_k + (2k+1)\, I_{k+1} = x^{-2k-1} e^{-\frac{1}{2}x^2},$$

and deduce that

$$1 - \Phi(x) = (2\pi)^{-\frac{1}{2}} \left[\left(\frac{1}{x} - \frac{1}{x^3} + \frac{3}{x^5} - \ldots - \frac{(-1)^k\, 1.3.5.\ldots.(2k-3)}{x^{2k-1}} \right) e^{-\frac{1}{2}x^2} + (-1)^k\, 1.3.5.\ldots.(2k-1) \int_x^\infty y^{-2k} e^{-\frac{1}{2}y^2} dy \right].$$

Show in particular that

$$(2\pi)^{-\frac{1}{2}} \left(\frac{1}{x} - \frac{1}{x^3} \right) e^{-\frac{1}{2}x^2} \leqslant 1 - \Phi(x) \leqslant (2\pi)^{-\frac{1}{2}} \frac{1}{x} e^{-\frac{1}{2}x^2} \quad (x > 0).$$

11.8 Convergence of random variables

Random variables have been defined as measurable functions from the underlying probability space $(\Omega, \mathscr{F}, \mathsf{P})$ into $(\mathbf{R}^*, \mathscr{B}^*)$. As measurable functions they have associated with them the modes of convergence discussed in Chapter 6. Three of these are of particular importance in probability theory†.

† Notice that the symbol $\rightsquigarrow$ used earlier in this chapter does not refer to convergence of random variables but to convergence of distributions of random variables.

(i) A sequence $\{X_n\}$ of random variables, regarded as a sequence of measurable functions, converges almost everywhere to a random variable X if

$$X_n(\omega) \to X(\omega)$$

for all ω except possibly for those lying in a set of P-measure zero. Probabilists generally use the terms 'convergence with probability one' or 'almost sure convergence' for this mode of convergence. Thus $X_n \to X$ almost surely if

$$\mathsf{P}\{\omega; \lim_{n\to\infty} X_n(\omega) = X(\omega)\} = 1.$$

(ii) A sequence $\{X_n\}$ of (almost certainly finite) random variables converges in measure to an (almost certainly finite) random variable X if, for any $\epsilon > 0$,

$$\mathsf{P}\{\omega; |X_n(\omega) - X(\omega)| > \epsilon\} \to 0$$

as $n \to \infty$. It is usual to say that X_n converges to X in probability; symbolically

$$\operatorname{p}\lim_{n\to\infty} X_n = X.$$

For finite random variables this is weaker than convergence with probability one, as was proved in §7.2.

(iii) A sequence $\{X_n\}$ converges in $\mathscr{L}_p$ to X if

$$\int_\Omega |X_n - X|^p \, d\mathsf{P} \to 0$$

as $n \to \infty$, i.e. if

$$\mathsf{E}\{|X_n - X|^p\} \to 0.$$

The case of importance in probability theory is $p = 2$; if

$$\mathsf{E}\{(X_n - X)^2\} \to 0$$

as $n \to \infty$ we say that X_n converges to X in mean square. This mode of convergence is stronger than (ii).

It will be seen that many of the elementary concepts in probability theory are just the same as concepts in measure theory, but are given different names. Indeed, we can set up a 'dictionary' as below, with measure-theoretic terms on the left, and the equivalent probability-theoretic terms on the right. (Notice that the words 'certain' and 'sure' are used interchangeably.)

Measurable set	*Event*
Empty set	*Impossible event*
Whole space	*Certain event*
Measure of a set	*Probability of an event*
Set of measure zero	*Event of probability zero*

Set of measure one	*Almost certain event*
Measurable function	*Random variable*
Integral	*Expectation*
Convergence a.e.	*Almost sure convergence*
Convergence in measure	*Convergence in probability*
$\mathscr{L}_2$ *convergence*	*Mean square convergence*

Exercises 11.8

1. Give examples to show that mean square convergence neither implies nor is implied by convergence with probability one.

2. If c is a constant, show that $X_n \to c$ in mean square if and only if

$$\mathsf{E}(X_n) \to c, \quad \operatorname{var}(X_n) \to 0.$$

3. If $X_1, X_2, \ldots$ are independent, each with mean μ and variance σ^2 $(0 \leqslant \sigma < \infty)$ show that, as $n \to \infty$,

$$\frac{X_1 + X_2 + \ldots + X_n}{n} \to \mu$$

in mean square.

12

CHARACTERISTIC FUNCTIONS

12.1 The space of distribution functions

A distribution function on $\mathbf{R}$ has been defined as a non-decreasing function, continuous from the right, and having limits 0 at $-\infty$ and 1 at $+\infty$. The class of all distribution functions will be denoted by $\mathscr{D}$, and this section will be devoted to a study of $\mathscr{D}$, and in particular to a discussion of the notion of convergence in $\mathscr{D}$.

It was shown in §11.7 that, if X_n is a random variable with distribution $B(n,p)$, and if

$$X_n^* = (X_n - np)/\{np(1-p)\}^{\frac{1}{2}},$$

then

$$X_n^* \rightsquigarrow N(0,1)$$

in the sense that, if F_n is the distribution function of X_n^* and F the distribution function corresponding to $N(0,1)$, then

$$\lim_{n\to\infty} F_n(x) = F(x)$$

for all x. This result suggests that an appropriate definition of convergence in $\mathscr{D}$ is pointwise convergence. This, however, is not quite right, as the following example shows.

For any real ξ, let D_ξ be the distribution function of the trivial random variable ξ, i.e.

$$\begin{aligned} D_\xi(x) &= 1 \quad (x \geqslant \xi) \\ &= 0 \quad (x < \xi). \end{aligned}$$

Then, if $\{\xi(n)\}$ is a sequence of numbers approaching 0 *from below*,

$$D_{\xi(n)}(x) \to D_0(x)$$

for all x. However, if $\{\xi(n)\}$ approaches 0 *from above*,

$$D_{\xi(n)}(x) \to D_0(x)$$

except at $x = 0$, where

$$D_{\xi(n)}(0) \to 0 \neq 1 = D_0(0).$$

The trouble arises from the discontinuity of D_0 at $x = 0$, which suggests that it is too exacting to require convergence at points at which the limiting distribution function is discontinuous. In fact, the following definition turns out to be the most useful.

A sequence $\{F_n\}$ of distribution functions is said to converge to a distribution function F if

$$F_n(x) \to F(x)$$

at all points of continuity x of F.

By a point of continuity of F is meant, of course, a point at which F is continuous. Since F has at most countably many discontinuities, points of continuity exist in profusion.

The definition of convergence in $\mathcal{D}$ has a number of useful alternative forms. Of these the one given by the following theorem will be important in what follows.

***Theorem* 12.1.** *Let F_n $(n = 1, 2, \ldots)$, F belong to $\mathcal{D}$. Then $F_n \to F$ if and only if, for every bounded continuous function $\phi: \mathbf{R} \to \mathbf{R}$,*

$$\lim_{n\to\infty} \int_{\mathbf{R}} \phi(x)\, dF_n(x) = \int_{\mathbf{R}} \phi(x)\, dF(x). \tag{12.1.1}$$

Proof. (i) Suppose that $F_n \to F$, and let D be the set of points at which F is discontinuous. Since D is countable, its complement is everywhere dense in $\mathbf{R}$. Let $\phi: \mathbf{R} \to \mathbf{R}$ be bounded and continuous, with

$$m = \sup |\phi(x)|,$$

and let ϵ be any positive number. Then we can choose a, b such that

$$a < b,$$
$$a \notin D, \quad F(a) < \epsilon/8m,$$
$$b \notin D, \quad F(b) > 1 - \epsilon/8m.$$

Since
$$F_n(a) \to F(a), \quad F_n(b) \to F(b),$$

there exists an integer n_1 such that, for all $n \geqslant n_1$,

$$F_n(a) < \epsilon/8m,$$
$$F_n(b) > 1 - \epsilon/8m.$$

It then follows that

$$\left|\int_{(b,\infty)} \phi\, dF_n - \int_{(b,\infty)} \phi\, dF\right| \leqslant m \int_{(b,\infty)} dF_n + m \int_{(b,\infty)} dF < \tfrac{1}{4}\epsilon,$$

and that
$$\left|\int_{(-\infty,a]} \phi\, dF_n - \int_{(-\infty,a]} \phi\, dF\right| < \tfrac{1}{4}\epsilon,$$

so that
$$\left|\int_{\mathbf{R}} \phi\, dF_n - \int_{\mathbf{R}} \phi\, dF\right| \leqslant \left|\int_{(a,b]} \phi\, d(F_n - F)\right| + \tfrac{1}{2}\epsilon,$$

for all $n \geqslant n_1$.

On the interval $(a, b]$, ϕ is uniformly continuous, and hence there exists $\delta > 0$ such that

$$x, y \in (a, b], \quad |x-y| < \delta \Rightarrow |\phi(x) - \phi(y)| < \tfrac{1}{8}\epsilon.$$

We may choose numbers

$$a = x_0 < x_1 < x_2 < \ldots < x_m = b$$

such that, for $1 \leqslant j \leqslant m$,

$$x_j \notin D, \quad |x_j - x_{j-1}| < \delta.$$

If a function $\psi: (a, b] \to \mathbf{R}$ is defined by

$$\psi(x) = \phi(x_j) \quad (x_{j-1} < x \leqslant x_j),$$

then

$$|\psi(x) - \phi(x)| < \tfrac{1}{8}\epsilon,$$

and so

$$\left|\int_{(a,b]} \phi\, d(F_n - F)\right| < \left|\int_{(a,b]} \psi\, d(F_n - F)\right| + \tfrac{1}{4}\epsilon.$$

Therefore, for $n \geqslant n_1$,

$$\left|\int_{\mathbf{R}} \phi\, dF_n - \int_{\mathbf{R}} \phi\, dF\right| < \left|\int_{(a,b]} \psi\, d(F_n - F)\right| + \tfrac{3}{4}\epsilon$$

$$= \left|\sum_{j=1}^{m} \phi(x_j)\{F_n(x_j) - F(x_j) - F_n(x_{j-1}) + F(x_{j-1})\}\right| + \tfrac{3}{4}\epsilon$$

$$\leqslant \sum_{j=0}^{m} A_j |F_n(x_j) - F(x_j)| + \tfrac{3}{4}\epsilon,$$

where A_j is independent of n. Since $x_j \notin D$,

$$\sum_{j=0}^{m} A_j |F_n(x_j) - F(x_j)| \to 0$$

as $n \to \infty$, and so there exists $n_0 \geqslant n_1$ such that this expression is less than $\tfrac{1}{4}\epsilon$ whenever $n \geqslant n_0$. Hence

$$\left|\int_{\mathbf{R}} \phi\, dF_n - \int_{\mathbf{R}} \phi\, dF\right| < \epsilon$$

for all $n \geqslant n_0$, and therefore (12.1.1) is satisfied.

(ii) Conversely, suppose that (12.1.1) holds. For rational $a < b$, take for ϕ the function ϕ_{ab} defined by

$$\begin{aligned} \phi_{ab}(x) &= b - a \quad (x < a) \\ &= b - x \quad (a \leqslant x \leqslant b) \\ &= 0 \qquad\quad (x > b). \end{aligned}$$

Then, for any $G \in \mathscr{D}$, integration by parts shows that

$$\int_{\mathbf{R}} \phi_{ab}\, dG = \int_a^b G(x)\, dx.$$

Thus (12.1.1) implies that

$$\lim_{n\to\infty} \int_a^b F_n(x)\, dx = \int_a^b F(x)\, dx.$$

Now let

$$F^*(x) = \limsup_{n\to\infty} F_n(x).$$

Then

$$(b-a)\, F^*(a) \leqslant \limsup_{n\to\infty} \int_a^b F_n(x)\, dx$$

$$= \int_a^b F(x)\, dx \leqslant (b-a)\, F(b),$$

so that

$$F^*(a) \leqslant F(b).$$

Letting $b \to a$ from above, $F^*(a) \leqslant F(a)$.

Hence, if x is real, and $a > x$ is rational,

$$F^*(x) \leqslant F^*(a) \leqslant F(a),$$

and, letting $a \to x$ from above,

$$F^*(x) \leqslant F(x).$$

An exactly similar argument yields

$$F_*(x) = \liminf_{n\to\infty} F_n(x) \geqslant F(x-),$$

so that

$$F(x-) \leqslant F_*(x) \leqslant F^*(x) \leqslant F(x).$$

In particular, whenever x is a point of continuity of F

$$F_*(x) = F^*(x) = F(x),$$

so that

$$\lim_{n\to\infty} F_n(x) = F(x).$$

It follows that $F_n \to F$.]

This theorem enables one to establish a number of properties of the convergence defined in $\mathscr{D}$. In particular, it shows that $\mathscr{D}$ can be made into a metric space in such a way that the convergence introduced above is identical with that associated with the metric.

***Theorem* 12.2.** *There exists a metric ρ on $\mathscr{D}$ such that $F_n \to F$ if and only if $\rho(F_n, F) \to 0$.*

Proof. It is clear from the proof that the 'if' part of the assertion of theorem 12.1 can be strengthened to read '$F_n \to F$ if (12.1.1) holds whenever $\phi = \phi_{ab}$ with $a < b$ rational'.

Now the set of functions $\{(b-a)^{-1}\phi_{ab}\}$ is countable; enumerate them as $\{\theta_1, \theta_2, \ldots\}$. Then $F_n \to F$ if and only if

$$\lim_{n\to\infty} \int_{\mathbf{R}} \theta_r dF_n = \int_{\mathbf{R}} \theta_r dF \quad (r = 1, 2, \ldots).$$

Moreover, the θ_r satisfy $\quad 0 \leqslant \theta_r(x) \leqslant 1.$

Now define, for any $F, G \in \mathscr{D}$,

$$\rho(F, G) = \sum_{r=1}^{\infty} 2^{-r} \left| \int_{\mathbf{R}} \theta_r dF - \int_{\mathbf{R}} \theta_r dG \right|.$$

Then easily
$$0 \leqslant \rho(F, G) \leqslant 1,$$
$$\rho(F, G) = \rho(G, F),$$
and
$$\rho(F, G) + \rho(G, H) \geqslant \rho(F, H).$$
Moreover,

$$\begin{aligned} \rho(F, G) = 0 &\Leftrightarrow \int_{\mathbf{R}} \theta_r dF = \int_{\mathbf{R}} \theta_r dG \quad (r = 1, 2, \ldots) \\ &\Leftrightarrow \int_{\mathbf{R}} \phi_{ab} dF = \int_{\mathbf{R}} \phi_{ab} dG \quad (a < b \text{ rational}) \\ &\Leftrightarrow \int_a^b F(x)\, dx = \int_a^b G(x)\, dx \\ &\Leftrightarrow F = G. \end{aligned}$$

Hence ρ is a metric on $\mathscr{D}$.

If F_n $(n = 1, 2, \ldots)$, $F \in \mathscr{D}$,

$$\rho(F_n, F) = \sum_{r=1}^{\infty} 2^{-r} \Delta_{r,n},$$

where
$$\Delta_{r,n} = \left| \int_{\mathbf{R}} \theta_r dF_n - \int_{\mathbf{R}} \theta_r dF \right|.$$

But $0 \leqslant \Delta_{r,n} \leqslant 1$, and so

$$\Delta_{r,n} \leqslant 2^r \rho(F_n, F).$$

Hence
$$\begin{aligned} \rho(F_n, F) \to 0 &\Rightarrow \Delta_{r,n} \to 0 \\ &\Rightarrow \int_{\mathbf{R}} \theta_r dF_n \to \int_{\mathbf{R}} \theta_r dF \quad (r = 1, 2, \ldots) \\ &\Rightarrow F_n \to F. \end{aligned}$$

Conversely, if $F_n \to F$, $\quad \Delta_{r,n} \to 0 \quad (n \to \infty)$,

and, since $\Delta_{r,n} \leqslant 1$, $\quad \rho(F_n, F) = \sum_{r=1}^{\infty} 2^{-r} \Delta_{r,n} \to 0,$

by the dominated convergence theorem (applied to counting measure on $\{1, 2, ...\}$).

The metric ρ therefore has the required properties. ∎

Thus $\mathscr{D}$ can be regarded as a metric space, and it becomes of interest to ask about its topological properties. For instance, it would be useful if $\mathscr{D}$ were compact, so that every sequence had a convergent subsequence. Unfortunately this is not the case; for instance if F_n is the distribution function of the normal distribution $N(0, n)$, it is easy to see that

$$\lim_{n\to\infty} F_n(x) = \tfrac{1}{2}$$

for all x, so that $\{F_n\}$ can have no convergent subsequence in $\mathscr{D}$. There is, however, a partial result in this direction, which is given in the following important theorem.

***Theorem* 12.3.** *Let $F_1, F_2, \ldots$ belong to $\mathscr{D}$. Then there exists an infinite sequence $n_1 < n_2 < \ldots < n_j < \ldots$ of integers and a non-decreasing, right-continuous function G with*

$$\lim_{x\to-\infty} G(x) \geqslant 0, \quad \lim_{x\to+\infty} G(x) \leqslant 1, \tag{12.1.2}$$

such that

$$\lim_{j\to\infty} F_{n_j}(x) = G(x),$$

whenever x is a point of continuity of G.

Proof. If $\mathbf{Q}$ denotes the set of rational points of $\mathbf{R}$, $\mathbf{Q}$ can be enumerated as

$$\mathbf{Q} = \{r_1, r_2, \ldots\};$$

write $f_{nk} = F_n(r_k)$.

We choose sequences $\sigma_0, \sigma_1, \sigma_2, \ldots$ of integers by induction as follows. Let $\sigma_0 = \{1, 2, 3, \ldots\}$. Now suppose that $\sigma_0, \sigma_1, \sigma_2, \ldots, \sigma_{k-1}$ have been chosen so that

(i)$_{k-1}$ $\sigma_0 \supset \sigma_1 \supset \sigma_2 \supset \ldots \supset \sigma_{k-1}$;

(ii)$_{k-1}$ for $j = 1, 2, \ldots, k-1$, the numbers f_{nj}, with n restricted to σ_j, converge to a limit f_j.

Then, for $n \in \sigma_{k-1}$, the f_{nk} form a bounded sequence of real numbers, which therefore has a convergent subsequence. There is therefore a sequence $\sigma_k \subset \sigma_{k-1}$ such that f_{nk} $(n \in \sigma_k)$ converges to a limit f_k. Then

(i)$_k$ $\sigma_0 \supset \sigma_1 \supset \sigma_2 \supset \ldots \supset \sigma_{k-1} \supset \sigma_k$;

(ii)$_k$ for $j = 1, 2, \ldots, k$, the numbers f_{nj} $(n \in \sigma_j)$, converge to f_j.

Hence by induction, there exist sequences $\sigma_0, \sigma_1, \sigma_2, \ldots$, satisfying (*i*)$_k$ and (ii)$_k$ for all k.

Now define a sequence $\{n_1, n_2, \ldots\}$ inductively by

$$n_1 = \min\{n \in \sigma_1\},$$
$$n_{j+1} = \min\{n \in \sigma_{j+1};\ n > n_j\} \quad (j = 1, 2, \ldots).$$

Then $n_j \in \sigma_j$, and so $n_j \in \sigma_k$ for $k \leqslant j$. Hence

$$\{n_k, n_{k+1}, n_{k+2}, \ldots\} \subset \sigma_k,$$

so that
$$\lim_{j\to\infty} f_{n_j k} = f_k.$$

Hence, if we define a function G_1: $\mathbf{Q} \to \mathbf{R}$ by

$$G_1(r_k) = f_k,$$

we have
$$\lim_{j\to\infty} F_{n_j}(r) = G_1(r)$$
for all $r \in \mathbf{Q}$.

Now define, for any $x \in \mathbf{R}$,

$$G(x) = \inf_{\substack{r\in\mathbf{Q}\\ r>x}} G_1(r).$$

Then G is non-decreasing, and is easily seen to be right-continuous. If $r > x$ is rational, then

$$\limsup_{j\to\infty} F_{n_j}(x) \leqslant \limsup_{j\to\infty} F_{n_j}(r) = G_1(r),$$

so that
$$\limsup_{j\to\infty} F_{n_j}(x) \leqslant G(x).$$

A similar argument gives

$$\liminf_{j\to\infty} F_{n_j}(x) \geqslant \sup_{\substack{r\in\mathbf{Q}\\ r<x}} G_1(r)$$
$$\geqslant G(y)$$

for any $y < x$, so that
$$\liminf_{j\to\infty} F_{n_j}(x) \geqslant G(x-).$$

In particular, if x is a point of continuity of G, then

$$\lim_{j\to\infty} F_{n_j}(x) = G(x).$$

Finally, since $0 \leqslant F_{n_j} \leqslant 1$, we have $0 \leqslant G \leqslant 1$, and, since G is non-decreasing, (12.1.2) follows. ∎

Corollary. *If* $\{F_n\}$ *is a sequence in* $\mathscr{D}$ *with the property that, given* $\epsilon > 0$, *there exists* Δ *with*

$$F_n(\Delta) - F_n(-\Delta) \geqslant 1 - \epsilon \tag{12.1.3}$$

for all sufficiently large n, *then* $\{F_n\}$ *has a convergent subsequence in* $\mathscr{D}$.

Proof. Letting $n \to \infty$ in (12.1.3),

$$G(\Delta) - G(-\Delta) \geqslant 1 - \epsilon,$$

so that

$$G(\infty) - G(-\infty) \geqslant 1 - \epsilon.$$

Since ϵ is arbitrary,

$$G(\infty) - G(-\infty) \geqslant 1,$$

which combined with (12.1.2) gives

$$G(-\infty) = 0, \quad G(\infty) = 1.$$

Hence $G \in \mathscr{D}$, and $F_{n_j} \to G$. ∎

We are now in a position to give a formal definition of the symbol $\rightsquigarrow$ used in §11.7. If random variables X_n have distribution functions F_n, and if $F_n \to F \in \mathscr{D}$, then we write

$$X_n \rightsquigarrow F.$$

In this notation F is generally replaced by a symbol (like $N(0, 1)$) representing the corresponding distribution.

Exercises 12.1

1. Let $F_1, F_2, \ldots, F$ be distribution functions of integer-valued random variables $X_1, X_2, \ldots, X$ with

$$\mathsf{P}(X_n = j) = p_j^{(n)}, \quad \mathsf{P}(X = j) = p_j.$$

Show that

$$F_n \to F$$

if and only if

$$p_j^{(n)} \to p_j$$

for all j.

2. Let X_n, X have distribution functions F_n, F. Show that, if $X_n \to X$ in probability, then $F_n \to F$. Show that the converse is in general false, but that it is true if, for some x, $\mathsf{P}(X = x) = 1$.

3. Give an example to show that $F_n \to F$ does not necessarily imply

$$\int_{\mathbf{R}} x\,dF_n \to \int_{\mathbf{R}} x\,dF.$$

How is this reconciled with theorem 12.1?

4. Paul Lévy has defined a metric Δ on $\mathscr{D}$ by

$$\Delta(F, G) = \inf\{\delta;\ F(x-\delta) - \delta \leqslant G(x) \leqslant F(x+\delta) + \delta \text{ for all } x\}.$$

Verify that Δ is a metric and that $F_n \to F$ if and only if $\Delta(F_n, F) \to 0$.

5. For $F, G \in \mathscr{D}$, let $D(F, G)$ be the total variation of $F - G$. Show that D is a metric on $\mathscr{D}$ and that $D(F_n, F) \to 0$ implies that $F_n \to F$. Show that $F_n \to F$ does not necessarily imply that $D(F_n, F) \to 0$.

12.2 Characteristic functions

One of the most powerful tools of the theory of probability is the characteristic function (together with its relatives the moment generating function, cumulant generating function and probability generating function), and the remainder of this chapter will be devoted to the use of this tool.

Characteristic function

The characteristic function of a (finite, real) random variable X is

$$\phi(t) = \mathsf{E}(e^{itX}), \tag{12.2.1}$$

where $i = \sqrt{(-1)}$.

In this definition t may be real or complex. Then e^{itX} is a *complex* random variable, and its expectation is defined in the obvious way by

$$\mathsf{E}(e^{itX}) = \mathsf{E}(\operatorname{Re} e^{itX}) + i\mathsf{E}(\operatorname{Im} e^{itX}).$$

These expectations exist when t is real, since then $|e^{itX}| = 1$, but they may or may not exist for complex t. Thus the characteristic function is always defined for real t, but not necessarily for all complex t.

***Theorem* 12.4.** *The characteristic function $\phi(t)$ of a random variable X has the following properties for real t;*

(i) $|\phi(t)| \leqslant 1$;

(ii) $\phi(0) = 1$;

(iii) *If for some real $\tau \neq 0$, $\phi(\tau) = 1$, then X is a discrete random variable, taking only values which are integral multiples of $2\pi/\tau$;*

(iv) *ϕ is a uniformly continuous function of t.*

Proof. (i) and (ii) are obvious.

(iii) If $\tau \neq 0$ and $\phi(\tau) = 1$, we have, on taking real parts,

$$\mathsf{E}(1 - \cos \tau X) = 0.$$

Since $1 - \cos \tau X \geqslant 0$ this implies that

$$\mathsf{P}(1 - \cos \tau X = 0) = 1,$$

i.e.

$$\mathsf{P}(X \text{ is divisible by } 2\pi/\tau) = 1.$$

(iv) If $t_1 - t_2 = t$, then

$$\begin{aligned} |\phi(t_1) - \phi(t_2)| &= |\mathsf{E}\{(e^{itX} - 1)\, e^{it_2X}\}| \\ &\leqslant \mathsf{E}|e^{itX} - 1| \\ &= \int_\Omega |e^{itX(\omega)} - 1| \mathsf{P}(d\omega) \\ &\to 0 \end{aligned}$$

as $t \to 0$ by bounded convergence. Hence

$$\phi(t_1) - \phi(t_2) \to 0$$

as $t_1 - t_2 \to 0$ uniformly in t_1, t_2. ▌

If a random variable X has distribution function F, theorem 11.3 shows that its characteristic function is

$$\phi(t) = \int_{\mathbf{R}} e^{itx}\, dF(x). \tag{12.2.2}$$

This expression is called the *Fourier–Stieltjes transform* of F. A knowledge of F enables one to calculate ϕ, and it will appear in the next section that, conversely, a knowledge of ϕ determines F. Thus F and ϕ are equivalent descriptions of the distribution of X.

One of the useful properties of the characteristic function is that it ties up very naturally with the addition of independent random variables. This is shown by the following result.

***Theorem* 12.5.** *Let X, Y be independent random variables with characteristic functions ϕ_X, ϕ_Y. Then the characteristic function ϕ_{X+Y} of $X+Y$ is given (for real t) by*

$$\phi_{X+Y}(t) = \phi_X(t)\,\phi_Y(t). \tag{12.2.3}$$

Proof.
$$\begin{aligned}
\phi_{X+Y}(t) &= \mathsf{E}\{\cos t(X+Y)\} + i\mathsf{E}\{\sin t(X+Y)\} \\
&= \mathsf{E}(\cos tX \cos tY) - \mathsf{E}(\sin tX \sin tY) \\
&\quad + i\mathsf{E}(\sin tX \cos tY) + i\mathsf{E}(\cos X \sin tY) \\
&= \mathsf{E}(\cos tX)\,\mathsf{E}(\cos tY) - \mathsf{E}(\sin tX)\,\mathsf{E}(\sin tY) \\
&\quad + i\mathsf{E}(\sin tX)\,\mathsf{E}(\cos tY) + i\mathsf{E}(\cos tX)\,\mathsf{E}(\sin tY) \\
&\qquad\qquad\qquad\qquad \text{(by theorem 11.6)} \\
&= \{\mathsf{E}(\cos tX) + i\mathsf{E}(\sin tX)\}\{\mathsf{E}(\cos tY) + i\mathsf{E}(\sin tY)\} \\
&= \phi_X(t)\,\phi_Y(t). ▌
\end{aligned}$$

Example. Suppose that X has the distribution $N(\mu, \sigma^2)$. Then its characteristic function is

$$\phi(t) = \frac{1}{\sigma(2\pi)^{\frac{1}{2}}} \int_{-\infty}^{\infty} \exp(itx) \exp\{-(x-\mu)^2/2\sigma^2\}\, dx.$$

This integral may be evaluated by the technique of contour integration. Making the substitution

$$z = (x-\mu)/\sigma - it\sigma,$$

we have $$\phi(t) = (2\pi)^{-\frac{1}{2}} \int \exp(-\tfrac{1}{2}z^2) \exp(-\tfrac{1}{2}\sigma^2 t^2 + i\mu t)\, dz,$$

where the integral is a line integral in the complex plane taken along the line $\operatorname{Im} z = -t\sigma$. (Such integrals have not been defined in this book, but should be taken in the ordinary elementary sense.) Cauchy's theorem shows that the integral is unchanged if the contour of integration is moved to the real axis, so that

$$\begin{aligned}\phi(t) &= (2\pi)^{-\frac{1}{2}} \exp\left(-\tfrac{1}{2}\sigma^2 t^2 + i\mu t\right) \int_{-\infty}^{\infty} e^{-\frac{1}{2}z^2}\, dz \\ &= \exp\left(-\tfrac{1}{2}\sigma^2 t^2 + i\mu t\right).\end{aligned}$$

Hence the characteristic function of $N(\mu, \sigma^2)$ is

$$\exp\left(i\mu t - \tfrac{1}{2}\sigma^2 t^2\right).$$

Now suppose that X and Y are independent, with distributions $N(\mu_1, \sigma_1^2)$ and $N(\mu_2, \sigma_2^2)$. Then $X + Y$ has characteristic function

$$\begin{aligned}\phi_{X+Y}(t) &= \exp\left(i\mu_1 t - \tfrac{1}{2}\sigma_1^2 t^2\right) \exp\left(i\mu_2 t - \tfrac{1}{2}\sigma_2^2 t^2\right) \\ &= \exp\left\{i(\mu_1 + \mu_2)\, t - \tfrac{1}{2}(\sigma_1^2 + \sigma_2^2)\, t^2\right\},\end{aligned}$$

which is just the characteristic function of $N(\mu_1 + \mu_2, \sigma_1^2 + \sigma_2^2)$. When, in the next section, it is proved that the characteristic function determines the distribution, it will follow that $X + Y$ has the distribution $N(\mu_1 + \mu_2, \sigma_1^2 + \sigma_2^2)$.

If in (12.2.1) we expand the exponential as a power series in t and then (formally) interchange E and Σ, we obtain

$$\phi(t) = \sum_{n=0}^{\infty} \frac{(it)^n}{n!} \mathsf{E}(X^n).$$

Hence, from the expansion of $\phi(t)$ as a power series in t, we can read off the 'moments' $\mathsf{E}(X^n)$ of X. For this reason ϕ is sometimes called the 'moment generating function' of X, although this name is more often reserved for the closely related function

$$\phi(-it) = \mathsf{E}(e^{tX}).$$

Some care is needed in justifying the expansion, and various results in this direction are known. For many purposes the most useful is the following.

***Theorem* 12.6.** *Suppose that, for some integer n,*

$$\mathsf{E}(|X|^n) < \infty.$$

Then

$$\phi(t) = \sum_{k=0}^{n} \frac{(it)^k}{k!} \mathsf{E}(X^k) + o(t^n) \qquad (12.2.4)$$

as $t \to 0$.

Proof. Elementary integration by parts yields the identity

$$e^z = \sum_{k=0}^{n} \frac{z^k}{k!} + \frac{z^n}{(n-1)!} \int_0^1 (1-y)^{n-1} (e^{zy} - 1)\, dy.$$

Putting $z = itX$ and taking expectations gives

$$\phi(t) = \sum_{k=0}^{n} \frac{(it)^k}{k!} \mathsf{E}(X^k) + R(t),$$

where $$R(t) = \frac{(it)^n}{(n-1)!} \int_{\mathbf{R}} \int_0^1 x^n (1-y)^{n-1} (e^{itxy} - 1)\, dy\, dF(x).$$

The integrand tends to zero with t, and is bounded in modulus by $2|x|^n (1-y)^n$. Hence, applying the dominated convergence theorem to the product measure $dy\,dF(x)$, and using the fact that

$$\int_{\mathbf{R}} \int_0^1 2|x|^n (1-y)^{n-1} dy\, dF(x) = \frac{2}{n} \mathsf{E}|X|^n < \infty,$$

the integral tends to zero, and so

$$R(t) = o(t^n) \quad (t \to 0).\ ▮$$

For example, in the case of the distribution $N(0, 1)$, $\mathsf{E}(|X^n|) < \infty$ for all n. Moreover,

$$\phi(t) = e^{-\frac{1}{2}t^2} = \sum_{m=0}^{\infty} \frac{(it)^{2m}}{2^m m!}.$$

Hence at once $$\mathsf{E}(X^n) = 0 \quad (n \text{ odd}),$$

$$= \frac{(2m)!}{2^m m!} \quad (n = 2m \text{ even}).$$

Exercises 12.2

1. Prove that (12.2.3) holds for any complex t for which $\phi_X(t)$ and $\phi_Y(t)$ exist.

2. Show that if, in the notation of § 11.7, X has the distribution

(i) $U(a, b)$, then
$$\phi(t) = \frac{e^{ibt} - e^{iat}}{i(b-a)t},$$

(ii) $E(\lambda)$, then
$$\phi(t) = \frac{\lambda}{\lambda - it},$$

(iii) Cauchy, then
$$\phi(t) = e^{-|t|},$$

(iv) $P(\mu)$, then
$$\phi(t) = \exp\{\mu(e^{it} - 1)\}.$$

Deduce the values of $\mathsf{E}(X^n)$ (where they exist) for these distributions.

3. If $\mathsf{E}(|X|^n) < \infty$, show that there exist numbers $\kappa_1, \kappa_2, \ldots, \kappa_n$ such that, as $t \to 0$,

$$\log \phi(t) = \sum_{k=1}^{n} \frac{(it)^k}{k!} \kappa_k + o(t^n).$$

The κ_j are called the *cumulants* of X. They have the property that, if X and Y are independent, then

$$\kappa_j(X+Y) = \kappa_j(X) + \kappa_j(Y).$$

Show that
$$\kappa_1 = \mathsf{E}(X), \quad \kappa_2 = \operatorname{var}(X).$$

Show that X has distribution $N(\mu, \sigma^2)$ if and only if κ_j exists for all j and

$$\kappa_1 = \mu, \quad \kappa_2 = \sigma^2, \quad \kappa_j = 0 \quad (j \geqslant 3).$$

4. If X is a non-negative random variable, show that its characteristic function $\phi(t)$ exists and is continuous in $\operatorname{Im} t \geqslant 0$ and is analytic in $\operatorname{Im} t > 0$. The function

$$\phi(is) = \int_{[0,\infty)} e^{-sx}\, dF(x)$$

is called the Laplace–Stieltjes transform of F. Notice that it is real, positive, and finite for real positive t.

12.3 The inversion and continuity theorems

It was asserted in the last section that a knowledge of the characteristic function $\phi(t)$ of a random variable X is sufficient to determine its distribution function $F(x)$. This fact is contained in the following theorem, which gives an almost explicit formula for F in terms of ϕ.

***Theorem* 12.7.** *Let F belong to $\mathscr{D}$, and define*

$$\bar{F}(x) = \tfrac{1}{2}\{F(x) + F(x-)\}.$$

If
$$\phi(t) = \int_{\mathbf{R}} e^{itx}\, dF(x),$$

then, for all a, b,

$$\bar{F}(b) - \bar{F}(a) = \lim_{A\to\infty} \frac{1}{2\pi} \int_{-A}^{A} \frac{e^{-iat} - e^{-ibt}}{it} \phi(t)\, dt. \qquad (12.3.1)$$

Proof. It clearly suffices to prove the result when $a < b$. Then

$$\begin{aligned} I_A &= \frac{1}{2\pi} \int_{-A}^{A} \frac{e^{-iat} - e^{ibt}}{it} \phi(t)\, dt \\ &= \frac{1}{2\pi} \int_{-A}^{A} \frac{e^{-iat} - e^{-ibt}}{it}\, dt \int_{\mathbf{R}} e^{itx}\, dF(x) \\ &= \frac{1}{2\pi} \int_{\mathbf{R}} dF(x) \int_{-A}^{A} \frac{e^{it(x-a)} - e^{it(x-b)}}{it}\, dt, \end{aligned}$$

by Fubini's theorem. Now

$$\begin{aligned}\int_{-A}^{A}\frac{e^{it(x-a)}-e^{it(x-b)}}{it}\,dt &= \int_{-A}^{A}dt\int_{x-b}^{x-a}e^{ity}\,dy\\ &= \int_{x-b}^{x-a}dy\int_{-A}^{A}e^{ity}\,dt\\ &= \int_{x-b}^{x-a}\frac{2\sin Ay}{y}\,dy\\ &= 2\int_{A(x-b)}^{A(x-a)}\frac{\sin z}{z}\,dz\\ &= 2\{K[A(x-a)]-K[A(x-b)]\}\end{aligned}$$

where
$$K(Z)=\int_0^Z\frac{\sin z}{z}\,dz.$$

Now K is an odd function, and it is well known (see exercise 12.3.(1)) that

$$K(Z)\to\tfrac{1}{2}\pi\quad(Z\to\infty).$$

Hence K is bounded, and

$$I_A=\frac{1}{\pi}\int_{\mathbf{R}}\{K[A(x-a)]-K[A(x-b)]\}\,dF(x).$$

As $A\to\infty$ the integrand converges to

$$\begin{cases}0 & (x<a)\\ \tfrac{1}{2}\pi & (x=a)\\ \pi & (a<x<b)\\ \tfrac{1}{2}\pi & (x=b)\\ 0 & (x>b).\end{cases}$$

Hence, by the dominated convergence theorem,

$$\begin{aligned}\lim_{A\to\infty}I_A &= \tfrac{1}{2}\{F(a)-F(a-)\}+\{F(b-)-F(a)\}+\tfrac{1}{2}\{F(b)-F(b-)\}\\ &= \bar{F}(b)-\bar{F}(a).\ \blacksquare\end{aligned}$$

Corollary. *If F, G in $\mathscr{D}$ have the same characteristic function, then $F=G$.*

Proof. From (12.3.1),

$$\bar{F}(b)-\bar{F}(a)=\bar{G}(b)-\bar{G}(a).$$

Let $a \to -\infty$ to give $$\bar{F}(b) = \bar{G}(b).$$
Hence, for all x,
$$F(x) = \lim_{b \to x+} \bar{F}(b) = \lim_{b \to x+} \bar{G}(b) = G(x). \;\rule{0.5em}{1em}$$

The inversion formula (12.3.1) is not in a very convenient form for actual calculation, and more useful formulae can be derived under more restrictive conditions. For example, suppose that
$$\int_{-\infty}^{\infty} |\phi(t)|\, dt < \infty.$$
Then, putting $a = 0$, (12.3.1) becomes
$$\begin{aligned}\bar{F}(b) - \bar{F}(0) &= \frac{1}{2\pi}\int_{-\infty}^{\infty} \frac{1-e^{-ibt}}{it}\,\phi(t)\,dt \\ &= \frac{1}{2\pi}\int_{-\infty}^{\infty} \phi(t)\,dt \int_0^b e^{-itx}\,dx \\ &= \frac{1}{2\pi}\int_0^b dx \int_{-\infty}^{\infty} \phi(t) e^{-itx}\,dt,\end{aligned}$$
by Fubini's theorem. Hence F has a probability density
$$f(x) = \frac{1}{2\pi}\int_{-\infty}^{\infty} \phi(t)\, e^{-itx}\, dt. \tag{12.3.2}$$

The corollary to theorem 12.7 shows that the characteristic function uniquely determines the distribution function. Thus there is a one-to-one correspondence between the class $\mathcal{D}$ and the class of all characteristic functions. Moreover, this correspondence is continuous, in a sense made precise by the following theorem.

***Theorem* 12.8.** *Let F_n $(n = 1, 2, \ldots)$ in $\mathcal{D}$ have characteristic function ϕ_n, and suppose that $F_n \to F$, where F in $\mathcal{D}$ has characteristic function ϕ, then $\phi_n(t) \to \phi(t)$ for all real t. Conversely, suppose that F_n has characteristic function ϕ_n, and that*
$$\phi(t) = \lim_{n \to \infty} \phi_n(t)$$
exists for all real t and is continuous at $t = 0$. Then ϕ is the characteristic function of some F in $\mathcal{D}$, and $F_n \to F$.

Proof: (i) If $F_n \to F$, it follows from theorem 12.1 (with $\phi(x) = \cos tx$ and $\phi(x) = \sin tx$) that $\phi_n(t) \to \phi(t)$.

(ii) To prove the converse, suppose that F_r in $\mathcal{D}$ is such that $\phi(t) = \lim_{n \to \infty} \phi_r(t)$ exists, and
$$\lim_{t \to 0} \phi(t) = \phi(0) = 1.$$

For any $a > 0$,

$$a^{-1}\int_0^a \operatorname{Re}\{1-\phi_n(t)\}\,dt = a^{-1}\int_0^a dt \int_{\mathbf{R}} (1-\cos tx)\,dF_n(x)$$

$$= a^{-1}\int_{\mathbf{R}} dF_n(x) \int_0^a (1-\cos tx)\,dt$$

$$= \int_{\mathbf{R}} \left(1-\frac{\sin ax}{ax}\right) dF_n(x),$$

using Fubini's theorem. Now

$$m = \inf_{|y|\geqslant 1} \left(1-\frac{\sin y}{y}\right)$$

is clearly strictly positive, and

$$1-\frac{\sin y}{y} \geqslant 0$$

for all y, and therefore

$$a^{-1}\int_0^a \operatorname{Re}\{1-\phi_n(t)\}\,dt \geqslant \int_{|ax|\geqslant 1} \left(1-\frac{\sin ax}{ax}\right) dF_n(x)$$

$$\geqslant m \int_{|ax|\geqslant 1} dF_n(x).$$

Now fix $\epsilon > 0$. Since $\phi(t) \to 1$ as $t \to 0$, there exists $a > 0$ such that

$$\operatorname{Re}\{1-\phi(t)\} < \epsilon m$$

for $0 \leqslant t \leqslant a$, and so

$$a^{-1}\int_0^a \operatorname{Re}\{1-\phi(t)\}\,dt < \epsilon m.$$

Hence, by bounded convergence,

$$a^{-1}\int_0^a \operatorname{Re}\{1-\phi_n(t)\}\,dt < \epsilon m$$

for all sufficiently large n. For such n,

$$\int_{|x|\geqslant a^{-1}} dF_n(x) < \epsilon,$$

so that

$$F_n(a^{-1}) - F_n(-a^{-1}) > 1-\epsilon.$$

It thus follows from the corollary to theorem 12.3 that there exists a subsequence $\{F_{n_j}\}$ of $\{F_n\}$ which converges to some F in $\mathscr{D}$. By the first part of the theorem, which has already been proved,

$$\phi_{n_j}(t) \to \int_{\mathbf{R}} e^{itx}\,dF(x),$$

so that ϕ is the characteristic function of F.

Now suppose that F_r does not converge to F. Then, if ρ denotes the metric of theorem 12.2,

$$\rho(F_n, F) \not\to 0.$$

Hence there exists $\delta > 0$ and a subsequence $\{F_{m_k}\}$ of $\{F_n\}$ such that

$$\rho(F_{m_k}, F) > \delta$$

for all k. Repeating the above argument with F_n replaced by F_{m_k} shows that $\{F_{m_k}\}$ has itself a subsequence which converges to some G in $\mathscr{D}$, and that

$$\phi(t) = \int_{\mathbf{R}} e^{itx}\, dG(x).$$

But $\rho(G, F) \geqslant \delta$, so that $G \neq F$, whilst at the same time

$$\int_{\mathbf{R}} e^{itx}\, dF(x) = \int_{\mathbf{R}} e^{itx}\, dG(x)$$

for all real t. This contradiction of the corollary to theorem 12.7 shows that $F_n \to F$. ▌

The theorem is very useful for proving results of the form $F_n \to F$, since in many cases it is easier to manipulate ϕ_n than F_n. This is particularly true when F_n is the distribution of a sum of independent random variables, and in the next chapter the characteristic function will be used extensively to prove results about such sums. As a simple illustration of the technique we give an alternative derivation of the normal approximation to the binomial distribution which was established in § 10.5.

Thus let X_n have distribution $B(n, p)$ $(0 < p < 1)$, and denote by X_n^* the standardized random variable

$$X_n^* = (X_n - np)/(npq)^{\frac{1}{2}}$$

(where, for convenience, we have written $q = 1-p$). The distribution function and characteristic function of X_n^* will be denoted by F_n and ϕ_n, respectively. Then

$$\begin{aligned}
\phi_n(t) = \mathsf{E}(e^{itX_n^*}) &= \mathsf{E}\left(\exp \frac{it(X_n - np)}{(npq)^{\frac{1}{2}}}\right) \\
&= \exp\left(-it\left(\frac{np}{q}\right)^{\frac{1}{2}}\right) \sum_{x=0}^{n} \binom{n}{x} p^x q^{n-x} \exp\{itx(npq)^{-\frac{1}{2}}\} \\
&= \exp\left(-it\left(\frac{np}{q}\right)^{\frac{1}{2}}\right) (q + p\exp\{it(npq)^{-\frac{1}{2}}\})^n \\
&= \exp\left[-it\left(\frac{np}{q}\right)^{\frac{1}{2}} + n\log(q + p\exp\{it(npq)^{-\frac{1}{2}}\})\right].
\end{aligned}$$

Holding p fixed and letting $n \to \infty$, we have

$$\phi_n(t) = \exp\left[-it\left(\frac{np}{q}\right)^{\frac{1}{2}} + n\log\left(1+it\left(\frac{p}{nq}\right)^{\frac{1}{2}} - \frac{t^2}{2nq} + O(n^{-\frac{3}{2}})\right)\right]$$

$$= \exp\left[-it\left(\frac{np}{q}\right)^{\frac{1}{2}} + it\left(\frac{np}{q}\right)^{\frac{1}{2}} - \frac{t^2}{2q} + \frac{t^2p}{2q} + O(n^{-\frac{1}{2}})\right]$$

$$\to \exp(-\tfrac{1}{2}t^2) = \phi(t), \text{ say.}$$

Now $\phi(t)$ is continuous at $t = 0$, and is hence the characteristic function of a unique distribution function, which is the limit of F_n as $n \to \infty$. In fact we have already seen that ϕ is the characteristic function of $N(0, 1)$, and so (since F is continuous)

$$F(x) = \lim_{n\to\infty} F_n(x) = (2\pi)^{-\frac{1}{2}} \int_{-\infty}^{x} e^{-\frac{1}{2}y^2}\, dy,$$

or equivalently $\quad X_n^* \rightsquigarrow N(0,1) \quad (n \to \infty).$

Exercises 12.3

1. By integrating $\sin z/z$ around a large semicircle standing on the real axis, prove the result used in the proof of theorem 12.7

$$\lim_{X\to\infty} \int_0^X \frac{\sin x}{x}\, dx = \tfrac{1}{2}\pi.$$

2. Verify directly that the inversion formula (12.3.2) is valid when
(i) $\phi(t) = e^{-|t|}$,
(ii) $\phi(t) = e^{-\frac{1}{2}t^2}$.

3. Use the method of characteristic functions to solve exercises 11.7 (6), 11.7 (7).

4. Show that, if $X_1, X_2, \ldots, X_n$ are independent and have distribution $E(1)$, then $X_1 + X_2 + \ldots + X_n$ has probability density

$$x^{n-1}e^{-x}/(n-1)! \quad (x \geqslant 0).$$

5. Use theorem 12.8 to show that $B(n, \mu/n) \to P(\mu)$ as $n \to \infty$.

6. Let ϕ_n be the characteristic function of $N(0, n)$. Show that

$$\phi(t) = \lim_{n\to\infty} \phi_n(t)$$

exists, but is not continuous at $t = 0$.

7. Show that F_n converges to some F in $\mathscr{D}$ if and only if there exists a function ϕ such that

$$\phi_n(t) \to \phi(t)$$

as $n \to \infty$ uniformly in every finite interval.

8. Let $Y_j (j = \pm 1, \pm 2, \ldots, \pm n^2)$ be independent variables having Poisson distributions with means $\mathsf{E}(Y_j) = cn^\alpha/|j|^{1+\alpha}$, where c, α are positive constants and $\alpha < 2$. Find the characteristic function ϕ_n of

$$\sum_{j=-n^2}^{n^2} \left(\frac{j}{n}\right) Y_j.$$

By considering the limit of ϕ_n as $n \to \infty$, show that, for any α in $0 < \alpha < 2$, there exists a distribution function F_α with

$$\int_{-\infty}^{\infty} e^{itx}\, dF_\alpha(x) = e^{-|t|^\alpha}.$$

Show also that F_α is stable in the sense of exercise 11.7 (7).

9. If $X_1, X_2, \ldots, X_n$ are independent, with distribution $U(-1, 1)$, show that $X_1 + X_2 + \ldots + X_n$ has probability density

$$f(x) = \frac{1}{\pi}\int_0^\infty \left(\frac{\sin t}{t}\right)^n \cos tx\, dt.$$

Show also that, in every interval of the form $(r, r+1)$ (where r is an integer), $f(x)$ is equal to a polynomial.

12.4 Generating functions

Suppose that X is a random variable which takes on only (positive and negative) integer values, and write

$$p_j = \mathsf{P}(X = j) \quad (j = 0, \pm 1, \pm 2, \ldots).$$

Then X has characteristic function

$$\phi(t) = \mathsf{E}(e^{itX}) = \sum_{j=-\infty}^{\infty} p_j e^{itj}$$
$$= P(e^{it}),$$

where
$$P(z) = \sum_{j=-\infty}^{\infty} p_j z^j.$$

The series for $P(z)$ converges at least when the complex variable z has $|z| = 1$; the resulting function is called the *probability generating function* of X.

Because the probability generating function is linked to the characteristic function by the simple relation

$$\phi(t) = P(e^{it}),$$

it has very similar properties. For convenience these are listed in the following theorem, the bulk of which consists of restatements (for the special case of integer-valued X) of theorems already proved.

Theorem 12.9. *Let* $P(z) = \sum_{j=-\infty}^{\infty} p_j z^j \quad (|z| = 1)$ *be the probability generating function of an integer-valued random variable X. Then*

(i) $P(1) = 1, \; |P(z)| \leqslant 1$;

(ii) *If there exists* $z \neq 1$ *with* $P(z) = 1$, z *must be of the form*

$$z = \exp(2\pi ip/q)$$

where p and q are integers with no common factor, and then

$$\mathsf{P}(q \textit{ divides } X) = 1;$$

(iii) $P: \{z \in \mathbf{C};\ |z| = 1\} \to \mathbf{C}$ *is continuous*;

(iv)
$$p_j = \frac{1}{2\pi i}\int_{|z|=1} \frac{P(z)}{z^{j+1}}\,dz = \frac{1}{2\pi}\int_0^{2\pi} P(e^{i\theta})\,e^{-ij\theta}\,d\theta. \tag{12.4.1}$$

If X and Y are independent, then

$$P_{X+Y}(z) = P_X(z)\,P_Y(z).$$

If, for each n,
$$P_n(z) = \sum_{j=-\infty}^{\infty} p_j^{(n)} z^j$$
is a probability generating function, and if

$$P(z) = \lim_{n\to\infty} P_n(z)$$

exists for all z in $|z| = 1$ *and is continuous at* $z = 1$, *then*

$$p_j = \lim_{n\to\infty} p_j^{(n)}$$

exists for all j, and
$$P(z) = \sum_{j=-\infty}^{\infty} p_j z^j.$$

Proof. The only assertion requiring proof is the inversion formula (12.4.1), which is of a slightly different form from that of theorem 12.7. Its proof follows on noting that

$$\int_{|z|=1} z^\nu\,dz = 0 \quad (\nu = 0, 1, 2, \ldots, -2, -3, \ldots)$$
$$= 2\pi i \quad (\nu = -1),$$

so that
$$p_j = \frac{1}{2\pi i}\sum_{k=-\infty}^{\infty} p_k \int_{|z|=1} z^{k-j-1}\,dz$$
$$= \frac{1}{2\pi i}\int_{|z|=1} \frac{P(z)}{z^{j+1}}\,dz,$$
using Fubini's theorem.]

Many of the simple distributions on the integers have explicit forms for their probability generating functions; a few of these are listed below, their verification being left to the reader.

Binomial distribution. $B(n,p): P(z) = (1-p+pz)^n$.

Poisson distribution. $P(\mu): P(z) = e^{\mu(z-1)}$.

Geometric distribution. $G(q): P(z) = (1-q)/(1-qz)$.

If the random variable is non-negative; its probability generating function
$$P(z) = \sum_{j=0}^{\infty} p_j z^j$$
exists in $|z| \leqslant 1$, and defines an analytic function in $|z| < 1$. Notice that, in this case,
$$p_j = \frac{1}{j!}\left(\frac{d^jP}{dz^j}\right)_{z=0}.$$

Exercises 12.4

1. Prove that, if X and Y are independent, with distributions $B(m,p)$ and $B(n,p)$, respectively, then $X+Y$ has distribution $B(m+n,p)$. Deduce the identity
$$\sum_{k=0}^{r}\binom{m}{k}\binom{n}{r-k} = \binom{m+n}{r}.$$

2. Show that
$$p_j = \binom{n+j-1}{j} q^j (1-q)^n$$
determines a probability distribution if $0 < q < 1$ $(n \geqslant 1)$, and show that
$$P(z) = \left(\frac{1-q}{1-qz}\right)^n.$$

3. If X and Y are independent with Poisson distributions $P(\lambda)$ and $P(\mu)$, show that
$$\mathsf{P}(X-Y=n) = \left(\frac{\lambda}{\mu}\right)^n e^{-(\lambda+\mu)}\frac{1}{2\pi}\int_0^{2\pi} e^{2\kappa\cos\theta}\cos n\theta\,d\theta,$$
where $\kappa = (\lambda\mu)^{\frac{1}{2}}$.

4. If X has probability generating function $P(z)$, prove that

(i) if $\mathsf{E}|X| < \infty$, then $\mathsf{E}(X) = P'(1)$,

(ii) if $\mathsf{E}(X^2) < \infty$, then $\operatorname{var}(X) = P''(1)+P'(1)-\{P'(1)\}^2$.

13

INDEPENDENCE

13.1 Sequences of independent trials

Let $(\Omega, \mathscr{F}, \mathsf{P})$ be a probability space. Then each ω in Ω can be thought of as a possible outcome of the random experiment described by this probability space. Now suppose that the experiment is repeated n times. Then the outcome of the compound experiment consisting of these n repetitions is the sequence

$$\omega = (\omega_1, \omega_2, \ldots, \omega_n),$$

where ω_j is the outcome of the jth experiment. In other words, the outcome of the compound experiment is an element of the n-fold Cartesian product

$$\Omega^n = \Omega \times \Omega \times \ldots \times \Omega$$

of Ω with itself.

Now let $A_1, A_2, \ldots, A_n$ belong to $\mathscr{F}$, and let $A = A_1 \times A_2 \times \ldots \times A_n$ be their product, which is of course a subset of Ω^n. The event $\omega \in A$ is just the intersection of the events $\omega_j \in A_j$ $(j = 1, 2, \ldots, n)$. If therefore the repetitions are regarded as independent, these last events are mutually statistically independent, and the probability of the event A is

$$\mathsf{P}^n(A) = \prod_{j=1}^{n} \mathsf{P}(A_j).$$

Now the results of §§6.1, 6.2 imply that the sets A (as the A_j vary) form a semi-ring which generates the product σ-ring

$$\mathscr{F}^{*n} = \mathscr{F} * \mathscr{F} * \ldots * \mathscr{F},$$

and P^n can be extended in a unique way to a probability measure on $\mathscr{F}^{*n}$. Thus, if $\mathscr{F}^n$ denotes the completion of $\mathscr{F}^{*n}$, with respect to P^n, then $(\Omega^n, \mathscr{F}^n, \mathsf{P}^n)$ is the probability space which describes the n-fold repetition of the experiment, on the assumption that the repetitions are independent of one another. This situation is often described as a sequence of n independent trials of the experiment described by $(\Omega, \mathscr{F}, \mathsf{P})$. P^n is of course the product measure $\mathsf{P} \times \mathsf{P} \times \ldots \times \mathsf{P}$.

For example, a single toss of a coin can, as we have seen, be described by a probability space Ω consisting of just two elements H and T. If this same coin is tossed n times, the resulting space Ω will contain the 2^n different possible outcomes. There are, however, situations in which one must go further than this, and consider an infinite sequence of repetitions. For example, suppose that it is decided to continue tossing a coin until heads have been obtained three times. Then, although it can be shown that this almost certainly requires only a finite number of tosses, it is impossible to set an *a priori* upper bound on this number, which may (though with small probability) be very large. It is thus necessary to consider an infinite sequence of tosses of the coin.

More generally, for any probability space $(\Omega, \mathscr{F}, P)$, one may consider an infinite sequence of independent trials or repetitions. Any such sequence will be of the form

$$\omega = (\omega_1, \omega_2, \ldots),$$

where ω_j is the outcome of the jth trial, and so the set of possible outcomes is just the product Ω^∞ of countably many copies of Ω. Moreover, any event A of the form

$$A = A_1 \times A_2 \times \ldots \times A_n \times \Omega \times \Omega \times \ldots \tag{13.1.1}$$

$$= \{\omega;\ \omega_j \in A_j\ (j = 1, 2, \ldots, n)\},$$

where $n \geq 1, A_j \in \mathscr{F}$, has a probability

$$\mathsf{P}^\infty(A) = \prod_{j=1}^{n} \mathsf{P}(A_j).$$

The class of all sets of the form (13.1.1) is a semi-ring and it generates the product σ-ring

$$\mathscr{F}^{*\infty} = \mathscr{F} * \mathscr{F} * \ldots.$$

Then P^∞ has a unique extension to $\mathscr{F}^{*\infty}$, and indeed to the completion $\mathscr{F}^\infty$ of $\mathscr{F}^{*\infty}$ with respect to P^∞. It follows that the appropriate probability space for an infinite sequence of independent trials is the infinite product space $(\Omega^\infty, \mathscr{F}^\infty, \mathsf{P}^\infty)$

An important special case arises when the outcome of a single trial is just a real random variable X. Then

$$\Omega = \mathbf{R}, \quad \mathscr{F} = \mathscr{L}_F, \quad \mathsf{P} = \mu_F,$$

where F is the distribution function of X, and μ_F is the corresponding Stieltjes measure. A sequence of independent trials results in an infinite sequence $(X_1, X_2, \ldots)$ of independent random variables each having the distribution function F, and this sequence is defined on the probability space $(\mathbf{R}^\infty, \mathscr{L}_F^\infty, \mu_F^\infty)$.

An important consequence of these remarks is that a theorem whose enunciation begins 'Let $X_1, X_2, \ldots$ be independent random variables with distribution function F' is not vacuous, since a probability space does exist on which such variables can be defined. One, but not the only, choice for this space is $(\mathbf{R}^\infty, \mathscr{L}_F^\infty, \mu_F^\infty)$.

Having seen how the concept of a sequence of independent trials can be formalised we now go on to look at some of the properties of independent events and independent random variables. It will not be necessary to refer explicitly to the underlying probability space, but it should be remembered that this will often be one constructed as indicated here by means of the theory of product measures.

Exercises 13.1

1. Let Ω be the unit interval, $\mathscr{F}$ the σ-field $\mathscr{L}$ of Lebesgue sets, and P Lebesgue measure. Define $X_n(\omega)$ to be the nth digit in the decimal expansion of ω. Show that $X_1, X_2, \ldots$ are independent, identically distributed, random variables, with

$$\mathsf{P}(X_n = r) = \tfrac{1}{10} \quad (r = 0, 1, 2, \ldots, 9).$$

2. With the same probability space as in the last example, define functions Y_n (the Rademacher functions) as follows:

$$Y_n(\omega) = 1 \quad \text{if the integer part of } 2^n x \text{ is even,}$$
$$= -1 \quad \text{if the integer part of } 2^n x \text{ is odd.}$$

Show that $Y_1, Y_2, \ldots$ are independent, identically distributed random variables, and find their distribution.

13.2 The Borel–Cantelli lemmas and the zero-one law

Let $A_1, A_2, \ldots$ be an infinite sequence of events. It is often of interest to know how many of these events occur. In particular, what is the probability that infinitely many occur? This question is often answered with the aid of two simple theorems, known as the Borel–Cantelli lemmas.

***Theorem* 13.1.** *Let $A_1, A_2, \ldots$ be events defined on a probability space $(\Omega, \mathscr{F}, \mathsf{P})$, and let*

$$A = \limsup_{n\to\infty} A_n = \bigcap_{N=1}^{\infty} \bigcup_{n=N}^{\infty} A_n \tag{13.2.1}$$

be the event that A_n occurs for infinitely many n. Then

$$\sum_{n=1}^{\infty} \mathsf{P}(A_n) < \infty \tag{13.2.2}$$

implies that $$\mathsf{P}(A) = 0.$$

Proof. For any N, $$A \subset \bigcup_{n=N}^{\infty} A_n,$$

so that $$\mathsf{P}(A) \leqslant \mathsf{P} \bigcup_{n=N}^{\infty} A_n \leqslant \sum_{n=N}^{\infty} \mathsf{P}(A_n).$$

Letting $N \to \infty$, $$\mathsf{P}(A) \leqslant 0,$$
and the theorem follows. ▌

Notice that in this theorem no assumption is made that the A_j are independent. In order to obtain a converse result some such assumption is necessary.

***Theorem* 13.2.** *Let $A_1, A_2, \ldots$ be events defined on the probability space $(\Omega, \mathscr{F}, \mathsf{P})$, and let $A = \limsup A_n$. Suppose moreover that the events A_n are independent. Then*
$$\sum_{n=1}^{\infty} \mathsf{P}(A_n) = \infty \tag{13.2.3}$$
implies that $$\mathsf{P}(A) = 1.$$

Proof. For any N, and any $M > N$

$$\begin{aligned}\mathsf{P} \bigcap_{n=N}^{\infty} (\Omega - A_n) &\leqslant \mathsf{P} \bigcap_{n=N}^{M} (\Omega - A_n) \\ &= \prod_{n=N}^{M} \mathsf{P}(\Omega - A_n) \\ &= \prod_{n=N}^{M} \{1 - \mathsf{P}(A_n)\} \\ &\leqslant \prod_{n=N}^{M} \exp\{-\mathsf{P}(A_n)\} \\ &= \exp\left\{-\sum_{n=N}^{M} \mathsf{P}(A_n)\right\}.\end{aligned}$$

Letting $M \to \infty$,

$$\begin{aligned}\mathsf{P} \bigcap_{n=N}^{\infty} (\Omega - A_n) &\leqslant \exp\left\{-\sum_{n=N}^{\infty} \mathsf{P}(A_n)\right\} \\ &= 0 \quad (\text{by } (13.2.3)).\end{aligned}$$

Hence, since $$\Omega - A = \bigcup_{N=1}^{\infty} \bigcap_{n=N}^{\infty} (\Omega - A_n),$$

$$1 - \mathsf{P}(A) \leqslant \sum_{N=1}^{\infty} \mathsf{P} \bigcap_{n=N}^{\infty} (\Omega - A_n) = 0,$$

so that $\mathsf{P}(A) = 1.$ ▌

It is important to note that theorem 13.1 applies to events which are not necessarily independent, while theorem 13.2 does not. If, however, the events A_n are independent, then the two theorems combine to imply that $\mathsf{P}(A)$ is 0 or 1 according as $\Sigma \mathsf{P}(A_n)$ converges or diverges.

It is worth while stressing the fact that, when the A_n are independent the event that infinitely many of them occur necessarily has probability either zero or one. This is an example of what is called a *zero-one law*, of which many exist in the theory of probability. In particular, it is a special case of the following result, which is often known as the zero-one law.

***Theorem* 13.3.** *Let $A_1, A_2, \ldots$ be events, and let $\mathscr{A}$ be the smallest σ-field containing each of these events. Suppose that E is an event in $\mathscr{A}$ with the property that, for any integers $j_1, j_2, \ldots, j_k$, the events*

$$E \quad \textit{and} \quad A_{j_1} \cap A_{j_2} \cap \ldots \cap A_{j_k}$$

are independent. Then $\mathsf{P}(E)$ is either 0 or 1.

Proof. Define Z_j to be equal to 1 on A_j and equal to 0 on $\Omega - A_j$. Then

$$\begin{aligned}\int_E Z_{j_1} Z_{j_2} \ldots Z_{j_k} d\mathsf{P} &= \mathsf{P}(A_{j_1} \cap A_{j_2} \cap \ldots \cap A_{j_k} \cap E) \\ &= \mathsf{P}(A_{j_1} \cap A_{j_2} \cap \ldots \cap A_{j_k}) \, \mathsf{P}(E) \\ &= \int_\Omega Z_{j_1} Z_{j_2} \ldots Z_{j_k} d\mathsf{P} \, \mathsf{P}(E).\end{aligned}$$

Hence
$$\int_E X \, d\mathsf{P} = \mathsf{E}(X) \, \mathsf{P}(E)$$

for any random variable X which is a linear combination of variables of the form $Z_{j_1} Z_{j_2} \ldots Z_{j_k}$. It is easy to check that the indicator of any event in the field $\mathscr{A}_0$ generated by the A_j is of this form, so that

$$\mathsf{P}(A \cap E) = \mathsf{P}(A) \, \mathsf{P}(E)$$

for all $A \in \mathscr{A}_0$. Theorem 4.2 shows that this holds also for all $A \in \mathscr{A}$, and in particular for $A = E$. Hence

$$\mathsf{P}(E) = \{\mathsf{P}(E)\}^2,$$

so that
$$\mathsf{P}(E) = 0 \quad \text{or} \quad 1.]$$

Corollary. *If $A_1, A_2, \ldots$ are independent events, and if E belongs, for all n, to the σ-field $\mathscr{A}_n$ generated by $A_n, A_{n+1}, A_{n+2}, \ldots$, then $\mathsf{P}(E)$ is either 0 or 1.*

Proof. It suffices to prove that

$$\mathsf{P}(A_{j_1} \cap A_{j_2} \cap \ldots \cap A_{j_k} \cap E) = \mathsf{P}(A_{j_1} \cap A_{j_2} \cap \ldots \cap A_{j_k}) \, \mathsf{P}(E).$$

Write $A = A_{j_1} \cap A_{j_2} \cap \ldots \cap A_{j_k}$, $n = \max(j_1, j_2, \ldots, j_k) + 1$.
Then, because the A_n are independent,

$$\mathsf{P}(A \cap A_{i_1} \cap A_{i_2} \cap \ldots \cap A_{i_p}) = \mathsf{P}(A)\,\mathsf{P}(A_{i_1} \cap A_{i_2} \cap \ldots \cap A_{i_p})$$

so long as $i_1, i_2, \ldots, i_p \geqslant n$. Hence, for all $B \in \mathscr{A}_n$,

$$\mathsf{P}(A \cap B) = \mathsf{P}(A)\,\mathsf{P}(B).$$

In particular, putting $B = E$, the result follows. ∎

An example of the situation in the corollary is given by the event

$$E = \limsup_{n \to \infty} A_n$$

discussed above, since $E = \bigcap_{N=n}^{\infty} \bigcup_{r=N}^{\infty} A_r \in \mathscr{A}_n$
for all n.

Exercises 13.2

1. A coin for which the probability p of falling heads satisfies $0 < p < 1$ is repeatedly tossed. Show that there is probability one that a run of 100 successive heads ultimately occurs if the tossing continues long enough.

2. Prove the following partial converse to theorem 13.1: if, for some $c > 0$,

$$\mathsf{P}(A_n) \geqslant c$$

for all n, then

$$\mathsf{P}(A) \geqslant c.$$

3. Let X be a random variable with distribution $U(0, 1)$, and let A_n be the event $\{X < 1/n\}$. Show that

$$\sum_{n=1}^{\infty} \mathsf{P}(A_n) = \infty,$$

but that

$$\mathsf{P}(\limsup_{n \to \infty} A_n) = 0.$$

Why does this not contradict theorem 13.2?

4. $X_1, X_2 \ldots$ are independent random variables, and E is the event that the series

$$\sum_{n=1}^{\infty} X_n$$

converges. Show that $\mathsf{P}(E)$ is either zero or one.

13.3 Sums of independent random variables

Much of the 'classical' theory of probability is concerned with the following problem. Let X_1, $X_2, \ldots$ be independent random variables having a common distribution function F, and write

$$S_0 = 0, \quad S_n = X_1 + X_2 + \ldots + X_n. \tag{13.3.1}$$

What can be said about the behaviour of the random variables S_n? This is a question which often arises in practice; the X_n may be small errors or random displacements, and then S_n is the cumulative error or the resultant displacement, and is often a quantity of interest. Of course, the methods of § 11.5, or those of Chapter 12, allow us, in principle, to compute the distribution of S_n. But even if such computations were practicable, they would not give us much intuitive idea of the way in which S_n behaves, particularly as n becomes large. The remainder of this chapter will therefore be devoted to different aspects of the behaviour of S_n for large values of n.

The simplest sort of result about S_n arises from the observation that S_n/n is just the average (i.e. the arithmetic mean) of $X_1, X_2, \ldots, X_n$. From the heuristic discussion of § 11.2 one might therefore expect that, for large n, S_n/n would tend to take values near $\mu = \mathsf{E}(X_n)$ (which of course, if it exists, is independent of n). This suggests the existence of results stating that, in some sense, S_n/n converges to μ as $n \to \infty$. Results of this sort are called *laws of large numbers*; they are precise formulations of the rough interpretation of the expectation of a random variable as the average of its values in a large number of independent trials.

The simplest result is concerned with mean square convergence, which is of course only meaningful when the random variables in question have finite variances.

***Theorem* 13.4.** *Let $X_1, X_2, \ldots$ be independent and identically distributed with $\mathsf{E}(X_n^2) < \infty$, and with $\mu = \mathsf{E}(X_n)$. Then*

$$\frac{X_1 + X_2 + \ldots + X_n}{n} \to \mu$$

in mean square as $n \to \infty$.

Proof.

$$\mathsf{E}\left(\frac{S_n}{n} - \mu\right)^2 = \frac{1}{n^2}\operatorname{var}(S_n)$$

$$= \frac{1}{n^2}\sum_{r=1}^{n}\operatorname{var}(X_r)$$

$$= \frac{1}{n}\operatorname{var}(X_1) \to 0$$

as $n \to \infty$.]

An immediate corollary, since mean square convergence implies convergence in probability, is that, if $\mathsf{E}(X_n^2) < \infty$, then

$$\operatorname*{p\,lim}_{n\to\infty} S_n/n = \mu.$$

This result is known as the *weak law of large numbers*. The restriction that the X_n have finite variance is, however, redundant; it suffices for the expectation μ to exist.

***Theorem* 13.5** (*Weak law of large numbers*). *Let $X_1, X_2, \ldots$ be independent and identically distributed, and suppose that the expectation $\mu = \mathsf{E}(X_n)$ exists and is finite. Then*

$$\text{(i)} \qquad \operatorname*{p\,lim}_{n\to\infty} \frac{X_1+X_2+\ldots+X_n}{n} = \mu$$

and (ii)
$$\frac{X_1+X_2+\ldots+X_n}{n} \rightsquigarrow D(\mu).$$

Proof. If
$$\phi(t) = \mathsf{E}(e^{itX_n}) = \int_{\mathbf{R}} e^{itx}\,dF(x),$$

the characteristic function of S_n/n is

$$\mathsf{E}(e^{itS_n/n}) = \prod_{r=1}^{n} \mathsf{E}(e^{itX_r/n})$$
$$= \{\phi(t/n)\}^n.$$

Theorem 12.6 shows that, for fixed t,

$$\phi(t/n) = 1+i\mu t/n+o(1/n),$$

so that
$$\mathsf{E}(e^{itS_n/n}) = \{1+i\mu t/n+o(1/n)\}^n$$
$$\to e^{i\mu t}$$

as $n\to\infty$. But $e^{i\mu t}$ is the characteristic function of $D(\mu)$, and so, by theorem 12.8,
$$S_n/n \rightsquigarrow D(\mu).$$

It follows that, for any $\epsilon > 0$,

$$\mathsf{P}(S_n/n < \mu-\epsilon) \to 0,$$

$$\mathsf{P}(S_n/n > \mu+\epsilon) \to 0,$$

so that
$$\mathsf{P}(|S_n/n-\mu| > \epsilon) \to 0$$

as $n\to\infty$. Hence
$$\operatorname*{p\,lim}_{n\to\infty} S_n/n = \mu,$$

and the theorem is proved. ▮

The last two theorems show that, for mean square convergence and for convergence in probability, the statement

$$S_n/n \to \mu$$

is true whenever it has meaning. It is therefore of interest to ask whether the same is true of almost sure convergence. The statement $S_n/n \to \mu$ with probability one is meaningful whenever μ exists, and is much stronger than the conclusion of theorem 13.5. It is in fact true, but is a much deeper result and requires a much more involved proof. The proof is based on an inequality due to Kolmogorov, which can be regarded as a generalisation of Tchebychev's inequality.

***Theorem* 13.6** (*Kolmogorov's inequality*). *Let $Y_1, Y_2, \ldots, Y_n$ be independent (but not necessarily identically distributed) random variables with* $\mathsf{E}(Y_r) = 0$, $\mathsf{E}(Y_r^2) < \infty$. *Then, for all $\epsilon > 0$,*

$$\mathsf{P}\{|Y_1+Y_2+\ldots+Y_r| > \epsilon \quad \textit{for some } r\} \leqslant \epsilon^{-2} \sum_{r=1}^{n} \operatorname{var}(Y_r). \qquad (13.3.2)$$

Proof. Let A_r denote the event

$$A_r = \{|Y_1+Y_2+\ldots+Y_s| \leqslant \epsilon\,(s = 1, 2, \ldots, r-1), |Y_1+Y_2+\ldots+Y_r| > \epsilon\},$$

so that the A_r are disjoint and

$$A = \bigcup_{r=1}^{n} A_r = \{|Y_1+Y_2+\ldots+Y_r| > \epsilon \quad \text{for some } r\}.$$

Write I_r for the random variable

$$\begin{aligned} I_r(\omega) &= 1 \quad (\omega \in A_r) \\ &= 0 \quad (\omega \notin A_r). \end{aligned}$$

Then, $I_r, Y_{r+1}, Y_{r+2}, \ldots, Y_n$ are independent, and

$$\begin{aligned}
\sum_{r=1}^{n} \operatorname{var}(Y_r) &= \operatorname{var}(Y_1+Y_2+\ldots+Y_n) \\
&= \mathsf{E}(Y_1+Y_2+\ldots+Y_n)^2 \\
&\geqslant \int_A (Y_1+Y_2+\ldots+Y_n)^2\, d\mathsf{P} \\
&= \sum_{r=1}^{n} \int_{A_r} (Y_1+Y_2+\ldots+Y_n)^2\, d\mathsf{P} \\
&= \sum_{r=1}^{n} \mathsf{E}\{I_r(Y_1+Y_2+\ldots+Y_n)^2\} \\
&= \sum_{r=1}^{n} \mathsf{E}\{I_r(Y_1+Y_2+\ldots+Y_r)^2 + I_r(Y_{r+1}+\ldots+Y_n)^2 \\
&\qquad + 2I_r(Y_1+Y_2+\ldots+Y_r)(Y_{r+1}+\ldots+Y_n)\} \\
&= \sum_{r=1}^{n} [\mathsf{E}\{I_r(Y_1+Y_2+\ldots+Y_r)^2\} + \mathsf{E}(I_r)\,\mathsf{E}(Y_{r+1}+\ldots+Y_n)^2 \\
&\qquad + 2\mathsf{E}\{I_r(Y_1+Y_2+\ldots+Y_r)\}\,\mathsf{E}(Y_{r+1}+\ldots+Y_n)].
\end{aligned}$$

The second term is non-negative and the third vanishes, so that

$$\sum_{r=1}^{n} \operatorname{var}(Y_r) \geqslant \sum_{r=1}^{n} \mathsf{E}\{I_r(Y_1+Y_2+\ldots+Y_r)^2\}$$

$$= \sum_{r=1}^{n} \int_{A_r} (Y_1+Y_2+\ldots+Y_r)^2 \, d\mathsf{P}$$

$$\geqslant \sum_{r=1}^{n} \int_{A_r} \epsilon^2 \, d\mathsf{P}$$

$$= \epsilon^2 \sum_{r=1}^{n} \mathsf{P}(A_r)$$

$$= \epsilon^2 \mathsf{P}(A).$$

Hence

$$\mathsf{P}(A) \leqslant \epsilon^{-2} \sum_{r=1}^{n} \operatorname{var}(Y_r). \rrbracket$$

***Theorem* 13.7** (*Strong law of large numbers*). *Let $X_1, X_2, \ldots$ be independent, with common distribution function F, and suppose that*

$$\mu = \int_{\mathbf{R}} x \, dF(x)$$

exists and is finite. Then with probability one,

$$\frac{X_1+X_2+\ldots+X_n}{n} \to \mu \tag{13.3.3}$$

as $n \to \infty$.

Proof. First define

$$\bar{X}_n = X_n \quad \text{if} \quad |X_n| \leqslant n,$$

$$= 0 \quad \text{if} \quad |X_n| > n.$$

Then

$$\sum_{n=1}^{\infty} \mathsf{P}(\bar{X}_n \neq X_n) = \sum_{n=1}^{\infty} \mathsf{P}(|X_n| > n)$$

$$= \sum_{n=1}^{\infty} \int_{|n|>n} dF(x)$$

$$= \int_{-\infty}^{\infty} \left(\sum_{n<|x|} 1 \right) dF(x)$$

$$\leqslant \int_{-\infty}^{\infty} |x| \, dF(x) < \infty,$$

so that, by theorem 13.1, there is probability one that $\bar{X}_n = X_n$ for all but finitely many n. Hence it suffices to prove (13.3.3) with X_n replaced by $\bar{X}_n$. Moreover,

$$\mathsf{E}(\bar{X}_n) = \int_{-n}^{n} x\,dF(x) \to \mu$$

as $n \to \infty$, so that

$$\{\mathsf{E}(X_1) + \mathsf{E}(X_2) + \ldots + \mathsf{E}(X_n)\}/n \to \mu$$

as $n \to \infty$. Hence it suffices to prove that

$$(Y_1 + Y_2 + \ldots + Y_n)/n \to 0$$

with probability one, where

$$Y_n = \bar{X}_n - \mathsf{E}(\bar{X}_n).$$

The object of truncating X_n is to use Kolmogorov's inequality, since Y_n has finite variance. This variance depends on n, but does not increase too fast. In fact,

$$\operatorname{var}(Y_n) = \operatorname{var}(\bar{X}_n) \leqslant \mathsf{E}(\bar{X}_n^2) = \int_{-n}^{n} x^2\,dF(x),$$

so that
$$\sum_{n=1}^{\infty} n^{-2} \operatorname{var}(Y_n) \leqslant \sum_{n=1}^{\infty} n^{-2} \int_{-n}^{n} x^2\,dF(x)$$
$$= \int_{-\infty}^{\infty} x^2 \Big(\sum_{n>|x|} n^{-2} \Big)\,dF(x)$$
$$\leqslant \int_{-\infty}^{\infty} x^2 \frac{1}{m-\frac{1}{2}}\,dF(x) \leqslant \int_{-\infty}^{\infty} 2|x|\,dF(x),$$

where m is the smallest integer greater than $|x|$.

Therefore
$$\sum_{n=1}^{\infty} n^{-2} \operatorname{var}(Y_n) < \infty. \tag{13.3.4}$$

We have to prove that $\frac{1}{n} \sum_{r=1}^{n} Y_r \to 0$

as $n \to \infty$. It is convenient first to prove this for a special sequence of values of n, namely, the powers of 2. Thus let

$$\eta_k = \sum_{n=1}^{2^k} Y_n,$$

fix $\epsilon > 0$, and let A_k be the event $\{|\eta_k 2^{-k}| > \epsilon\}$. Then, by Tchebychev's inequality,

$$\begin{aligned} \mathsf{P}(A_k) &\leqslant 2^{-2k}\epsilon^{-2}\operatorname{var}(\eta_k) \\ &= 2^{-2k}\epsilon^{-2}\sum_{n=1}^{2^k}\operatorname{var}(Y_k). \end{aligned}$$

Therefore

$$\begin{aligned} \sum_{k=0}^{\infty}\mathsf{P}(A_k) &\leqslant \epsilon^{-2}\sum_{k=0}^{\infty}\sum_{n=1}^{2^k} 2^{-2k}\operatorname{var}(Y_n) \\ &= \epsilon^{-2}\sum_{n=1}^{\infty}\operatorname{var}(Y_n)\sum_{k=k(n)}^{\infty} 2^{-2k}, \end{aligned}$$

where

$$k(n) = \min\{k;\ 2^k \geqslant n\}.$$

Thus

$$\begin{aligned} \sum_{k=0}^{\infty}\mathsf{P}(A_k) &\leqslant \epsilon^{-2}\sum_{n=1}^{\infty}\operatorname{var}(Y_n)\tfrac{4}{3}2^{-2k(n)} \\ &\leqslant \tfrac{4}{3}\epsilon^{-2}\sum_{n=1}^{\infty} n^{-2}\operatorname{var}(Y_n) \\ &< \infty \quad (\text{by } (13.3.4)). \end{aligned}$$

Hence, by theorem 13.1, the event

$$B(\epsilon) = \{|\eta_k 2^{-k}| > \epsilon \quad \text{for infinitely many } k\}$$

has probability zero, and therefore so has

$$\bigcup_{m=1}^{\infty} B\left(\frac{1}{m}\right) = \{\eta_k 2^{-k} \nrightarrow 0\}.$$

Therefore

$$\frac{\eta_k}{2^k} \to 0 \quad (k \to \infty)$$

with probability one.

It now remains only to 'fill up the gaps' between the powers of 2. For any k, consider the random variable

$$\zeta_k = \max\left\{\left|\sum_{r=2^{k-1}+1}^{n} Y_r\right|;\ 2^{k-1} < n \leqslant 2^k\right\}.$$

By Kolmogorov's inequality, for any $\epsilon > 0$,

$$\begin{aligned} \mathsf{P}(\zeta_k 2^{-k} > \epsilon) &\leqslant \epsilon^{-2}2^{-2k}\sum_{r=2^{k-1}+1}^{2^k}\operatorname{var}(Y_r) \\ &\leqslant \epsilon^{-2}\sum_{r=2^{k-1}+1}^{2^k} r^{-2}\operatorname{var}(Y_r), \end{aligned}$$

and so

$$\begin{aligned} \sum_{k=0}^{\infty}\mathsf{P}(\zeta_k 2^{-k} > \epsilon) &\leqslant \epsilon^{-2}\sum_{r=2}^{\infty} r^{-2}\operatorname{var}(Y_r) \\ &< \infty \quad (\text{by } (13.3.4)). \end{aligned}$$

Hence, by theorem 13.1,

$$\mathsf{P}(\zeta_k 2^{-k} > \epsilon \quad \text{for infinitely many } k) = 0,$$

and so, as before, $\zeta_k 2^{-k} \to 0 \quad (k \to \infty)$

with probability one.

But, if $2^{k-1} < n \leqslant 2^k$,

$$\left|\frac{Y_1 + Y_2 + \ldots + Y_n}{n}\right| = \left|\frac{\eta_{k-1} + \sum_{r=2^{k-1}+1}^{n} Y_r}{n}\right| \leqslant \frac{\eta_{k-1} + \zeta_k}{2^{k-1}}$$

$$\to 0$$

with probability one, as n(and so k) tends to infinity. ∎

The difference between the weak and the strong laws of large numbers can be seen if they are written in the following equivalent forms:

weak law $\mathsf{P}\left(\left|\frac{S_n}{n} - \mu\right| > \epsilon\right) \to 0 \quad (n \to \infty, \quad \text{all} \quad \epsilon > 0),$

strong law $\mathsf{P}\left(\left|\frac{S_r}{r} - \mu\right| > \epsilon \quad \text{for some} \quad r \geqslant n\right) \to 0$

$$(n \to \infty, \quad \text{all} \quad \epsilon > 0).$$

The difference lies in the 'for some $r \geqslant n$', which means that we are not only demanding that S_r/r be close to μ but that it remain close to μ as r increases. It is this that makes the strong law the deeper result and necessitates the much more difficult proof. An alternative proof of theorem 13.7, depending on the point-wise ergodic theorem, will be given in Chapter 15.

Exercises 13.3

1. With the notation of theorem 13.7, show that,

$$\text{if} \quad \mu > 0, \quad \lim_{n \to \infty} S_n = +\infty,$$

$$\text{if} \quad \mu < 0, \quad \lim_{n \to \infty} S_n = -\infty.$$

2. Show that the strong law of large numbers remains true even if the X_n are not identically distributed, if $\mu = \mathsf{E}(X_n)$ for all n, and

$$\sum_{n=1}^{\infty} n^{-2}\, \mathsf{E}\{\min(X_n^2, n^2)\} < \infty.$$

3. Use the result of exercise 13.1 (1), together with theorem 13.7, to show that almost all real numbers (i.e. all but a set of zero Lebesgue measure) have the property that the frequency of any possible digit among the first n digits of their decimal expansions tends to $\frac{1}{10}$ as $n \to \infty$. Any such number is called *normal to base* 10; normality to base r is similarly defined for $r = 2, 3, 4, \ldots$. Prove that almost all numbers are normal to every base.

4. Use the distribution $P(n)$ to show that

$$1+n+\frac{n^2}{2!}+\frac{n^3}{3!}+\ldots+\frac{n^n}{n!} \sim \tfrac{1}{2}e^n$$

as $n \to \infty$.

5. A sequence $E_1, E_2, \ldots$, of independent events satisfies $\mathsf{P}(E_n) = p$ for all n. Use the random variables χ_{E_n} to show that, if

$$r_n(\omega) = \text{number of } j \leqslant n \text{ with } \omega \in E_j,$$

then

$$\frac{r_n(\omega)}{n} \to p$$

as $n \to \infty$ for almost all ω.

13.4 The central limit theorem

If $X_1, X_2, \ldots$, are independent and identically distributed, and if $S_n = X_1 + X_2 + \ldots + X_n$, the results of the last section show that, for large n, there is high probability that S_n/n is near to the expectation μ (if it exists). Just how near it is likely to be is the subject of this section.

Suppose that the X_n have finite mean μ and variance σ^2. The case of zero variance is trivial (for then $\mathsf{P}(S_n = n\mu) = 1$), and we therefore take $\sigma > 0$. Then the results of § 11.5 show that

$$\mathsf{E}(S_n) = n\mu, \quad \operatorname{var}(S_n) = n\sigma^2, \tag{13.4.1}$$

so that the *standardised* variable

$$Z_n = \frac{S_n - n\mu}{\sigma n^{\frac{1}{2}}} \tag{13.4.2}$$

has

$$\mathsf{E}(Z_n) = 0, \quad \operatorname{var}(Z_n) = 1$$

for all n. The remarkable and celebrated 'central limit theorem' associated with the names of De Moivre and Laplace states that, whatever the distribution of the X_n, the distribution of Z_n converges, as $n \to \infty$, to the normal distribution $N(0, 1)$.

The proof of this result is quite simple, but it illustrates the power of the technique of characteristic functions.

***Theorem* 13.8** (*Central limit theorem*). *Let $X_1, X_2, \ldots$, be independent and identically distributed, with finite mean μ and finite variance σ^2 (where $\sigma > 0$). Then, if Z_n is defined by* (13.4.2),

$$\mathsf{P}(Z_n \leqslant z) \to (2\pi)^{-\frac{1}{2}} \int_{-\infty}^{z} e^{-\frac{1}{2}x^2}\, dx, \tag{13.4.3}$$

or symbolically

$$Z_n \rightsquigarrow N(0,1) \tag{13.4.4}$$

as $n \to \infty$.

Proof. Consider the random variable

$$X'_n = (X_n - \mu)/\sigma.$$

Then X'_n has mean zero and variance one, and so, by theorem 12.6,

$$\mathsf{E}(e^{itX'_n}) = 1 - \tfrac{1}{2}t^2 + o(t^2)$$

as $t \to 0$. But

$$Z_n = n^{-\frac{1}{2}}(X'_1 + X'_2 + \dots + X'_n),$$

and so, for fixed t,

$$\begin{aligned} \mathsf{E}(e^{itZ_n}) &= \prod_{j=1}^{n} \mathsf{E}(e^{itn^{-\frac{1}{2}}X'_j}) \\ &= \{\mathsf{E}(e^{itn^{-\frac{1}{2}}X'_1})\}^n \\ &= \left\{1 - \frac{t^2}{2n} + o\left(\frac{1}{n}\right)\right\}^n \\ &\to e^{-\frac{1}{2}t^2} \end{aligned}$$

as $n \to \infty$. But $e^{-\frac{1}{2}t^2}$ is the characteristic function of $N(0,1)$, and so, by theorem 12.8,

$$Z_n \rightsquigarrow N(0,1).$$

Because the distribution function of $N(0,1)$, which is the right-hand side of (13.4.3), is continuous, this is equivalent to (13.4.3). ∎

This theorem, since it gives information about the distribution of S_n as $n \to \infty$, is a strengthening of the weak law of large numbers, but one which is only achieved by replacing the assumption $\mathsf{E}|X_n| < \infty$ by the more stringent one that $\mathsf{E}(X_n^2) < \infty$. There is a corresponding strengthening of the strong law, which will be discussed in the next section.

As a simple example, suppose that X_n takes only the two values 0 and 1, and set $p = \mathsf{P}(X_n = 1)$. Then S_n is just the number of $r \leqslant n$ for which $X_r = 1$, and so has the binomial distribution $B(n,p)$. Then (13.4.4) asserts that

$$\frac{S_n - np}{\{np(1-p)\}^{\frac{1}{2}}} \rightsquigarrow N(0,1),$$

which is just a different way of writing the result (10.5.7) which was earlier proved directly.

The central limit theorem only of course applies when the X_n have finite variance, and it might be asked what happens when this is not the case. Suppose, for example, that the X_n have the Cauchy distribution, with probability density

$$\frac{1}{\pi(1+x^2)}.$$

Then (exercise 12.2.(2)) (iii))

$$\mathsf{E}(e^{itX_n}) = e^{-|t|},$$

and so

$$\mathsf{E}(e^{itS_n}) = \{e^{-|t|}\}^n = e^{-n|t|}.$$

It follows that

$$\mathsf{E}(e^{itS_n/n}) = e^{-|t|},$$

so that S_n/n has a Cauchy distribution. Thus the weak law of large numbers is violated. Moreover S_n, when normalised so that it has a non-trivial limiting distribution, has a distribution which converges not to a normal distribution but to a Cauchy distribution.

The Cauchy distribution is an example of a class of probability distributions known as the *symmetric stable laws*. These are distributions whose characteristic functions are of the form

$$\exp(-c|t|^\alpha), \tag{13.4.5}$$

where $c > 0$. It is easy to show (exercise 12.3 (8)) that such distributions exist if (and only if) $0 < \alpha \leqslant 2$. The Cauchy distribution has $\alpha = 1$; when $\alpha = 2$ the distributions are normal distributions. If the X_n have characteristic function (13.4.5), it is trivial to verify that, for any n, $S_n/n^{1/\alpha}$ has the same distribution as X_1.

In general, however, if $\mathsf{E}(X_n^2) = \infty$, it is not possible to find sequences of numbers a_n, b_n such that the distribution of

$$\frac{S_n - a_n}{b_n}$$

converges to a non-degenerate distribution. The reader interested in the extensive work that has been done on this problem, and in the general topic of sums of independent random variables, should refer to B. V. Gnedenko and A. N. Kolmogorov, *Limit distributions for Sums of Independent Random Variables*, Addison-Wesley, 1954, or to M. Loève, *Probability Theory*, van Nostrand, 1963 (Chapter vi).

Exercises 13.4

1. The cumulants $\kappa_r(X)$ of a random variable X were defined in exercise 12.2.(3). Show that, if $\kappa_r = \kappa_r(X_n)$ exists, then, if Z is given by (13.4.2)

$$\kappa_r(S_n) = n\kappa_r,$$

and

$$\kappa_r(Z_n) = n^{-\frac{1}{2}(r-2)}\kappa_r\,\sigma^{-r} \quad (r \geqslant 2).$$

Deduce that, if $\mathsf{E}|X_n|^k < \infty$, then as $n \to \infty$,

$$\mathsf{E}(Z_n^{2k}) \to \frac{(2k)!}{k!2^k}.$$

13.5 The law of the iterated logarithm

Let $X_1, X_2, \ldots$ be independent and identically distributed with mean μ and variance σ^2 $(\sigma > 0)$, and let S_n, Z_n be defined as in the previous section. The central limit theorem shows that, when n is large, the distribution of Z_n approximates to $N(0,1)$, Thus, for a given large value of n, Z_n is unlikely to be large. For instance, since

$$(2\pi)^{-\frac{1}{2}} \int_{10}^{\infty} e^{-\frac{1}{2}x^2} dx \simeq 10^{-23},$$

the probability that $Z_n > 10$ is, for large n, of the order of 10^{-23}, which is for most purposes negligible. If, however, we look at all Z_n for n between say, 10^{24} and 10^{25}, it is plausible that a few, at least, of these values will happen to rise above 10. Similarly, one might expect that a few values of Z_n, as n increases, might be arbitrarily large.

Just how large these outlying values of Z_n are likely to be is the content of the *law of the iterated logarithm*. This states that, with probability one, the inequality

$$Z_n > (c \log\log n)^{\frac{1}{2}} \tag{13.5.1}$$

is satisfied for infinitely many n if $c < 2$, but for only finitely many if $c > 2$. In other words,

$$\limsup_{n\to\infty} Z_n/(2\log\log n)^{\frac{1}{2}} = 1. \tag{13.5.2}$$

Similarly, with probability one,

$$\liminf_{n\to\infty} Z_n/(2\log\log n)^{\frac{1}{2}} = -1. \tag{13.5.3}$$

In its full generality, this result was proved by P. Hartman and A. Wintner ('On the law of the iterated logarithm', *American Journal of Mathematics* **63** (1941) 169–76). The details are complicated, and will not be reproduced here. In order, however, to illustrate the main ideas, we give the proof for the very special case in which the X_n have a normal distribution.

***Theorem* 13.9.** *Let $X_1, X_2, \ldots$ be independent random variables with the distribution $N(0, 1)$. Then with probability one,*

$$\limsup_{n\to\infty} \frac{X_1+X_2+\ldots+X_n}{(2n\log\log n)^{\frac{1}{2}}} = 1. \tag{13.5.4}$$

* *Proof.* As in the strong law of large numbers, we first prove the result for a suitably chosen subsequence $\{n_j\}$ of values of n, and then show that the full result follows. Three points of notation: as usual

$$S_n = X_1+X_2+\ldots+X_n, \quad Z_n = S_n/n^{\frac{1}{2}},$$

ψ will denote the function

$$\begin{aligned}\psi(x) &= (2\log\log x)^{\frac{1}{2}} \quad (x \geqslant 3)\\ &= \psi(3) \qquad\qquad (x < 3),\end{aligned}$$

and Φ will denote the distribution function

$$\Phi(x) = (2\pi)^{-\frac{1}{2}}\int_{-\infty}^{x} e^{-\frac{1}{2}y^2}\,dy$$

of $N(0, 1)$. We recall (exercise 11.7.(9)) that

$$1-\Phi(x) \sim (2\pi)^{-\frac{1}{2}}x^{-1}e^{-\frac{1}{2}x^2} \tag{13.5.5}$$

as $x \to +\infty$. Notice also that S_n has distribution $N(0, n)$, so that, in particular,

$$\mathsf{P}(S_n > 0) = \tfrac{1}{2}.$$

Exactly similarly, if $m < n$,

$$\mathsf{P}(X_{m+1}+X_{m+2}+\ldots+X_n > 0) = \tfrac{1}{2},$$

i.e.

$$\mathsf{P}(S_n > S_m) = \tfrac{1}{2}.$$

For any x, let A_n, B_n denote the events

$$A_n = \{S_n > x\},$$

$$B_n = \{S_1, S_2, \ldots, S_{n-1} \leqslant x, S_n > x\}.$$

Then A_n can be expressed as the disjoint union

$$A_n = \bigcup_{k=1}^{n} (B_k \cap A_n),$$

and therefore

$$\mathsf{P}(A_n) = \sum_{k=1}^{n} \mathsf{P}(B_k \cap A_n).$$

But B_k, together with $\{S_n > S_k\}$, implies A_n, and so

$$B_k \cap A_n \supset B_k \cap \{S_n > S_k\}.$$

Hence
$$\begin{aligned}\mathsf{P}(B_k \cap A_n) &\geqslant \mathsf{P}(B_k \cap \{S_n > S_k\})\\ &= \mathsf{P}(B_k)\,\mathsf{P}(S_n > S_k)\\ &= \tfrac{1}{2}\mathsf{P}(B_k),\end{aligned}$$
and therefore
$$\begin{aligned}\mathsf{P}(A_n) &\geqslant \sum_{k=1}^{n} \tfrac{1}{2}\mathsf{P}(B_k)\\ &= \tfrac{1}{2}\mathsf{P}\bigcup_{k=1}^{n} B_k\\ &= \tfrac{1}{2}\mathsf{P}(S_k > x \quad \text{for some} \quad k \leqslant n).\end{aligned}$$
Using the fact that S_n has distribution $N(0, n)$, it follows that
$$\mathsf{P}(S_k > x \quad \text{for some} \quad k \leqslant n) \leqslant 2\{1 - \Phi(n^{-\frac{1}{2}}x)\}. \qquad (13.5.6)$$

After these preliminary remarks, we can now proceed to the proof of the theorem. It suffices to prove that the following two assertions hold with probability one.

(I) If $\lambda > 1$, there are at most finitely many n with
$$S_n > \lambda n^{\frac{1}{2}}\psi(n).$$
(II) If $\lambda < 1$, there are infinitely many n with
$$S_n > \lambda n^{\frac{1}{2}}\psi(n).$$

To prove (I), fix $\lambda > 1$, and let n_j be the integer part of λ^j. Then
$$\begin{aligned}&\mathsf{P}(S_n > \lambda n^{\frac{1}{2}}\psi(n) \quad \text{for some } n \text{ in} \quad n_j < n \leqslant n_{j+1})\\ &\quad\leqslant \mathsf{P}(S_n > \lambda n_j^{\frac{1}{2}}\psi(n_j) \quad \text{for some } n \text{ in} \quad n_j < n \leqslant n_{j+1})\\ &\quad\leqslant \mathsf{P}(S_n > \lambda n_j^{\frac{1}{2}}\psi(n_j) \quad \text{for some} \quad n \leqslant n_{j+1})\\ &\quad\leqslant 2\{1 - \Phi[n_{j+1}^{-\frac{1}{2}}\lambda n_j^{\frac{1}{2}}\psi(n_j)]\}, \quad \text{by} \quad (13.5.6),\\ &\quad= P_j \quad \text{(say)}.\end{aligned}$$
Now, from (13.5.5),
$$\begin{aligned}P_j &\sim (2/\pi)^{\frac{1}{2}}(n_{j+1}/n_j)^{\frac{1}{2}}\lambda^{-1}\{\psi(n_j)\}^{-1}\exp[-\tfrac{1}{2}\lambda^2(n_j/n_{j+1})\{\psi(n_j)\}^2]\\ &\sim (2/\lambda\pi)^{\frac{1}{2}}\{\psi(n_j)\}^{-1}\exp[-\lambda^2(n_j/n_{j+1})\log\log n_j].\end{aligned}$$
Also $\psi(n_j) \geqslant \psi(3)$, and $(n_j/n_{j+1}) \to \lambda^{-1}$ as $j \to \infty$. Hence, for sufficiently large j,
$$(n_j/n_{j+1}) > \lambda^{-\frac{3}{2}},$$
and so
$$\begin{aligned}P_j &< (2/\lambda\pi)^{\frac{1}{2}}\{\psi(3)\}^{-1}\exp[-\lambda^{\frac{1}{2}}\log\log n_j]\\ &= c\exp[-\lambda^{\frac{1}{2}}\log\log n_j], \text{ say},\\ &= c(\log n_j)^{-\lambda^{\frac{1}{2}}}\\ &\sim dj^{-\lambda^{\frac{1}{2}}}, \quad 0 < d < \infty.\end{aligned}$$

Hence, since $\lambda^{\frac{1}{2}} > 1$,
$$\sum_{j=1}^{\infty} P_j < \infty,$$
and so
$$\sum_{j=1}^{\infty} \mathsf{P}(S_n > \lambda n^{\frac{1}{2}}\psi(n) \quad \text{for some } n \text{ in} \quad n_j < n \leqslant n_{j+1}) < \infty.$$
It then follows from theorem 13.1, that, with probability one, there are at most finitely many intervals $(n_j, n_{j+1}]$ which contain a value of n for which $S_n > \lambda n^{\frac{1}{2}}\psi(n)$. Hence there are at most finitely many such values of n, and (I) holds with probability one.

To prove (II), fix $\lambda < 1$, and let ν be an integer so large that
$$\lambda\nu^{\frac{1}{2}} + 2 < (\nu - 1)^{\frac{1}{2}}. \tag{13.5.7}$$
Write
$$\Sigma_r = S_{\nu^r}, \quad \sigma_r = \Sigma_r - \Sigma_{r-1},$$
so that the σ_r are independent, and σ_r has distribution $N(0, \nu^r - \nu^{r-1})$. In the assertion (I) that has already been proved, replace X_n by $-X_n$ (for all n) and set $\lambda = 2$, $n = \nu^{r-1}$. Then, with probability one,
$$\Sigma_{r-1} \geqslant -2\nu^{\frac{1}{2}(r-1)}\psi(\nu^{r-1})$$
for all but finitely many r. Hence, for all but finitely many r,
$$\Sigma_r \geqslant \sigma_r - 2\nu^{\frac{1}{2}(r-1)}\psi(\nu^{r-1}),$$
and in order to prove (II) it suffices to show that, with probability one,
$$\sigma_r > \lambda\nu^{\frac{1}{2}r}\psi(\nu^r) + 2\nu^{\frac{1}{2}(r-1)}\psi(\nu^{r-1}),$$
for infinitely many r.

Using (13.5.7) and the fact that ψ is non-decreasing, we see that it suffices to prove that
$$\sigma_r > (\nu - 1)^{\frac{1}{2}}\nu^{\frac{1}{2}(r-1)}\psi(\nu^r),$$
for infinitely many r. Moreover, since the σ_r are independent we can use theorem 13.2; it therefore suffices to show that
$$\sum_{r=1}^{\infty} \mathsf{P}\{\sigma_r > (\nu - 1)^{\frac{1}{2}}\nu^{\frac{1}{2}(r-1)}\psi(\nu^r)\} = \infty,$$
i.e. that
$$\sum_{r=1}^{\infty} \{1 - \Phi[\psi(\nu^r)]\} = \infty.$$
This is true since, as $r \to \infty$,
$$\begin{aligned} 1 - \Phi[\psi(\nu^r)] &\sim (2\pi)^{-\frac{1}{2}}[\psi(\nu^r)]^{-1}\exp\{-\tfrac{1}{2}[\psi(\nu^r)]^2\} \\ &= [4\pi \log\log \nu^r]^{-\frac{1}{2}}\exp\{-\log\log \nu^r\} \\ &\sim (4\pi)^{-\frac{1}{2}}(\log r)^{-\frac{1}{2}}r^{-1}, \end{aligned}$$
and
$$\sum_{r=2}^{\infty} (\log r)^{-\frac{1}{2}}r^{-1} = \infty.$$
Thus (II) holds with probability one.

To complete the proof of the theorem, note that (I) implies that

$$\mathsf{P}\left\{\limsup_{n\to\infty} \frac{S_n}{(2n \log\log n)^{\frac{1}{2}}} > 1 + \frac{1}{k}\right\} = 0$$

for $k = 1, 2, \ldots$, so that

$$\mathsf{P}\left\{\limsup_{n\to\infty} \frac{S_n}{(2n \log\log n)^{\frac{1}{2}}} > 1\right\} = 0,$$

and that similarly (II) implies that

$$\mathsf{P}\left\{\limsup_{n\to\infty} \frac{S_n}{(2n \log\log n)^{\frac{1}{2}}} < 1\right\} = 0. \blacksquare$$

The proof in the general case follows similar lines, and uses the fact that, for large n, Z_n is approximately normally distributed. It requires however, a stronger version of this result than is contained in the central limit theorem, and in particular a good estimate for

$$\mathsf{P}(Z_n > \theta\psi(n))$$

for fixed θ and large n.

Exercises 13.5

1. A positive increasing function ϕ is said to belong to the *lower class* for the sequence $\{S_n\}$ if

$$\mathsf{P}(S_n > n^{\frac{1}{2}}\phi(n) \text{ for infinitely many } n) = 1,$$

and to the *upper class* if this probability is zero. The zero-one law shows that every ϕ belongs to one or other of these classes. The law of the iterated logarithm then states that $(c \log\log n)^{\frac{1}{2}}$ belongs to the upper class if $c > 2$, and to the lower class if $c < 2$. Modify the proof of theorem 13.9 to show that, under the conditions of that theorem, ϕ belongs to the upper class if and only if

$$\int_1^\infty \frac{1}{t\phi(t)} e^{-\frac{1}{2}\{\phi(t)\}^2}\, dt < \infty.$$

In particular,

$$(2 \log\log n)^{\frac{1}{2}}$$

belongs to the lower class. For what values of α does

$$(2 \log\log n + \alpha \log\log\log n)^{\frac{1}{2}}$$

belong to the lower class?

14

FINITE COLLECTIONS OF RANDOM VARIABLES

14.1 Joint distributions

In the last chapter the emphasis was on collections of random variables which had the very special property of statistical independence. In general, a collection of random variables does not have this property, and this chapter is concerned with the machinery for describing and manipulating such collections. The present section is devoted to an analogue for finite collections of the theory of the probability distribution which was developed in § 11.3 for a single random variable.

Let $(\Omega, \mathscr{F}, \mathsf{P})$ be a probability space on which is defined a collection $X = (X_1, X_2, \ldots, X_n)$ of random variables. We shall assume for simplicity that each X_r is almost certainly finite, so that X_r is a measurable function from $(\Omega, \mathscr{F})$ into $(\mathbf{R}, \mathscr{B})$. It follows that the function X which sends $\omega \in \Omega$ into the point

$$(X_1(\omega), X_2(\omega), \ldots, X_n(\omega)) \in \mathbf{R}^n$$

is a measurable function from $(\Omega, \mathscr{F})$ into $(\mathbf{R}^n, \mathscr{B}^n)$.

The measure P now induces a measure P' on $(\mathbf{R}^n, \mathscr{B}^n)$ by

$$\mathsf{P}'(B) = \mathsf{P}\{\omega;\ X(\omega) \in B\} \quad (B \in \mathscr{B}^n). \tag{14.1.1}$$

If $\mathscr{F}'$ is the completion of $\mathscr{B}^n$ with respect to P', then $(\mathbf{R}^n, \mathscr{F}', \mathsf{P}')$ is a probability space. The measure P' is called the *joint probability distribution* of the collection X. In the case $n = 1$ dealt with in § 11.3, P' was identified with the Stieltjes measure induced by the distribution function of the random variable. A similar identification is possible in the general case, but it is much more complicated and correspondingly less useful, and will not be used here.

The following theorem is an immediate consequence of theorem 6.8.

***Theorem* 14.1.** *If the collection $X = (X_1, X_2, \ldots, X_n)$ has joint distribution P' and if $g\colon \mathbf{R}^n \to \mathbf{R}$ is Borel measurable, then*

$$\mathsf{E}\{g(X_1, X_2, \ldots, X_n)\} = \int_{\mathbf{R}^n} g(x_1, x_2, \ldots, x_n)\, d\mathsf{P}', \tag{14.1.2}$$

in the sense that, if either side exists, then so does the other, and they are then equal.

As in the case of a single random variable there are two simple types of joint distribution which often occur. The first is the *discrete* distribution, which is the distribution of a collection $X = (X_1, X_2, \ldots, X_n)$ with the property that each X_r takes on only a countable number of possible values. Then X takes on only a countable number of possible values $\xi_1, \xi_2, \ldots$ in $\mathbf{R}^n$, and if $p_j = \mathsf{P}(X = \xi_j)$, the measure P' is given by

$$\mathsf{P}'(B) = \sum_{\{j;\, \xi_j \in B\}} p_j. \tag{14.1.3}$$

The points ξ_j are completely arbitrary; the probabilities p_j must, of course, satisfy

$$p_j \geqslant 0, \quad \sum_{j=1}^{\infty} p_j = 1. \tag{14.1.4}$$

The second type is the *continuous* distribution, which is that in which P' is absolutely continuous with respect to n-dimensional Lebesgue measure. When this happens, it follows from the Radon–Nikodym theorem that P' is determined by its derivative f with respect to Lebesgue measure, in fact

$$\mathsf{P}(X \in B) = \mathsf{P}'(B) = \int_B f(x)\, dx \tag{14.1.5}$$

(where $B \in \mathscr{B}^n$ and $\int \ldots dx$ denotes integration with respect to Lebesgue measure), and (14.1.2) becomes

$$\mathsf{E}\{g(X_1, X_2, \ldots, X_n)\} = \int_{\mathbf{R}^n} g(x) f(x)\, dx. \tag{14.1.6}$$

The collection X is then said to have the *joint probability density* f. Notice that if f_1 and f_2 are densities for X, then $f_1 = f_2$ a.e. Clearly a Lebesgue integrable function f is a joint probability density if and only if

$$f(x) \geqslant 0 \quad \text{a.e.}, \quad \int_{\mathbf{R}^n} f(x)\, dx = 1. \tag{14.1.7}$$

A particular case of a joint distribution arises when the X_r are independent with distribution functions F_r. Then, by theorem 11.6,

$$\mathsf{P}'(B) = \int_{x \in B} \cdots \int dF_1(x_1)\, dF_2(x_2) \ldots dF_n(x_n),$$

so that P' is just the completion of the Cartesian product of the Stieltjes measures μ_{F_r} $(r = 1, 2, \ldots, n)$. If X_r has probability density f_r, then a joint probability density for X is given by

$$f(x_1, x_2, \ldots, x_n) = f_1(x_1) f_2(x_2) \ldots f_n(x_n).$$

The converse is also clearly true; if P' is the product of Stieltjes probability measures μ_{F_r}, then $X_1, X_2, \ldots, X_n$ are independent.

Suppose that we know the joint distribution, which by a change of notation we denote by P_n, of $\mathbf{X}_n = (X_1, X_2, \ldots, X_n)$. Then *a fortiori* we know the joint distribution P_{n-1} of the subcollection

$$\mathbf{X}_{n-1} = (X_1, X_2, \ldots, X_{n-1}).$$

In fact, if $B \in \mathscr{B}^{n-1}$,

$$\begin{aligned} \mathsf{P}_{n-1}(B) &= \mathsf{P}(\mathbf{X}_{n-1} \in B) \\ &= \mathsf{P}(\mathbf{X}_n \in B \times \mathbf{R}) \\ &= \mathsf{P}_n(B \times \mathbf{R}). \end{aligned}$$

P_{n-1} is called a *marginal distribution* obtained from P_n. An important case is that in which $\mathbf{X}_n$ has a joint probability density.

***Theorem* 14.2.** *If $(X_1, X_2, \ldots, X_n)$ has a joint probability density $f(x_1, x_2, \ldots, x_n)$, then $(X_1, X_2, \ldots, X_{n-1})$ has a joint probability density*

$$g(x_1, x_2, \ldots, x_{n-1}) = \int_{-\infty}^{\infty} f(x_1, x_2, \ldots, x_{n-1}, \xi)\, d\xi. \qquad (14.1.8)$$

(Roughly speaking, we 'integrate out' the unwanted variable x_n.)

Proof.

$$\begin{aligned} \mathsf{P}_{n-1}(B) &= \mathsf{P}_n(B \times \mathbf{R}) \\ &= \int_{B \times \mathbf{R}} f(\mathbf{x})\, d\mathbf{x} \\ &= \int_B \int_{\mathbf{R}} f(x_1, x_2, \ldots, x_{n-1}, \xi)\, dx_1, dx_2 \ldots dx_{n-1}\, d\xi \\ &= \int_B g(x_1, x_2, \ldots, x_{n-1})\, dx_1\, dx_2 \ldots dx_{n-1}. \rbrack \end{aligned}$$

Corollary. *Under the conditions of the theorem, if $0 < m < n$, then $(X_1, X_2, \ldots, X_m)$ has joint probability density*

$$\int_{-\infty}^{\infty} \cdots \int_{-\infty}^{\infty} f(x_1, x_2, \ldots, x_m, \xi_{m+1}, \xi_{m+2}, \ldots, \xi_n)\, d\xi_{m+1}\, d\xi_{m+2} \ldots d\xi_n.$$

Exercises 14.1

1. Let X have distribution $U(0,1)$, and let $X_1 = X_2 = X$. Describe the joint distribution of (X_1, X_2) and show that it is neither discrete nor continuous. (In fact such distributions, which are singular with respect to Lebesgue measure, are much less pathological in n dimensions than in one dimension.)

2. Let X, Y be random variables taking only integer values, and let

$$p_{ij} = \mathsf{P}(X = i, Y = j).$$

Show that the marginal distributions of X and Y are given by

$$p_{i.} = \sum_j p_{ij}, \quad p_{.j} = \sum_i p_{ij},$$

respectively. Show that X and Y are independent if and only if there exist numbers a_i, b_j such that, for all i and j,

$$p_{ij} = a_i b_j.$$

14.2 Conditioning with respect to a random variable

Suppose that on a probability space $(\Omega, \mathscr{F}, \mathsf{P})$ there is defined a random variable Y having a discrete distribution. Then Y takes on values from a countable set $\{\eta_1, \eta_2, \ldots,\}$, and we may clearly take

$$p_j = \mathsf{P}(Y = \eta_j) > 0.$$

Also

$$p_1 + p_2 + \ldots = 1.$$

Now, if $A \in \mathscr{F}$ is any event, we may want to talk about the probability of A conditional on the value of Y. If $Y = \eta_j$, this is given by

$$\mathsf{P}(A \mid Y = \eta_j) = \mathsf{P}(A \cap \{Y = \eta_j\})/p_j \tag{14.2.1}$$

as in §10.4.

It is interesting to look at this in a slightly different way. If A is held fixed, equation (14.2.1) defines a function

$$\phi(y) = \mathsf{P}(A \mid Y = y) \tag{14.2.2}$$

for all values of Y for which $\mathsf{P}(Y = y) > 0$, a function which indicates the way in which the probability of A is affected by a knowledge of the value of Y. The fact that ϕ is undefined for other values of y is irrelevant, for Y takes such values with probability zero.

With ϕ defined by (14.2.2) we can consider the random variable $\phi(Y)$, which is defined with probability one. This random variable is conventionally (if confusingly) denoted by

$$\mathsf{P}(A \mid Y).$$

Thus $\mathsf{P}(A \mid Y)$ is a function from Ω into $\mathbf{R}$, defined by

$$\mathsf{P}(A \mid Y)(\omega) = \mathsf{P}(A \mid Y = Y(\omega)).$$

The function $\phi(y)$ has a simple characteristic property. Let B be any Borel subset of $\mathbf{R}$. Then, if G is the distribution function of Y,

$$\begin{aligned}\int_B \phi(y)\, dG(y) &= \sum_{\{j;\, \eta_j \in B\}} \phi(\eta_j) p_j \\ &= \sum_{\{j;\, \eta_j \in B\}} \mathsf{P}(A \cap \{Y = \eta_j\}) \\ &= \mathsf{P}(A \cap \{Y \in B\}).\end{aligned}$$

Hence, for any $B \in \mathscr{B}$,

$$\mathsf{P}(A \cap \{Y \in B\}) = \int_B \phi(y)\, dG(y). \tag{14.2.3}$$

Alternatively, by theorem 11.4, this can be written as

$$\mathsf{P}(A \cap \{Y \in B\}) = \int_{\{Y \in B\}} \phi(Y)\, d\mathsf{P},$$

so that (14.2.3) is equivalent to the property that, whenever E is an event of the form $\{\omega;\ Y(\omega) \in B\}$ for $B \in \mathscr{B}$,

$$\mathsf{P}(A \cap E) = \int_E \phi(Y)\, d\mathsf{P}. \tag{14.2.4}$$

All this is based on the assumption that Y has a discrete distribution, so that it almost certainly takes on a value y for which $\mathsf{P}(Y = y) > 0$. When Y has a more general distribution, this simple analysis breaks down. For example, if Y has a continuous distribution,

$$\mathsf{P}(A \mid Y = y)$$

is not defined for any y, since $\mathsf{P}(Y = y) = 0$. At the expense of some complication, however, it is possible to salvage something of the idea of conditional probability.

Thus let Y be any (finite) random variable on $(\Omega, \mathscr{F}, \mathsf{P})$, and let G be its distribution function. Let $A \in \mathscr{F}$ be an event, which will be held fixed during the discussion. For any $B \in \mathscr{B}$, define

$$\lambda(B) = \mathsf{P}(A \cap \{Y \in B\}).$$

Then it is trivial to verify that λ is a measure on $(\mathbf{R}, \mathscr{B})$, which is finite since

$$\lambda(\mathbf{R}) = \mathsf{P}(A).$$

Now let μ_G be the Stieltjes measure determined by G, so that

$$\mu_G(B) = \mathsf{P}(Y \in B).$$

Then, for any $B \in \mathscr{B}$,

$$\lambda(B) = \mathsf{P}(A \cap \{Y \in B\}) \leqslant \mathsf{P}(Y \in B) = \mu_G(B),$$

so that

$$\mu_G(B) = 0 \Rightarrow \lambda(B) = 0,$$

i.e. λ is absolutely continuous with respect to μ_G. It now follows from the Radon–Nikodym theorem that there exists a measurable function $\phi: \mathbf{R} \to \mathbf{R}$ such that, for any $B \in \mathscr{B}$,

$$\lambda(B) = \int_B \phi(y)\, \mu_G(dy) = \int_B \phi(y)\, dG(y).$$

Moreover, any function ϕ_1 satisfying this equation has

$$\mu_G\{y;\ \phi_1(y) \neq \phi(y)\} = 0.$$

Thus we have proved the following result.

***Theorem* 14.3.** *Let Y be a random variable with distribution function G defined on a probability space $(\Omega, \mathscr{F}, \mathsf{P})$, and let $A \in \mathscr{F}$. Then there exists a Borel measurable function $\phi: \mathbf{R} \to \mathbf{R}$ such that, for all $B \in \mathscr{B}$,*

$$\mathsf{P}(A \cap \{Y \in B\}) = \int_B \phi(y)\, dG(y). \tag{14.2.5}$$

Moreover, a function ϕ_1 satisfies (14.2.5) *if and only if $\phi_1 = \phi$ except on a set of μ_G-measure zero.*

Any such function $\phi(y)$ is called *a version of the conditional probability* $\mathsf{P}(A \mid Y = y)$. The conditional probability $\mathsf{P}(A \mid Y = y)$ can be regarded as an element of $\mathscr{L}_1(\mathbf{R}, \mathscr{B}, \mu_G)$; it is then an equivalence class of functions differing only on sets of μ_G-measure zero.

If ϕ is a version of $\mathsf{P}(A \mid Y = y)$, the random variable $\phi(Y)$ is called *a version of the conditional probability* $\mathsf{P}(A \mid Y)$. If ϕ_1 is another version, $\phi_1(Y) = \phi(Y)$ with probability one, so that $\mathsf{P}(A \mid Y)$ is determined up to differences on events of probability zero. If $E = \{\omega;\ Y(\omega) \in B\}$, then

$$\mathsf{P}(A \cap E) = \int_E \mathsf{P}(A \mid Y)\, d\mathsf{P}. \tag{14.2.6}$$

As an important example of this analysis, let X and Y be two random variables which have a joint probability density $f(x, y)$. Then, by theorem 14.2, X has a probability density

$$f_1(x) = \int_{\mathbf{R}} f(x, y')\, dy'$$

and Y has a probability density

$$f_2(y) = \int_{\mathbf{R}} f(x', y)\, dx'.$$

Let $B_1, B_2 \in \mathscr{B}$, and put $A = \{\omega;\ X(\omega) \in B_1\}$. Then any version ϕ of $\mathsf{P}(A \mid Y = y)$ satisfies

$$\begin{aligned}\int_{B_2} \phi(y) f_2(y)\, dy &= \mathsf{P}(A \cap \{Y \in B_2\}) \\ &= \mathsf{P}(X \in B_1,\ Y \in B_2) \\ &= \int_{B_1} \int_{B_2} f(x, y)\, dx\, dy.\end{aligned}$$

Since this holds for all B_2, we have

$$\phi(y) = \int_{B_1} f(x,y)\,dx \Big/ f_2(y)$$

for almost all y (where 'almost all' refers to the measure with density f_2). Hence one version of $\mathsf{P}(A \mid Y = y)$ is

$$\int_{B_1} f(x,y)\,dx \Big/ f_2(y).$$

If therefore we write $\quad f(x|y) = f(x,y)/f_2(y),$

then
$$\mathsf{P}(X \in B_1 \mid Y = y) = \int_{B_1} f(x|y)\,dx. \tag{14.2.7}$$

Thus $f(x|y)$ can be interpreted as the probability density of X conditional on $Y = y$. For this reason

$$f(x|y) = f(x,y) \Big/ \int_{\mathbf{R}} f(x',y)\,dx' \tag{14.2.8}$$

is called the *probability density of X conditional on Y*, and

$$f(y|x) = f(x,y) \Big/ \int_{\mathbf{R}} f(x,y')\,dy' \tag{14.2.9}$$

is called the *probability density of Y conditional on X*.

As an example to make these ideas more concrete, suppose that the height X and the weight Y of a man drawn at random from the population have a joint probability density $f(x,y)$. Then $f(x|y)$ is, as a function of x, the probability density of the height of men with weight y, while $f(y|x)$ is the probability density of weight among men of a given height x.

Notice that
$$f(y|x) = f(x|y)\frac{f_2(y)}{f_1(x)}, \tag{14.2.10}$$

a result sometimes called the *generalised Bayes theorem.*

Exercises 14.2

1. If A and Y are independent, show that

$$\mathsf{P}(A \mid Y) = \mathsf{P}(A)$$

with probability one.

2. The joint distribution of (X, Y) is uniform over the disc

$$D = \{(x,y);\ x^2 + y^2 \leqslant 1\}, \text{ i.e.}$$

$$\mathsf{P}((X,Y) \in B) = 1/\pi \times \text{Lebesgue measure of } B \cap D \quad (B \in \mathscr{B}^2).$$

Show that
$$\mathsf{P}(|X| < \tfrac{1}{2}|Y = y) = 1$$
for almost all y in $\tfrac{1}{2}\sqrt{3} < |y| < 1$, and that
$$\mathsf{P}(|X| < \tfrac{1}{2}|Y = y) = \tfrac{1}{2}(1-y^2)^{-\frac{1}{2}}$$
for almost all y in $|y| < \tfrac{1}{2}\sqrt{3}$.

3. Let X, Y be independent random variables with distribution $N(0,1)$, and define $R \geqslant 0$ and $\Theta \in [0, 2\pi)$ by
$$X = R\cos\Theta, \quad Y = R\sin\Theta.$$
Show that, if $0 < \theta_1 < \theta_2 < 2\pi$, then with probability one,
$$\mathsf{P}(\theta_1 < \Theta < \theta_2|R) = (\theta_2 - \theta_1)/2\pi.$$

4. Let $h\colon \mathbf{R} \to \mathbf{R}^+$ be Borel measurable. Show that
$$\int_A h(Y)\,d\mathsf{P} = \mathsf{E}\{h(Y)\,\mathsf{P}(A\,|\,Y)\}.$$

14.3* Conditioning with respect to a σ-field

The construction of $\mathsf{P}(A|Y)$ in the previous section is a particular case of a more general construction which is sometimes useful, and which will now be described. Let $(\Omega, \mathscr{F}, \mathsf{P})$ be a probability space, and let $A \in \mathscr{F}$ be fixed. Then
$$\mathsf{P}_A(E) = \mathsf{P}(A \cap E), \quad E \in \mathscr{F},$$
clearly defines a measure P_A on $\mathscr{F}$, which is absolutely continuous with respect to P. Hence, by the Radon–Nikodym theorem, there exists an $\mathscr{F}$-measurable function $\phi\colon \Omega \to \mathbf{R}$ with
$$\mathsf{P}_A(E) = \int_E \phi(\omega)\,\mathsf{P}(d\omega), \quad E \in \mathscr{F}.$$
Moreover, if ϕ_1 is another function satisfying this equation, $\phi_1 = \phi$ almost everywhere. In other words, there is a unique function (or more accurately equivalence class of functions) ϕ in $\mathscr{L}_1(\Omega, \mathscr{F}, \mathsf{P})$ satisfying the above equation.

This result is, however, trivial, since a possible choice of ϕ is the indicator function χ_A of A.

But now let $\mathscr{G}$ be a sub-σ-field of $\mathscr{F}$, not necessarily containing A. Then the restrictions P' and P'_A of P and P_A to $\mathscr{G}$ are measures on $\mathscr{G}$, and P'_A is absolutely continuous with respect to P'. Hence there exists a unique function ϕ in $\mathscr{L}_1(\Omega, \mathscr{G}, \mathsf{P}')$ such that
$$\mathsf{P}'_A(E) = \int_E \phi\,d\mathsf{P}' \quad (E \in \mathscr{G}),$$
i.e.
$$\mathsf{P}_A(E) = \int_E \phi\,d\mathsf{P} \quad (E \in \mathscr{G}). \tag{14.3.1}$$

This function is called the *probability of A conditional on the σ-field $\mathscr{G}$*, and is denoted by $\mathsf{P}(A|\mathscr{G})$. Each function in the equivalence class ϕ is called a *version* of $\mathsf{P}(A|\mathscr{G})$.

It is important once again to stress that each version of $\mathsf{P}(A|\mathscr{G})$ is measurable not only with respect to $\mathscr{F}$ but also with respect to $\mathscr{G}$. If Y is a random variable, and $\mathscr{G} = Y^{-1}\mathscr{B}$, it is easy to verify that $\mathsf{P}(A|\mathscr{G})$ is just the conditional probability $\mathsf{P}(A|Y)$ defined in the previous section.

Suppose that $(\Omega, \mathscr{F}, \mathsf{P})$ is a probability space, and that $\mathscr{G} \subset \mathscr{F}$ is a σ-field. We have seen how, for any $A \in \mathscr{F}$, there is defined an equivalence class of $\mathscr{G}$-measurable functions $\mathsf{P}(A|\mathscr{G})$ satisfying

$$\mathsf{P}(A \cap E) = \int_E \mathsf{P}(A|\mathscr{G})\, d\mathsf{P} \quad (E \in \mathscr{G}). \tag{14.3.2}$$

It is easy to show that the assertions

$$0 \leqslant \mathsf{P}(A|\mathscr{G}) \leqslant 1, \quad \mathsf{P}(\Omega|\mathscr{G}) = 1,$$

$$\mathsf{P}\left(\bigcup_{n=1}^{\infty} A_n \middle| \mathscr{G}\right) = \sum_{n=1}^{\infty} \mathsf{P}(A_n|\mathscr{G}) \quad (A_n \in \mathscr{F} \text{ disjoint}),$$

hold with probability one, regardless of the choice of version of the $\mathsf{P}(A|\mathscr{G})$. It does not follow, however, that $\mathsf{P}(\cdot|\mathscr{G})$ is a measure on $(\Omega, \mathscr{F})$ because the exceptional sets of probability zero will depend on the choice of the A_n, and may cover an event of positive probability. This is an example of the sort of difficulty which arises in the manipulation of conditional probabilities; the methods by means of which these difficulties may sometimes be overcome may be found, for example, in J. L. Doob, *Stochastic Processes*, Wiley, 1953, or M. Loève, *Probability theory*, van Nostrand, 1963.

Exercises 14.3

1. Show that, if $A \in \mathscr{G}$, then almost certainly
$$\mathsf{P}(A|\mathscr{G}) = \chi_A.$$

2. Show that, if A and $\mathscr{G}$ are independent (i.e. A and E are independent for all $E \in \mathscr{G}$), then almost certainly
$$\mathsf{P}(A|\mathscr{G}) = \mathsf{P}(A).$$

3. If $\mathscr{Q}$ is the σ-field containing only the sets $\varnothing$ and Ω, show that, almost certainly
$$\mathsf{P}(A|\mathscr{Q}) = \mathsf{P}(A).$$

4. If $\mathscr{F}(B)$ is the ϕ-field generated by a set $B \in \mathscr{F}$, show that, almost certainly,
$$\mathsf{P}(A|\mathscr{F}(B)) = \mathsf{P}(A|B)\,\chi_B + \mathsf{P}(A|\Omega - B)\,\chi_{\Omega - B}.$$

14.4 Moments

If $X_1, X_2, \ldots, X_k$ are random variables, expressions of the form

$$\mathsf{E}(X_1^{n_1} X_2^{n_2} \ldots X_k^{n_k}),$$

where the n_j are (usually) non-negative integers, are called *moments* of the X_j. They may or may not exist, depending on whether the expectations are defined. If $n_1 + n_2 + \ldots + n_k = N$, the above expression is called an Nth-order moment.

The first-order moments ($N = 1$) are, of course, just the expectations

$$\mu_i = \mathsf{E}(X_i) \tag{14.4.1}$$

of the random variables. The second-order moments are of the form

$$\mu_{ij} = \mathsf{E}(X_i X_j), \tag{14.4.2}$$

where i and j may be equal or unequal. Notice that, by the Schwarz inequality (§ 7.4),

$$\mu_{ij}^2 \leqslant \mu_{ii} \mu_{jj}, \tag{14.4.3}$$

so that, if $\mathsf{E}(X_i^2) < \infty$ for all i, then all the second-order moments exist.

It is often convenient, before computing the moments, to subtract from the random variables their expectations μ_i. The resulting expressions are called *central moments*. The most important case arises when $N = 2$, and gives rise to the concept of the covariance of two random variables.

Covariance

If X and Y are random variables with finite means and variances, their covariance is defined by

$$\operatorname{cov}(X, Y) = \mathsf{E}\{[X - \mathsf{E}(X)][Y - \mathsf{E}(Y)]\}. \tag{14.4.4}$$

The Schwarz inequality shows that this expectation exists, and indeed that

$$\{\operatorname{cov}(X, Y)\}^2 \leqslant \mathsf{E}\{[X - \mathsf{E}(X)]^2\} \, \mathsf{E}\{[Y - \mathsf{E}(Y)]^2\}$$
$$= \operatorname{var}(X) \operatorname{var}(Y).$$

It follows that, if $\operatorname{var}(X)$ and $\operatorname{var}(Y)$ are positive, the quantity

$$\rho(X, Y) = \operatorname{cov}(X, Y)/\{\operatorname{var}(X) \operatorname{var}(Y)\}^{\frac{1}{2}} \tag{14.4.5}$$

lies in $[-1, 1]$; it is called the *correlation coefficient* of X and Y. Observe that

$$\operatorname{cov}(X, X) = \operatorname{var}(X),$$

so that

$$\rho(X, X) = 1.$$

Similarly,

$$\rho(X, -X) = -1.$$

A useful expression for the covariance is obtained by noting that (14.4.4) yields

$$\operatorname{cov}(X, Y) = \mathsf{E}\{XY - \mathsf{E}(X)\,Y - X\mathsf{E}(Y) + \mathsf{E}(X)\,\mathsf{E}(Y)\}$$
$$= \mathsf{E}(XY) - \mathsf{E}(X)\,\mathsf{E}(Y) - \mathsf{E}(X)\,\mathsf{E}(Y) + \mathsf{E}(X)\,\mathsf{E}(Y).$$

Hence
$$\operatorname{cov}(X, Y) = \mathsf{E}(XY) - \mathsf{E}(X)\,\mathsf{E}(Y). \tag{14.4.6}$$

If X and Y are independent, then by theorem 11.6.

$$\operatorname{cov}(X, Y) = 0.$$

Two random variables whose covariance is zero are said to be *uncorrelated*, and therefore independent variables are uncorrelated. For this reason the (dimensionless) quantity $\rho(X, Y)$ is often used as a measure of the dependence between X and Y. However, it is possible (see exercise 14.4.(5)) for two variables to have $\rho(X, Y) = 0$, and so to be uncorrelated, without being independent.

***Theorem* 14.4.** *Let $X_1, X_2, \ldots, X_n$ be random variables with $\mathsf{E}(X_i^2) < \infty$, and write*

$$v_{ij} = \operatorname{cov}(X_i, X_j).$$

Then, if $c_1, c_2, \ldots, c_n$ are any real numbers, the random variable

$$Z = c_1 X_1 + c_2 X_2 + \ldots + c_n X_n$$

has variance
$$\operatorname{var}(Z) = \sum_{i=1}^{n} \sum_{j=1}^{n} c_i c_j v_{ij}. \tag{14.4.7}$$

Proof. Let $\mu_i = \mathsf{E}(X_i)$, $X_i' = X_i - \mu_i$. Then

$$\mathsf{E}(Z) = \sum_{i=1}^{n} c_i \mu_i,$$

so that
$$\operatorname{var}(Z) = \mathsf{E}\left\{\sum_{i=1}^{n} c_i X_i'\right\}^2$$
$$= \mathsf{E}\left\{\sum_{i=1}^{n} \sum_{j=1}^{n} c_i c_j X_i' X_j'\right\}$$
$$= \sum_{i=1}^{n} \sum_{j=1}^{n} c_i c_j \mathsf{E}(X_i' X_j')$$
$$= \sum_{i=1}^{n} \sum_{j=1}^{n} c_i c_j v_{ij}. \blacksquare$$

***Corollary* 1.**

$$\operatorname{var}(X + Y) = \operatorname{var}(X) + \operatorname{var}(Y) + 2\operatorname{cov}(X, Y). \tag{14.4.8}$$

(This is the form which (11.5.6) takes when X and Y are not independent.)

Corollary 2. *The matrix*

$$V = (v_{ij};\ i,j = 1,2,\ldots,n)$$

is positive-semi-definite.

Proof. For any real c_i,

$$\sum_{i=1}^{n}\sum_{j=1}^{n} c_i c_j v_{ij} = \operatorname{var}\left(\sum_{i=1}^{n} c_i X_i\right) \geqslant 0. \;\rrbracket$$

Notice that, if the matrix V is positive-semi-definite but not positive-definite, so that c_i exist, not all zero, with

$$\sum_{i=1}^{n}\sum_{j=1}^{n} c_i c_j v_{ij} = 0,$$

then it must follow that

$$\operatorname{var}\left(\sum_{i=1}^{n} c_i X_i\right) = 0,$$

so that there is a constant c with

$$\mathsf{P}\left(\sum_{i=1}^{n} c_i X_i = c\right) = 1.$$

Hence the random point $X = (X_1, X_2 \ldots, X_n)$ in $\mathbf{R}^n$ lies with probability one on the hyperplane

$$\sum_{i=1}^{n} c_i x_i = c.$$

Covariances are important in a number of contexts; they will reappear in § 14.6 and in Chapter 15. Higher-order moments, however, appear to be of less importance.

Exercises 14.4

1. Show that, if a collection $(X_1, X_2, \ldots, X_k)$ of random variables satisfies $E|X_i|^N < \infty$ for all i, then all the Nth order moments of this collection exist.

2. Give examples to show that, for any $\rho \in [-1, 1]$, there exist random variables X, Y with $\rho(X, Y) = \rho$.

3. If $0 < \operatorname{var}(X) < \infty$, show that

$$\begin{aligned} \operatorname{cov}(X, aX+b) &= 1 \quad \text{if} \quad a > 0, \\ &= -1 \quad \text{if} \quad a < 0. \end{aligned}$$

4. Show that, if $\rho(X, Y) = 1$, then there exists constants a, b with $a > 0$ such that

$$\mathsf{P}(Y = aX + b) = 1.$$

What happens if $\rho(X, Y) = -1$?

5. Let X have distribution $N(0, 1)$, and let $Y = X^2$. Show that X and Y are uncorrelated but not independent.

6. Let X and Y have

$$0 < \mathrm{var}(X) = \mathrm{var}(Y) < \infty.$$

Show that $X+Y$ and $X-Y$ are uncorrelated. Are they independent?

14.5 The multinomial distribution

In Chapter 10 we considered the problem of a sequence of n independent trials, each of which could result in either a success or a failure, with respective probabilities p and $1-p$. The total number of successes was shown to have the binomial distribution $B(n, p)$.

Now consider a more general situation, in which a trial may result in any one of k outcomes, labelled 1, 2, ..., k, the probability of outcome j being p_j. If X_j is the number of times in n independent trials for which the outcome j occurs, we can ask for the joint distribution of the collection $(X_1, X_2, ..., X_k)$. The answer is given by the following theorem.

***Theorem* 14.5.** *For the situation described above, let*

$$P_n(x_1, x_2, ..., x_k) = \mathsf{P}(X_1 = x_1, X_2 = x_2, ..., X_k = x_k).$$

Then, if $x_1, x_2, ..., x_k$ are non-negative integers with $x_1 + x_2 + ... + x_k = n$,

$$P_n(x_1, x_2, ..., x_k) = \frac{n!}{x_1!\,x_2!\ldots x_k!} p_1^{x_1} p_2^{x_2} \ldots p_k^{x_k}; \tag{14.5.1}$$

otherwise $$P_n(x_1, x_2, ..., x_k) = 0.$$

Proof. If the outcome of the nth trial is Y, we have

$$\begin{aligned} P_n(x_1, x_2, ..., x_k) &= \sum_{i=1}^{k} p_i \mathsf{P}(X_1 = x_1, X_2 = x_2, ..., X_k = x_k \mid Y = i) \\ &= \sum_{i=1}^{k} p_i P_{n-1}(x_1 - \delta_{i1}, x_2 - \delta_{i2}, ..., x_k - \delta_{ik}).\dagger \end{aligned}$$

Now, if (14.5.1) is true with n replaced by $(n-1)$, then

$$\begin{aligned} P_n(x_1, x_2, ..., x_k) &= \sum_{i=1}^{k} p_i \frac{(n-1)!}{(x_1-\delta_{i1})! \ldots (x_k - \delta_{ik})!} p_1^{x_1-\delta_{i1}} \ldots p_k^{x_k-\delta_{ik}} \\ &= \frac{(n-1)!}{x_1!\,x_2! \ldots x_k!} p_1^{x_1} p_2^{x_2} \ldots p_k^{x_k} \sum_{i=1}^{k} x_i \\ &= \frac{n!}{x_1!\,x_2! \ldots x_k!} p_1^{x_1} p_2^{x_2} \ldots p_k^{x_k}. \end{aligned}$$

† $\delta_{ij} = 1$ if $i = j$, 0 if $i \neq j$.

Hence, since (14.5.1) is clearly true for $n = 1$, it is true for all n by induction. ▮

A collection of random variables which have the joint distribution described in theorem 14.5 is said to have the *multinomial distribution* $M(n; p_1, p_2, \ldots, p_k)$. This is clearly a discrete distribution, since X_i can only take the values $0, 1, 2, \ldots, n$. Some of the properties of this distribution are given in the following theorem.

***Theorem* 14.6.** *If* $(X_1, X_2, \ldots, X_k)$ *have the joint distribution*

$$M(n; p_1, p_2, \ldots, p_k),$$

then

(i) *for any real numbers* $z_1, z_2, \ldots, z_k$,

$$\mathsf{E}(z_1^{X_1} z_2^{X_2} \ldots z_k^{X_k}) = (p_1 z_1 + p_2 z_2 + \ldots + p_k z_k)^n; \qquad (14.5.2)$$

(ii) X_i *has distribution* $B(n, p_i)$, *so that*

$$\mathsf{E}(X_i) = np_i, \quad \operatorname{var}(X_i) = np_i(1-p_i);$$

and (iii) *if* $i \neq j$, $$\operatorname{cov}(X_i, X_j) = -np_ip_j. \qquad (14.5.3)$$

Proof. For any r, define a random variable Y_r which is equal to z_j if the outcome of the rth trial is j. Then the Y_r are independent, and clearly

$$Y_1 Y_2 \ldots Y_n = z_1^{X_1} z_2^{X_2} \ldots z_k^{X_k},$$

so that

$$\begin{aligned} \mathsf{E}(z_1^{X_1} z_2^{X_2} \ldots z_k^{X_k}) &= \mathsf{E}(Y_1 Y_2 \ldots Y_n) \\ &= \mathsf{E}(Y_1)\,\mathsf{E}(Y_2) \ldots \mathsf{E}(Y_n) \\ &= \{\mathsf{E}(Y_1)\}^n \\ &= (p_1 z_1 + p_2 z_2 + \ldots + p_k z_k)^n, \end{aligned}$$

and (i) is proved. To prove (ii) put $z_i = z$, $z_j = 1$ $(j \neq i)$. Then

$$\mathsf{E}(z^{X_i}) = (1 - p_i + p_i z)^n,$$

showing that X_i has distribution $B(n, p_i)$. Finally, in (14.5.2) put

$$z_i = 1 + s, \quad z_j = 1 + t, \quad z_\alpha = 1 \quad (\alpha \neq i, j).$$

Then $$\mathsf{E}\{(1+s)^{X_i}(1+t)^{X_j}\} = (1 + p_i s + p_j t)^n.$$

Equating coefficients of st

$$\mathsf{E}(X_i X_j) = n(n-1)\,p_i p_j,$$

so that, by (14.4.6),

$$\begin{aligned} \operatorname{cov}(X_i, X_j) &= n(n-1)\,p_i p_j - np_i np_j \\ &= -np_i p_j. \end{aligned}$$ ▮

Exercises 14.5

1. Let $X_1, X_2, \ldots, X_k$ be independent random variables on a probability space $(\Omega, \mathscr{F}, \mathsf{P})$ such that X_j has distribution $P(\mu_j)$, and write

$$S = X_1 + X_2 + \ldots + X_k.$$

For any positive integer n, define a probability measure P_n on $(\Omega, \mathscr{F})$ by

$$\mathsf{P}_n(A) = \mathsf{P}(A \mid S = n) \quad (A \in \mathscr{F}),$$

and let $\mathscr{F}_n$ be the completion of $\mathscr{F}$ with respect to P_n. Prove that, on the probability space $(\Omega, \mathscr{F}_n, \mathsf{P}_n)$, the random variables $X_1, X_2, \ldots, X_k$ have the joint distribution $M(n; p_1, p_2, \ldots, p_k)$, where $p_j = \mu_j/(\mu_1 + \mu_2 + \ldots + \mu_k)$.

2. Let $X_1, X_2, \ldots, X_n$ have joint distribution $M(n; p_1, p_2, \ldots, p_k)$ $(p_i > 0)$, and define a new probability measure P' by

$$\mathsf{P}'(A) = \mathsf{P}(A \mid X_1 = 3) \quad (A \in \mathscr{F}).$$

Find the distribution of X_2 under P'.

3. If P_n is defined by (14.5.1), and if A is any positive number, use Stirling's formula (§ 10.5) to show that, as $n \to \infty$,

$$P_n(x_1, x_2, \ldots, x_k) \sim (2\pi n)^{-\frac{1}{2}(k-1)} (p_1 p_2 \ldots p_k)^{-\frac{1}{2}} \exp\left[-\tfrac{1}{2}\sum_{j=1}^{k} \left(\frac{x_j - np_j}{np_j}\right)^2\right]$$

uniformly in $|x_j - np_j|^2 < np_j A$.

14.6 The multinormal distribution†

Of all possible joint distributions, by far the most important is the *multinormal* (or *multivariate normal*) distribution. It arises in many applications, and has a number of special properties which make its manipulation particularly simple.

If A is any $(n \times n)$ symmetric matrix, consider the quadratic form

$$Q(\boldsymbol{x}) = \boldsymbol{x}^T A \boldsymbol{x} = \sum_{i=1}^{n} \sum_{j=1}^{n} a_{ij} x_i x_j,$$

(where $\boldsymbol{x}$ denotes the point in $\mathbf{R}^n$ with coordinates x_j, regarded where necessary as a column vector, and $\boldsymbol{x}^T$ denotes its transpose), and write

$$f(\boldsymbol{x}) = e^{-\frac{1}{2}Q(\boldsymbol{x})}.$$

There exists an orthogonal matrix P with the property that $P^T A P$ is diagonal, so that, if $\boldsymbol{y} = P^T \boldsymbol{x}$,

$$Q(\boldsymbol{x}) = \sum_{i=1}^{n} \alpha_i y_i^2,$$

† In this section a familiarity with the theory of matrices such as may be found in Birkhoff and Maclane, *A Survey of Modern Algebra*, Macmillan, 1953, or L. Mirsky, *Linear Algebra*, Oxford, 1955 will be assumed.

where $\alpha_1, \alpha_2, \ldots, \alpha_n$ are the eigenvalues of A. Then, since Lebesgue measure is invariant under rotation,

$$\int_{\mathbf{R}^n} f(x)\,dx = \int_{\mathbf{R}^n} \exp\{-\tfrac{1}{2}(\alpha_1 y_1^2 + \ldots + \alpha_n y_n^2)\}\,dy_1\,dy_2 \ldots dy_n$$
$$= \prod_{j=1}^{n} \int_{\mathbf{R}} \exp(-\tfrac{1}{2}\alpha_j y_j^2)\,dy_j.$$

Hence
$$\int_{\mathbf{R}^n} f(x)\,dx = \infty$$

unless $\alpha_j > 0$ for all j, i.e. unless A is positive-definite. If A is positive-definite, then, since

$$\int_{\mathbf{R}} e^{-\frac{1}{2}\alpha y^2} dy = (2\pi/\alpha)^{\frac{1}{2}},$$

we have
$$\int_{\mathbf{R}^n} f(x)\,dx = \prod_{j=1}^{n} (2\pi/\alpha_j)^{\frac{1}{2}}$$
$$= (2\pi)^{\frac{1}{2}n} (\det A)^{-\frac{1}{2}},$$

where $\det(A)$ is the determinant of A. It follows that, if A is positive-definite, the function

$$(2\pi)^{-\frac{1}{2}n} (\det A)^{\frac{1}{2}} \exp(-\tfrac{1}{2} x^T A x)$$

is a probability density on $\mathbf{R}^n$. For a reason which will become apparent, it is usual to put $V = A^{-1}$ (as that V is also positive-definite and symmetric).

Multinormal distribution

A collection $(X_1, X_2, \ldots, X_n)$ which has the joint probability density

$$(2\pi)^{-\frac{1}{2}n} (\det V)^{-\frac{1}{2}} \exp(-\tfrac{1}{2} x^T V^{-1} x) \qquad (14.6.1)$$

is said to have the multinormal distribution $N(0, V)$.

If $(X_1, X_2, \ldots, X_n)$ has distribution $N(0, V)$, and if $\mu_1, \mu_2, \ldots, \mu_n$ are real numbers, the collection $(X_1 + \mu_1, X_2 + \mu_2, \ldots, X_n + \mu_n)$ has joint probability density

$$(2\pi)^{-\frac{1}{2}n} (\det V)^{-\frac{1}{2}} \exp\{-\tfrac{1}{2}(x - \boldsymbol{\mu})^T V^{-1} (x - \boldsymbol{\mu})\}. \qquad (14.6.2)$$

A collection having this density is said to have the multinormal distribution $N(\boldsymbol{\mu}, V)$. Since it is clear that $N(\boldsymbol{\mu}, V)$ can be reduced to $N(0, V)$ by subtracting the numbers μ_i, we shall restrict attention to the, only apparently more special, case $N(0, V)$.

The basic property of the multinormal distribution is given by the following theorem.

***Theorem* 14.7.** *Let* $(X_1, X_2, \ldots, X_n)$ *have the multinormal distribution* $N(0, V)$. *Let* C *be an* $(m \times n)$ *matrix with* $m \leqslant n$, *and suppose that the rank of* C *is* m. *Then the collection of random variables* $(Y_1, Y_2, \ldots, Y_m)$ *given by*

$$Y = CX$$

has the multinormal distribution $N(0, CVC^T)$.

Proof. Consider first the case $m = n$. Then C is a non-singular linear mapping of $\mathbf{R}^n$ onto itself. For $B \in \mathscr{B}^n$.

$$\begin{aligned} \mathsf{P}(Y \in B) &= \mathsf{P}(X \in C^{-1}B) \\ &= (2\pi)^{-\frac{1}{2}n} (\det V)^{-\frac{1}{2}} \int_{C^{-1}B} \exp(-\tfrac{1}{2}x^T V^{-1} x)\, dx \\ &= (2\pi)^{-\frac{1}{2}n} (\det V)^{-\frac{1}{2}} \int_B \exp\{-\tfrac{1}{2}y^T (C^{-1})^T V^{-1} C^{-1} y\} \det(C^{-1})\, dy \\ &= (2\pi)^{-\frac{1}{2}n} (\det CVC^T)^{-\frac{1}{2}} \int_B \exp\{-\tfrac{1}{2}y^T (CVC^T)^{-1} y\}\, dy, \end{aligned}$$

where we have used the result of exercise 6.5 (4). Hence, Y has joint probability density

$$(2\pi)^{-\frac{1}{2}n} (\det CVC^T)^{-\frac{1}{2}} \exp\{-\tfrac{1}{2}y^T (CVC^T)^{-1} y\},$$

and therefore has distribution $N(0, CVC^T)$.

Now turn to the more difficult case $m < n$. First, note that we can find an $(n-m) \times n$ matrix C_0 such that the $(n \times m)$ matrix

$$C_1 = \begin{pmatrix} C \\ \cdots \\ C_0 \end{pmatrix}$$

is non-singular. Then, if we extend the set $(Y_1, Y_2, \ldots, Y_m)$ to the larger set $(Y_1, Y_2 \ldots, Y_n)$ by

$$Y = C_1 X,$$

it follows from the result just proved that $(Y_1, Y_2, \ldots, Y_n)$ has distribution $N(0, V_1)$, where

$$V_1 = C_1 V C_1^T.$$

Hence, by the corollary to theorem 14.2, $(Y_1, Y_2, \ldots, Y_m)$ has a joint probability density

$$f(y_1, y_2, \ldots, y_m) = \int_{-\infty}^{\infty} \cdots \int_{-\infty}^{\infty} (2\pi)^{-\frac{1}{2}n} (\det V_1)^{-\frac{1}{2}}$$

$$\times \exp(-\tfrac{1}{2}y^T V_1^{-1} y)\, dy_{m+1}\, dy_{m+2} \ldots dy_n.$$

To evaluate this integral write y and V_1^{-1} in partitioned form as

$$y = \begin{pmatrix} \eta \\ \cdots \\ \zeta \end{pmatrix}, \quad V_1^{-1} = \begin{pmatrix} U_1 & U_2^T \\ U_2 & U_3 \end{pmatrix},$$

where $\boldsymbol{\eta}$ is $(m \times 1)$ and U_1 is $(m \times m)$. Then the identity

$$\mathbf{y}^T V_1^{-1} \mathbf{y} = (\boldsymbol{\zeta} + U_3^{-1} U_2 \boldsymbol{\eta})^T U_3 (\boldsymbol{\zeta} + U_3^{-1} U_2 \boldsymbol{\eta}) + \boldsymbol{\eta}^T (U_1 - U_2^T U_3^{-1} U_2) \boldsymbol{\eta},$$

which is easily verified, leads, after the substitutions

$$\boldsymbol{\zeta}_1 = \boldsymbol{\zeta} + U_3^{-1} U_2 \boldsymbol{\eta},$$

$$U = U_3 - U_2^T U_3^{-1} U_2,$$

to

$$f(\boldsymbol{\eta}) = \int_{\mathbf{R}^{n-m}} (2\pi)^{-\frac{1}{2}n} (\det V_1)^{-\frac{1}{2}} \exp\left(-\tfrac{1}{2}\boldsymbol{\zeta}_1^T U_3 \boldsymbol{\zeta}_1 - \tfrac{1}{2}\boldsymbol{\eta}^T U \boldsymbol{\eta}\right) d\boldsymbol{\zeta}_1$$

$$= c \exp\left(-\tfrac{1}{2}\boldsymbol{\eta}^T U \boldsymbol{\eta}\right)$$

for some c. It follows that $(Y_1, Y_2, \ldots, Y_m)$ has joint distribution $N(0, U^{-1})$. It remains only to find U^{-1}.

We have

$$V_1 = \begin{pmatrix} C \\ \hdashline C_0 \end{pmatrix} V \,(C^T \mid C_0^T)$$

$$= \begin{pmatrix} CVC^T & CVC_0^T \\ C_0VC^T & C_0VC_0^T \end{pmatrix}$$

$$= \begin{pmatrix} W_1 & W_2^T \\ W_2 & W_3 \end{pmatrix}, \quad \text{say.}$$

Then, since

$$\begin{pmatrix} U_1 & U_2^T \\ U_2 & U_3 \end{pmatrix} \begin{pmatrix} W_1 & W_2^T \\ W_2 & W_3 \end{pmatrix} = \begin{pmatrix} I_m & 0 \\ 0 & I_{n-m} \end{pmatrix},$$

we have

$$U_1 W_1 + U_2^T W_2 = I_m,$$

$$U_2 W_1 + U_3 W_2 = 0.$$

The second equation gives $W_2 = -U_3^{-1} U_2 W_1$, which, substituted into the first, yields

$$(U_1 - U_2^T U_3^{-1} U_2) W_1 = I_m,$$

i.e.

$$U W_1 = I_m.$$

Hence

$$U^{-1} = W_1 = CVC^T,$$

and we have proved that $(Y_1, Y_2, \ldots, Y_m)$ has joint distribution

$$N(0, CVC^T). \blacksquare$$

***Corollary* 1.** *If $(X_1, X_2, \ldots, X_n)$ has joint distribution $N(0, V)$, then $(X_1, X_2, \ldots, X_m)$ (where $m < n$) has joint distribution $N(0, V_1)$, where V_1 is the $(m \times m)$ principal submatrix of V. In particular, X_1 has distribution $N(0, v_{11})$, (where v_{ij} is the (i,j)-th element of V).*

***Corollary* 2.** *If $(X_1, X_2, \ldots, X_n)$ have joint distribution $N(0, V)$, the random variable*

$$c_1 X_1 + c_2 X_2 + \ldots + c_n X_n$$

has the normal distribution $N(0, \sigma^2)$, *where*

$$\sigma^2 = \sum_{i=1}^{n} \sum_{j=1}^{n} c_i c_j v_{ij}.$$

***Corollary* 3.** *If* $(X_1, X_2, \ldots, X_n)$ *has joint distribution* $N(0, V)$, *then*

$$\mathsf{E}(X_i) = 0,$$

$$\operatorname{var}(X_i) = v_{ii},$$

$$\operatorname{cov}(X_i, X_j) = v_{ij}.$$

(For this reason V is often called the *variance–covariance matrix.*)

Proof. From corollary 2, for any c_i,

$$0 = \mathsf{E}\left(\sum_{i=1}^{n} c_i X_i\right) = \sum_{i=1}^{n} c_i \mathsf{E}(X_i),$$

and $$\sum_{i=1}^{n} \sum_{j=1}^{n} c_i c_j v_{ij} = \operatorname{var}\left(\sum_{i=1}^{n} c_i X_i\right) = \sum_{i=1}^{n} \sum_{j=1}^{n} c_i c_j \operatorname{cov}(X_i, X_j).$$

The result follows on equating coefficients of c_i and $c_i c_j$. ∎

***Corollary* 4.** *If the distribution of* (X, Y) *is multinormal, then* X *and* Y *are independent if and only if* $\operatorname{cov}(X, Y) = 0$.

Proof. The 'only if' assertion is trivial. Conversely, if $\operatorname{cov}(X, Y) = 0$, the matrix V, and therefore the matrix V^{-1}, is diagonal, and the probability density of (X, Y) is of the form $f(x)\,g(y)$ which, as remarked in §14.1, implies that X and Y are independent. ∎

***Corollary* 5.** *Let* $X_1, X_2, \ldots, X_n$ *be independent and have the normal distribution with mean* 0 *and variance* σ^2, *and let* $P = (p_{ij})$ *be an orthogonal* $(n \times n)$ *matrix. Then the variables* $Y_1, Y_2, \ldots, Y_n$ *given by*

$$Y_i = \sum_{j=1}^{n} p_{ij} X_j$$

are also independent with distribution $N(0, \sigma^2)$.

Proof. Clearly $(X_1, X_2, \ldots, X_n)$ has joint distribution $N(0, \sigma^2 I)$, where I is the $(n \times n)$ identity matrix. Hence $(Y_1, Y_2, \ldots, Y_n)$ has joint distribution $N(0, V)$, where $V = P\sigma^2 I P^T = \sigma^2 I$. ∎

Exercises 14.6

1. Let (X, Y) have a multinormal distribution $N(0, V)$. Show that there exists a number λ such that for any $B \in \mathscr{B}$, there is probability one that

$$\mathsf{P}(X \in B \mid Y) = \sigma^{-1}(2\pi)^{-\frac{1}{2}} \int_B \exp\left\{-\frac{1}{2\sigma^2}(x - \lambda Y)^2\right\} dx.$$

(In other words, the distribution of X given Y is $N(\lambda Y, \sigma^2)$.)

2. Let $X_1, X_2, \ldots, X_n$ be independent and have distribution $N(0,1)$, and let $S = X_1+X_2+\ldots+X_n$. Show that it is possible to find an orthogonal matrix P so that, if $\mathbf{Y} = P\mathbf{X}$, then

$$Y_1 = S/n^{\frac{1}{2}}.$$

Show further that, for any r, $X_r - X_1$ is a linear function of $Y_2, Y_3, \ldots, Y_n$. Deduce that the collection $(X_2-X_1, X_3-X_1, \ldots, X_n-X_1)$ is independent of S.

If $\bar{X} = S/n$ is the average of the X_j, and $F: \mathbf{R}^n \to \mathbf{R}$ is any Borel function, show that

$$F(X_1-\bar{X}, X_2-\bar{X}, \ldots, X_n-\bar{X})$$

is independent of $\bar{X}$. In particular, $\bar{X}$ and

$$\sum_{j=1}^{n} (X_j-\bar{X})^2$$

are independent, a result of some importance in statistics.

3. If $\mathbf{X} \in \mathbf{R}^n$ has joint distribution $N(0, V)$ and the $(m \times n)$ matrix C has rank less than m, the joint distribution of $\mathbf{Y} = C\mathbf{X}$ is called the *singular multinormal distribution* $N(0, CVC^T)$. Prove that

(i) the definition is unambiguous, in that the distribution of $C\mathbf{X}$ depends only on CVC^T, and not on C and V separately,

(ii) CVC^T is a singular positive-semi-definite matrix,

(iii) for any such matrix W there is a collection of random variables with joint distribution $N(0, W)$,

(iv) the distribution is concentrated on a linear subspace of $\mathbf{R}^m$ whose dimension is equal to the rank of C, and is absolutely continuous with respect to Lebesgue measure on that subspace,

(v) if $\mathbf{X}$ has distribution $N(0, V)$, where V may or may not be singular, then $C\mathbf{X}$ has distribution $N(0, CVC^T)$.

4. State and prove analogues for the distribution $N(\boldsymbol{\mu}, V)$ of theorem 14.7, its corollaries, and exercises 14.6 (1, 2, 3).

15

STOCHASTIC PROCESSES

15.1 Renewal processes

Let $X_1, X_2, \dots$ be independent finite random variables with the same distribution function F, and suppose that X_n is (almost certainly) positive, so that
$$F(0) = \mathsf{P}(X_n \leqslant 0) = 0.$$
Then, if
$$S_0 = 0, \quad S_n = X_1 + X_2 + \dots + X_n,$$
the points
$$S_0, S_1, S_2, S_3, \dots$$
form a strictly increasing sequence in $[0, \infty)$.

This mathematical model is often used to represent practical situations in which a sequence of events occurs in time in an irregular fashion, S_n being regarded as the time of occurrence of the nth event after $t = 0$. For example, the times of arrival of calls at a telephone exchange can often be at least approximately represented by a sequence of this type.

Another example in which this model may be appropriate is that in which a piece of equipment contains a certain item that periodically needs replacing. Suppose, for instance, that it contains a light bulb. At $t = 0$ a bulb is installed, and remains in use until, after a time X_1 (say), it fails and is replaced by a second bulb. This fails after a time X_2 (say) and is itself replaced. If the process continues, and if the lifetime of the nth bulb is X_n, then S_n is the instant at which it fails and is replaced. Because the lifetime of a light bulb depends on circumstances in its manufacture and use which can be regarded as random, the quantities X_n will be random variables. It is plausible to assume that they are moreover independent and identically distributed, and we then have exactly the model described.

Because of this application to the 'theory of renewals', the sequence
$$S_0, S_1, S_2, \dots$$
is often called a *renewal process*; F is then called the *lifetime distribution function.*

It is, of course, possible to apply to this model the theory of sums of independent random variables developed in Chapter 13. Note that
$$\mu = \mathsf{E}(X_n) = \int_0^\infty x\,dF(x) \tag{15.1.1}$$

exists and is strictly positive, though it may equal $+\infty$. If $\mu < \infty$, the strong law of large numbers shows that, with probability one,

$$\lim_{n\to\infty} S_n/n = \mu. \tag{15.1.2}$$

This result is also true if $\mu = \infty$, as can be seen as follows. For any integer N, let

$$X_n^{(N)} = \min(X_n, N).$$

Then $X_1^{(N)}, X_2^{(N)}, \ldots$ are independent and identically distributed, with finite mean

$$\mu^{(N)} = \int_0^\infty \min(x, N)\, dF(x).$$

Hence, with probability one,

$$\lim_{n\to\infty} (X_1^{(N)} + X_2^{(N)} + \ldots + X_n^{(N)})/n = \mu^{(N)},$$

and since

$$S_n \geqslant X_1^{(N)} + X_2^{(N)} + \ldots + X_n^{(N)},$$

there is probability one that

$$\liminf_{n\to\infty} S_n/n \geqslant \mu^{(N)}.$$

Now let $N \to \infty$, so that

$$\mu^{(N)} \to \int_0^\infty x\, dF(x) = \infty.$$

Then

$$\liminf_{n\to\infty} S_n/n = \infty.$$

It follows that (15.1.2) holds whether μ is finite or infinite.

An important corollary of this result is that, since $\mu > 0$, there is probability one that

$$\lim_{n\to\infty} S_n = +\infty. \tag{15.1.3}$$

This means that the number of points S_n falling in any finite interval is almost certainly finite.

In particular, the number $N(t)$ falling in the interval $(0, t]$ (where $0 \leqslant t < \infty$) is almost certainly finite. Clearly it is given by

$$N(t) = \max\{n \geqslant 0;\ S_n \leqslant t\}. \tag{15.1.4}$$

If the random variables X_n are defined on a measure space $(\Omega, \mathscr{F}, \mathsf{P})$, then $N(t)$ is a function from Ω into $\{0, 1, 2, 3, \ldots\}$, and, for any n,

$$\begin{aligned}\{\omega;\ N(t) = n\} &= \{\omega;\ S_n \leqslant t < S_{n+1}\} \\ &= \{\omega;\ S_n \leqslant t\} - \{\omega;\ S_{n+1} \leqslant t\} \in \mathscr{F},\end{aligned}$$

so that $N(t)$ is a non-negative integer-valued random variable.

Thus, starting from the independent random variables X_n, we have arrived at an uncountable collection

$$\{N(t);\, t \geqslant 0\}$$

of random variables. It is not difficult to see that they are not independent (using, for instance, the fact that $N(t_1) \leqslant N(t_2)$ if $t_1 < t_2$). A collection of random variables like $\{N(t)\}$, indexed by a parameter t (which often, but not always, represents time, but may not even be real), is called a *stochastic process*. In the next section this concept will be more formally defined and analysed. Note that $\{N(t)\}$ can be regarded as a *random function* of t; in this case it is a step function with jumps of $+1$ at the random points $S_1, S_2, \ldots$.

Let F_n denote the distribution function of S_n. In principle, the F_n can be successively computed from the obvious recurrence relation

$$F_1(x) = F(x), \quad F_{n+1}(x) = \int_0^x F_n(x-y)\, dF(y) \quad (x \geqslant 0), \tag{15.1.5}$$

(cf. §11.5), although in practice other methods (usually involving characteristic functions) may be more convenient. Then, if n is a non-negative integer,

$$\mathsf{P}\{N(t) = n\} = \mathsf{P}(S_n \leqslant t) - \mathsf{P}(S_{n+1} \leqslant t),$$

so that the distribution of $N(t)$ is given by

$$\mathsf{P}\{N(t) = n\} = F_n(t) - F_{n+1}(t). \tag{15.1.6}$$

It is also possible to find the joint distributions of collections of the form $\{N(t_1), N(t_2), \ldots, N(t_k)\}$. For instance

$$\mathsf{P}\{N(t_1) = n_1, N(t_2) = n_2\} = \mathsf{P}(S_{n_1} \leqslant t_1 < S_{n_1+1}, S_{n_2} \leqslant t_2 < S_{n_2+1}),$$

and the latter probability can be expressed in a straightforward but complicated way in terms of the known distribution of the X_n.

As an example, suppose that X_n has distribution function

$$F(x) = 1 - e^{-\lambda x} \quad (x \geqslant 0), \tag{15.1.7}$$

for some $\lambda > 0$. Then X_n has probability density $\lambda e^{-\lambda x}\,(x \geqslant 0)$, and

$$\mu = \mathsf{E}(X_n) = \int_0^\infty x \,.\, \lambda e^{-\lambda x} dx = \lambda^{-1}.$$

It is easy to show, by induction using (15.1.5), that S_n has probability density

$$f_n(x) = \lambda^n x^{n-1} e^{-\lambda x}/(n-1)! \quad (x \geqslant 0),$$

so that
$$\mathsf{P}\{N(t)=n\} = \int_0^t \frac{\lambda^n x^{n-1} e^{-\lambda x}\,dx}{(n-1)!} - \int_0^t \frac{\lambda^{n+1} x^n e^{-\lambda x}\,dx}{n!}$$
$$= \frac{\lambda^n}{n!}\int_0^t (n x^{n-1} e^{-\lambda x} - \lambda x^n e^{-\lambda x})\,dx$$
$$= \frac{\lambda^n}{n!} t^n e^{-\lambda t}.$$

In other words, $N(t)$ has a Poisson distribution with mean λt. For this reason a renewal process for which F is given by (15.1.7) is called a *Poisson process*.

Renewal processes in general, and Poisson processes in particular, have a number of important properties. The interested reader will find a detailed discussion in D. R. Cox, *Renewal Theory*, Methuen, 1962.

Exercises 15.1

1. The *renewal function* $H(t)$ is defined by
$$H(t) = \mathsf{E}\{N(t)\}.$$
Show that
$$H(t) = \sum_{n=1}^{\infty} F_n(t).$$
By using the inequality
$$F_n(t) \leqslant \mathsf{E}(e^{\alpha t - \alpha S_n}),$$
prove that $H(t)$ is finite for all t.

2. If $\mu < \infty$, use the strong law of large numbers to prove that, with probability one,
$$\lim_{t\to\infty} \frac{N(t)}{t} = \frac{1}{\mu}.$$
(In other words, the *density* of the points S_n on the line is $1/\mu$.)

Deduce that
$$\lim_{t\to\infty} \frac{H(t)}{t} = \frac{1}{\mu}.$$

3. If X_n has probability density $x^{k-1}e^{-x}/(k-1)!$, where k is a positive integer, show that
$$\mathsf{P}\{N(t)=n\} = \sum_{r=kn}^{kn+k-1} \frac{t^r e^{-t}}{r!}.$$

4. Define $\tilde{X}(t) = S_{N(t)+1} - t$. Show that, for a Poisson process, $N(t)$ and $\tilde{X}(t)$ are independent, and that the distribution of $\tilde{X}(t)$ is independent of t. Deduce that the numbers of events of a Poisson process in disjoint intervals are independent. Is this true of any other renewal process?

5. If $\theta > 0$, show that

$$\int_0^\infty e^{-\theta t}\, dH(t) = \frac{\phi(\theta)}{1-\phi(\theta)},$$

where

$$\phi(\theta) = \int_0^\infty e^{-\theta x}\, dF(x).$$

15.2 The general theory of stochastic processes

In the previous section we encountered a collection $\{N(t)\}$ of random variables defined on the same probability space, and called a stochastic process. Such collections arise in many applications of probability theory, and a detailed theory has been built up to describe them.

Stochastic process

If $(\Omega, \mathscr{F}, \mathsf{P})$ is a probability space, and T any set, a collection $\{X(t); t \in T\}$ of (real) random variables $X(t)$ on Ω is called a (real) stochastic process with parameter set T.

In many applications, like that of the previous section, t represents a time parameter, so that T is the real line or some subset of it. This is not always the case, however; there are important examples in which T has more complex structure.

If $\{X(t); t \in T\}$ is a stochastic process on Ω, then $X(t)$, being a random variable on Ω, has a probability distribution. More generally, if $t_1, t_2, \ldots, t_n$ are distinct elements of T, the random variables

$$X(t_1), X(t_2), \ldots, X(t_n)$$

will have a joint distribution, which will be a probability measure

$$\mathsf{P}_{t_1 t_2 \ldots t_n}$$

on $\mathbf{R}^n$. These measures, for different values of $n, t_1, t_2, \ldots, t_n$, are called the *finite-dimensional distributions* of the stochastic process. The complete collection of finite-dimensional distributions clearly contains a great deal of information about the process: just how much will appear later.

It is rather clear (cf. §6.6) that the probability measures

$$\mathsf{P}_{t_1 t_2 \ldots t_n}$$

must satisfy the following two conditions, which are known as the *Kolmogorov consistency conditions.*

$\mathscr{K}_1$: If $B \in \mathscr{B}^n$, then

$$\mathsf{P}_{t_1 t_2 \ldots t_n t_{n+1}}(B \times \mathbf{R}) = \mathsf{P}_{t_1 t_2 \ldots t_n}(B).$$

$\mathscr{K}_2$: If π is a permutation of $(1, 2, \ldots, n)$, and if the function

$\phi: \mathbf{R}^n \to \mathbf{R}^n$ is defined by

$$\phi(x_1, x_2, \ldots, x_n) = (x_{\pi(1)}, x_{\pi(2)}, \ldots, x_{\pi(n)}),$$

then
$$\mathsf{P}_{t_{\pi(1)} t_{\pi(2)} \ldots t_{\pi(n)}} = \mathsf{P}_{t_1 t_2 \ldots t_n} \phi^{-1}.$$

Condition $\mathscr{K}_1$ says that the joint distribution of

$$\{X(t_1), X(t_2), \ldots, X(t_n)\}$$

can be obtained from that of $\{X(t_1), X(t_2), \ldots, X(t_n), X(t_{n+1})\}$ by 'integrating out' the last variable, while $\mathscr{K}_2$ says that permuting the variables alters their joint distribution in the obvious way.

The conditions $\mathscr{K}_1$ and $\mathscr{K}_2$ are the only restrictions on the collection of finite-dimensional distributions. This is the content of the following theorem, which is fundamental to the theory.

***Theorem* 15.1.** *Let T be any set, and suppose that, to every finite subset $\{t_1, t_2, \ldots, t_n\}$ of T there is assigned a probability measure*

$$\mathsf{P}_{t_1 t_2 \ldots t_n}$$

on $(\mathbf{R}^n, \mathscr{B}^n)$. Suppose further that the conditions $\mathscr{K}_1$ and $\mathscr{K}_2$ are satisfied. Then there exists a probability space $(\Omega, \mathscr{F}, \mathsf{P})$ and a collection

$$\{X(t);\ t \in T\}$$

of random variables on this space such that the joint distribution of

$$\{X(t_1), X(t_2) \ldots, X(t_n)\} \quad \textit{is} \quad \mathsf{P}_{t_1 t_2 \ldots t_n}.$$

Proof. It follows from theorem 6.10 that there exists a probability measure μ on $(\mathbf{R}^T, \mathscr{B}^T)$ such that for $B \in \mathscr{B}^n$,

$$\mu\{x;\ (x(t_1), x(t_2), \ldots, x(t_n)) \in B\} = \mathsf{P}_{t_1 t_2 \ldots t_n}(B).$$

Taking $\Omega = \mathbf{R}^T$, $\mathscr{F}$ the completion of $\mathscr{B}^T$ with respect to μ, and $\mathsf{P} = \mu$, we obtain a probability space $(\Omega, \mathscr{F}, \mathsf{P})$. If $X(t): \Omega \to \mathbf{R}$ is the function which maps x onto $x(t)$, the $X(t)$ are random variables and

$$\begin{aligned}\mathsf{P}\{(X(t_1), X(t_2), \ldots, X(t_n)) \in B\} &= \mu\{x;\ (x(t_1), x(t_2), \ldots, x(t_n)) \in B\} \\ &= \mathsf{P}_{t_1 t_2 \ldots t_n}(B).\ \rrbracket\end{aligned}$$

The theorem shows that, given any consistent family of finite-dimensional distributions, we can construct a stochastic process having these distributions. For this reason, much of the theory of stochastic processes is really a study of the implications of the consistency conditions $\mathscr{K}_1$ and $\mathscr{K}_2$.

Let $\{X(t);\, t \in T\}$ be a stochastic process defined on a probability space $(\Omega, \mathscr{F}, \mathsf{P})$. Then, as in the proof of theorem 15.1, the finite-dimensional distributions define a probability measure P' on $(\mathbf{R}^T, \overline{\mathscr{B}^T})$, where $\overline{\mathscr{B}^T}$ is the completion of $\mathscr{B}^T$ with respect to P'. Let $X(t,\omega)$ denote the value of $X(t)$ at the point $\omega \in \Omega$, and let $X: \Omega \to \mathbf{R}^T$ be the mapping which sends ω into the function $\{X(t,\omega);\, t \in T\}$. Then $(\mathbf{R}^T, X^{-1}\mathscr{F}, \mathsf{P}X^{-1})$ is a probability space, often called the *space of sample functions* because its points are typical realisations of the stochastic process. From the definition of the finite-dimensional distributions, if $A \subset \mathbf{R}^T$ is a cylinder set with finite-dimensional base, $A \in X^{-1}\mathscr{F}$ and

$$\begin{aligned}(\mathsf{P}X^{-1})\,A &= \mathsf{P}(X^{-1}A)\\ &= \mathsf{P}(X(\omega) \in A)\\ &= \mathsf{P}'A.\end{aligned}$$

Hence this holds for all A in the σ-field $\mathscr{B}^T$ generated by the cylinder sets, and hence for all $A \in \overline{\mathscr{B}^T}$. Thus

$$\overline{\mathscr{B}^T} \subset X^{-1}\mathscr{F},$$

and
$$\mathsf{P}X^{-1} = \mathsf{P}'$$
on $\overline{\mathscr{B}^T}$.

In other words, $\mathsf{P}X^{-1}$ is an extension of P' to the larger σ-field $X^{-1}\mathscr{F}$. However, as was shown in §6.7, extensions beyond the completion of the σ-field (i.e. beyond $\overline{\mathscr{B}^T}$) are necessarily non-unique, and this gives rise to difficulties which considerably complicate the theory.

More precisely, the probability of an event of the form

$$\{\omega;\; X(\omega) \in A\}$$

can be calculated *from the finite-dimensional distributions* of the process so long as A belongs to $\overline{\mathscr{B}^T}$. If it does not so belong, then the probability of an event, or even whether its probability is defined, will depend on the choice of $(\Omega, \mathscr{F}, \mathsf{P})$. An example of this phenomenon is given in exercise (15.2.(5)). General techniques for avoiding these difficulties have been developed but will not be described here; for a full discussion of them the reader may turn to Doob, *Stochastic processes*, Wiley, 1953.

It is very common to regard a stochastic process as being specified by its finite-dimensional distributions. This is satisfactory so long as we confine our attention to events of the form $\{\omega;\, X(\omega) \in A\}$ for $A \in \overline{\mathscr{B}^T}$. The trouble is that, when T is uncountable, the class $\overline{\mathscr{B}^T}$ is often uncomfortably small (cf. §6.6).

Exercises 15.2

1. Let p_i, p_{ij} $(i,j = 0, 1, 2, \ldots)$ be non-negative numbers satisfying

$$\sum_{j=0}^{\infty} p_j = \sum_{j=0}^{\infty} p_{ij} = 1$$

for all i. Show that there exists a stochastic process $X(t)$ with parameter set $\{0, 1, 2, \ldots\}$ for which

$$\mathsf{P}\{X(0) = x_0, X(1) = x_1, X(2) = x_2, \ldots, X(n) = x_n\}$$
$$= p_{x_0} p_{x_0 x_1} p_{x_1 x_2} \cdots p_{x_{n-1} x_n}$$

for any non-negative integers $x_0, x_1, x_2, \ldots, x_n$.

2. Let p_i be as in exercise 1, and let $p_{ij}(\cdot)$ $(i,j = 0, 1, 2, \ldots)$ be non-negative functions of a positive real variable with

$$\sum_{j=0}^{\infty} p_{ij}(t) = 1$$

for all i, t. Show that there exists a stochastic process $X(t)$ with parameter set $[0, \infty)$ for which

$$\mathsf{P}\{X(0) = x_0, X(t_1) = x_1, X(t_2) = x_2, \ldots, X(t_n) = x_n\}$$
$$= p_{x_0} p_{x_0 x_1}(t_1)\, p_{x_1 x_2}(t_2 - t_1) \ldots p_{x_{n-1} x_n}(t_n - t_{n-1})$$

$(0 < t_1 < t_2 < \ldots < t_n$, x_j non-negative integers), so long as the $p_{ij}(\cdot)$ satisfy

$$p_{ij}(u+v) = \sum_{k=0}^{\infty} p_{ik}(u)\, p_{kj}(v)$$

for all $u, v > 0$.

3. Let $X(t)$ be a stochastic process with parameter set T. For any subset S of T, let $\mathscr{F}_S$ denote the smallest σ-field of events with respect to which the function $X(s): \Omega \to \mathbf{R}$ is measurable for all $s \in S$. If T is a subset of $\mathbf{R}$, X is said to be a *Markov process* if, whenever $t \in T$, $A \in \mathscr{F}_{\{s \in T;\, s<t\}}$, $B \in \mathscr{F}_{\{s \in T;\, s>t\}}$, we have

$$\mathsf{P}(A \cap B \mid X(t)) = \mathsf{P}(A \mid X(t))\, \mathsf{P}(B \mid X(t))$$

with probability one. Show that the processes of the two previous examples are Markov processes. (They are called *Markov chains*, and are of considerable practical importance. A comprehensive account of the theory of these processes is given in K. L. Chung, *Markov Chains with Stationary Transition Probabilities*, Springer, 1960.)

4. Let T be the class of Borel subsets of $[0, 1]$. Show that there exists a stochastic process $X(t)$ with parameter set T such that

(i) $X(t)$ has a Poisson distribution whose mean is the Lebesgue measure of t,

(ii) if $t_1, t_2, \ldots, t_n$ are disjoint, $X(t_1), X(t_2), \ldots, X(t_n)$ are independent, and $X(t_1 \cup t_2 \cup \ldots \cup t_n) = X(t_1) + X(t_2) + \ldots + X(t_n)$.

5. Let $T = \mathbf{R}$, and define finite-dimensional distributions $\mathsf{P}_{t_1 t_2 \ldots t_n}$ which concentrate probability one at the origin in $\mathbf{R}^n$. Show that these satisfy the Kolmogorov consistency conditions. If P' is the corresponding measure on $(\mathbf{R}^T, \mathscr{B}^T)$, and if

$$E = \{x \in \mathbf{R}^T;\ x(t) = 0 \quad \text{for all} \quad t\},$$

show that both E and $\mathbf{R}^T - E$ have P'-outer measure 1. Use theorem 6.12 to deduce that there exists a stochastic process $X(t)$ for which

$$\mathsf{P}\{X(t) = 0\} = 1$$

holds for all t, but for which

$$\mathsf{P}\{X(t) = 0 \quad \text{for all} \quad t\} = 0.$$

Can you give a direct construction for such a process?

15.3 Gaussian processes

In this section we illustrate the general theory by considering an important class of stochastic processes, whose manipulation is relatively simple.

Let T be any set (usually infinite, and possibly uncountable), and let $v(\cdot,\cdot): T \times T \to \mathbf{R}$ be a function with the two properties:

$\mathscr{D}_1$: $v(t_1, t_2) = v(t_2, t_1)\ (t_1, t_2 \in T)$,

$\mathscr{D}_2$: for any finite subset $\{t_1, t_2, \ldots, t_n\} \subset T$ and any real numbers $z_1, z_2, \ldots, z_n$, not all zero,

$$\sum_{i=1}^{n} \sum_{j=1}^{n} v(t_i, t_j)\, z_i z_j > 0. \tag{15.3.1}$$

Any such function is called a *positive-definite* function on T.

With any positive-definite function there can be associated a stochastic process in the following way. Let S be a finite sequence $\{t_1, t_2, \ldots, t_n\}$ of distinct elements of T, and write V_S for the $(n \times n)$ matrix whose (i,j)th element is $v(t_i, t_j)$. Then $\mathscr{D}_1$ and $\mathscr{D}_2$ imply that V_S is a positive-definite symmetric matrix. Hence, there exists a probability measure

$$\mathsf{P}_S = \mathsf{P}_{t_1 t_2 \ldots t_n}$$

on $\mathbf{R}^n$ with density

$$f_S(\mathbf{x}) = (2\pi)^{-\frac{1}{2}n} (\det V_S)^{-\frac{1}{2}} \exp\left(-\tfrac{1}{2} \mathbf{x}^T V_S^{-1} \mathbf{x}\right); \tag{15.3.2}$$

the multinormal distribution $N(0, V_S)$. Moreover, it is an easy consequence of theorem 14.7 that the finite-dimensional distributions P_S satisfy the consistency conditions $\mathscr{K}_1$ and $\mathscr{K}_2$, and hence there exists a stochastic process $X(t)$ with these finite-dimensional distributions. Thus we have proved the following result.

***Theorem* 15.2.** *Let v be a positive-definite function on a set T. Then there exists a stochastic process $\{X(t);\ t \in T\}$ with the property that all its finite-dimensional distributions are multinormal, and*

$$\mathsf{E}(X_t) = 0, \quad \mathsf{E}(X_s X_t) = v(s,t). \tag{15.3.3}$$

A stochastic process satisfying the conditions of the theorem is called a *Gaussian process* with *autocovariance function* v. Notice that two Gaussian processes with the same autocovariance function have the same finite-dimensional distributions, given by (15.3.2).

The effect of this theorem is largely to reduce the study of Gaussian processes to a study of positive-definite functions. This is too large a subject to pursue here; the interested reader may refer to Loève, *Probability Theory* (Chapter x).

Perhaps the most important example of a Gaussian process is the *Wiener process*. This is a process $X(t)$ defined for real positive t (with the convention that $X(0) = 0$).

Wiener process

A Gaussian process is said to be a Wiener process if $T = (0, \infty)$ and

$$v(s,t) = \min(s,t). \tag{15.3.4}$$

The function (15.3.4) clearly satisfies $\mathscr{D}_1$; it satisfies $\mathscr{D}_2$ since, if $0 < t_1 < t_2 < \ldots < t_n$,

$$\begin{aligned}\sum_{i,j=1}^{n} v(t_i,t_j)\, z_i z_j &= \sum_{i=1}^{n} t_i \left\{ z_i^2 + 2z_i \sum_{j=i+1}^{n} z_j \right\} \\ &= \sum_{i=1}^{n} t_i \left\{ \left(\sum_{j=i}^{n} z_j \right)^2 - \left(\sum_{j=i+1}^{n} z_j \right)^2 \right\} \\ &= \sum_{i=1}^{n} (t_i - t_{i-1}) \left(\sum_{j=i}^{n} z_j \right)^2 > 0.\end{aligned}$$

Hence the function is positive-definite, and Wiener processes exist. All Wiener processes have the same finite-dimensional distributions.

It was shown in the previous section that a consistent family of finite-dimensional distributions determines (in a unique way) a probability measure P' on $(\mathbf{R}^T, \mathscr{B}^T)$; the measure corresponding to Wiener processes is called *Wiener measure* and is denoted by W. The completion of $\mathscr{B}^T$ with respect to W will be denoted by $\mathscr{W}$, so that $(\mathbf{R}^T, \mathscr{W}, W)$ is the space of sample functions for the Wiener process. The next theorem shows that this is the same measure as that defined on pp. 161–2 by a somewhat different method.

***Theorem* 15.3.** *Let $X(t)$ be a Wiener process, and let $X(s,t) = X(t) - X(s)$ be the increment of the process on an interval (s,t). Then*

(i) *$X(s,t)$ has distribution $N(0, t-s)$, and*

(ii) *if (s_1, t_1) and (s_2, t_2) are disjoint, then $X(s_1, t_1)$ and $X(s_2, t_2)$ are independent.*

Proof. (i) Since $X(t)$ is Gaussian, the joint distribution of $X(s)$ and $X(t)$ is multinormal, and so, by theorem 14.7, $X(s,t)$ has a normal distribution with zero mean and variance

$$\begin{aligned}\operatorname{var} X(s,t) &= \mathsf{E}\{X(t) - X(s)\}^2 \\ &= \mathsf{E}\{X(t)\}^2 + \mathsf{E}\{X(s)\}^2 - 2\mathsf{E}\{X(s)\,X(t)\} \\ &= v(t,t) + v(s,s) - 2v(s,t) \\ &= t + s - 2s = t - s.\end{aligned}$$

(ii) It is sufficient to deal with the case $s_1 < t_1 \leqslant s_2 < t_2$. Then $X(s_1, t_1)$, $X(s_2, t_2)$ have a multinormal joint distribution, and their covariance is

$$\begin{aligned}\mathsf{E}\{X(s_1,t_1)\,X(s_2,t_2)\} &= \mathsf{E}\{X(t_1) - X(s_1)\}\{X(t_2) - X(s_2)\} \\ &= \mathsf{E}\{X(t_1)X(t_2) - X(s_1)\,X(t_2) - X(t_1)\,X(s_2) \\ &\qquad + X(s_1)\,X(s_2)\} \\ &= v(t_1,t_2) - v(s_1,t_2) - v(t_1,s_2) + v(s_1,s_2) \\ &= t_1 - s_1 - t_1 + s_1 = 0.\end{aligned}$$

Hence by corollary 4 to theorem 14.7, $X(s_1, t_1)$ and $X(s_2, t_2)$ are independent. ▌

The Wiener process has been extensively used as a model for situations in which a particle wanders erratically about on the real line. Property (ii) then states that the displacements of the particle in successive time intervals are independent random variables, while property (i) implies that the distribution of any such displacement depends only on the length of the time interval.

Notice that, when $t - s$ is small, the distribution of $X(t) - X(s)$ is concentrated near the origin, so that the particle is unlikely to move a large distance in a short time. More precisely for any t, the convergence

$$\lim_{h \to 0} X(t+h) = X(t)$$

holds in mean square, and therefore in probability. In this sense, then, the particle executes a continuous path in $\mathbf{R}$.

It would be much more satisfactory, however, if we could prove that $X(t)$ was, with probability one, a continuous function of t. More precisely, if C denotes the subset of $\mathbf{R}^{(0,\infty)}$ consisting of the continuous functions from $(0,\infty)$ into $\mathbf{R}$, with limit 0 at 0, we would like to be able to show that $C \in \mathscr{W}$ and that $W(C) = 1$. Unfortunately, however, C does not lie in $\mathscr{W}$ and its Wiener measure is not defined. In fact (see exercise 15.3 (4)) both C and $\mathbf{R}^{(0,\infty)} - C$ have outer Wiener measure 1. Hence a statement such as 'Wiener processes are almost certainly continuous' cannot be made.

It is, however, possible to make a weaker statement. Let $\mathbf{Q}_2$ denote the set of binary rationals, i.e. the set of numbers x for which $2^n x$ is an integer for some n. Denote by C' the set of functions from T into $\mathbf{R}$ whose restriction to $(0, A) \cap \mathbf{Q}_2$ is uniformly continuous (and has limit 0 at 0) for all A. Clearly $C \subset C' \subset \mathbf{R}^{(0,\infty)}$.

***Theorem* 15.4.** *With the above notation $C' \in \mathscr{B}^{(0,\infty)} \subset \mathscr{W}$, and $W(C') = 1$.*

Thus a Wiener process is almost certainly continuous on $\mathbf{Q}_2$, and uniformly continuous on any bounded subset of it. At the expense of some complication, $\mathbf{Q}_2$ can be replaced by any countable dense subset of $(0, \infty)$.

Proof. For any $x = \{x(t);\ t > 0\} \in \mathbf{R}^{(0,\infty)}$, and any interval (a, b), define

$$M(a, b; x) = \sup |x(t) - x(s)|,$$

where the supremum is taken over all $s, t \in \mathbf{Q}_2$ with $a \leqslant s < t \leqslant b$, and x is extended to $[0, \infty)$ by setting $x(0) = 0$. For any integer k, define

$$M_k(x) = \max\{M((r-1)\,2^{-k}, r2^{-k}; x);\ 1 \leqslant r \leqslant 2^k\}.$$

Since $\mathbf{Q}_2$ is countable, the function $M_k: \mathbf{R}^{(0,\infty)} \to \mathbf{R}$ is $\mathscr{B}^{(0,\infty)}$-measurable. It is very easy to see that $M_k(x)$ is non-increasing in k, and that $M_k(x) \to 0$ as $k \to \infty$ if and only if x is uniformly continuous on $[0, 1] \cap \mathbf{Q}_2$. Thus the class C_1 of functions in $\mathbf{R}^{(0,\infty)}$ uniformly continuous on $[0, 1] \cap \mathbf{Q}_2$ can be written

$$C_1 = \{x \in \mathbf{R}^{(0,\infty)};\ \lim_{k\to\infty} M_k(x) = 0\},$$

and so belongs to $\mathscr{B}^{(0,\infty)}$.

Now let X be a Wiener process, and consider the random variable $M(a, b; X)$ where $a = (r-1)\,2^{-k}$, $b = r2^{-k}$. Then

$$\begin{aligned} M(a, b; X) &\leqslant 2 \lim_{n\to\infty} \max_{1 \leqslant m \leqslant 2^{n-k}} |X(a, a + m2^{-n})| \\ &= 2 \lim_{n\to\infty} M^{(n)}, \quad \text{say.} \end{aligned}$$

For fixed n, write

$$\xi_m = X(a + (m-1)2^{-n}, a + m2^{-n})\,2^{\frac{1}{2}n}.$$

Then $$\xi_1, \xi_2, \ldots, \xi_{2^{n-k}}$$
are independent with distribution $N(0, 1)$, and
$$M^{(n)} = 2^{-\frac{1}{2}n} \max_m |\xi_1 + \xi_2 + \ldots + \xi_m|.$$

Now recall the inequality (13.5.6) which was used in the proof of the law of the iterated logarithm, and which shows that
$$\mathsf{P}(\xi_1 + \xi_2 + \ldots + \xi_m > x \quad \text{for some} \quad m \leqslant N) \leqslant 2\{1 - \Phi(N^{-\frac{1}{2}} x)\},$$
and
$$\mathsf{P}(\xi_1 + \xi_2 + \ldots + \xi_m < -x \quad \text{for some} \quad m \leqslant N) \leqslant 2\{1 - \Phi(N^{-\frac{1}{2}} x)\}$$
where Φ is the distribution function of $N(0, 1)$. Hence, if $y > 0$, we have (putting $N = 2^{n-k}$, $x = 2^{\frac{1}{2}n} y$)
$$\mathsf{P}(M^{(n)} > y) \leqslant 4\{1 - \Phi(2^{\frac{1}{2}k} y)\}.$$

Letting $n \to \infty$, and using the countable additivity of P,
$$\mathsf{P}\{M(a, b; X) > 2y\} \leqslant 4\{1 - \Phi(2^{\frac{1}{2}k} y)\}.$$
By the inequality of exercise 11.7 (9), this implies that
$$\mathsf{P}\{M((r-1)\, 2^{-k}, r2^{-k}; X) > 2y\} \leqslant 4(2\pi)^{-\frac{1}{2}}\, 2^{-\frac{1}{2}k}\, y^{-1} \exp(-2^{k-1} y^2)$$
$$= p_k(y), \text{ say.}$$

Now, for fixed k, the random variables
$$M((r-1)\, 2^{-k}, r2^{-k}; X)$$
for different values of r, being functions of the increments of X on disjoint intervals, are independent, and so
$$\mathsf{P}\{M_k(X) \leqslant 2y\} = \prod_{r=1}^{2^k} \mathsf{P}\{M((r-1)\, 2^{-k}, r2^{-k}; X) \leqslant y\}$$
$$\geqslant \{1 - p_k(y)\}^{2^k}.$$

Since M_k does not increase with k,
$$\mathsf{P}\{\lim_{k \to \infty} M_k(X) \leqslant 2y\} \geqslant \{1 - p_k(y)\}^{2^k}$$
for all k. Since $2^k p_k(y) \to 0$ as $k \to \infty$, the right-hand side converges to 1, and hence
$$\mathsf{P}\{\lim_{k \to \infty} M_k(X) \leqslant 2y\} \geqslant 1.$$

Since this holds for all $y > 0$,
$$\mathsf{P}(X \in C_1) = 1,$$
i.e.
$$W(C_1) = 1.$$

Exactly the same argument shows that, for any integer N, the class C_N of functions in $\mathbf{R}^{(0,\infty)}$ which are uniformly continuous in $(0, N) \cap \mathbf{Q}_2$ and converge to 0 at the origin satisfies

$$C_N \in \mathscr{B}^{(0,\infty)}, \quad W(C_N) = 1.$$

Since
$$C' = \bigcap_{N=1}^{\infty} C_N,$$

we have
$$C' \in \mathscr{B}^{(0,\infty)}, \quad W(C') = 1. \blacksquare$$

This result shows that any Wiener process has the property that it is almost certainly continuous on $\mathbf{Q}_2$. The theorem can also be used to construct a Wiener process which is almost certainly continuous everywhere.

Let x belong to C'. Then it is very easy to see that the limit

$$y(t) = \lim_{\substack{\tau \to t \\ \tau \in \mathbf{Q}_2}} x(\tau)$$

exists, and defines a continuous function y. If we put $y = \theta x$, θ is a function from C' into C, and clearly $\theta x = x$ for all $x \in C$. (In other words, θ is a projection from C' onto its subspace C.)

Now let X be any Wiener process, defined on a probability space $(\Omega, \mathscr{F}, \mathsf{P})$. Since $X \in C'$ for almost all ω, the function $Y = \theta X$ is defined with probability one. Since $\mathbf{Q}_2$ is countable, $Y(t)$ is a random variable, so that Y is a stochastic process. Since

$$Y(t) = \operatorname{p} \lim_{\substack{\tau \to t \\ \tau \in \mathbf{Q}_2}} X(\tau),$$

the finite-dimensional distributions P^Y of Y are connected with those of X by

$$\mathsf{P}^Y_{t_1 t_2 \ldots t_n} = \lim_{\substack{\tau_j \to t_j \\ \tau_j \in \mathbf{Q}_2}} \mathsf{P}^X_{\tau_1 \tau_2 \ldots \tau_n},$$

and it follows that Y is a Wiener process. By construction $Y \in C$ with probability one. Hence Y is a Wiener process which is almost certainly continuous.

As described before, Y induces a probability measure $\mathsf{P}Y^{-1}$ on $\mathbf{R}^{(0,\infty)}$, which is defined on a σ-field $Y^{-1}\mathscr{F}$ which contains $\mathscr{W}$ as a sub-σ-field. We have seen that $C \in Y^{-1}\mathscr{F}$ and that $\mathsf{P}Y^{-1}C = 1$. Moreover, $\mathsf{P}Y^{-1}$ coincides with W on $\mathscr{B}^{(0,\infty)}$, and hence on $\mathscr{W}$. In other words, we have constructed an extension $\overline{W}$ of Wiener measure W for which C is measurable and has $\overline{W}(C) = 1$.

One consequence of this is that, whenever $E \supset C$ is in $\mathscr{B}^{(0,\infty)}$,

$$W(E) = \overline{W}(E) \geqslant \overline{W}(C) = 1,$$

so that the outer Wiener measure of C satisfies

$$W^*(C) = \inf_{E \supset C} W(E) \geqslant 1.$$

Thus C has outer Wiener measure 1. Exercise 15.3 (4) shows that $\mathbf{R}^{(0,\infty)} - C$ also has outer measure 1, so that C is not Wiener measurable.

The above discussion illustrates the difficulties which arise when considering stochastic processes with uncountable parameter sets T. It also contains the germ of the general method for resolving such difficulties; if in a suitable topology T has a countable dense subset T_0 (like $\mathbf{Q}_2$), then the process can often be studied more easily on T_0 and then extended by continuity to the whole of T.

Exercises 15.3

1. Let $X(t)$ be any stochastic process with $\mathsf{E}X(t)^2 < \infty$ for all t, and let $v(s,t)$ be the covariance of $X(s)$ and $X(t)$. Show that $v(s,t)$ is positive-semi-definite in the sense that (15.3.1) holds with '$>$' replaced by '$\geqslant$'.

2. If $X(t)$ is a Wiener process, show that

$$Y(t) = e^{-t}X(e^{2t}) \quad (-\infty < t < \infty),$$

is a Gaussian process with autocovariance function

$$v_Y(s,t) = e^{-|t-s|}.$$

3. If $X(t)$ is a Wiener process, show that

$$Z(t) = X(t) - tX(1) \quad (0 < t < 1),$$

is a Gaussian process with autocovariance function

$$v_Z(s,t) = s(1-t) \quad (s \leqslant t).$$

(Z is called the *tied-down Wiener process* on $(0, 1)$.)

4. Let $X(t)$ be a Wiener process, and let τ be a random variable, independent of X, and having a continuous distribution on $(0, \infty)$. Define a new process $\hat{X}(t)$ by

$$\hat{X}(t) = X(t) \quad (t \neq \tau),$$
$$\hat{X}(\tau) = 0.$$

Show that $\hat{X}$ is a Wiener process, and that

$$\mathsf{P}(\hat{X} \in C) = 0.$$

Deduce that

$$W^*(\mathbf{R}^{(0,\infty)} - C) = 1.$$

5. Show that theorem 15.4 can be strengthened to read

$$W\{x; \lim_{k\to\infty} M_k(x)\, 2^{\alpha k} = 0\}$$

for any $\alpha < \frac{1}{2}$. Deduce that, if L_α is the class of functions x for which, given t, there exists δ with

$$|x_{t+h} - x_t| < |h|^\alpha$$

for all $|h| < \delta$, then

$$\overline{W}(L_\alpha) = 1.$$

6. Prove, using the law of the iterated logarithm, that

$$\overline{W}\{x;\ \limsup_{t\to\infty} X(t)/(2t \log\log t)^{\frac{1}{2}} = 1\} = 1.$$

15.4 Stationary processes

Let $X(t)$ be a stochastic process whose parameter set T is a subset of the real line. Then t is a real variable, and it is intuitively convenient to think of it, as indeed is so in many applications, as measuring time. It may be that, although $X(t)$ varies with t, its probability distribution does not. More generally, the joint distribution of $X(t_1), X(t_2), \ldots, X(t_n)$ may depend only on the differences $t_j - t_1$. This is a property which is possessed by many processes which represent systems in some sort of statistical equilibrium, and is known as *stationarity*.

Stationary process

A stochastic process $X(t)$ with parameter set $T \subset \mathbf{R}$ is said to be stationary if whenever $t_1, t_2, \ldots, t_n, t_1 + h, t_2 + h, \ldots, t_n + h$ all belong to T, its finite-dimensional distributions satisfy

$$\mathsf{P}_{t_1+h,\, t_2+h,\, \ldots,\, t_n+h} = \mathsf{P}_{t_1 t_2 \ldots t_n}. \tag{15.4.1}$$

In other words, a process is stationary if its finite-dimensional distributions are not affected by a shift in the time axis.

In particular, suppose that T consists of the positive integers, so that the stochastic process $X(t)$ is a sequence $\{X_1, X_2, \ldots,\}$ of random variables. Then this sequence is stationary if and only if, for each n,

$$\mathsf{P}_{2,3,4,\ldots,n+1} = \mathsf{P}_{1\,2\,3,\ldots,n}. \tag{15.4.2}$$

If $\{X_1, X_2, \ldots\}$ is a stationary sequence of random variables, then, as was shown in § 15.2, it induces a probability measure P' on $(\mathbf{R}^T, \mathscr{B}^T)$ (where $T = \{1, 2, \ldots\}$) by

$$\mathsf{P}'(B) = \mathsf{P}(\{X_1, X_2, \ldots\} \in B).$$

Now let $S: \mathbf{R}^T \to \mathbf{R}^T$ be the function defined by

$$S\{x_1,, x_2, \ldots\} = \{x_2, x_3, \ldots\}; \tag{15.4.3}$$

S is called the *shift function*. Then S and P' have an important property. To see this, let $B \in \mathscr{B}^T$ be a cylinder set, i.e. a set of the form

$$B = \{x \in \mathbf{R}^T;\ (x_1, x_2, \ldots, x_n) \in A\}$$

for $A \in \mathscr{B}^n$. Then

$$\begin{aligned} \mathsf{P}'(S^{-1}B) &= \mathsf{P}(\{X_1, X_2, \ldots\} \in S^{-1}B) \\ &= \mathsf{P}(S\{X_1, X_2, \ldots\} \in B) \\ &= \mathsf{P}(\{X_2, X_3, \ldots\} \in B) \\ &= \mathsf{P}((X_2, X_3, \ldots, X_{n+1}) \in A) \\ &= \mathsf{P}_{2,3,\ldots,n+1}(A) \\ &= \mathsf{P}_{1,2,\ldots,n}(A) \\ &= \mathsf{P}((X_1, X_2, \ldots, X_n) \in A) \\ &= \mathsf{P}(\{X_1, X_2 \ldots\} \in B) \\ &= \mathsf{P}'(B). \end{aligned}$$

Thus the probability measure $\mathsf{P}'S^{-1}$ coincides with P' on all cylinder sets, and hence on the σ-field $\mathscr{B}^T$ generated by the cylinder sets. Thus we have proved the following result, which we state in the terminology of Chapter 7.

***Theorem* 15.5.** *Let the stationary sequence*

$$\{X(t); \quad t \in T = \{1, 2, \ldots\}\} = \{X_1, X_2, \ldots\}$$

induce the measure P' *on* $(\mathbf{R}^T, \mathscr{B}^T)$. *Then the shift function* S *defined by* (15.4.3) *is a measure-preserving transformation of* $(\mathbf{R}^T, \mathscr{B}^T, \mathsf{P}')$ *into itself.*

This result enables us to use the results of §7.5, and in particular Birkhoff's theorem (theorem 7.9). This theorem, when translated into the language of stationary sequences of random variables, reads as follows:

***Theorem* 15.6.** *Let* $\{X_1, X_2, \ldots,\}$ *be a stationary sequence of random variables with* $\mathsf{E}|X_1| < \infty$. *Then*

$$Z = \lim_{n \to \infty} \frac{1}{n}(X_1 + X_2 + \ldots + X_n) \tag{15.4.4}$$

exists with probability one, and

$$\mathsf{E}(Z) = \mathsf{E}(X_1). \tag{15.4.5}$$

Proof. Let $f: \mathbf{R}^T \to \mathbf{R}$ be the mapping which sends $x = \{x_1, x_2, \ldots\}$ onto x_1. Then

$$f(S^i x) = x_{i+1}.$$

Since
$$\int_{\mathbf{R}^T} |f(x)| \, d\mathsf{P}' = \int_\Omega |X_1| \, d\mathsf{P} = \mathsf{E}|X_1| < \infty,$$

f is integrable on $(\mathbf{R}^T, \mathscr{B}^T, \mathsf{P}')$, and the result follows on applying theorem 7.9. ∎

This theorem is often called the *strong law of large numbers* for stationary sequences, because of its similarity to the strong law theorem 13.7. Notice, however, that it is not necessarily true that Z is almost certainly equal to $\mathsf{E}(X_1)$; it is possible for Z to be a non-degenerate random variable and all we can say is that its expectation must be $E(X_1)$.

The strong law for independent random variables (theorem 13.7) can be deduced from theorem 15.6. If $X_1, X_2, \ldots$ are independent and identically distributed with finite mean μ, the sequence $\{X_1, X_2, \ldots\}$ is clearly stationary, and $\mathsf{E}|X_1| < \infty$. Hence, by theorem 15.6, the limit

$$Z = \lim_{n\to\infty} (X_1+X_2+\ldots+X_n)/n$$

exists with probability one, and

$$E(Z) = \mu.$$

Theorem 13.7 will then follow if we can prove that

$$\mathsf{P}(Z = \mu) = 1.$$

Now consider the event $E = \{Z > \mu\}$. For any n, E is independent of $X_1, X_2, \ldots, X_n$. A trivial modification of theorem 13.3 (the zero-one law) then shows that $\mathsf{P}(E)$ must be either zero or one. Now

$$\mathsf{P}(Z > \mu) = 1$$

would imply $\mathsf{E}(Z) > \mu$ and lead to a contradiction, so that $\mathsf{P}(Z > \mu) = 0$. Similarly, $\mathsf{P}(Z < \mu) = 0$ and so finally $\mathsf{P}(Z = \mu) = 1$, and theorem 13.7 is proved.

This is a very typical application of theorem 15.6: Z is proved to be degenerate, and so to equal $\mathsf{E}(X_1)$ with probability one.

Exercises 15.4

1. Show that the process Y of exercise 15.3 (2) is stationary.

2. Prove the following converse of theorem 15.5: if P' is a probability measure on $(\mathbf{R}^T, \mathscr{B}^T)$ which is preserved by S, then there exists a stationary process which induces the measure P'.

3. A stationary sequence $\{X_1, X_2, \ldots\}$ has $\mathsf{E}(X_1^2) < \infty$. Show that $\mu = \mathsf{E}(X_n)$ is independent of n, and that the covariance of X_m and X_n takes the form

$$\operatorname{cov}(X_m, X_n) = v(n-m)$$

for some even function v. Show that, as $n \to \infty$,

$$(X_1+X_2+\ldots+X_n)/n \to \mu$$

in mean square if and only if

$$\sum_{r=1}^{n}\left(1-\frac{r}{n}\right)v(r) = o(n)$$

as $n \to \infty$. Show in particular that this is true if $v(n) \to 0$ as $n \to \infty$.

4. A process $\{X(t);\, t \in \mathbf{R}\}$ is said to be *second-order stationary* if $\mathsf{E}\{X(t)\}$ is independent of t and if $\mathsf{E}\{X(s)\,X(t)\}$ depends only on $(t-s)$. Prove that

(i) a stationary process is second-order stationary,

(ii) a second-order stationary process need not be stationary,

(iii) a second-order stationary Gaussian process is stationary.

5. Let X be a non-degenerate random variable with finite mean, and define a sequence $\{X_1, X_2, \ldots\}$ by $X_n = X$ for all n. Show that this is a stationary sequence for which the limit Z of theorem 15.6 does not satisfy

$$\mathsf{P}\{Z = \mathsf{E}(X_1)\} = 1.$$

6. Show that a sequence $\{X_1, X_2, \ldots\}$ is stationary if and only if there exists a sequence $\{Y_1, Y_2, \ldots\}$ such that for any n, the joint distribution of $(X_1, X_2, \ldots, X_n)$ is the same as that of $(Y_n, Y_{n-1}, \ldots, Y_1)$.

INDEX OF NOTATION

† Note that the symbol $\boldsymbol{c}$ has two distinct uses, which should not be confused

The symbol ❚ is used to signal the end of a proof.

GENERAL INDEX